AMERICAN VACUUM SOCIETY CLASSICS

THE PHYSICAL BASIS OF ULTRAHIGH VACUUM

P. A. Redhead
J. P. Hobson
E. V. Kornelsen

AMERICAN VACUUM SOCIETY CLASSICS

THE PHYSICAL BASIS OF ULTRAHIGH VACUUM

AMERICAN VACUUM SOCIETY
CLASSICS

THE PHYSICAL BASIS OF ULTRAHIGH VACUUM

P. A. Redhead
J. P. Hobson
E. V. Kornelsen

Radio & Electrical Engineering Division
National Research Council
Ottawa, Canada

American Institute of Physics   **New York**

© 1993 by American Institute of Physics
All rights reserved
Printed in the United States of America

American Institute of Physics
335 East 45th Street
New York, NY 10017-3483

Library of Congress Cataloging-in-Publication Data

Redhead, P. A.
    The physical basis of ultrahigh vacuum/P. A. Redhead, J. P. Hobson, E. V. Kornelsen.
      p. cm.—(AVS classics of vacuum science and technology)
    Originally published: London: Chapman and Hall, 1968.
    Includes bibliographical references and index.
    ISBN 1-56396-122-9
    1. Vacuum. 2. Vacuum technology. I. Hobson, J. P. II. Kornelsen, E. V.
III. Title. IV. Series.
    QC166.R38   1993
    533.5–dc20   92-46643
                                                                                                                  CIP

TO
our families

ACKNOWLEDGEMENTS

We are specially indebted to Mrs. J. Baker who typed the manuscript and corrected many errors in grammar, spelling and other matters. Thanks are also due to the staff of the library, drafting office and photographic section of the National Research Council for their constant help. The assistance of many authors in supplying unpublished information is also gratefully acknowledged.

P. A. REDHEAD
J. P. HOBSON
E. V. KORNELSEN

OTTAWA, 1967

Series Preface

The science of producing and measuring high and ultrahigh vacuum environments has fundamental interest for basic research in addition to a wide variety of important practical applications. Basic research involving particle physics, atomic and molecular physics, plasma physics, physical chemistry, and surface science often involves careful production, control, and measurement of a vacuum environment in order to perform experiments. Practical applications of vacuum science and technology are found in many materials processing techniques used for microelectronic, photonic, and magnetic materials and in the simulation of space and rarefied gas environments.

As in most of modern science, the rapid development of the field has been accompanied by a parallel growth in the related technical literature including specialized journals, monographs, and textbooks. There exist early publications in vacuum science and technology which have attained the status of indispensable references among practitioners, lecturers, and students of the field. Many of these "classic" publications have gone out of print and are currently unavailable to newcomers to the field. The present series, commissioned by the American Vacuum Society, and published within the Book Program of the American Institute of Physics to celebrate the 40th anniversary of the Society in 1993, is entitled the **"American Vacuum Society Classics."**

The American Vacuum Society Classics will reprint important books from the last four decades that continue to have significant impact on the modern development of the field. It is the goal of the American Vacuum Society Classics to reprint these books in a high quality and affordable format to ensure wide availability to the technical community, individual researchers, and students.

H. F. Dylla
Continuous Electron Beam
Accelerator Facility
Newport News, VA
and
College of William and Mary
Williamsburg, VA

Author's Preface to the Reprint Edition

The *Physical Basis of Ultrahigh Vacuum* was written at a time (1966–68) when ultrahigh vacuum (UHV) was in use in many laboratories for studies in surface science and when very pure gases were needed; it was also being used on a larger scale in space simulators, storage rings, and in plasma containment devices in fusion research. The commercial availability of UHV components was quite limited and UHV systems were frequently constructed by the experimenter from a mixture of commercially available components and home-made parts; there were no manufacturers concerned with the production of UHV components and systems only, nor was there any commercial use of UHV in manufacturing. The book was written for the researcher who wished to use UHV and who might have to design and construct his own UHV components and systems because of the lack of suitable commercial equipment or, in some cases, the resources to buy such equipment. This book was designed to explain both the design and use of UHV systems and components, as well as the underlying physical and chemical principles on which the performance of the equipment depended. It was this juxtaposition of the practical and the phenomenological which made this book unique in its field.

While most of the physical and chemical phenomena occurring in UHV were known, many of their quantitative aspects were not, particularly their limits at low pressures. This book reviewed extensive research data bearing on these matters, much of which have not been updated since.

Although there have been major improvements in the availability and quality of commercial UHV equipment since this book was written, the underlying principles are unchanged, though better understood in some cases, and the descriptions of components and systems are sufficiently general to still be of use. The limits to the performance of UHV equipment can still best be understood in terms of the same basic physical phenomena. The continuing value of this book is the close association of the underlying physical principles with the practical problems of the design and use of UHV components and systems.

CONTENTS

Acknowledgements *page* vi

1 INTRODUCTION 1

Index of Symbols 4

PART A: PHYSICAL PROCESSES

2 MOLECULE-MOLECULE AND MOLECULE-SURFACE INTERACTIONS 11

   2.1 Energies between Molecules 11
      2.1.1 Van der Waals Energies and Valence Energies, 12
      2.1.2 Induction Energies and Electrostatic Energies, 16

   2.2 Intermolecular Collisions 18
   2.3 Energies between a Molecule and a Surface 22
      2.3.1 Energies of Physisorption, 22
      2.3.2 Energies of Chemisorption, 29

   2.4 Dynamic Interactions between a Molecule and a Surface 37
      2.4.1 Vapour Pressure, 37
      2.4.2 Adsorption Isotherms, 39
      2.4.3 Impact between a Molecule and a Surface: Accommodation, Condensation and Sticking, 54
      2.4.4 Desorption of a Molecule from a Surface, 73
      2.4.5 Adsorption of Mixtures, Cryotrapping and Replacement, 83

   2.5 Diffusion, Solution and Permeation of Gases in Solids 88
      2.5.1 Solution and De-solution of Gas from a Semi-infinite Solid, a Slab, a Cylinder, and a Sphere, 92
      2.5.2 De-solution of Gas from Vacuum Chamber Wall and Permeation, 99
      2.5.3 Modifications to Idealized Solutions for Diffusion, 100
      2.5.4 Specific Examples of Diffusion Mechanisms in Ultrahigh Vacuum Practice, 105

3 COLLISION PROCESSES IN GASES 110

   3.1 Elastic Collisions of Electrons with Atoms ($e0/e0$) 111

Contents

3.2 Inelastic Electron-Atom Collisions — 114
  3.2.1 Ionization, 114
  3.2.2 Excitation, 117
3.3 Elastic Atom-Atom and Ion-Atom Collisions (00/00, 10/10) — 120
3.4 Inelastic Effects in Ion-Atom Collisions — 123
  3.4.1 Charge Transfer (10/01), 123
  3.4.2 Excitation (10/10), 125
  3.4.3 Ionization (10/11), 125
3.5 Photo-absorption and Photo-ionization — 128
3.6 Positive Ion Recombination — 131

4 INTERACTION OF CHARGED PARTICLES WITH SURFACES — 134

4.1 Electron Scattering from Surfaces — 134
  4.1.1 The Energies of Scattered Electrons, 134
  4.1.2 Elastic Reflection of Electrons, 135
  4.1.3 Characteristic Energy Losses and Rediffused Electrons, 152
  4.1.4 Secondary Electron Emission, 159
4.2 Electron-Impact Desorption — 167
  4.2.1 Electron-Impact Desorption from Adsorbed Layers, 168
  4.2.2 Electron-Impact Desorption from Thick Layers, 172
  4.2.3 Electron-Impact Desorption from Surfaces of Unknown State, 174
  4.2.4 Theory of Electron-Impact Desorption, 176
4.3 Impact of Energetic Atoms and Ions on Surfaces — 181
  4.3.1 Back-Scattering, 188
  4.3.2 Sputtering, 192
  4.3.3 Radiation Damage, 204
  4.3.4 Entrapment and Re-emission, 209
  4.3.5 Penetration and Channelling, 216
  4.3.6 Electron Ejection, 222
4.4 Emission of Ions from Hot Surfaces — 229
  4.4.1 Thermionic Emission of Ions, 229
  4.4.2 Surface Ionization, 230

5 INTERACTION OF RADIATION WITH SURFACES — 234

5.1 Photo-electric Emission — 234
5.2 Interaction of Photons with Adsorbed Gases — 239

6 MECHANICAL PROPERTIES OF MATERIALS AT VERY LOW PRESSURES — 243

6.1 Adhesion or Cold Welding — 243
6.2 Friction, Lubrication and Wear — 246
6.3 Creep, Fatigue and Fracture — 248

Contents

PART B: PRESSURE MEASUREMENT

7 GENERAL CONSIDERATIONS OF PRESSURE MEASUREMENT — 253

7.1 Calibration of UHV Gauges — 253
7.1.1 Absolute Pressure Measurement, 253
7.1.2 The McLeod Gauge, 254
7.1.3 Extensions to Lower Pressure, 257
7.1.4 Relative Sensitivity for Different Gases, 262

7.2 Measuring Systems as Sinks and Sources — 263
7.2.1 Pumping and Re-emission in Gauges, 263
7.2.2 Gas Interactions at Hot Surfaces, 275
7.2.3 Sinks and Sources in Tubing (Blears effect), 280

7.3 Pressure Measurements in Non-Uniform Environments — 281

7.4 Residual Currents — 287
7.4.1 Soft X-ray Photo-emission, 287
7.4.2 Electron-Impact Desorption, 290

7.5 Methods of Current Measurement — 294
7.5.1 General Problems of Current Collection in UHV, 294
7.5.2 Measurement with dc Amplifiers, 295
7.5.3 Electron Multipliers, 295
7.5.4 Conversion-Scintillation Detectors, 298

7.6 Cathodes and Cathode Effects — 299

8 TOTAL PRESSURE GAUGES — 304

8.1 Hot-Cathode Gauges — 304
8.1.1 The Bayard-Alpert Gauge, 305
8.1.2 Suppressor Gauge, 313
8.1.3 Extractor Gauge, 315
8.1.4 Orbitron Gauge, 317
8.1.5 Hot-Cathode Gauges with Magnetic Field, 319
8.1.6 Other Hot-Cathode Gauges, 322
8.1.7 Measurement of Residual Currents, 323

8.2 Crossed-Field, Cold-Cathode Gauges — 329
8.2.1 Penning Gauge, 329
8.2.2 Magnetron Gauge, 331
8.2.3 Inverted-Magnetron Gauge, 333
8.2.4 Current-Pressure Characteristics and Oscillatory Behaviour, 334

8.3 Comparison of Total Pressure Gauges — 336

Contents

9 PARTIAL PRESSURE GAUGES — 340

9.1 Mass-Spectrometers — 340
    9.1.1 General Characteristics, 342
    9.1.2 Examples of Ultrahigh Vacuum Mass-Spectrometers, 345
    9.1.3 Measurement Problems in Partial Pressure Analysis, 352
    9.1.4 Mass Numbers Typically Observed, 354

9.2 Desorption Spectrometers — 355
    9.2.1 Chemical Desorption Spectrometers, 355
    9.2.2 Physical Desorption Spectrometers, 361

9.3 Gauges using Field Emission or Work-Function Changes — 365

PART C: PRODUCTION OF ULTRAHIGH VACUUM

10 PROCESSING TECHNIQUES FOR ULTRAHIGH VACUUM — 369

10.1 Pre-treatment of Materials — 371
10.2 Baking Procedures — 373
10.3 Degassing Procedures after Baking — 378

11 PUMPS FOR ULTRAHIGH VACUUM — 379

11.1 Molecular Drag and Turbo-Molecular Pumps — 379
11.2 Diffusion Pumps — 380
11.3 Cryopumps — 383
    11.3.1 Cryosorption Pumps, 385
    11.3.2 Cryogenic Pumps, 391

11.4 Getter Pumps — 399
    11.4.1 Getter Materials, 399
    11.4.2 Types of Getter Pumps, 403

11.5 Ion Pumps — 404
    11.5.1 Sputter-Ion Pumps, 406
    11.5.2 Getter-Ion Pumps, 411

11.6 Comparison of Properties of UHV Pumps — 416

12 EXAMPLES OF ULTRAHIGH VACUUM SYSTEMS — 417

12.1 Small Glass and Metal Systems — 417
    12.1.1 Apparatus, 417
    12.1.2 Residual Conditions Achieved, 417
    12.1.3 Residual Processes, 420
    12.1.4 Admission of Gas, 427

Contents

12.2 Large UHV Systems   430
    12.2.1 Conventional Systems of Moderate Size, 431
    12.2.2 Cryogenic Systems, 432
    12.2.3 Accelerator Vacuum Systems, 433
    12.2.4 Systems for Plasma Physics Research, 433
    12.2.5 Space Simulators, 434

References   435

Appendix A. Recent References   467

Subject Index   477

Author Index   485

CHAPTER ONE
INTRODUCTION

Ultrahigh vacuum (uhv) has been defined as the region of pressure below 10^{-9} torr. Prior to 1950, no means existed for measuring pressures below about 10^{-8} torr and thus there is no direct evidence that uhv had been obtained before 1950. However, there is considerable indirect evidence that uhv was obtained by some earlier experimenters. For example, Anderson (1935) almost certainly obtained uhv conditions in his experiments; he estimated pressures from the rate of change of work function of a clean tungsten surface. Nottingham (1947) was the first to predict that electrons in an ionization gauge would release soft x-rays on striking a surface and that these soft x-rays would in turn release photo-electrons from the ion-collector. The photocurrent would be indistinguishable from the ion current in the measuring circuit and would constitute a pressure-independent residual current. Ionization gauges available at the time (1947) were incapable of measuring pressure below about 10^{-8} torr because of this 'x-ray effect'.

Several new designs of ionization gauge were developed, shortly after Nottingham's comment, in which the limitation in pressure measurement caused by soft x-rays was reduced (Bayard and Alpert 1950; Lander 1950; Metson 1951). The design of Bayard and Alpert was simple and effective, permitting pressure measurement to about 10^{-10} torr, and has been almost universally accepted. Once a simple means for pressure measurement in the uhv range was available, a rapid advance in all aspects of uhv technology occurred. By about 1960 the production and measurement of pressures down to about 10^{-10} torr had become routine in many laboratories and commercial uhv equipment was available.

Since 1960 there has been a steady decrease in the lowest pressures produced and measured (at the time of writing (1967) the current lower limit at room temperature is about 10^{-14} torr) and there has been a continuing improvement in understanding of problems relevant to uhv production and measurement. In the same period there has been a marked improvement in commercially available uhv equipment and the use of uhv techniques has become world-wide.

The advances in understanding of phenomena in uhv systems and in uhv technology have required contributions from many different disciplines (surface physics and chemistry, thermodynamics, electron and ion-impact physics, cryogenics, electronics; to name a few). In this book an attempt is made to correlate information from different disciplines and to examine the diverse physical and chemical processes relevant to the production, measurement and application of ultrahigh vacuum. We have taken a similar approach in writing this book to that we used earlier in a review article on ultrahigh vacuum appearing in *Advances in Electronics and Electron Physics* (Redhead *et al.* 1962b). The emphasis throughout is on the nature and magnitude of underlying physical and chemical phenomena which govern the behaviour of uhv systems and devices; the description of practical devices and systems is kept to the minimum necessary to exemplify the principles under discussion. Questions still exist on the relative importance of some processes in uhv systems and in these cases we have usually included a discussion of the relevant processes.

This book is intended for the graduate student, scientist or engineer who has to make use of uhv technology in his work, and who needs a fuller understanding of the principles governing the behaviour of his uhv equipment. It might also serve as an introduction to research on the physical processes and technology of uhv.

The book is divided, like Caesar's Gaul, into three parts; part A is concerned with the physical processes that govern phenomena in uhv systems. The production and measurement of uhv is dominated by effects not occurring in the gas phase; namely, surface effects (adsorption, desorption, secondary emission, etc.) and bulk effects (permeation, solution, ion penetration, etc.). In Chapter 2 we deal with the interaction of gas molecules with surfaces and solids; in Chapters 4 and 5 we deal with the interaction of charged particles and radiation with surfaces and their penetration into the solid. Chapter 3 contains a brief discussion of the interaction of charged particles, radiation and energetic neutrals with gases and emphasizes those aspects which are relevant to problems of uhv production and instrumentation.

Part B is concerned with the measurement of pressure, both total and partial, in the uhv range. Emphasis is placed on the general processes and problems involved rather than a detailed engineering description of various types of gauge.

Part C contains a discussion of the production of uhv and is concerned with processing, pumping methods, and equipment. Once again, the emphasis is on the principles of the methods and devices rather than on engineering detail.

The manifold applications of uhv are not discussed in detail in this book. Applications in the sciences include the physics and chemistry of surfaces; experiments requiring the maintenance of high gas purity at low pressures; the study of mechanical properties of solids with control of the surface conditions. In brief, any experiment which requires the production of an atomically clean solid surface or high gas purity can best be done with the use of uhv techniques. In engineering we find uhv being used for the simulation of extra-terrestrial conditions, for the production of thin-film and other solid-state devices and many other applications.

In choosing references we have attempted to be critical, rather than inclusive, throughout. For the book as a whole, references published after about the end of 1966 have not been included in the text. Additional references which would have been included if they had been available at the time of writing are listed with titles, in the appendix.

In those areas of science and engineering relevant to uhv the units are more than usually chaotic. The most confused discipline appears to be vacuum technology, where the English-speaking countries use a strange mixture of the metric and British systems (e.g. the speed of a diffusion pump is usually given in litres per second, while its diameter is given in inches). In this book the metric system is used throughout but no attempt has been made to use a consistent set of units within the metric system, rather we have used those units which have most widespread acceptance in the particular field. In some cases we have used different units for the same quantity, depending on the area of science in which it is used. For example, activation energies in adsorption are given in calories per mole, whereas activation energies in electronic processes are given in electron volts per particle. The following table lists the most frequently used symbols and the corresponding units.

INDEX OF SYMBOLS

Listed below are the most frequently used symbols.

A	mass number, number of nucleons	
$b(T)$	Henry's Law constant at fixed T	
c	condensation coefficient	
C	conductance	litre sec^{-1}
C_0	solubility of a gas in a solid	Quantity of gas (cm^3 at STP) in 1 cm^3 of solid at 1 atmosphere external pressure.
D	diffusion constant	cm^2 sec^{-1}
D_s	surface diffusion constant	cm^2 sec^{-1}
e	electronic charge	$1 \cdot 602 \times 10^{-19}$ coulomb
E	potential energy of interaction of two molecules	kcal mole^{-1}
E	activation energy	kcal mole^{-1} or electron volt
E_a	activation energy of adsorption in chemisorption	kcal mole^{-1}
E_d	activation energy of desorption	kcal mole^{-1}
E_D	activation energy for diffusion	cal mole^{-1} or kcal mole^{-1}
E_K	activation energy for permeation	kcal mole^{-1}
E_k	kinetic energy	electron volt
E_m	energy minimum in inter-molecular potential	kcal mole^{-1} or electron volt
E_p	activation energy of desorption for physical adsorption	cal mole^{-1} or kcal mole^{-1}
E_s	activation energy of surface diffusion	cal mole^{-1}
E^*	heat of dissociation	electron volt or kcal mole^{-1}
F	outgassing rate	cm^3 (STP) cm^{-2} sec^{-1}, or torr litre cm^{-2} sec^{-1}, or molecules cm^{-2} sec^{-1}
h	Planck's constant	$6 \cdot 6 \times 10^{-27}$ erg sec
I^+	ion current	ampere
I^-	electron current	ampere
k	Boltzmann's constant	$1 \cdot 38 \times 10^{-16}$ erg °K^{-1}
k	re-emission constant	

Index of Symbols

K	gauge sensitivity factor $$K = \frac{I^+}{I^-}\frac{1}{p}$$	torr^{-1}
K	permeation constant	cm^3(STP)sec^{-1}cm^{-1}atmos^{-1}
L	leak rate	molecules sec^{-1}
m_e	electron mass	$9{\cdot}038 \times 10^{-31}$ kilogram
m	atomic mass	gram
n	molecular density in a gas	molecules cm^{-3}
n_0	number of molecules in a torr litre at 295°K	$3{\cdot}27 \times 10^{19}$ molecules litre^{-1} torr^{-1}
N_A	total number of molecules adsorbed	
N_0	Avogadro's number	$6{\cdot}023 \times 10^{23}$
p	partial pressure	torr
p_0	vapour pressure	torr
P	total pressure	torr
q	heat of adsorption	kcal mole^{-1} or electron volt
q_c	heat of chemisorption	kcal mole^{-1}
q_A	heat of adsorption of atoms in chemisorption	kcal mole^{-1}
q_i	energy of interaction between adsorbed molecules	cal mole^{-1}
q_l	latent heat of vaporization	kcal mole^{-1}
q_p	heat of physisorption	kcal mole^{-1}
Q	collision cross-section	cm^2
r_m	position of energy minimum in intermolecular potential	angstrom or cm
R	gas constant	$8{\cdot}31 \times 10^7$ erg mole^{-1} deg^{-1} or $1{\cdot}9865$ cals. deg^{-1} mole^{-1}
r_0	range constant in Lennard-Jones potential	angstrom or cm
r	distance apart of two molecular centres	angstrom or cm
s	sticking probability	
S	pumping speed	litre sec^{-1}
t	time	sec
T	temperature	°K
T_g	gas temperature	°K
T_r	temperature of molecules reflected from a surface	°K
T_s	surface temperature	°K
v_1	first-order desorption rate constant	sec^{-1}
v_2	second-order desorption rate constant	sec^{-1} cm^2
V	volume	litre

Index of Symbols

V_i	ionization potential	volts
v	molecular velocity	cm sec^{-1}
$\tilde{V}$	molar volume of gas	cm^3 mole^{-1}
Y	sputtering yield	atoms/ion
Z	atomic number	
z	distance from a surface to a molecule	angstrom or cm
z_c	distance from surface to minimum of chemisorption energy	angstrom or cm
z_p	distance from a surface to energy minimum for physically adsorbed molecule	angstrom or cm
α	recombination coefficient	cm^3 sec^{-1}
α	accommodation coefficient	
α	polarizibility of a molecule	cm^3
α-phase	more loosely bonded phase in chemisorption	
β-phase	more strongly bonded phase in chemisorption	
ε	energy constant in Lennard-Jones potential	kcal mole^{-1} or electron volt
ε	Polyani potential $\varepsilon = -RT(\ln p/p_0)$	kcal mole^{-1}
ε_0	permittivity of free space	$8 \cdot 85 \times 10^{-12}$ farad/meter
ζ	relative coverage for a local isotherm	
η	charge-to-mass ratio	coulomb/kg
θ	scattering angle	
θ	relative surface coverage	fraction of a monolayer
λ	electron wavelength $\lambda = \left(\dfrac{150}{V}\right)^{\frac{1}{2}}$	angstrom, with V in volts
λ	molecular mean free path	cm
μ	electric dipole moment of a molecule	esu
ν	specific arrival rate of molecules at a surface	molecules cm^{-2} sec^{-1} torr^{-1}
σ	molecules adsorbed per unit area	molecules cm^{-2}
σ_m	monolayer capacity of an adsorbed layer	molecules cm^{-2}

Σ	total electron back-scattering coefficient	
τ	time of sojourn of a molecule on a surface	sec
τ	time constant	sec
τ_0	nominal vibration time of an adsorbed molecule	sec
τ_s	time of sojourn of a molecule at a surface site before moving to an adjacent site by surface diffusion	sec
τ'	average time of sojourn of a molecule on a surface measured in many experiments	sec
ϕ	electron work-function	volt
χ	electric quadrupole moment of a molecule	esu

Part A

PHYSICAL PROCESSES

CHAPTER TWO
MOLECULE-MOLECULE AND MOLECULE-SURFACE INTERACTIONS

The most frequent spontaneous event in an ultrahigh vacuum (uhv) system is a collision between a gas molecule and a surface. The subsequent behaviour of the impinging molecule, which is controlled by the forces between the molecule and the surface, is a subject of primary concern in the physics of uhv systems. However, many features of the interaction of molecule A and a surface consisting of molecules B can be predicted from a study of the interaction of two isolated molecules A and B. Thus we begin our study of the physics of uhv with a study of the potential energy of interaction (E) of two molecules without relative kinetic energy (Section 2.1). Next we allow the two molecules to collide with kinetic energy (Section 2.2) and find that the first-order deviations from the perfect gas law are simply related to the intermolecular potential. Next we construct the potential between a surface and a single molecule (Section 2.3), then allow the molecule to impact on the surface (Section 2.4). In Section 2.5 the molecule penetrates the surface. The general procedure is to examine processes of progressively greater complexity until we have included most of those processes which occur in normal practice or have been the subject of particular study in uhv systems. It is useful to separate discussions of systems in thermodynamic equilibrium for which general thermodynamic relations may be written, from those in a transient state or in steady state equilibrium, for which the most useful description is often in terms of individual molecular events.

This chapter is restricted primarily to thermal energies. Discussion of energies higher than thermal may be found in Chapter 3 and Chapter 4.

2.1 ENERGIES BETWEEN MOLECULES

Normally, both atoms and molecules will be referred to as molecules, and primary emphasis will be on the ground state of the molecule. As is customary, two molecules with centres (defined in some convenient way) distance r apart, are allowed to approach each other from infinity and the nature and magnitude of the potential energy of interaction (E) between them is examined as a function of r. The relation between potential energy and force (F) is that of the conservative force field:

$$F(r) = -\frac{dE}{dr}; \qquad E(r) = \int_r^\infty F(r)\,dr. \qquad (2.1)$$

Spherical symmetry has been assumed for simplicity. A detailed discussion of the following may be found in Hirschfelder *et al.* (1964), or Beattie and Stockmayer (1951). For molecules in the ground state there are four main contributions to $E(r)$:

$$E = E^{dis} + E^{ind} + E^{es} + E^{val}. \qquad (2.2)$$

E^{dis} is known as the dispersion energy and is present for every pair of molecules. It is the energy responsible for the liquefaction of all gases at sufficiently low temperatures, and is known as the van der Waals energy. E^{ind} is non-zero only if either or both of the interacting molecules have a permanent electric moment (either dipole, μ, or quadrupole, χ). The electric moment of one molecule induces an electric moment in the other molecule and an energy of induction results. E^{es} is non-zero only if both molecules have a permanent electric moment, and arises because of the electrostatic energy of interaction of the two permanent moments. E^{val} is the valence energy responsible both for the binding of atoms into molecules and the short range repulsion of all molecules at sufficiently small values of r. For a sizeable fraction of molecules likely to be encountered in uhv studies $\mu = 0$ and $\chi = 0$; for pairs of these molecules $E^{ind} = 0$, $E^{es} = 0$, $E = E^{dis} + E^{val}$. Thus dispersion and valence energies are of major importance in uhv studies, and form the basis of cryogenic and getter pumps. These energies have no classical analogue and can only be calculated quantum mechanically, while the induction and electrostatic energies can be obtained classically from the appropriate electric moments and polarizibilities. Even E^{dis} and E^{val} can be obtained classically once the charge distribution of the two interacting molecules is known according to the Hellmann-Feynman theorem (Hirschfelder *et al.* 1964); however, the charge distribution or wave function of the system must be obtained from wave mechanics. There is a fifth contribution to E known as the resonance energy which is non-zero only if one or both molecules are in an electronically excited state. The majority of events in uhv systems do not involve excited states and we do not consider E^{res}.

Measured energies will, of course, be quantized values resulting from the insertion of $E(r)$ into a suitable wave equation.

2.1.1 Van der Waal's Energies and Valence Energies

Because of the underlying importance of E^{dis} and E^{val} in uhv studies we have chosen to give a brief discussion of the derivation of these two energies for the hydrogen molecule which illustrates the essential distinction between E^{dis} and E^{val}. Details of the derivation may be found in Schiff (1949) or Herzberg (1950). The hydrogen molecule is first regarded as two separate hydrogen atoms, with nuclei A and B, and electrons 1 and 2, respectively. Thus the initial, so-called unperturbed, energy E^0 of the pair, for use in the Schroedinger equation, consists only of the sum of the separate coulomb energies and is given by

$$E^0 = 3\cdot 32 \times 10^{-8} \left(-\frac{1}{r_{1A}} - \frac{1}{r_{2B}} \right) \text{ kcal/mole,} \qquad (2.3)$$

where r_{1A} and r_{2B} are the electron to proton distances in centimetres for atoms A and B respectively. The ground state electron wave function for the whole system is simply the product of the ground state electron wave functions of the separate atoms $\psi_A(1)$, $\psi_B(2)$ or linear combinations thereof. When the atoms are moved sufficiently close together to interact, the electron spins will be required to become antiparallel ($^1\Sigma$ state) or parallel ($^3\Sigma$ state) and this will restrict the only wave functions of interest to be

$^1\Sigma \qquad \psi_s = \psi_A(1)\psi_B(2) + \psi_A(2)\psi_B(1) \qquad (2.4)$

$^3\Sigma \qquad \psi_a = \psi_A(1)\psi_B(2) - \psi_A(2)\psi_B(1). \qquad (2.5)$

ψ_s is symmetric in the exchange of the two electrons so that multiplication by the antisymmetric spin function of the $^1\Sigma$ state will yield an overall, antisymmetric result. Similarly ψ_a is antisymmetric in the exchange of the two electrons so that multiplication by the symmetric spin function of $^3\Sigma$ state will yield an overall antisymmetric result. When the two atoms are far apart the energy eigen values of the $^1\Sigma$ and $^3\Sigma$ states are the same. When there is weak interaction between the atoms (long range) *both* states show approximately the same reduction in energy (i.e. the atoms attract). This attractive energy which is just E^{dis} is shown by second-order perturbation theory to vary as r^{-6} where r is the distance between the nuclei. The result is illustrated at the right of Fig. 2(1). The attractive nature of E^{dis} is *universal* for all approaching molecules at long range. As the distance between the two hydrogen nuclei is reduced, the energies corresponding to the two eigen functions given by eqns. (2.4) and (2.5) behave quite differently as may be seen from Fig. 2.1. These energies initially of unlike sign arise essentially from the symmetry properties of the wave functions and are defined as E^{val}.

For the $^3\Sigma$ state, E^{val} is positive, and soon overcomes the attractive effect of E^{dis} to yield a shallow minimum of some 7 cal/mole at $r = 4\cdot 5$ Å. The $^3\Sigma$ state of the hydrogen molecule is characteristic of two atoms which do not form a stable chemical compound. On the other hand, for the $^1\Sigma$ state E^{val} is at first negative as r is reduced, thus adding to the attractive effect of E^{dis} and forming a much deeper minimum of 109 kcal/mole at $r = 0\cdot 741$ Å. At even smaller values of r E^{val} is positive for both states and the nuclei are repulsive. The deep minimum at a relatively small value of r is characteristic of two atoms which form a stable molecule. The qualitative basis of the binding energy is that an unpaired electron in each atom can be formed into a pair of electrons with opposing spins in the molecule. The greater the number of unpaired electrons in the atoms the greater the binding energy. In this spin valence theory the

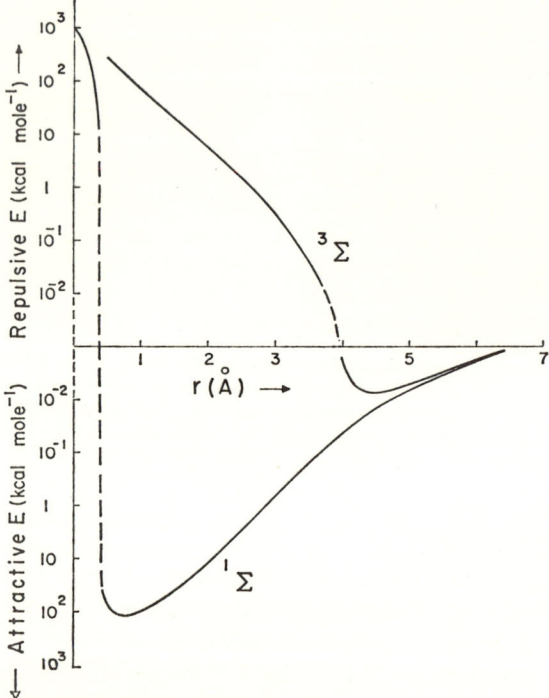

Fig. 2.1 Potential energy of interaction of two hydrogen atoms as function of distance r between nuclei. Figure constructed from values given by Hirschfelder *et al*. (1964) pp. 1061–1063.

valence of an atom is just the number of unpaired electrons in its incomplete outer shell. A considerable body of chemical binding data can be correlated on this basis.

Detailed calculations of E may be made quite simply at long range from the well-known Lennard-Jones potential which actually covers all ranges but is most useful at long ranges:

$$E(r) = 4\varepsilon \left[\left(\frac{r_0}{r}\right)^{12} - \left(\frac{r_0}{r}\right)^{6} \right]. \qquad (2.6)$$

Note that E varies as $-1/r^6$ at long range. Table 2.1 gives values of ε and r_0 for insertion into eqn. (2.6) for interaction between two like molecules. r_0 is the separation distance at which E becomes zero and may be considered as the collision diameter for low energy collisions. Equation (2.6) always generates a minimum $E_m = -\varepsilon$ at $r_m = 2^{1/6} r_0 = 1 \cdot 12 \, r_0$. Values of r_m are given in Table 2.1. As an example of the application of eqn. (2.6) the attractive part of E has been calculated from it in Fig. 2.2 for the interaction of two argon atoms. Equation (2.6) also provides repulsive energies but more accurate values are provided by the machine results of Abrahamson (1963) using a Thomas-

Molecule-Molecule and Molecule-Surface Interactions

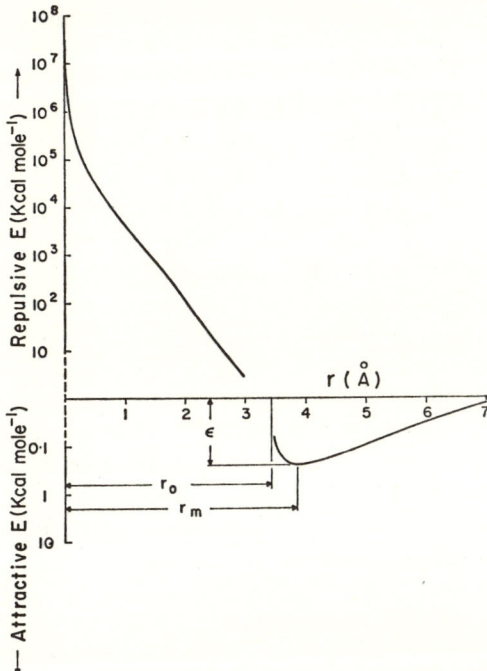

Fig. 2.2 Potential energy of interaction of two argon atoms as function of distance r between nuclei. Figure constructed from values given by Abrahamson (1963) for repulsive energies and Lennard-Jones constants of Table 2.1 for attractive energies.

TABLE 2.1
Constants for the Lennard-Jones potential between gas molecules

$$E = 4\varepsilon\left[\left(\frac{r_0}{r}\right)^{12} - \left(\frac{r_0}{r}\right)^6\right]$$

1	2	3	4	5	6
	ε				1 Atm
Gas molecules	(eV)	(kcal/mole)	r_0 (Å)	$r_m = 1 \cdot 12 r_0$	boiling point (°K)
He—He	0·00088	0·0203	2·556	2·863	4·2
H$_2$—H$_2$	0·00318	0·0735	2·928	3·279	20·4
Ne—Ne	0·00307	0·0709	2·789	3·124	27·3
N$_2$—N$_2$	0·00789	0·182	3·681	4·123	77·4
CO—CO	0·00950	0·219	3·590	4·021	81·9
Ar—Ar	0·0107	0·246	3·418	3·828	87·4
O$_2$—O$_2$	0·00976	0·225	3·433	3·845	90·2
CH$_4$—CH$_4$	0·0118	0·272	3·822	4·281	111·7
Kr—Kr	0·0163	0·377	3·61	4·043	120·3
Xe—Xe	0·0197	0·455	4·055	4·542	165·1
CO$_2$—CO$_2$	0·0163	0·377	3·996	4·476	194·7

Values taken from Hirschfelder *et al.* (1964) p. 1110.

Fermi-Dirac model of the argon atom, and Abrahamson's tabulated results have been used for the repulsive part of the interaction in Fig. 2.2. Equation (2.6) and Abrahamson's results match at $r = 3$ Å but eqn. (2.6) gives higher positive values of E at $r < 3$ Å. A minimum of potential energy analogous to that of the $^1\Sigma$ state of two hydrogen atoms illustrated in Fig. 2.1 is formed for all atoms forming a stable diatomic molecule. Table 2.2 lists the depth of the potential well (dissociation energy) and the magnitude of r at the minimum (internuclear distance) for the common diatomic gases.

A comparison of Tables 2.1 and 2.2 illustrates two general points underlying much of what follows in this chapter.

(1) The values of ε in Table 2.1 are much smaller than the values of the dissociation energies of Table 2.2 In Section 2.2 it will be found that the former are roughly representative of energies of binding of molecules on cryosorption and cryogenic pumps, while the latter are roughly representative of energies of binding of chemically active molecules on getters. Desorption from surfaces in general varies as $\exp(-E/RT)$ and the basic reason why getters operate at room temperature while cryogenic pumps require lower temperatures is already established.

TABLE 2.2
Dissociation energies and internuclear distances for common diatomic gases

1	2	3	4
Gas	Dissociation Energy (eV)	(kcal/mole)	Internuclear Distance (Å)
H_2	4·476	103·2	0·7416
D_2	4·553	105·0	0·7416
CO	11·108	256·2	1·1281
N_2	7·373	170·0	1·094
NO	5·29	122·0	1·1508
O_2	5·080	117·1	1·20739
F_2	<2·75	<63·4	1·435
Cl_2	2·475	57·07	1·988
Br_2	1·971	45·45	2·283
I_2	1·5417	35·55	2·666

Values from Herzberg (1950), Table 39.

(2) If the gases of Table 2.1 are ordered with ascending values of ε then the list will also be approximately an ordering of ascending boiling temperatures, which is also the order of descending difficulty of pumping cryogenically.

2.1.2 Induction Energies and Electrostatic Energies

The induction energy E^{ind} may be derived from classical electrostatics. It can be shown that a dipole polarizes a nearby neutral body and produces an induced dipole which interacts with the field of the original dipole to yield an energy of attraction (averaged over all orientations of the dipole),

$$E^{\text{ind}} = 1\cdot 44 \times 10^{13} \left(-\frac{\mu^2 \alpha_2}{r^6}\right) \text{kcal/mole}, \tag{2.7}$$

where μ is the electric moment of the first molecule in esu and α_2 the polarizibility of the second in cm^3, r being the separation in cm between the two. Values of μ and α are given in Table 2.3.

TABLE 2.3

Electric dipole and quadrupole moment, and polarizibility for common molecules

1 Molecule	2 Dipole moment $\mu \times 10^{18}$ (esu)	3 Quadrupole moment $X \times 10^{26}$ (esu)	4 Polarizability $\alpha \times 10^{24}$ (cm^3)
He	0[a]	0	0·204[d]
H$_2$	0[b]	1·25[c]	0·79[d]
Ne	0[a]	0	0·392[d]
N$_2$	0[b]	1·5[b]	1·76[d]
CO	0·10[a]	1·6[b]	1·95[c]
Ar	0[a]	0	1·63[d]
O$_2$	0[b]	0·55[b]	1·60[c]
CH$_4$	0[a]		2·60[d]
Kr	0[a]	0	2·465[d]
Xe	0[a]	0	4·01[d]
CO$_2$	0[a]	3·1[c]	2·65[c]
H$_2$O	1·85[a]		1·48[d]

[a] Maryott and Buckley (1953).
[b] Townes and Schawlow (1955)
[c] Hirschfelder et al. (1964)
[d] Beattie and Stockmayer (1951)

The electrostatic energy E^{es} can also be obtained from classical electrostatics and is the interaction energy of the permanent charge distributions of both interacting molecules. For large separations an average weighted over all relative orientation yields:

For two dipoles

$$E^{\text{es}} = -1\cdot 44 \times 10^{13} \left(\frac{2}{3kT} \frac{\mu_1^2 \mu_2^2}{r^6}\right) \text{kcal/mole}. \tag{2.8}$$

Note that both eqns. (2.7) and (2.8) yield energies varying as r^{-6}, i.e. in the same way as the attractive portion of E^{dis}.

For one dipole and one quadrupole

$$E^{\text{es}} = -1\cdot 44 \times 10^{13} \frac{1}{kT} \frac{\mu^2 \chi^2}{r^8} \text{kcal/mole}. \tag{2.9}$$

For two quadrupoles

$$E^{es} = -1.44 \times 10^{13} \frac{7}{40kT} \frac{\chi_1^2 \chi_2^2}{r^{10}} \text{ kcal/mole} \qquad (210)$$

where the units of μ, χ are esu as given in Table 2.3, r is in cm, and kT in ergs. Note that these energies depend upon temperature.

It is clear from Table 2.3 that the only molecule common in uhv studies with a significant dipole moment is water.

To compare the magnitudes of the energies given by eqns. (2.6), (2.7), (2.8), (2.9) and (2.10) we have selected combinations from Table 2.3 for explicit calculation at room temperature with $r = 3.828 \times 10^{-8}$ cm, which is the value of r_m for two argon atoms shown in Fig. 2.2.

From eqn. (2.6)	Ar-Ar	E^{dis}	$= -0.246$ kcal/mole
From eqn. (2.7)	H_2O-Ar	E^{ind}	$= -0.025$ kcal/mole
From eqn. (2.8)	H_2O-H_2O	E^{es}	$= -0.86$ kcal/mole
From eqn. (2.9)	H_2O-CO	E^{es}	$= -0.066$ kcal/mole
From eqn. (2.10)	CO-CO	E^{es}	$= -0.006$ kcal/mole

Note that the only energy which exceeds E^{dis} for Ar-Ar is the interaction energy of the two water dipoles, which is the highest such interaction energy likely to be found in uhv studies. The quadrupole moment of the water molecule has not been included in Table 2.3 because it is a true tensor and cannot be expressed as a scalar, but the components of this tensor are comparable in magnitude to χ for CO.

2.2. INTERMOLECULAR COLLISIONS

First we present briefly an important theorem for a gas in thermodynamic equilibrium, which expresses the second virial coefficient of a gas in terms of the intermolecular potential. This formula is not of quantitative importance for uhv studies in the gas phase, but the analogous formula for gas surface interactions is a measure of physical adsorption on surfaces in the system, which may be of considerable importance.

The equation of state of a gas may be expressed as

$$\frac{p\tilde{V}}{RT} = 1 + \frac{B(T)}{\tilde{V}} + \ldots\ldots\ldots\ldots \qquad (2.11)$$

which is known as the virial equation of state, $\tilde{V}$ being the molar volume (cm^3/mole), p the pressure in dynes/cm^2 (1 torr = 1333 dyne/cm^2), R the universal gas constant in ergs/mole deg (8.31×10^7), T the absolute temperature. $B(T)$ is the second virial coefficient, which is a function of temperature and represents the first-order deviation from the perfect gas law. It can be shown by statistical mechanics (Hirschfelder *et al.* 1964) for molecules whose energy of interaction is a function of the intermolecular separation r,

$$B(T) = -2\pi N_0 \int_0^\infty \left[\exp\left(-\frac{E(r)}{RT}\right) - 1\right] r^2 \, dr \qquad (2.12)$$

where N_0 is Avogadro's number ($6 \cdot 023 \times 10^{23}$). From eqn. (2.12) $B(T)$ may be found by integration using $E(r)$ as given for example in Fig. 2.2. For Xe, Hirschfelder et al. (1964) present Table 2.4 comparing experimental values of $B(T)$ with those calculated from the Lennard-Jones potential using the constants given in Table 2.1. It is seen in Table 2.4 that the agreement between theory and experiment is excellent. In general, the value of $B(T)/\tilde{V}$ is too small to be interesting in uhv studies, being only about 0·5% at S.T.P. and even lower at lower pressures.

TABLE 2.4
Experimental and calculated values of $B(T)$ for Xenon

1 Temperature °K	2 $B(T)$ (Experimental) cc/mole	3 $B(T)$ Calc.
298·2	—130·2	—128·4
348·2	— 94·5	— 94·5
373·2	— 81·2	— 81·5
498·2	— 39·1	— 38·9
548·3	— 28·0	— 28·0
573·3	— 23·5	— 23·4

From Hirschfelder et al. (1964) p. 167.

For systems not in thermodynamic equilibrium, it is necessary to examine the details of intermolecular collisions. While the considerations of Section 2.1 rest on firm grounds they do not predict immediately the results of any experiments. The simplest experiments which test the theory involve intermolecular collisions in the gas phase.

Extensive studies have been made of such collisions with crossed molecular beams (Bernstein 1963). The quantities measured are the differential and total elastic scattering cross-sections, the total inelastic cross-sections and the total cross-sections for reaction in the case where chemical reaction can take place. An elastic collision is defined as one in which the internal quantum numbers of the reactants are unchanged, an inelastic collision is one in which these numbers are changed and a reactive collision is one in which the chemical identity of the reactants has changed. The cross-section for an inelastic collision is less important than the other types and may frequently be neglected (Greene et al. 1966).

The information to be derived from elastic scattering of atoms and molecules in the thermal energy range (< 1 eV) has been reviewed by Bernstein (1963). It has been established for the simple systems (K-Hg, Cs-Hg and Li-Hg) that the differential cross-section $I(\theta)$ at small scattering angles (θ) varies as $\theta^{-7/3}$. This relationship is to be expected if the attractive potential $E(r)$ at long range varies as $-1/r^6$ was was found in Section 2.1 for the dispersion energy. It has also been found experimentally for these systems that the total elastic scattering cross-section Q is related to the relative velocity of impact v by the simple relation (Massey and Mohr 1934),

$$Q = \left(1\cdot 76 \times 10^{29} \frac{C}{v}\right)^{2/5} \text{cm}^2 \tag{2.13}$$

where C is the attractive force constant in the Lennard-Jones (6-12) potential and is given by (see eqn. (2.6))

$$C = 4\varepsilon r_0^6. \text{ erg cm}^6 \tag{2.14}$$

Values of Q, using the Lennard-Jones constants of Table 2.1 have been obtained at room temperature ($T = 295°K$) from eqn. (2.13) and are shown in Table 2.5 for a number of gases. These values of Q represent cross-sections for measurable deflections of one gas molecule by another, and a dominant fraction of these collisions are for very small angles of deflection. This is the reason these cross-sections are much larger than cross-sections associated mainly with the repulsive part of the intermolecular potential, namely the mean free path, thermal conductivity, viscosity and gaseous diffusion, where the magnitude of the appropriate cross-section is given approximately by πr_0^2 (Table 2.5). Values of r_0 are given in Table 2.1.

The simple relation for the mean free path λ which arises from the kinetic theory of gases is given by Dushman and Lafferty (1962)

$$\lambda = \frac{1}{2^{\frac{1}{2}} n \pi r_0^2} \text{ cm} \tag{2.15}$$

TABLE 2.5
Collision cross-sections and mean free paths for simple gases
(calculated for $T = 295°K$)

1 Gas	2 $Q \times 10^{15}$ (cm²)	3 $\pi r_0^2 \times 10^{15}$ (cm²)	4 P(torr) *for* $\lambda = 10$ cm
He	5·45	2·05	$1\cdot05 \times 10^{-3}$
H₂	4·40	2·69	$8\cdot03 \times 10^{-4}$
Ne	15·4	2·44	$8\cdot85 \times 10^{-4}$
N₂	46·7	4·26	$5\cdot07 \times 10^{-4}$
CO	47·3	4·05	$5\cdot33 \times 10^{-4}$
Ar	47·2	3·67	$5\cdot88 \times 10^{-4}$
O₂	44·1	3·70	$5\cdot84 \times 10^{-4}$
CH₄	53·6	4·59	$4\cdot70 \times 10^{-4}$
Kr	74·1	4·09	$5\cdot28 \times 10^{-4}$
Xe	115	5·17	$4\cdot18 \times 10^{-4}$
CO₂	83·2	5·02	$4\cdot30 \times 10^{-4}$

where n is the number of molecules/cm^3 and is therefore linearly related to the pressure.

Expressing eqn. (2.15) in terms of the pressure p in torr gives

$$p = \frac{7 \cdot 3T}{10^{20} \lambda \pi r_0^2} \text{ torr} \qquad (2.16)$$

where λ is in cm.

Equation 2.16 has been calculated in Table 2.5 setting $T = 295°K$, $\lambda = 10$ cm which is a typical dimension for a small vacuum system, and using the values of r_0 as given in Table 2.1. At values of p greater than those given in Table 2.5 in a system with a typical dimension of 10 cm, effects of mean free path will be clearly in evidence. About an order in pressure below those of Table 2.5 the first effects of intermolecular collisions will become measurable. While these pressures are high relative to the uhv range, nevertheless calibration of uhv gauges is carried out in this range and hence this range of pressures is important for quantitative work at lower pressures. Deviations from a linear relation between ion current and pressure for an ion gauge may be expected in this pressure range.

Direct measurements of ε (eqn. (2.6)) can be obtained from the magnitude of the so-called 'rainbow angle' (θ_r) in elastic scattering with crossed molecular beams (Bernstein 1963). At a certain scattering angle $\theta_r (\sim 20°)$ the elastic differential scattering cross-section has a maximum. Values of ε may be obtained by this method and are considered accurate to $\pm (5-10)\%$.

For systems of low reduced mass the elastic differential cross-section shows periodic variations with angle which arise from quantum mechanical interferences. The distance r_m of Table 2.1 can be deduced from the periodicity of this variation and are considered accurate to $\pm (10-20)\%$.

Greene et al. (1966) have shown that the basic characteristics of elastic scattering from a spherically symmetric potential can occur also in chemically reactive systems (K+HCl, K+HBr, K+HI, K+CH$_3$Br), but in addition, of course, inelastic and reactive cross-sections are present.

From the extent to which the elastic scattering in these systems differs from elastic scattering in non-reactive systems, conclusions concerning reaction cross-sections may be drawn. Greene et al. (1966) give the following total collision cross-sections at a relative energy of impact of 2 kcal/mole (0·087 eV): K+HCl 4×10^{-16} cm^2; K+HBr 31×10^{-16} cm^2; K+HI 24×10^{-16} cm^2; K+CH$_3$Br 23×10^{-16} cm^2; K+CBr$_4$ 236×10^{-16} cm^2. These cross-sections are relatively insensitive to impact energy near 2 kcal/mole but have the following threshold energies: K+HCl 0·59 kcal/mole; K+HBr 0·15 kcal/mole; K+HI 0·2 kcal/mole. Reactions of this type in the gas phase may thus be of importance when the mean free path for elastic collisions becomes commensurate with the size of the apparatus. Crossed beams of the alkalis and halogens (Datz and Minturn 1964, Cs+Br$_2$; Wilson et al. 1964, K, Rb, Cs+Br$_2$, I$_2$, ICl, IBr) are more reactive than those mentioned above and the main characteristics of elastic scattering are not found. The reaction cross-sections are $\geq 100 \times 10^{-16}$ cm^2, the alkali product recoils forward with respect to the incident alkali-atom and most of the chemical energy released appears as vibrational excitation of the alkali halide.

The components for reactions such as these are not unknown in uhv systems, alkali atoms and ions being emitted frequently from gauge filaments (Riddoch and Leck 1958; Lichtman 1965b) and the halogens being found as minor constituents of residual spectra (Lichtman and McQuistan 1965) but it does not seem likely that these gas phase reactions play a major role in uhv systems.

2.3 ENERGIES BETWEEN A MOLECULE AND A SURFACE

2.3.1 Energies of Physisorption

The equilibrium spacings between two molecules influenced only by dispersion and repulsive valence forces are, in general, larger than the interatomic spacings found in solids and liquids. Further, dispersion forces are additive (Margenau 1939). Thus, in the first approximation, the potential between a molecule and a surface is the sum of the potentials between the molecule and all the atoms of the solid. This summation may be simplified to an integration over a semi-infinite continuum. Figure 2.3 shows the required co-ordinate system for calculation of the energy of physical adsorption or physisorption.

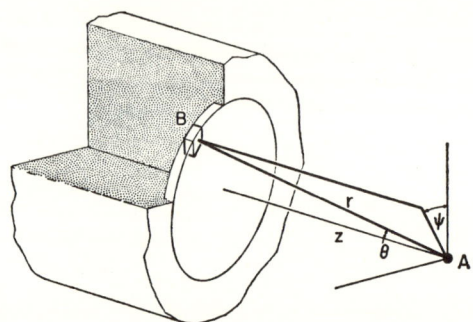

Fig. 2.3 Co-ordinate system for integrating intermolecular potential energy to yield potential energy between molecule and surface. A is molecule, B is the solid.

For a single molecule at A the volume element at B supplies a potential energy from eqn. (2.6)

$$dE = N4\varepsilon \left[\left(\frac{r_0}{r} \right)^{12} - \left(\frac{r_0}{r} \right)^{6} \right] r^2 \sin\theta \, dr \, d\theta \, d\psi \qquad (2.17)$$

where N is the number of atoms/cm^3 in the semi-infinite continuum and the Lennard Jones 6–12 potential has been chosen to represent the intermolecular potential. This integration may be carried out by elementary methods to yield

$$E = 4\pi\varepsilon N r_0^3 \left[\frac{1}{45}\left(\frac{r_0}{z}\right)^9 - \frac{1}{6}\left(\frac{r_0}{z}\right)^3\right] \tag{2.18}$$

where z is the distance from the centre of the first layer of atoms in the solid to the centre of the single molecule.

It may be noted that: The 6–12 intermolecular potential has been reduced to a 3–9 molecule-surface potential; the maximum energy of binding has been increased from ε to $2 \cdot 2\ N\varepsilon r_0^3$; the location of the zero of E has been reduced from r_0 to $r_0/1 \cdot 4$; and the location of the minimum in E (i.e. E_p) has been reduced from $1 \cdot 12\ r_0$ to $z = r_0/1 \cdot 16$. If the atoms of the solid are spaced differently in a direction parallel to the surface rather than in a direction perpendicular to it, then there may be simpler integrations than that given above. For example, on the basal face of graphite the interlayer spacing is $3 \cdot 35$ Å, while the average interatomic spacing in a layer is $1 \cdot 6$ Å. In this case it is more appropriate to integrate eqn. (2.17) over the surface plane only. This leads to the expression

$$E = 4\pi\varepsilon N_A r_0^2 \left[\frac{1}{6}\left(\frac{r_0}{z}\right)^{10} - \frac{1}{3}\left(\frac{r_0}{z}\right)^4\right] \tag{2.19}$$

where N_A is the number of atoms cm^{-2} in the solid surface. This is known as the 4–10 molecule-surface potential. The forms of eqns. (2.18) and (2.19) have been widely used to determine empirically the values of the coefficients, the calculated values being too approximate. From experimental data it is frequently not possible to establish whether eqn. (2.18) or eqn. (2.19) (or any one of other similar equations) should be used. Values obtained from experimental data on a highly homogeneous adsorbent, assuming the validity of eqn. (2.18) and eqn. (2.19), are shown in Table 2.6.

TABLE 2.6

Values of E_p obtained from experimental data assuming 3-9 and 4-10 molecule-surface interaction potentials for physical adsorption on a homogeneous surface

1	2	3	4
Adsorbent	Gas	$-E_p$ (3-9) cal/mole	$-E_p$ (4-10) cal/mole
Graphitized Carbon Black (homogeneous)	Neon	758[a]	729[a]
	Argon	2200[a]	2130[a]
	Krypton	2900[a]	2790[a]
	Xenon	3800[a]	3690[a]
	Hydrogen	1141[b]	1097[b]
	Deuterium	1151[b]	1109[b]
	Methane	2947[b]	2846[b]
	Deutero-methane	2881[b]	2846[b]

[a]Sams et al. 1960)   [b]Constabaris et al. (1961)

The variations of E_p (minimum value of physisorption energy) obtained using one model or the other are not great, but the values of E_p are several times the value of ε given in column 3 of Table 2.1. Representative values of z_p (distance from surface to energy minimum) are given in Table 2.7. They are, in general, smaller than the values of r_m given in column 5 of Table 2.1. Theoretical questions arise (Young and Crowell 1962) when the integration of (2.17) is used for the energy between an adsorbed molecule and a metal since the electrons in the latter are mobile, but most theories give the z^{-3} dependence of eqn. (2.18) for the attractive energy. Pierotti and Halsey (1959) have used this dependence in a comparison between the various theoretical values for the coefficient in eqn. (2.18) and experimental data for krypton adsorption on iron, copper, sodium and tungsten.

TABLE 2.7
Values for distance from surface to energy minimum (z_p) for physical adsorption

1	2	3
Adsorbed molecule	Adsorbing surface	
	P^{33} Graphitized carbon z_p (A)	Porous glass z_p (A)
H_2	2·62	1·86
D_2	2·62	
CH_4	3·71	
CD_4	3·74	
O_2		1·89
N_2		1·99
He		1·87
Ne	3·62	1·93
Ar	3·67	2·20
Kr	3·76	
Xe	3·99	

from Halsey (1961)
The reader should refer to Halsey (1961) for detailed derivation of these distances.

Spatial variations of E will exist even on perfect crystal faces and summations over individual atoms are clearly required to handle cases of this type. A number of such calculations have been made Young and Crowell 1962), particularly for adsorption on graphite for which the result has been obtained that the variations are relatively minor as shown in Table 2.8. Values of z_p are also given in Table 2.8. Note the good agreement between the calculated value of z_p for Ar on graphite with that deduced from experiment given in column 2 of Table 2.7. For the adsorption of argon on the ionic solid sodium chloride, however, the variations over the surface are greater (Table 2.8). Ricca (1967), in a paper contrasting integration of eqn. (2.17) with atom by atom summation, has calculated large spatial variations for the adsorption of all the noble gases on noble-gas crystals. The atom by atom approach has been

Molecule-Molecule and Molecule-Surface Interactions 25

used by Ehrlich and Hudda (1959) in analysing their field emission microscope data for the adsorption of Xe on different crystal faces of W. Least binding is obtained for the (110) face and greatest binding for the (116) face in semiquantitative agreement with the observations. Goodman (1967), however, does not consider the principle of atomic summation to be firmly established.

Van der Waals energies between a gas molecule and an atomically smooth surface (Table 2.6 and Table 2.8) are not greatly different from the energies between the same gas molecule and its own liquid (energy of vaporization). This may be seen by comparing Table 2.9 with Tables 2.6 and 2.8. The energies of vaporization of course determine the ordering of the boiling temperatures (Table 2.1). Energies of sublimation of common metals are also given in Table 2.9.

TABLE 2.8
Calculated variations of adsorption energy with location on single crystals for physical adsorption

1 Gas	2 Adsorbent	3 Description of site	4 z_p (A)	5 E_p (cal/mole)	6 Reference
Argon	Graphite	Above centre of lattice hexagon	3·55	1803	Crowell & Young (1953)
		Above midpoint between two carbon atoms	3·60	1778	,,
		Above a carbon atom	3·60	1771	,,
Argon	Sodium Chloride	Centre of lattice cell		1875	Hayakawa (1957)
		Midpoint of lattice edge		1624	,,
		Over sodium ion		1978	,,
		Over chlorine ion		1510	,,

Based on Tables of Young and Crowell (1962)

TABLE 2.9
Energies of vaporization of the common gases and energies of sublimation of common metals

Gas	He	H_2	Ne	N_2	CO	Ar	O_2	CH_4	Kr	Xe	CO_2
Energy of Vaporization (kcal/mole)	0·020	0·215	0·431	1·333	1·444	1·558	1·630	1·955	2·158	3·021	4·041

Metal	Ba	Ag	Cu	Ni	Ti	Pt	Mo	C	Ta	W
Energy of Sublimation (kcal/mole)	42	68	81	101	113	122	155	171	180	202

When the surface is not flat there will be a tendency for an adsorbed molecule to migrate to the various defects, steps, fissures, cracks, etc. where the adsorbed molecule can 'make contact' with a greater number of surface atoms and hence be bound to the surface with a higher energy. A discussion of surface defects has been given by Dunning (1967). Local fields capable of polarizing the adsorbed molecule may arise. When, in addition, the surface is made up of a series of unlike atoms (for example, a glass surface) then the calculations of the adsorption energy from first principles is a very complex task. Such surfaces are classed in general as 'heterogeneous' and it is not possible to characterize the interaction of a molecule with such surfaces by a single valued energy, except at high temperature and low coverage where an effectively single value E_p is obtained. Table 2.10 shows a series of such average energies obtained by Steele and Halsey (1954, 1955). They are higher than those of Table 2.6 for the same molecule adsorbed by about a factor 2. A theoretical analysis of low pressure isotherms by Hobson (1965b) has given the distribution of surface energies for argon on glass, shown in Fig. 2(4) by the curve $q_i = 0$, representing no energy of interaction between adsorbed molecules. The analysis also suggests that the distribution of energies on many heterogeneous surfaces for the same adsorbed molecule may be semiquantitatively the same as that shown in Fig. 2.4. The same general distribution of energies has been found (Hobson 1965b) for N_2 and He on a heterogeneous surface although the magnitudes of the adsorption energies are different. At very low coverages adsorbed argon atoms fill the high energy sites of Fig. 2.4 and their mean energy is high corresponding roughly to the value in Table 2.10. It is possible

TABLE 2.10
Values of E_p for physical adsorption of common gases on heterogeneous surfaces at high temperatures and low coverages
(kcal/mole)

Solid	Gas							
	He	H_2	Ne	N_2	Ar	O_2	CH_4	Kr
Porous glass	0·68	1·97	1·54	4·26	3·78	4·09		
Saran Charcoal	0·63	1·87	1·28	3·70	3·66		4·64	
Carbon black	0·60		1·36		4·34			
Alumina					2·80			3·46

(from Steele and Halsey 1954, 1955)

that even at very low coverages the adsorbed molecules are grouped together on the surface and hence energies of interaction between adsorbed molecules add to the energy of binding to the surface. The experiments upon which Fig. 2.4 is based do not distinguish between these contributions. The curve labelled $q_i = 1558$ cal/mole represents the distribution of surface energies in the event that the adsorbed molecules are bound to each other with the full heat of vaporization (Table 2.9). The curve $q_i = 500$ cal/mole represents an intermediate case which is more likely in practice. As the coverage increases the mean binding energy decreases until, when the surface is covered with several layers of adsorbate, the binding energy passes over into the energy of vaporization or sublimation, depending on the temperature. This transition is important in uhv systems since the coverage on surfaces under ambient background conditions corresponds usually to a state of very low coverage, with energies of adsorption similar to those of Table 2.10, whereas after exposure to large gas loads found, for example, in cryogenic pumps, a sufficient quantity of adsorbate is present to change the heat of adsorption into the heat of vaporization or sublimation (Table 2.9). This transition is general for all gases.

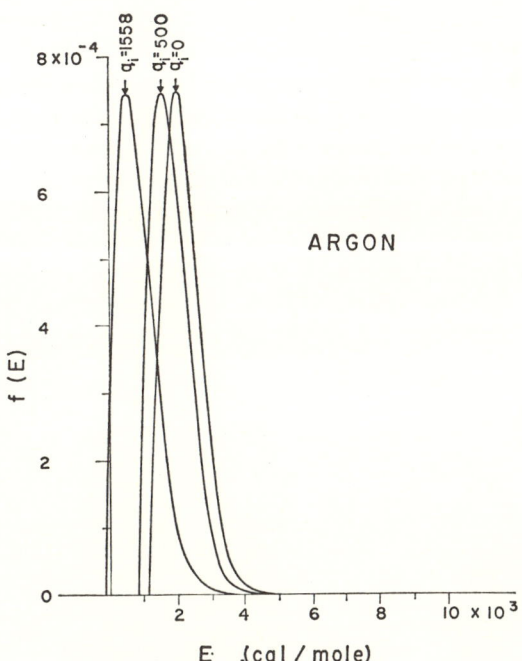

Fig. 2.4 Distribution of energies of physical adsorption for argon on a heterogeneous surface. $f(E)$ is fraction of surface area with energy between E and $E+dE$. q_i (cal/mole) is energy of interaction between adsorbed atoms. (From Hobson 1965b).

As noted above, local fields on a heterogeneous surface may polarize adsorbed molecules and contribute a term to the adsorption energy analogous to E_{ind} (Section 2.1.2). The presence of polarization can be observed directly as the change of surface potential, or work function, when a gas is adsorbed on a surface (Eberhagen 1960; Tompkins 1967). Most studies of this subject have been made for chemically adsorbed gases or in conventional vacuum systems, but in the field emission microscope under uhv conditions it has been found that physical adsorption of the inert gases reduces the work function of W (Ne, Ar, Xe on W, Gomer 1958; Ar, Kr, Xe on W, Ehrlich and Hudda 1959). Tompkins (1967), following Ehrlich and Hudda (1959), estimates that 20% of the adsorption energy of Xe on W could be contributed by polarization of the Xe atom by local fields.

The interactions between the molecules adsorbed on a surface are complex, involving all the considerations of intermolecular energies presented in Section 2.1 in addition to the consideration that the molecule polarized by the surface becomes, in effect, polar for purposes of intermolecular calculations. A recent example of this type of calculation, predicting isotherms in the uhv range, has been performed by Taylor et al. (1965) for He on porous glass. Despite all these complexities of physical adsorption on heterogeneous surfaces the empirical results, as will be seen later, are average results and present a much simpler picture.

An important quantity in the determination of the non-equilibrium properties of adsorption is the saddle of potential energy existing between adjacent adsorption sites. This saddle is called the activation energy of surface diffusion, E_s. It may be calculated for simple situations from the considerations leading to Table 2.8 (see also Ricca (1967) and Goodman (1967)) but

TABLE 2.11
Energies of surface diffusion (E_s) and desorption (E_p) in physical adsorption

1 Absorbed molecule	2 Adsorbent	3 E_s (cal/mole)	4 E_p (cal/mole)	5 E_s/E_p	6 Reference
N_2	Porous Glass	1600	3200	0·50	Barrer (1954b)
O	,, ,,	1800	3000	0·60	
Ar	,, ,,	1400	2600	0·54	
Kr	,, ,,	1800	3300	0·54	
CH_4	,, ,,	1500	3000	0·50	
C_2H_6	,, ,,	3200	5300	0·60	
Ne	Saran Charcoal	1250	1650	0·76	Dacey (1961)
H_2	,, ,,	1850	2350	0·79	
D_2	,, ,,	1850	2400	0·77	
Ar	,, ,,	3800	4900	0·78	
CH_4	,, ,, Single crystal tungsten	4450	5900	0·75	
Ar	(310) → (100)	>1100	1900	>0·58	Ehrlich & Hudda (1959) Ehrlich (1963b)
Kr	(310) → (100)	600	4500	0·13	
Xe	(111) → (111)	2000	8000	0·25	
Xe	(210) → (100)	3800	8000	0·47	
CO_2	CO_2/W	2400	5500	0·43	Gomer (1959)
CO	CO/W	[900]	[2300]	0·39	

The symbols X/W refer to an X covered tungsten surface. Quantities in brackets refer to preliminary values or rough estimates.

Molecule-Molecule and Molecule-Surface Interactions

in general the situation is too complex for calculation, Table 2.11 lists energies of surface diffusion obtained from experimental data. Desorption energies which may be identified with E_p for physical adsorption are included for comparison. Reference will be made later to the ratio E_s/E_p.

2.3.2 Energies of Chemisorption

The basic distinction between physisorption and chemisorption energies has been illustrated in Fig. 2.1 by the $^3\Sigma$ and $^1\Sigma$ states of the hydrogen molecule respectively. Chemisorption or valence energies are larger and of shorter range and for this reason the integration of an equation like eqn. (2.17) is not valid and indeed chemisorption energies between a molecule and a surface retain more of the characteristics of the chemical bond between two molecules than do physisorption energies.

The main distinctions between chemisorption and physisorption are now compared:

(i) The energies of chemisorption are larger than those of physisorption. An arbitrary, and sometimes poorly defined, dividing energy is set at 10 kcal/mole. Chemisorption energies range up to some 200 kcal/mole. As a result the range of absolute temperatures over which chemisorption is significant is some twenty times that for physisorption.

(ii) Equilibrium distances between the surface and an adsorbed atom or molecule are shorter in chemisorption, being typically between 1 and 3Å, whereas in physisorption they are typically nearer 4Å (see Tables 2.2 and 2.7).

(iii) Chemisorption is highly specific to the nature of the adsorbent whereas physisorption is not. The magnitude of the heat of adsorption depends not only on the nature of the adsorbent but also may differ markedly from one crystal face to another on the same material. For example, Delchar and Ehrlich (1965) found that nitrogen adsorbs on tungsten (100) with a heat of 75 kcal/mole at 300°K, whereas no adsorption at all was observed on the (110) surface at 300°K. The specificity of various metals towards the chemisorption of simple gases at 300°K is indicated in Table 2.12 which is taken from Bond (1962). The experimental magnitude of the heats of adsorption decreases in the following sequence

$$O_2 > C_2H_2 > C_2H_4 > Co > H_2 > CO_2 > N_2$$

from measurements on evaporated films of metals.

The heats of chemisorption of the seven gases (H_2, O_2, N_2, CO, CO_2, C_2H_4 and NH_3) on different metals follow the order

Ti,Ta > Nb > W, Cr > Mo > Fe > Mn > Ni, Co > Rh > Pt, Pd > Cu, Au.

(iv) Chemisorption ceases when the adsorbate molecule can no longer interact with the unsatisfied valence bonds of the surface atom. Thus after the formation of one or two layers of a given gas on a surface, the surface appears to additional impinging molecules as if it were the condensed phase of

TABLE 2.12

Classification of metals and semi-metals based on adsorption properties at 300° K
(*A* indicates adsorption, NA no adsorption)

Group	Metals	Gases						
		O_2	C_2H_2	C_2H_4	CO	H_2	CO_2	N_2
A	Ca, Sr, Ba, Ti, Zr, Hf, V, Nb, Ta, Cr, Mo, W, Fe[a], Re[c]	A	A	A	A	A	A	A
B_1	Ni, Co[c]	A	A	A	A	A	A	NA
B_2	Rh, Pd, Pt, Ir[c]	A	A	A	A	A	NA	NA
C	Al, Mn, Cu, Au[b]	A	A	A	A	NA	NA	NA
D	K	A	A	NA	NA	NA	NA	NA
E	Mg, Ag[a], Zn, Cd, In, Si, Ge, Sn, Pb, As, Sb, Bi	A	NA	NA	NA	NA	NA	NA
F	Se, Te	NA	NA	NA	NA	NA	NA	NA

[a]The adsorption of N_2 on Fe is activated, as is the adsorption of O_2 on Ag films sintered at 273°K.
[b]Au does not adsorb O_2.
[c]Metal probably belongs to this group, but the behaviour of films is not known.
(Table from Bond 1962).

the adsorbate with binding energies similar to enegies of vaporization (Table 2.9). In uhv systems this sudden change of energy appears as the saturation of a getter.

(v) Chemisorption is often, but not always, accompanied by dissociation of a diatomic (or polyatomic) adsorbate molecule and may exhibit an activation energy. This situation is depicted in Fig. 2.5 for a diatomic molecule. The line marked PA_2 refers to the energy of physisorption of a diatomic molecule containing 2 atoms A. E_p and z_p have the same meaning as in Section 2.3.1 with numerical values given in Tables 2.6 and 2.7. If, however, there is a net gain in attractive energy when the molecule is dissociated ($A_2 \rightarrow 2A$) far from the surface by supplying it an energy E^* (Table 2.2) and then bringing it to the surface along line C as two atoms, then clearly this dissociative adsorption will be thermodynamically more stable than molecular adsorption. The energy condition for dissociative chemisorption is therefore (see Fig. 2.5)

$$q_c = 2q_A - E^* > E_p. \qquad (2.20)$$

q_c is expressed per mole of undissociated gas, q_A is the binding energy per mole of atoms to the surface. A mechanism for dissociation of the molecule far from the surface is usually absent and dissociative chemisorption will occur only if some mechanism is present for supplying the energy E_a, the activation energy of adsorption, if this should happen to be greater than zero. It is clear from Fig. 2.5 that the value of E_a will depend upon the detailed shapes of the curves P and C and may range from zero to E^*.

Gases that do not dissociate on adsorption may also exhibit an activation energy of adsorption because electrons within the molecule must rearrange before bonding to the surface is possible.

When a chemisorbed molecule desorbs from the surface it must receive an activation energy of desorption E_d

$$E_d = E_a + q_c. \tag{2.21}$$

The net energy required to desorb a molecule is, however, q_c. Thus E_d is measured in desorption rate experiments (Section 2.4.4) while q_c is measured as a heat in measurements near equilibrium, and this is the reason the symbol E has been changed to q. In the majority of cases of chemisorption of simple gases on clean metals the activation energy of adsorption is very small and the heat of adsorption equals the activation energy of desorption. This is always true for physical adsorption; thus $E_p = q_p$ in Fig. 2.5.

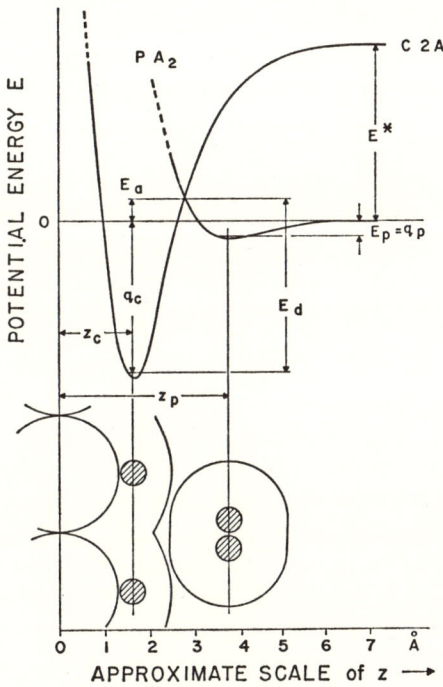

Fig. 2.5 Potential energy diagram illustrating dissociative chemisorption of a diatomic gas molecule. Cross-hatched circles symbolize nuclei of atoms. (Based on diagram from Bond 1962).

There are two main methods for measuring q, from isotherm data and from calorimetric measurements. From isotherm data the isosteric heat of adsorption is obtained

$$q_{st} = RT^2 \left(\frac{\delta \ln p}{\delta T}\right)_\theta. \tag{2.22}$$

If the heats are measure isothermally at particular values of θ, when no external work is done during adsorption, the differential heat q_{diff} is obtained. It can be shown that

$$q_{diff} = q_{st} - RT. \qquad (2.23)$$

When heats are measured calorimetrically the differential heat so measured is probably intermediate between q_{st} and q_{diff}. In chemisorption the difference between the two heats is usually less than the errors of measurement and can be ignored.

Since the bonding in chemisorption occurs via the unsatisfied valence bonds of the surface atoms we might expect the heats of adsorption to be similar in magnitude to the heats of formation of the corresponding chemical compounds. Table 2.13 shows the heats of adsorption at low coverage (the initial heats), measured calorimetrically on evaporated metal film, for O_2, CO

TABLE 2.13

Initial heats of adsorption of oxygen bearing gases on evaporated films

1	2	3	4	5	6	7	8	9
Metal	$q(O_2)$	Stable oxide	Heat of formation of oxide	$q(CO)$	Carbonyl	Heat of formation of carbonyl	$q(CO_2)$	$q(CO_2)$ calc.
Al	211	Al_2O_3	266					
Ti	236	Ti_3O_2	235	153			191	203
Cr	174	Cr_2O_3	183					
Mn	150	Mn_2O_3	153	78			63	85
Fe	136	Fe_3O_4	134	46	$Fe(CO)_2$	28	68	46
Co	100	Co_3O_4	102	47			37	29
Ni	107	NiO	116	42	$Ni(CO)_4$	35	54	28
Nb	208	Nb_2O_5	182	132			152	168
Mo	172	MoO_2	131	74	$Mo(CO)_6$	36	109	92
Rh	102	Rh_2O	48	46			—	29
Pd	68	PdO	42	43			—	49
Ta	212	Ta_2O_5	193	134			182	172
W	194	WO_2	134	126	$W(CO)_6$	42	122	155
Pt	70	PtO	34	48			—	15

All values in kcal/mole.
Compiled from Brennan *et al.* (1960), Brennan and Hayes (1965) and Brennan and Hayward (1965).

and CO_2. The heats of adsorption for oxygen (Brennan *et al.* 1960) are compared with the heats of formation of the corresponding oxide (columns 3, 3 and 4) and the agreement is seen to be reasonable. When we compare the initial heats of adsorption of carbon monoxide with the heats of formation of metal carbonyls (Brennan and Hayes 1965), for the few instances where the heats of formation are known, we see (columns 5 and 7) that there is no real correspondence. One would not expect the energy of the bridged surface bond for adsorbed CO (the commonest CO bond configuration) to be similar to that of a molecule, which is one of many linearly bonded to a common metal atom. Column 8 lists the heats of adsorption of CO_2 measured by Brennan and Hayward (1965) who suggest that CO_2 is dissociatively adsorbed as CO and O and that the theoretical value of the heat of adsorption would then be given by

$$q(CO_2)_{calc} = q(CO) + \tfrac{1}{2}q(O_2) - 68. \qquad (2.24)$$

For simplicity q_c has been written as q.

These calculated values are given in column (9). It can be seen that there is a reasonable correspondence between $q(CO_2)$ and the calculated value except for a few cases. This indicates that in most cases dissociative adsorption occurs.

The heat of adsorption of most gases on polycrystalline adsorbates is observed to decrease markedly as the coverage increases. This behaviour is exemplified by the data of Table 2.14 for hydrogen on various evaporated metal films. Figure 2.6 shows the heats of adsorption of CO on W as a function of coverage derived from different experimental methods.

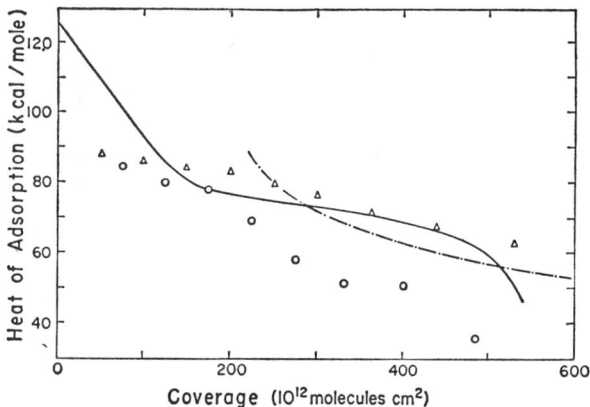

Fig. 2.6 Heats of adsorption of CO on W as a function of coverage. Solid line, calorimetric measurements (Brennan and Hayes 1965); Δ, O the integral and differential heats from flash desorption (Ehrlich 1961b); dashed line, differential heats from field emission microscopy (Gomer 1961). (from Brennan and Hayes 1965).

This decrease of heat with coverage is discussed in detail in Bond (1962) and Hayward and Trapnell (1964). Variation of q with θ leads to isotherms of the Freundlich type (when $dq/d\theta = -q_m/\theta$) or the Temkin type (when $dq/d\theta = -\alpha$). The decrease of q with θ can be accompanied by an increase of the activation energy of adsorption with θ. This type of behaviour is indicated schematically in Fig. 2.7.

The specificity apparent in the data on heats of chemisorption presented above has encouraged considerable efforts to correlate the surface phenomena

TABLE 2.14
Heats of adsorption of hydrogen (kcal/mole) on evaporated metal films

Metal	$q(H_2)$ at $\theta = 0$	$q(H_2)$ at $\theta = 0\cdot 2$
Cr	45	
Ni	29	27
Mo	40	
Rh	26	25
Pd	27	
Ta	45	
W	52	39

(from Beeck 1950; Stevenson 1955; Wahba and Kemball 1953).

with the bond structure of the solid or with other parameters of the adsorbent or adsorbate. It has been widely held that chemisorption occurs via a covalent bond and a hole in the d-band of the metal. This postulate has been elaborated by Dowden (1950), Dilke *et al.* (1948) and Beeck (1950). Ehrlich (1963b) and (1964a) has demonstrated that this postulate, though attractively elegant, does not fit the observations when closely examined.

Other attempts have been made to calculate heats of adsorption; these methods are reviewed in some detail by Hayward and Trapnell (1964) and

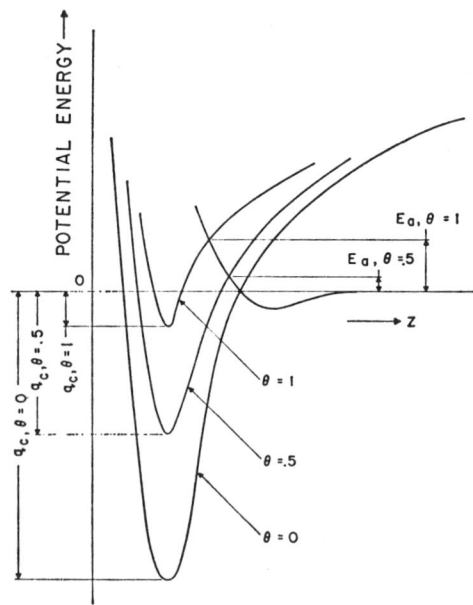

Fig. 2.7 Schematic variation of activation energy of adsorption, E_a, as heat of adsorption, q_c, declines with increasing coverage, θ. (Based on diagram from Bond (1962)).

Bond (1962). Most of these methods, which attempt to relate surface behaviour to bulk properties, are based on inadequate theoretical foundations. Some useful empirical correlations have been determined, for example Brennan has shown that the initial heats of adsorption of O_2, CO, CO_2 and C_2H_4 on evaporated metal films are apparently correlated to the metallic radius as shown in Figs. 2.8 (a, b, c, d). These curves show two branches with the group VIII metals on the lower branch.

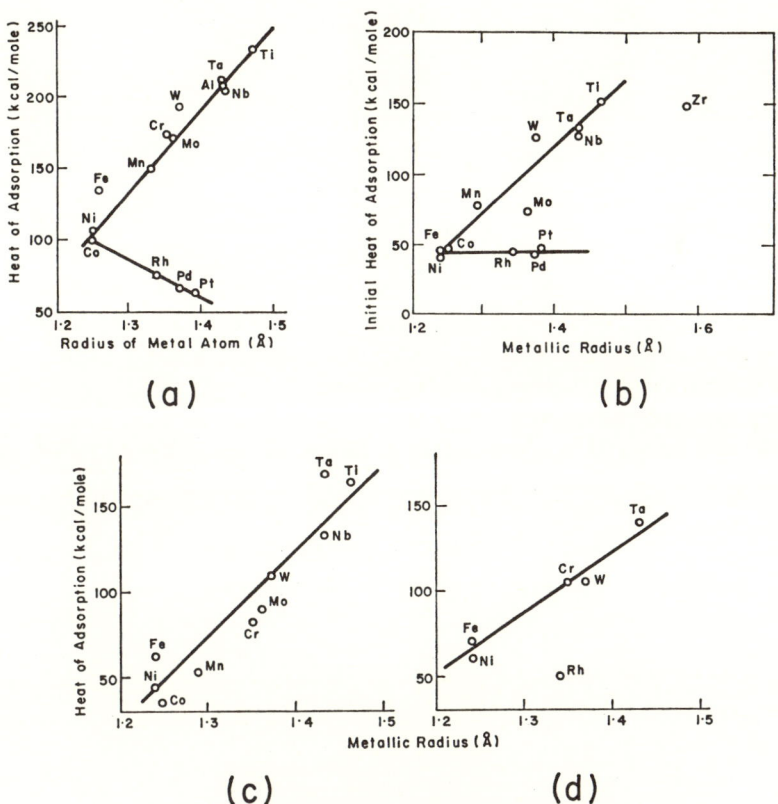

Fig. 2.8 Initial heats of adsorption as a function of metallic radius:
    (a) O_2 (Brennan et al. 1960)
    (b) CO (Brennan and Hayes 1965, Fig. 4).
    (c) CO_2 (Brennan and Hayward 1965, Fig. 5).
    (d) ethylene (Brennan and Hayward 1965).

Measurements of the kinetics of chemisorption on metal surfaces, other than evaporated films, by methods such as flash desorption, field emission and field-ion microscopy have clearly demonstrated the existence of multiple binding states on the same surface. Figure 2.9 shows the desorption spectra for CO on W, for various amounts adsorbed, as an example of multiple binding

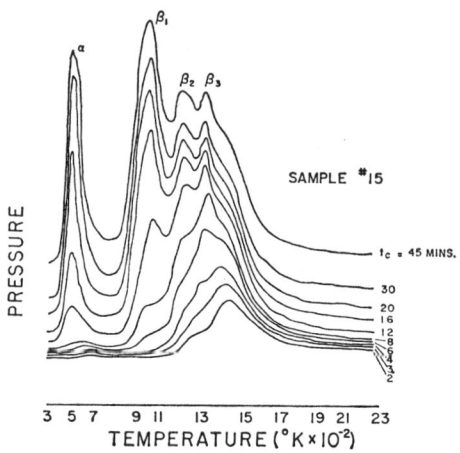

Fig. 2.9 Desorption spectra for variation adsorption times of CO on W at a temperature of 300°K. The zero level of these curves has been shifted to prevent overlap (Redhead 1961).

states. These desorption spectra were obtained after the desorption time indicated at the right of each curve. At least four adsorbed phases are clearly differentiated, the α-phase with low binding energy (∼25 kcal/mole) and the three β-phases with binding energies of about 60, 70 and 76 kcal/mole.

The multiple states of binding observed result both from surface heterogeneity and from the formation of different bonding configurations on the same crystal surface. The assignment of the various observed binding states to heterogeneity or other causes is still a matter of some controversy in several cases. However, recent work using field-ion and field emission microscopy has greatly advanced our knowledge of the various adsorption states on refractory metals. For example, Rootsaert et al. (1962) have shown by field emission microscopy that the adsorption states of CO on W (see Fig. 2.9) were associated with the different crystal faces as shown in Table 2.15. The only adsorbed state which occurs on all crystal faces is the α-state. It has been proposed by several authors that the α-state is linearly bonded via the C atom to one

TABLE 2.15
Adsorption states of CO on W

State	q_c (kcal/mole)	Surface dipole
α	20	Positive, all faces
β_1	52	Negative, (411) (310) (210)
β_2	70	Negative, (111) and others
β_3	100	Negative, (111) zone

(from Rootsaert et al. 1962).

tungsten atom whereas the β states are bonded to two tungsten atoms. Measurements of isotopic exchange, which occurs in β states only (Madey *et al.* 1965) suggest that the β-CO lies flat with both the C and O bonding to W. Other evidence from electron energy-loss measurements also suggest that the β-CO is a lying down configuration.

The detailed arrangements of the adsorbed states and the adsorption kinetics of the simple diatomic gases (N_2, CO, H_2, NO) on refractory metals has been studied in some detail and the reader is directed to the review articles of Ehrlich (1963a, 1963b, 1964a, 1964b) and Gomer (1966).

2.4 DYNAMIC INTERACTIONS BETWEEN A MOLECULE AND A SURFACE

As in the case of intermolecular collisions, equilibrium (constant T, no mass transfer) phenomena are independent of the details of the collision between a gas molecule and a surface, depending only on the potential discussed in the previous sections. Some of these equilibrium phenomena, particularly the vapour pressure and physical adsorption isotherms, are of practical interest in uhv systems. Equilibrium phenomena are discussed in Sections 2.4.1 and 2.4.2. followed by a discussion of phenomena in which there is a transfer of mass and/or the temperature is a function of time or of position (Sections 2.4.3, 2.4.4., 2.4.5).

2.4.1. Vapour Pressure

The Clausius-Clapeyron equation

$$\frac{dp_0}{dT} = \frac{q_l}{T\Delta V} \tag{2.25}$$

gives the relation between vapour pressure p_0 and the latent heat of vaporization q_l. T is the absolute temperature, and ΔV is the molar change in volume during a transition from condensed phase to saturated vapour. Generally the volume of vapour is large compared to the volume of the same mass of liquid, i.e. $\Delta V = RT/p_0$. Thus

$$\frac{dp_0}{dT} = \frac{p_0 q_l}{RT^2}. \tag{2.26}$$

This may be integrated to yield

$$p_0 = C \exp\left(-\frac{q_l}{RT}\right) \tag{2.27}$$

if q_l is assumed constant. For values of p_0 at temperatures near the normal boiling point (Table 2.1) this equation may be used in conjunction with values of q_l from Table 2.9. C may be obtained from a known combination of values of p_0, q_l and T. Equation 2.27 is based on several assumptions which are not

TABLE 2.16
Vapour pressure data for some common gases

Symbol	Compound	Data temp. range, °K	Temperatures (°K) for vapour pressures (torr)														
			10^{-13}	10^{-12}	10^{-11}	10^{-10}	10^{-9}	10^{-8}	10^{-7}	10^{-6}	10^{-5}	10^{-4}	10^{-3}	10^{-2}	10^{-1}	1	10^1	10^2	10^3
Ar	Argon	82– 88	20·3	21·3	22·5	23·7	25·2	26·8	28·6	30·6	33·1	35·9	39·2	43·2	48·2	54·4	62·5	73·4	ⓜ89·9
Br₂	Bromine	253–331	102·0	106·5	111·0	116·5	122·0	128·5	135·5	143·5	152·5	163·0	174·5	188·5	204·	224·	248.	ⓜ282·	339·
CH₄	Methane	48–112	24·0	25·3	26·7	28·2	30·0	32·0	34·2	36·9	39·9	43·5	47·7	52·9	59·2	67·3	77·7	91·7	115·0
CO	Carbon Monoxide	56–133	20·5	21·5	22·6	23·8	25·2	26·7	28·4	30·3	32·5	35·0	38·0	41·5	45·8	51·1	57·9	ⓜ67·3	ⓜ84·1
CO₂	Carbon Dioxide	107–196	59·5	62·2	65·2	68·4	72·1	76·1	80·6	85·7	91·5	98·1	106·0	114·5	125·0	137·5	153·5	173·0	198·0
COS	Carbonyl Sulphide	162–224													(124·5)	ⓜ139·5	159·5	187·0	229·
CS₂	Carbon Disulphide	194–319												(160·0)	177·5	199·5	228·	269·	329·
Cl₂	Chlorine	162–420	66·1	69·1	72·4	76·0	80·0	84·4	89·4	95·1	101·5	109·0	117·5	127·5	140·0	155·0	ⓜ173·0	201·	245·
F₂	Fluorine	54– 89														(52·2)	ⓜ59·5	70·5	87·5
H₂	Hydrogen	14– 21	2·67	2·83	3·01	3·21	3·45	3·71	4·03	4·40	4·84	5·38	6·05	6·90	8·03	9·55	11·70	15·10	21·4
HBr	Hydrogen Bromide	120–205	51·8	54·3	57·1	60·2	63·7	67·6	72·1	77·1	82·9 – 89·6	97·5	107·0	118·5	132·5	151·0	175·0	ⓜ209·
HCl	Hydrogen Chloride	132–195	49·7	52·1	54·6	57·5	60·6	64·1	68·1	72·5	77·6	83·4	90·1	98·1	108·5	121·0	137·0	158·5	ⓜ193·0
HF	Hydrogen Fluoride	240–290														(179·0)	ⓜ207·	245·	301·
H₂O	Water	175 380	113·0	118·5	124·0	130·0	137·0	144·5	153·0	162·0	173·0	185·0	198·5	215·	233·	256·	ⓜ284·	325·	381·
H₂S	Hydrogen Sulphide	153–213	57·1	59·8	62·7	65·9	69·5	73·5	78·0	83·1	89·0	95·7	103·5 – 113·5	124·5	138·5	156·5	180·5	ⓜ218·
He	Helium	0·9–5·2													0·980	1·268	1·738	2·634	4·518
I₂	Iodine	298–456	141·5	147·5	154·0	161·5	169·5	178·5	188·5	199·5	212·	226·	243·	262·	285·	312·	345·	ⓜ389·	471·
Kr	Krypton	63–121	27·9	29·4	30·9	32·7	34·6	36·8	39·3	42·2	45·5	49·4	53·9	59·4	66·3	74·8	85·9	101·0	ⓜ123·5
N₂	Nitrogen	54–128	18·1	19·0	20·0	21·1	22·3	23·7	25·2	27·0	29·0	31·4	34·1	37·5	41·7	47·0	54·0	ⓜ63·4	80·0
N₂O	Nitrous Oxide	103–186	55·8	58·3	61·1	64·2	67·6	71·3	75·5	80·3	85·7	91·9	99·0	107·5	117·5	129·5	144·0	162·5	ⓜ189·5
NO	Nitric Oxide	73–180	37·7	39·4	41·3	43·4	45·6	48·1	50·9	54·0	57·6	61·6	66·3	71·7	78·1	85·7	95·0	106·5	ⓜ123·5
NH₃	Ammonia	145–240	70·9	74·1	77·6	81·5	85·8	90·6	95·9	102·0	108·5	116·5	125·5	136·0	148·0	163·0	181·0	ⓜ206·	245·
Ne	Neon	15– 45	5·50	5·79	6·11	6·47	6·88	7·34	7·87	8·48	9·19	10·05	11·05	12·30	13·85	15·80	18·45	22·1	ⓜ27·5
O₂	Oxygen	57–154	21·8	22·8	24·0	25·2	26·6	28·2	29·9	31·9	34·1	36·7	39·8	43·3	48·1	54·1	ⓜ62·7	74·5	92·8
SO₂	Sulphur Dioxide	178–263	78·9	82·4	86·3	90·4	95·1	100·0	106·0	112·5	119·5	128·0	137·5	148·5	161·5	177·0	195·5	ⓜ225·	269·
Xe	Xenon	110–166	38·5	40·5	42·7	45·1	47·7	50·8	54·2	58·2	62·7	68·1	74·4	82·1	91·5	103·5	118·5	139·5	ⓜ170·0

ⓜ Melting Point.
⎯ Transition Point in Solid

From Honig and Hook (1960).

Molecule-Molecule and Molecule-Surface Interactions

exactly true. For values of p_0 at temperatures far removed from the normal boiling point the reader should make reference to Nesmayanov (1963) for the vapour pressures of all but the common gases and to Honig and Hook (1960) for the vapour pressures of the common gases. A useful table from the paper by Honig and Hook is given as Table 2.16. Two useful vapour pressures are: N_2 at 20°K is $2 \cdot 2 \times 10^{-11}$ torr; H_2 at 4·2°K is $3 \cdot 5 \times 10^{-7}$ torr (Borovik et al. 1960; Hengevoss 1965).

2.4.2 Adsorption Isotherms

Since the potential energy near a solid surface is more negative (attractive) than it is some distance away the average density of gas molecules near the surface is always, in equilibrium, greater than it is in the gas phase. This qualitative statement has been given simple mathematical form by Steele and Halsey (1955) and their formulation is now presented. This formulation has been applied to physical adsorption but is applicable in principle to chemisorption as well.

If the geometric volume of the system V_{geo} is defined as that volume enclosed by a surface passing through the centres of all the surface atoms, then the volume available to the centres of gas molecules (considered as spheres) is $V_{geo} - Az_p$, where A is the internal surface area of the system and z_p is the mean distance of a surface atom from an adsorbed atom. $V_{geo} - Az_p$ would be the volume measured from the perfect gas law (expressed here in cgs units).

$$V_{geo} - Az_p = \frac{nRT}{p} \qquad (2.28)$$

if n, T and p were measured precisely and if there were no tendency for gas to concentrate near the surfaces. This tendency, however, always exists and gives rise to an apparent volume (V_{app}) which is in excess of that given by eqn. (2.28). This excess volume V_{ex} may be written

$$V_{ex} = V_{app} - (V_{geo} - Az_p) = \int^{V_{geo} - Az_p} \left[\exp\left(-\frac{E}{RT}\right) - 1 \right] dV. \qquad (2.29)$$

Equation (2.29) is analogous to eqn. (2.12) for gas-gas collisions and expresses the result that every element of volume must be multiplied by its Boltzmann factor $\exp(-E/RT)$, if the total apparent volume of the system is to be calculated. Now V_{ex} may be considered as the amount adsorbed, i.e. it is the quantity of matter in the system, in excess of that expected on purely geometric grounds

We digress briefly at this point to define certain quantities which are widely used in adsorption literature.

V_{ex} is given as a volume at p,T, but may readily be converted into the total number of molecules adsorbed N_A.

$$N_A = (9\cdot 67 \times 10^{18}) \frac{V_{ex} p}{T} \text{ molecules} \qquad (2.30)$$

where V_{ex} is in cm^3, p in torr, T in °K.

A more used quantity is the number of molecules adsorbed per unit area

$$\sigma = \frac{N_A}{A} \text{ molecules/cm}^2. \qquad (2.31)$$

In turn this may be written as a fraction (θ) of the number of molecules per cm^2 in a complete monolayer, σ_m.

$$\theta = \frac{\sigma}{\sigma_m}. \qquad (2.32)$$

To convert σ to θ it is necessary to know σ_m for the gas in question. There has been much discussion in the literature concerning the magnitudes of σ_m (see, for example, Kodera and Onishi (1959); Young and Crowell (1962); Brennan and Graham (1965)), but for the present purposes the number of molecules/cm^2 in the surface of the liquid appears as useful as any other value and these are shown for the common gases in Table 2.17. Note that all values of σ_m lie between 5×10^{14} and 10×10^{14} molecules cm^{-2}. Note also that the areas per molecule are less than the values of Table 2.5.

We now return to the discussion of eqn. (2.29). If E at every point in the volume could be defined, then eqn. (2.29) could be used to calculate the physical adsorption isotherm which relates θ to p at fixed T in the form we will use it. In general, E is not known everywhere because of the effects of heterogeneity and the interaction between adsorbed molecules (Section 2.3.1) and hence eqn. (2.29) is of practical value only in certain special cases. One of these cases is that of a homogeneous surface at a temperature sufficiently high that there is no interaction between adsorbed molecules. This case can be realized experimentally using the adsorbent P33(2700) which is a graphitized carbon black with a highly homogeneous surface (Polley *et al.* 1953). Since the graphite surface has only minor variations in E in the plane of the surface (Table 2.8) the integration variable in eqn. (2.29) is the distance normal to the surface. Under these conditions eqn. (2.29) becomes an expression of Henry's Law ($\theta = b(T)p$ at fixed T), and this result is found experimentally (Constabaris *et al.* 1959). The experimental results of Constabaris *et al.* (1959) are shown in Fig. 2.10a which is a composite figure, on a log-log plot assembled to demonstrate the type of adsorption isotherm obtained on various types of surfaces over a wide range of pressures and temperatures. From the Henry's Law data in Fig. 2.10a an experimental value of $b(T)$ may be obtained. From eqn. (2.29) assuming some suitable variation of $E(z)$ with z (distance normal to the surface) (see Section 2.3.1) a theoretical value of $b(T)$ may be obtained which can be compared with the experimental result. This procedure has been carried out for several gas-surface potentials by Sams *et al.* (1960) who have

TABLE 2.17

Molecules per cm² and molecular area in the liquid surface of the common gases at their normal boiling temperatures

Gas	He	H_2	Ne	N_2	CO	Ar	O_2	CH_4	Kr	NO	Xe	N_2O	CO_2	NH_3	H_2O
Molecules cm⁻² × 10⁻¹⁴ $\sigma_m \times 10^{-14}$	7·22	6·95	10·0	6·14	5·95	6·94	7·09	5·52	6·14	8·00	5·33	5·95	5·88	7·75	9·5
Area per molecule × 10¹⁵ cm²	1·38	1·44	1·0	1·62	1·68	1·44	1·41	1·81	1·62	1·25	1·88	1·68	1·70	1·29	1·05

Calculated for plane of closest packing in hexagonal close packing

$$\sigma_m = 0{\cdot}2883 \left(\frac{4\sqrt{2}\,N_0\,\delta}{M} \right)^{2/3} \text{ molecules cm}^{-2}$$

N_0 = Avogadro's Number = 6·023 × 10²³, δ = density (gm cm⁻³), M = Molecular Weight.

been able to deduce values of E_p (those of Table 2.6) and values of the adsorption area and the distance z_p. In general the results obtained from each gas-surface interaction potential are comparable and it is not possible to choose a particular interaction potential as superior to the others.

While these procedures of Halsey and his colleagues are undoubtedly the more correct way to handle adsorption isotherms on a homogeneous surface, it is an interesting and useful fact that, once the energy E_p is known, the isotherm can be predicted to a precision adequate for most practical purposes, by the simple kinetic arguments of Langmuir (1918). These arguments equate the arriving flux assuming unity condensation coefficient, to the desorbing flux

$$\frac{3{\cdot}5\times 10^{22}p}{(MT)^{\frac{1}{2}}} = \frac{\sigma}{\tau_0}\exp\left(-\frac{E_p}{RT}\right) \qquad (2.33)$$

where τ_0 is the vibration time of an adsorbed molecule. Desorption will be discussed in Section 2.4.4. below. Setting $\tau_0 = 10^{-12}$ sec (following Kiselev and Poshkus (1963)) and using eqn. (2.32)

$$\theta = b(T)p = p\,\frac{3{\cdot}5\times 10^{10}}{\sigma_m(MT)^{\frac{1}{2}}}\exp(E_p/RT). \qquad (2.34)$$

From eqn. (2.34) the adsorption isotherm can be calculated from the values of E_p of Table 2.6 and the values of σ_m of Table 2.17. This calculation is shown dashed in Fig. 2.10a and the agreement with the experimental result is good. Similar results are obtained with other gases. We will return to this kinetic view of adsorption in Section 2.4.4 below.

If the adsorbing surface is heterogeneous and contains a number of adsorption energies E_i so that the fraction of the surface area containing sites with energy E_i is $f(E_i)$ then the relative coverage θ for the whole surface will be

$$\theta = \sum_i b_i(T)p \qquad (2.35)$$

provided the temperature is high enough to permit Henry's Law to hold on each individual patch of the surface. It is important to note that if Henry's Law holds for each energy E_i then it will hold for the whole surface as well. The essential conditions for Henry's Law to hold for each individual patch is that the relative coverage be very low. This will be true if $T\to\infty$ and/or $p\to 0$, and in these limits Henry's Law is to be expected for every physical adsorption isotherm. The limit $T\to\infty$ has been studied by Steele and Halsey (1954, 1955) who found Henry's Law for all the gases of Table 2.10 at temperatures mainly above room temperature. The energies of Table 2.10 were obtained by applying the analysis above as if the surface were homogeneous. The energies of Table 2.10 are therefore mean effective energies. The experimental result of Steele and Halsey (1954) for argon on porous glass is shown in Fig. 2.10a.

The case $p\to 0$ is of particular interest in uhv because the new techniques have permitted this limit to be approached for the adsorption of vapours for the first time. A vapour is defined as a gas near its condensation temperature. It turns out, however, that in many cases of practical interest the pressures that have been achieved in uhv systems are still too high for the appearance of Henry's Law. This is shown in Fig. 2.10a by the shape of the isotherms for argon on rutile, zirconium, and Pyrex at temperatures in the vicinity of the normal boiling point of argon (87·4°K). All isotherms except those at the lower right of Fig. 2.10a have slopes less than 45° to the abscissa indicating failure of Henry's Law. It is anticipated that lowering the pressure sufficiently far below 10^{-10} torr should generate Henry's Law for all isotherms of Fig. 2.10a, and we will return to this point below. At very high coverages (several monolayers) the characteristics of the adsorbent become obliterated and the pressure above the adsorbed layer becomes the vapour pressure of the adsorbate, and the heat of adsorption becomes the heat of vaporization. This is the situation for argon on rutile at 85°K at high pressures, shown in Fig. 2.10a. Thus the adsorption isotherm on a heterogeneous surface may be expected to pass from a Henry's Law region at very low pressures and coverages, through a region concave downward (effect of heterogeneity) at intermediate coverages, with a limiting pressure at the vapour pressure at very

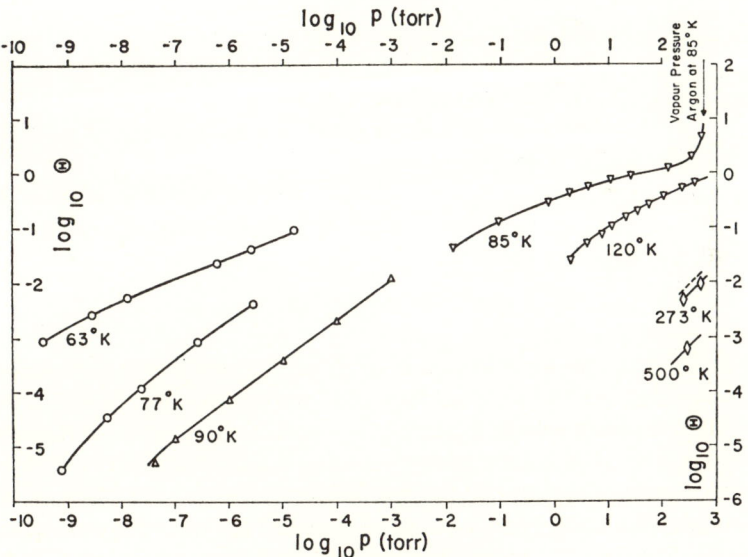

Fig. 2.10a Isotherm data for argon illustrating ranges of pressure, coverage, and temperature that have been measured.

◇   P33(2700)   Constabaris et al. (1959)

Dashed line has been calculated from eqn. 2.34 with $E_p = 2200$ kcal/mole, $\sigma_m = 6.94 \times 10^{14}$, $M = 40$, $T = 273.2°K$.

□	Porous Glass	Steele and Halsey (1954)
▽	Rutile (T$_2$O$_2$)	Drain and Morrison (1953)
△	Zirconium	Hansen (1962)
○	Pyrex	Hobson and Armstrong (1963)

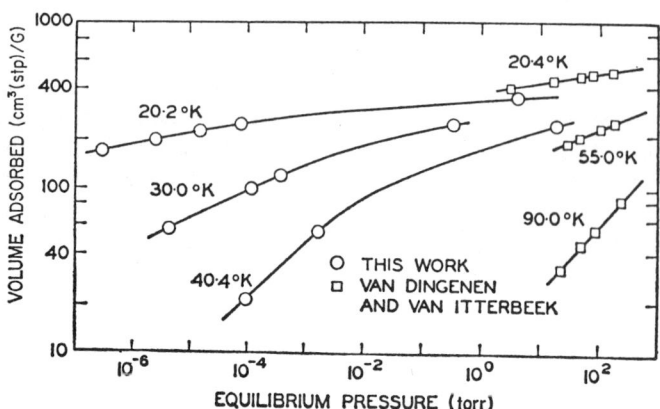

Fig. 2.10*b* Adsorption isotherms for hydrogen on coconut charcoal at various temperatures.
(From Stern *et al*. 1965; also showing data of Van Dingenen and Van Itterbeek (1939). 1cm^3 S.T.P. = 2.68×10^{19} molecules).

high coverages. These isotherms are those found most commonly in the uhv range and are typical of isotherms measured in sorption pumps. As an example Fig. 2.10*b* shows isotherms of hydrogen on activated charcoal measured by Stern *et al*. (1965) during a study of sorption pump materials. Isotherms on powdered or dispersed adsorbents however, are difficult to interpret because the adsorbent is difficult to cool in the uhv range, and the results may not be characteristic of the desired temperature. Stern *et al*. (1965) comment as follows about their isotherm measurements for H_2 on a variety of dispersed adsorbents '... it appears likely that true adsorption equilibrium was not reached in any of the present experiments. Therefore the isotherms ... must be considered as "practical" isotherms; these are entirely satisfactory for the purpose of the present research'. A detailed discussion of the difficulties of obtaining thermal equilibrium with dispersed adsorbents has been given by Hobson (1967).

An increasing body of isotherm data obtained in uhv systems is accumulating, for which there are no doubts about the thermal uniformity of the adsorbent, since this was the wall of the vacuum vessel immersed in a suitable liquid bath. A sampling of such data is shown in Fig. 2.11 for liquid nitrogen temperature. Data similar to those shown in Fig. 2.11 have been measured for He on Pyrex (Hobson 1959); Kr, Xe on Pyrex (Ricca *et al*. 1967); Kr, Xe on films of W, Mo (Ricca and Bellardo 1967); Kr on Pyrex, Ni (Schram 1966); Kr, Xe on Pyrex and Mo films (Endow and Pasternak 1966). From explicit tests it was found that thermodynamic equilibrium was present for all the data of Fig. 2.11. A thermal transpiration correction was made in all these data so that the pressures given are those above the adsorbed layer at the temperature of the adsorbent. The data are therefore suitable for analysis by the methods of equilibrium thermodynamics. The assignment of θ for these data will be discussed below.

A general examination of Fig. 2.11 shows that there is good agreement between independent workers and that the gas rather than the adsorbent tends to be the main determinant of the isotherm (see, for example, the lower three curves of Fig. 2.11 for Ar). All isotherms of Fig. 2.11 have approximately the same shape, being curved gently downward on the log-log plot. The adsorbents in these measurements were not atomically clean, having interacted with the residual gases of the various systems, but were nevertheless fully reproducible experimentally in each system.

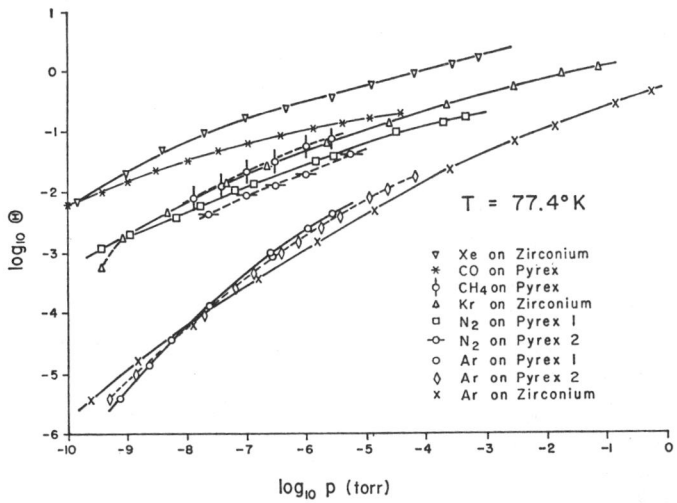

Fig. 2.11 Physical adsorption isotherms for various gases on various adsorbents at $T = 77.4°K$. References: Ar, Kr, Xe on Zr, Hansen (1962); Ar, N_2 on Pyrex 1, Hobson and Armstrong (1963); CH_4 on Pyrex, N_2 on Pyrex 2, Ricca and Medana (1964); Ar on Pyrex 2, Ricca *et al.* (1967); CO on Pyrex, Ricca *et al.* (1966).

Physical adsorption isotherms have been the subject of much study (Brunauer 1945; Young and Crowell 1962) in the conventional pressure range ($p > 10^{-4}$ torr) but the results of Fig. 2.11 taken with uhv apparatus were nevertheless rather surprising and have led to a new method of analysis which will now be presented.

This analysis rests upon three empirical observations found to be satisfied by the great majority of the isotherms of Fig. 2.11 and associated data:

(i) All isotherms failed to reach Henry's Law ($\theta \alpha p$) at the lowest pressures measured.

(ii) The temperature dependence of the isotherms was that of the Polanyi potential theory (Brunauer 1945). This means that all isotherms for a given adsorbent-adsorbate combination followed a relation of the form

$$\theta = g(\varepsilon) \tag{2.36}$$

where ε is given by

$$\varepsilon = -RT\ln p/p_0 \tag{2.37}$$

p_0 being the vapour pressure of the adsorbate at the temperature of the measurement. ε in eqn. (2.37) is not the same as ε in eqn. (2.6).

(iii) The shape of each isotherm was provided by specifying $g(\varepsilon)$ according to the isotherm of Dubinin and Radushkevich (1947)

$$\ln\theta = -B\varepsilon^2 \tag{2.38}$$

with B constant.

Property (i) above is exhibited in Fig. 2.11 and properties (ii) and (iii) may be tested for experimental data by plotting the quantity adsorbed against ε^2 on a log-linear plot as in Fig. 2.12. Note that the data for all temperatures lie on a single straight line. Kaganer (1957) suggests that the intercept of this straight line with the vertical axis in this plot be interpreted as the number of molecules in a monolayer σ_m. For the data of Fig. 2.12 for N_2 $\sigma_m = 6.4 \times 10^{14}$ molecules cm^{-2} and for Ar $\sigma_m = 7 \times 10^{13}$ atoms cm^{-2}. Thus the Dubinin-Radushkevich equation (2.38) provides a means of finding the monolayer coverage from isotherm data at very low pressures, just as the Brunauer-Emmett-Teller equation (1938) provided a means of finding the monolayer coverage from isotherm data in the range $0.01 < p/p_0 < 1$. The interpretation of this coverage is, however, not clear at present. Reference to Table 2.17 shows that σ_m for N_2 from Fig. 2.12 corresponds closely to packing in a liquid layer, while σ_m for Ar from Fig. 2.12 is approximately an order of magnitude *lower* than packing in a liquid layer. Surface roughness would make σ_m *higher* than that for a liquid layer. Ricca and Medana (1964) obtain a low value of $\sigma_m = 1.45 \times 10^{14}$ molecules cm^{-2} also for CH_4 on Pyrex and find discrepancies of the same type (more Xe adsorbed than Kr) when comparing the adsorption of Kr and Xe on films of W and Mo (Ricca and Bellardo 1967). Ricca *et al.* (1967) suggests that σ_m applies only to molecules adsorbed on 'active' parts of the surface and this suggestion is at least plausible, and is supported by the results of Endow and Pasternak (1966) who found that σ_m decreased about a factor 2 during the 10–15 days following deposition of a film of Mo. Also Ross and Roberts (1965) have found values of σ_m about four times normal for particular samples of 'superactive' glass. Despite difficulties in the physical interpretation of σ_m the intercept from plots like that of Fig. 2.12 provides a convenient means of defining $\theta = \sigma/\sigma_m$, and for most of the data of Fig. 2.11 θ was obtained in this way. Some questions arise about the value of p_0 to be used in eqn. (2.37) and these are discussed by Ricca *et al.* (1967). The choice of p_0 influences the values of σ_m and B but since the exact interpretation of these quantities is in some doubt we do not discuss this point further. The slopes in Fig. 2.12 provide a measure of B in eqn. (2.38) whose inverse square root $B^{-\frac{1}{2}}$

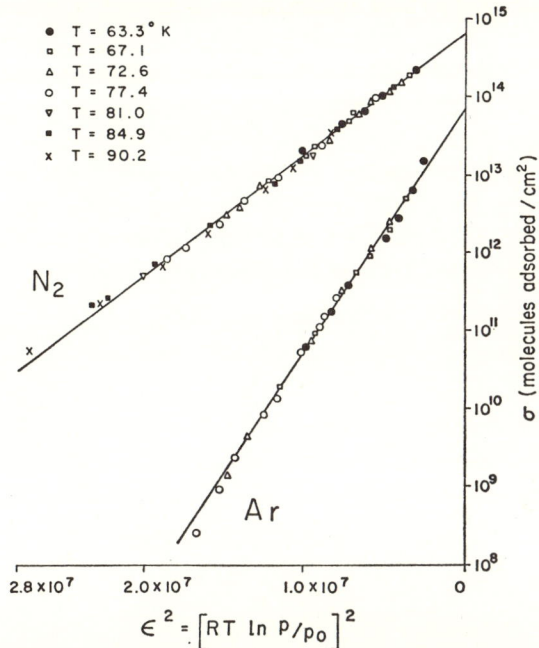

Fig. 2.12 Physical adsorption isotherms of N_2 and Ar on Pyrex plotted as a function of $\varepsilon^2 = (RT \ln p/p_0)^2$. (Hobson and Armstrong 1963).

TABLE 2.18
Values of $B^{-\frac{1}{2}}$ (cal/mole) obtained by fitting isotherms with Dubinin-Radushkevich equation

Adsorbent	Gas							
	He	H_2	N_2	CO	Ar	CH_4	Kr	Xe
Glass	158[a]	528[b]	1673[c]	1984[e]	1179[c]	1445[d]	1379[f]	1543[f]
			1555[d]		1165[f]		1340[g]	1470[g] (mean)
							1123[h]	
Mo film							2304[i]	2950[i]
W film							2650[i]	3409[i]
Zr film					1220[j]		1360[j]	1257[j] (mean)
Ni							1010[h]	

[a] Hobson (1959); [b] Van Itterbeek et al. (1964); [c] Hobson and Armstrong (1963); [d] Ricca and Medana (1964); [e] Ricca et al. (1966); [f] Ricca et al. (1967); [g] Endow and Pasternak (1966); [h] Schram (1966); [i] Ricca and Bellardo (1967); [j] Hansen (1962).

is expressed as an energy and has a value roughly that of the heat of vaporization (Table 2.9). Table 2.18 is a collection of values of $B^{-\frac{1}{2}}$ obtained by various workers. For a given gas the values of $B^{-\frac{1}{2}}$ for fresh metal films is higher than that for glass, although Endow and Pasternak (1966) found that as their films of Mo aged over a period of 10–15 days their values of $B^{-\frac{1}{2}}$ for Kr approached those of glass. The values for glass are in good agreement.

The value of B may be obtained from measurements at a single temperature. Hobson (1965a, 1965b, 1966) has developed a theoretical model for physical adsorption on a heterogeneous surface, in which a single value of B is used to derive the distribution of adsorption energies and to generate isotherms over a very wide range of p, T, θ. The first step in the analysis was the assumption of a local isotherm equation on each patch of energy E_i so that the relative coverage for the whole surface was given by

$$\theta(p,T) = \int_0^\infty \zeta(E_i,p,T) f(E_i) \, dE_i \tag{2.39}$$

where $\zeta(E_i,p,T)$ was the local isotherm on a patch with adsorption energy E_i at pressure p. Equation (2.39) has been used many times for adsorption on a heterogeneous surface (see Honig 1954) and the only novel feature in the present analysis was the choice of a particularly simple mathematical form for $\zeta(E_i,p,T)$ which enabled an explicit solution for $f(E_i)$ to be obtained. Adamson and Ling (1961) and Ross and Olivier (1964) have used eqn. (2.39) but their choice of $\zeta(E_i,p,T)$ was sufficiently complex to necessitate the use of a computer for the finding of $f(E)$. For application of the method of Ross and Olivier (1964) to uhv isotherm data see Schram (1966). Figure 2.4 shows $f(E_i)$ as found by Hobson (1965b) for Ar on Pyrex for several local isotherms having different degrees of adsorbate-adsorbate interaction (q_i). These distribution functions were obtained from isotherm data at one temperature only. Figures 2.13a and 2.13b show the calculated isotherms for argon and helium obtained only from the experimental data for adsorption on Pyrex at 77·4°K and 4·2°K respectively. Similar calculations for N_2 are not shown. Several points in Figs. 2.13a and 2.13b may be noted: (1) not only do the calculations predict isotherms correctly for adsorption on Pyrex, they also predict isotherms on other adsorbents quite accurately, collecting the data of Fig. 2.10a into a cohesive unit. This suggests that the distribution of adsorption energies for a given adsorbate on a variety of heterogeneous surfaces is similar, and this result is of value in predicting isotherms for unknown heterogeneous adsorbents, (2) a transition to Henry's Law is predicted for each isotherm although isotherm data at pressures below those available are required to test the prediction, (3) over the range of available data there is little difference in the calculated results for different degrees of adsorbate-adsorbate interaction postulated in the local isotherm, (4) the pressure above a helium layer at $\theta \sim 10^{-5}$ is extremely low ($p \sim 10^{-20}$ torr). It is relatively easy to obtain $\theta \sim 10^{-5}$ in an uhv system.

Theoretical calculations on low pressure isotherms for physical adsorption have also been made by Huang (1965) for He and N_2 on Pyrex and by Taylor et al. (1965) for He on porous glass. These authors start with an ideal surface and calculate the binding energy of the adsorbate molecule according to the

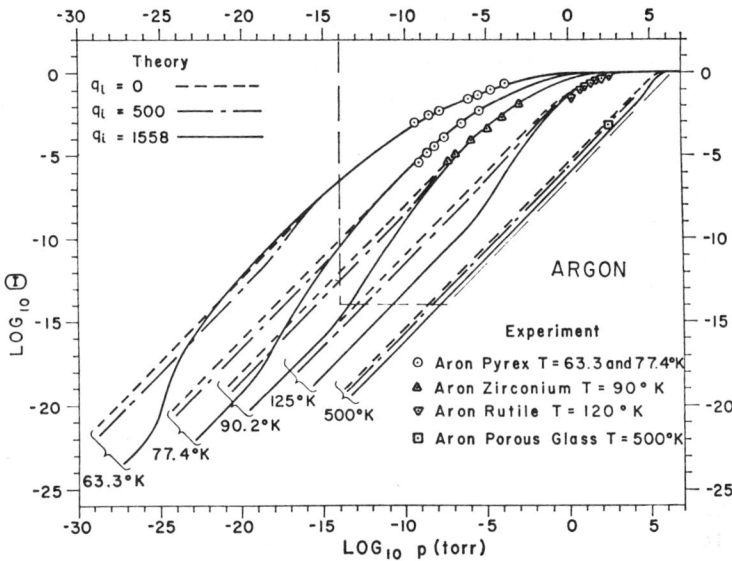

Fig. 2.13(a) Calculated isotherms (lines) for the physical adsorption of Ar on a heterogeneous surface. q_i is adsorbate-adsorbate interaction energy (cal mole^{-1}). Area within the dashed zone at upper right represents region in which measurements might be possible with modern techniques. Experimental data (points): Ar on Pyrex, Hobson and Armstrong (1963); Ar on Zr, Hansen (1962); Ar on rutile, Drain and Morrison (1953); Ar on porous glass, Steele and Halsey (1954). (from Hobson 1966).

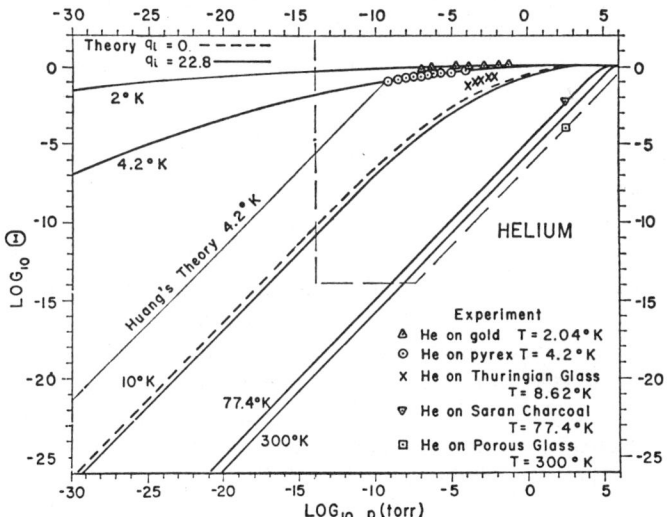

Fig. 2.13(b) Calculated isotherms (lines) for the physical adsorption of He on a heterogeneous surface. q_i is adsorbate-adsorbate interaction energy (cal mole^{-1}). Area within the dashed zone at upper right represents region in which measurements might be possible with modern techniques. Experimental data (points): He on gold, Meyer (1956); He on Pyrex, Hobson (1959); He on thuringian glass, Keesom and Schweers (1941); He on saran charcoal and porous glass, Steele and Halsey (1955).
Line labelled Huang's theory 4·2°K is the isotherm predicted theoretically by Huang (1965) on the basis of a homogeneous surface. (from Hobson 1966).

general procedures outlined in Section 2.3. This approach leads them to Henry's Law isotherms for coverages well below a monolayer. The result of Huang's calculation for He at 4·2°K is shown in Fig. 2.13*b*. It is not in serious disagreement with experimental data currently existing but does diverge from the calculations of the heterogeneous model at lower pressures. Further experiments are required to resolve the issue. The onset of Henry's Law predicted by Taylor *et al.* (1965) for He on porous glass at 4·2°K is at lower pressures still and it may not be possible to check the prediction of Taylor *et al.* (1965) experimentally. Taylor *et al.* (1965) also calculate energies at coverages above a monolayer and at present their isotherm calculation is of more interest at pressures above the uhv range.

In practice the physical adsorption isotherm represents the lower working limit of the cryosorption pump, although in normal operation a cryosorption pump will operate at pressures above this because it will operate under dynamic conditions rather than at equilibrium. The amount of gas which can be adsorbed on a cryosorption pump is limited by the amount of cooled sorbent available. The form of sorbents varies widely. Table 2.19 has been assembled to illustrate the surface areas available with various sorbents. When sorbents are dispersed as small particles the adsorbing area is measured per gram (traditionally in metre^2/gm). This area may be converted to molecules/gm by using Table 2.17. When a sorbent is bonded to a flat substrate whose area can be measured then a roughness factor may be defined as the apparent increase in adsorbing area caused by the presence of the sorbent. For evaporated films both measures of sorbing area are useful and are included in Table 2.19. Roughness factors less than unity as found for glass represent examples, noted above, where apparent adsorption areas are less than the geometric area.

Because of elevated temperatures most chemisorption isotherms are of rather less interest in uhv systems than physisorption isotherms and less data is available. Isotherms for coverages near a monolayer at temperatures near room temperature are, however, of importance (Robinson 1960; Hobson and Earnshaw 1967) but have not been systematically studied.

The three most important isotherms observed in the chemisorption of gases on metals are known as Langmuir, Freundlich and Tempkin isotherms respectively. The Langmuir isotherm is based on the simplest model and assumes that the rate of adsorption is proportional to the number of empty states. For non-dissociative adsorption

$$\theta = \frac{Ap}{1+Ap}, \tag{2.40}$$

where A is a constant. If adsorption is dissociative and immobile then

$$\theta = \frac{(Ap)^{\frac{1}{2}}}{1+(Ap)^{\frac{1}{2}}}. \tag{2.41}$$

TABLE 2.19
Surface areas of adsorbents

Material	Form	Gas used	Area/gm (metre²/gm)	Roughness factor	Reference
Silica Gel Type R	Granule	N_2	784		Stern et al. (1965)
„ „ „ ID	„	„	311		„
Alumina	Pellet	„	287		„
Hydrated Ferric Oxide	Granule	„	194		„
Coconut Charcoal	Granule	„	889		„
Linde molecular sieve type 13X	Powder	„	514		„
Linde molecular sieve type 5A	Pellet	„	600		„
Linde molecular sieve type 5A	Bonded to Al plate	H_2	390	$1 \cdot 3 \times 10^5$	Stern et al. (1966)
Sanan charcoal S-85	Powder	N_2	1170		Steele & Halsey (1955)
Porous glass		N_2	117		„
P-33 (2700) graphite	„	N_2	13		Constabaris & Halsey (1957)
Co	Film	Kr	4·9	10	Brennan & Graham (1965)
Ni	„		4·3	9	„
Pt	„		6·5	7	„
Fe	„		22·6	60	„
Mo	„		71	173	„
Ta	„		26·2	38	„
W	„		39·2	40	„
Ti	„		13·6	15	„
Zr	„		33·2	17	„
Single crystal Cu 110 face	Solid	N_2		1·42	Rhodin (1950)
Single crystal Cu 100 face	„	„		1·35	„
Single crystal Cu 111 face	„	„		1·20	„
Glass	„	N_2		1·37	Ricca & Medana (1964)
		CH_4		0·26	„
		Ar		0·14	Ricca et al. (1967)
		Kr		0·16	„
		Xe		0·05	„

The Freundlich isotherm

$$\theta = Bp^{1/n}, \qquad (2.42)$$

where B and n are constants at a given temperature, can be derived if it is assumed that the heat of adsorption decreases exponentially with coverage. The Temkin isotherm

$$\theta = \frac{RT}{q_0 \alpha} . \ln(cp) \qquad (2.43)$$

can be derived from the Langmuir isotherm if the heat decreases linearly with coverage, viz: $q = q_0(1-\alpha\theta)$. In the above, $c = a\exp(q_0/RT)$. These isotherms, and methods of testing whether experimental data fit the isotherms, are considered in detail by Hayward and Trapnell (1964).

The application of these isotherms to experimental data of chemisorption of gases on metals is frequently of doubtful value. In most cases the situation is too complex to be described by the very simple models of the above isotherms. In particular, most gases when chemisorbed on clean metal surfaces do so in several different adsorbed phases with different heats of adsorption. Except in some very simple cases, it is probably best to treat isotherm data empirically.

We now give a few examples of simple isotherms of gases chemisorbed on metals under uhv conditions when the cleanliness of the surface was assured. Hickmott (1960a) has measured the isotherm for the adsorption of hydrogen on a tungsten wire in the temperature range 77–1940°K. In the range of coverage of Hickmott's experiment (0·4 to 0·7) an excellent agreement with the Temkin isotherm is observed, as can be seen from the data of Fig. 2.14. The heat of adsorption, derived from these isotherms, decreases linearly with coverage as required by a Temkin isotherm. The heat of adsorption was found to drop from 29 kcal/mole at low coverage to 12 kcal/mole at full coverage.

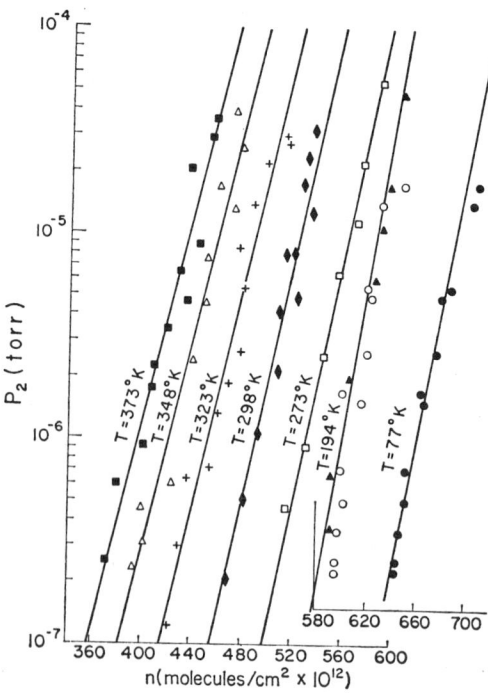

Fig. 2.14 Hydrogen isotherms on tungsten. Filled circles: $n_\alpha + n_\beta$ at 77°K; open circles: n_β at 77°K. α and β refer to different adsorbed phases of hydrogen on tungsten. (Hickmott 1960a).

The isotherms for chemisorption of nitrogen on tungsten wires has been measured by Kisliuk (1959a) in the temperature range 1330 to 1540°K, the results are shown in Fig. 2.15. p' is the pressure corrected by the factor $(T_s/T_g)^{\frac{1}{2}}$; T_s is the temperature of the solid and T_g of the gas. It can be seen that these data are a good fit to a Langmuir isotherm for dissociative adsorption

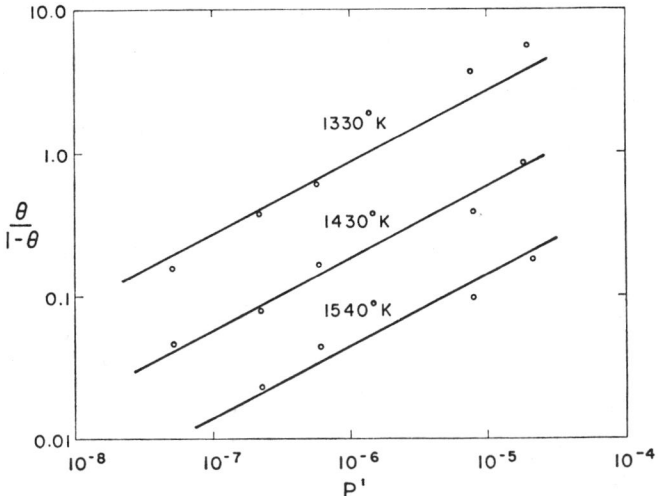

Fig. 2.15 Nitrogen isotherms on tungsten. p' is the ambient pressure multiplied by $(T_s/T_g)^{\frac{1}{2}}$, where T_s is the temperature of the solid and T_g of the gas. The lines are drawn with a slope of 0·5 (Kisliuk 1959a).

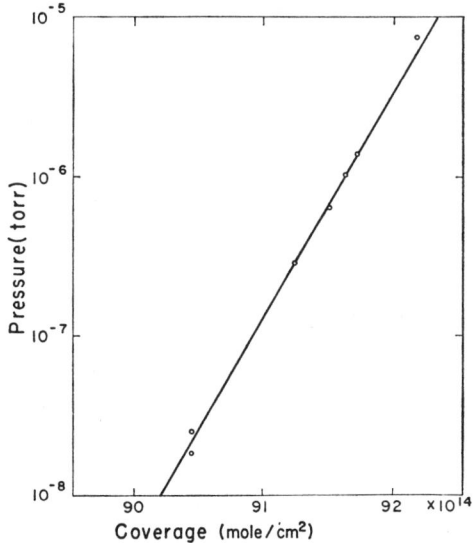

Fig. 2.16 Test of the Temkin Isotherm for H_2 on a Mo film at 78°K. (Hayward et al. 1966).

(giving a slope of 0·5 on a plot of $\theta/(1-\theta)$ against log p). The isosteric heats derived from the isotherms, by means of the Clausius-Clapeyron equation (2.22), give a constant heat of 116 kcal/mole.

Many measurements of adsorption isotherms on evaporated metal films and powders are not representative of adsorption on clean metal surfaces as would be observed in uhv.

Measurements of the isotherms of hydrogen on evaporated films of molybdenum at 78°K have been made under uhv conditions and are shown in Fig. 2.16 (Hayward *et al.* 1966). It can be seen that a Temkin type of isotherm was observed.

2.4.3 Impact between a Molecule and a Surface: Accommodation, Condensation and Sticking

Equilibrium properties do not depend upon the details of a single collision between a gas molecule and a wall, because the situation can be described by an equation of the form of eqn. (2.29), which includes no details of either the collision or desorption process. But there are many phenomena in uhv systems which do depend upon the separate processes of collision and desorption. In particular, the speeds of getters and of cryosorption and cryogenic pumps depend upon the sticking or condensation probability upon impact. The conductances of all passages from one portion of an apparatus to another depend upon the degree of sticking of molecules on the walls; the readings of a gauge located in a system in which temperature gradients exist depend upon a complex combination of gas-wall collisions. Even the reading of an ionization gauge usually depends upon local temperature or flow distributions. Some gauges, for example thermal conductivity gauges (Leck 1964), operate as the result of the energy exchange when a gas molecule strikes a surface at a different temperature.

There have been numerous studies, both experimental and theoretical, of the collision between a molecule and a surface. Many of the experimental workers have considered that they were studying the collision between a known molecule and a known surface, and hence were contributing data which could be analysed theoretically in a basic way. However, it seems more likely in most of these studies that the surface was unknown and hence the data could not be analysed as desired. Despite this qualification, however, the data are of interest here since most surfaces in an uhv system are unknown in the same way. On the other hand, the theories that have been proposed for the gas surface collision apply only in specially prepared circumstances and are not applicable in detail to the vast majority of gas surface collisions in an uhv system. The gap between this abundance of experimental data for poorly defined surfaces and idealized theories for the gas-surface collision has narrowed in recent years, but our presentation below remains a subjective assessment of the present situation.

There have been four main fields of study, the first three being primarily experimental (1) the angular distribution of the scattered flux, (2) the exchange of energy between the molecule and the surface (accommodation coefficients), (3) the condensation coefficient or probability that the molecule will become adsorbed on the surface, (4) theoretical studies of the gas surface collision. Clearly these four subdivisions are related and indeed a complete theory of the gas surface collision would provide answers to the first three subdivisions.

2.4.3.1 The angular distribution of the scattered flux. The earliest studies were those of Wood (1915) who used atomic beams of Hg and Cd impinging on glass, and of Knudsen (1915) who used an atomic beam of Hg on glass. Both found that the intensity of the scattered flux varied as the cosine of the angle of reflection. This cosine law of reflection has been found many times since the early work. Figure 2.17 shows the scattering of N_2 by surfaces of steel, aluminum, and glass obtained by Hurlbut (1957). Points lying on the circle represent the cosine law of reflection, as is the case for most of the data. Impact of glass at a large angle of incidence (0) represents the main deviation from the cosine law. Knudsen (1934) has pointed out that if the cosine law applies for each individual encounter then the flux incident on each unit area of a spherical surface is uniform everywhere. This property of a sphere, coupled with the cosine law, is finding increased modern application in the calibration of gauges (Moore 1965) and production of particular uhv components (Kornelsen and Domeij 1966). The cosine law of reflection is a central assumption in Clausing's (1932) calculation of the conductance of a pipe in the pressure range where the mean free path is large relative to the diameter of the pipe. The essential correctness of pipe conductance calculations based on the cosine law of reflection has been confirmed recently by Levenson *et al.* (1963). There is no doubt that the cosine law of reflection occurs in many cases of the gas-solid impact. It would be convenient if it were universally true, but it is not.

Knauer and Stern (1929), Stern (1929), Estermann and Stern (1930a, b) studied the scattering of molecular beams of He and H_2 by a single crystal of LiF and NaCl, and found specularly reflected and diffracted beams. Figure 2.18 shows the reflection of He by a single crystal of NaCl. The specularly reflected beam is shown in Fig. 2.18*a* together with first order diffraction maxima in Fig. 2.18*b*. The results could be handled quantum mechanically by assigning a de Broglie wavelength to the incident molecule

$$\lambda = \frac{h}{mv} = \frac{h}{(2mE)^{\frac{1}{2}}}, \qquad (2.44)$$

h being Planck's constant, m the atomic mass, v the molecular velocity and E the kinetic energy (all in cgs units).

A quantum mechanical analysis (see Section 2.4.3.4 below) could even provide a quantitative description of second-order anomalies in the shapes of diffraction maxima. The necessary physical condition (Fraser 1931) requiring

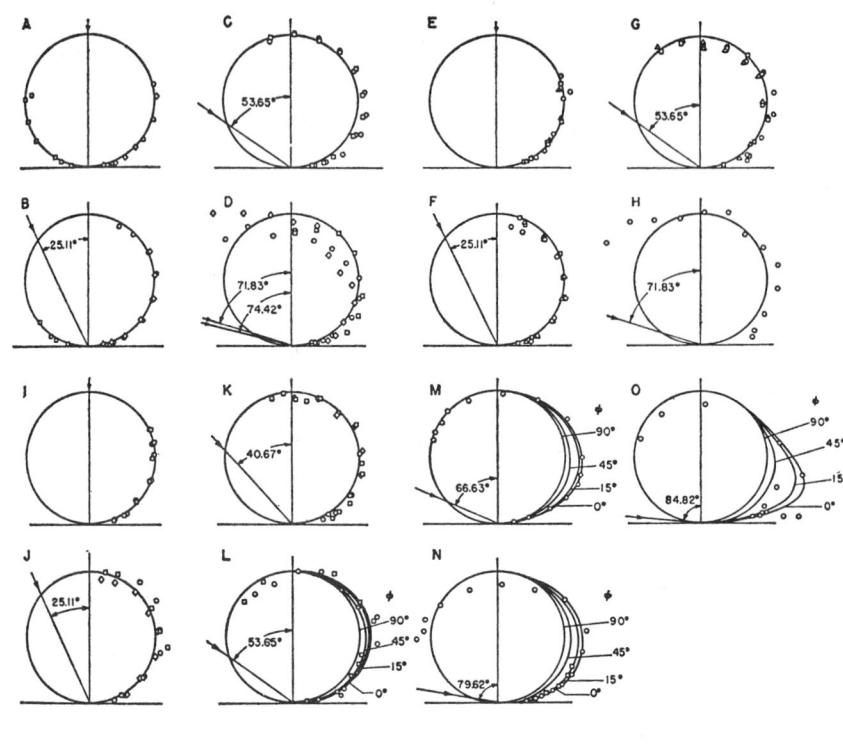

Fig. 2.17 Polar plots of scattering data of N_2 at room temperature from various surfaces. Arrows represent incident beam. Azimuth zero is plane of incidence. (Hurlbut 1957).

the consideration of the wave properties of the incident molecule is that the roughness of the surface resolved along the direction of the incident beam shall be small relative to the particle wavelength. If ψ is the glancing angle (i.e. the angle between the incident beam and the plane of the surface) and if l is the magnitude of surface roughness normal to the surface, then for specular reflection and diffraction

$$l \sin \psi \leqq \lambda. \qquad (2.45)$$

For a gas at room temperature $E = 3/2\,kT = 6 \times 10^{-14}$ ergs. Thus the wavelength decreases with the mass of the molecule and specular reflection accompanied by diffraction is more likely for the lighter gases and at lower temperatures. For helium at room temperature the wavelength is 0.73×10^{-8} cm. Even for an ideally smooth surface l in eqn. (2.45) cannot be reduced below the magnitude of the thermal vibrations which are of the order of 10^{-8} cm but which fall with surface temperature. Thus for eqn. (2.45) to be satisfied ψ must be relatively small ($\lesssim 30°$), i.e. reflection and diffraction are to be expected only at glancing angles of incidence. Equation (2.45), while necessary, is not sufficient to ensure specular reflection and/or diffraction. It is also required that inelastic processes involving energy transfer cannot take place. These will be discussed below. It is clear from the magnitudes arising from eqns. (2.44) and (2.45) that specular reflection accompanied by diffraction should be relatively rare in practical uhv systems.

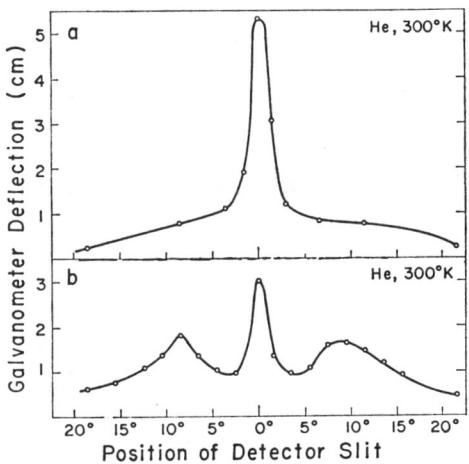

Fig. 2.18 Reflection of He atoms from face of single crystal of NaCl. Angle of incidence 78.5°. Position of detector slit is in degrees about the surface normal measured from the plane of reflection.
(a) incident beam in 100 plane
(b) incident beam in 110 plane.
(from Stern 1929)

However, specular reflection can take place under conditions in which the diffraction equations (2.44) and (2.45) are not satisfied. Zahl and Ellett (1931) measured nearly specular beams when Hg atoms impinged on single crystals of NaCl, KBr, and KCl, even when the incoming beam approached normal incidence. Qualitatively similar results were obtained by Hancox (1932) for beams of Hg and Cd impinging on crystals of LiF, LiCl, NaF. The more modern work (Datz et al. 1962 (He on Pt); Smith and Fite 1962 (H_2 on Ni, H_2 on W, Cl_2 on Ni); Smith 1964 (Kr, Ar, Ne, D_2 on Ni) and Smith and Saltsburg 1964 (He, Ar on Au)), have decisively confirmed that specular beams may frequently be found when the diffraction equations (2.44) and (2.45) are not satisfied. Hurlbut and Beck (1959) have even demonstrated that there is a strong specular component when Ar atoms impinge on liquid Ga. Smith and Saltsburg (1964) show, in their study of the reflection of He and Ar from films of gold grown epitaxially on mica that the presence of specular reflection depends on several factors which are of particular interest in uhv systems. It was found (*a*) smooth single crystal surfaces (111 surface of gold) were required for specular reflection of He and for nearly specular reflection of argon, (*b*) the adsorption of background gases reduced the degree of specular reflection for helium but not for argon, (*c*) the reflected beam for argon was displaced toward the surface normal by 20° from the specular direction, (*d*) a reversible transition from diffuse to directed scattering was observed for both He and Ar at a target temperature between 150 and 200°C. This result was attributed to the physical adsorption of some background gas in this temperature range. A similar result obtained by Datz *et al.* (1962) is shown in Fig. 2.19. In an extension of their work to the heavier inert gases, Smith and Saltsburg (1964) (He, Ne, H_2, Ar, Xe on Au) have shown that the heavier the gas the greater the tendency to diffuse reflection and the further off the specular direction the reflected beam. It may be noted that Edmonds and Hobson (1965) suggest variations in the degree of specular reflection of molecules on glass surfaces at different temperatures to explain an anomalous value of the thermal transpiration ratio in an uhv system.

In view of this relatively wide observation of specular or nearly specular reflection it might be wondered why the cosine law has been observed so

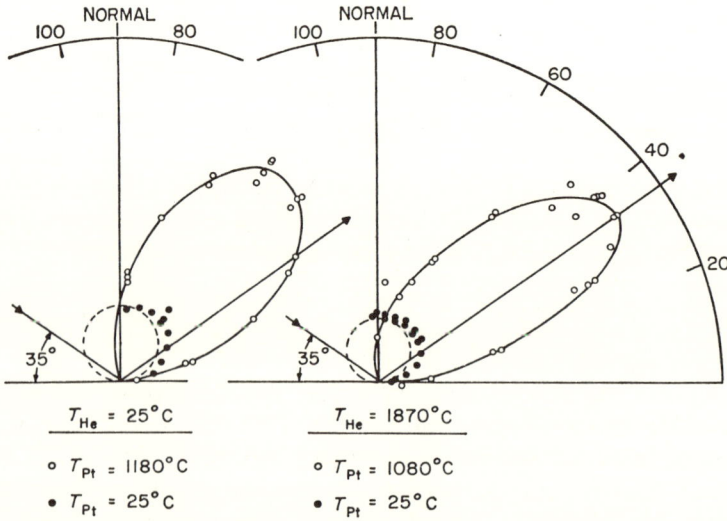

Fig. 2.19 Angular distribution of scattering of He from Pt. Arrow approaching from left is incident direction and arrow leaving toward right is direction of specular reflection. (Datz *et al.* 1962).

frequently. There are two main reasons for this and these may be distinguished by measurements of the energy and momentum transferred to the surface. First, if the surfaces are not smooth but expose a variety of faces inclined at a variety of angles then, even if individual collisions result in specular reflection the overall scattered beam will have a distribution in angle determined by the topography of the surface. If this topography is random then the reflected beam will tend to have a cosine distribution. Since most surfaces typically found in an uhv system will be rough to some degree, a cosine-like distribution is to be expected, even with specular reflection at individual collisions.

Second, if at least a portion of the incident kinetic energy is lost to the surface, i.e. the collision is inelastic, then the conditions for specular reflection are destroyed and the scattering will tend toward cosine. Energy accommodation is measured by the accommodation coefficient and it will be found below that many gas-solid impacts are inelastic.

2.4.3.2 Accommodation coefficients (Kaminsky 1965; Wachman 1962). If a group of thermal molecules approaches a surface with average kinetic incident energy E_g, and as a result of the collision are reflected from the surface with average kinetic energy E_r, which is different from the average energy they would have had at the temperature of the surface E_s, then for a gas without internal degrees of freedom the accommodation coefficient is defined as

$$\alpha = \frac{E_r - E_g}{E_s - E_g} = \frac{T_r - T_g}{T_s - T_g} \tag{2.46}$$

where $E_r, E_g, E_s = 2kT_r, 2kT_g, 2kT_s$, respectively, i.e. T_r, T_g, T_s are temperatures equivalent to the energy of the molecules. For the important case where $\Delta T = T_s - T_g \to 0$, eqn. (2.46) transforms to

$$\alpha = \lim_{\Delta T \to 0} \frac{T_r - T_g}{\Delta T}. \tag{2.47}$$

Several other forms of the accommodation coefficient have been proposed and measured. For example, for molecules having rotational internal energy it is possible to define the rotational accommodation coefficient

$$\alpha_R^* = \frac{E_r^* - E_g^*}{E_s^* - E_g^*} = \frac{T_r^* - T_g^*}{T_s^* - T_g^*} \tag{2.48}$$

where E_r^* etc. are now rotational energies and T_r^* etc. are temperatures corresponding to these rotational energies. However, in uhv studies, it is the translational accommodation coefficient defined by eqn. (2.47) that is of primary interest and we restrict our remarks to this accommodation coefficient. For a value of α between zero and unity, for example $\alpha = 0.5$, it is not possible to tell whether all the molecules halved the energy difference between themselves and the surface or whether half of them were reflected without loss of energy and the other half accommodated fully to the surface. The distinction is important in uhv studies for in the former interpretation it would be concluded that the condensation coefficient (c) or probability of the molecule sticking to the surface was zero, whereas in the second interpretation it would be concluded that $c = 0.5$. The value of c determines the pumping speed of the surface in question and will be discussed explicitly below. The state of an impinging molecule which has accommodated fully to the surface is independent of the impact. Its subsequent behaviour, in particular its rate

TABLE 2.20
Selected results on thermal accommodation coefficient (α)

Surface	Condition	Surface temp. $T_s(°K)$	Incident gas	Gas temp. °K	$\Delta T = T_s - T_g$ °K	Accommodation coefficient	Ref.
W	Initially degassed but probably slightly gas covered during the run	320	He	303	17	0·016	Thomas & Schofield (1955)
W	,,	300	Ne	(see ΔT)	20	0·057	Morrison & Roberts (1939)
W	,,	1860	Ar	Near room temp.		0·091	De Poorter & Searcy (1963)
Pt	,,	405	Kr	(see ΔT)	100	0·69	Thomas & Brown (1950)
Fe	H_2-covered	303·1—333·1	Ne			0·099	Eggleton & Tompkins (1952)
Fe	O_2-covered	303·1—333·1	Ne			0·268	,,
Fe	N_2-covered	303·1—333·1	Ne			0·438	,,
Ni	Surface covered with gases of unknown composition	298	He			0·385	Amdur & Guildner (1957)
Ni	,,	298	Ne			0·824	,,
Ni	,,	298	Ar			0·935	,,
W	,,	298	H_2			0·357	,,
W	,,		N_2			0·868	,,
W	,,		O_2			0·905	,,
Pt	,,	283	H_2			0·287	,,
Pt	,,		N_2			0·816	,,
Pt	,,		O_2			0·853	,,
Ni	,,	298	H_2			0·294	,,
Ni	,,		N_2			0·824	,,
Ni	,,		O_2			0·862	,,

(from Kaminsky 1965).

of desorption, is of interest and will be discussed in detail below. Here, however, the fully accommodated molecule is of interest only as a contributor to the overall accommodation coefficient.

There have been many measurements of the accommodation coefficient for a great variety of gas-surface combinations. Table 2.20 gives a selection from an extensive compilation of values of α for gases on metals from Kaminsky (1965). Three results appear to be true in general: (1) The cleaner the metal surface the smaller the accommodation coefficient; (2) the lighter the molecular weight of the gas the smaller the accommodation coefficient; (3) all values of α lie between 0·01 and 1 with all values for gas covered surfaces lying above 0·1. Such gas covered surfaces are probably typical of routine surfaces in uhv systems.

Massey and Burhop (1952) note that for a rough surface on which an impinging molecule makes n collisions before leaving the surface the apparent accommodation coefficient α_{app} is larger than α for a single collision

$$\alpha_{app} = 1 - (1-\alpha)^n. \tag{2.49}$$

There are thus several mechanisms likely to be found in practical systems causing $\alpha > 0·1$. High values of α will usually be associated with the cosine law of reflection.

Figure 2.20 shows values of α for the common gases on a glass surface as a function of temperature. Note that the accommodation coefficients increase in the same order as the boiling temperatures of Table 2.1 and that α generally increases as the temperature falls. This behaviour of α with temperature is more extensively documented by Thomas and Olmer (1943) and is the type of variation most generally found. However, it may be noted that the only exception to this temperature variation in Fig. 2.20 is helium. The accommo-

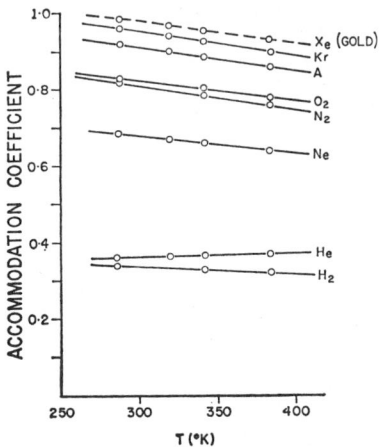

Fig. 2.20 Accommodation coefficients of common gases on glass as a function of temperature. (Schafer and Gertstacker 1956).

dation coefficient of helium on tungsten and of hydrogen on hydrogen-covered tungsten has been recently studied by Wachman (1966) and he confirms the qualitative temperature variations shown for these gases in Fig. 2.20, although the absolute values of α are lower. Figure 2.21 compares experimental values for $\alpha(T)$ for He on clean tungsten with modern theories (see Section 2.4.3.4). Accommodation coefficients of vapours measured in uhv apparatus are given in Table 2.21.

An experiment providing a new view of accommodation is that of Meyer and Gomer (1958) who have studied the energy exchange between relatively

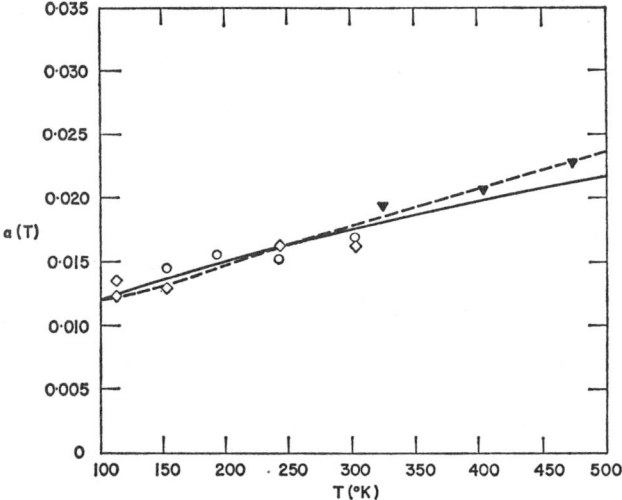

Fig. 2.21 Temperature dependence of accommodation coefficient of helium on clean tungsten (experiments and theories). Solid line, Goodman-Wachman formula (1966); dashed line, Goodman's lattice theory (1964); ○ Thomas (1965) experimental data; ◇ Silvernail (1955) experimental data; ▼ Wachman (1966) experimental data. (from Wachman 1966).

TABLE 2.21
Translational accommodation coefficients on Pyrex glass

Substance	p(torr)	α	Temperature region (°K)
Bromine	7.8×10^{-6}	0.55 ± 0.06	195–460
Iodine	8.0×10^{-6}	0.64 ± 0.06	245–460
Benzene	1.5×10^{-5}	0.37 ± 0.05	195–465
Napthalene	9.8×10^{-6}	0.37 ± 0.05	250–465
n-hexane	7.4×10^{-5}	0.47 ± 0.05	195–465
Ethylacetate	6.0×10^{-5}	0.46 ± 0.05	195–500

(from Morrison and Tuzi 1965).

cold gas molecules of He, Ne, Kr, H_2 and CH_4 and a hot graphite surface (T_s = 1100 to 1800°K). They present a series of measurements of α for various gas and surface temperatures from which they conclude that the cold gas molecules ($T_g \sim 300°K$) accommodate to the surface until they reach a critical temperature T^* (see Table 2.22) with which they desorb. If $T_g > T^*$ then α is calculated in the usual way from

$$\dot{q} = \alpha \dot{n} \bar{q}(T_s - T_g) \tag{2.50}$$

where $\dot{q}$ is the rate of energy transfer to the surface, $\dot{n}$ is the number of impacts on the hot surface per second, and $\bar{q}$ is the average specific heat (q_v) of the gas in the temperature range being studied plus $R/2$. They find α 0·1 for H_2 and $\alpha \approx 0\cdot2$ for He. The results imply that for surface temperatures $T_s < T^*$ values of α should approach unity. This conclusion is in contradiction to some of the results of Table 2.20 and Fig. 2.20 although the surfaces are not, of course, the same.

TABLE 2.22

Critical temperature T^* derived from the break in the curves q versus T_g (after Meyer and Gomer 1958). Surface graphite

	$\bar{q} = q_v + R/2$ (cal/mole)	T^* (°K)
He	4	435
H_2	6	470
Ne	4	590
Kr	4	810
CH_4	10·5	680

The velocity distribution of the scattered molecules is closely related to the accommodation coefficient, but detailed work on reflected velocity distributions appears to be at an early stage (Kaminsky 1965).

2.4.3.3. Condensation and sticking. After a molecule has collided with a surface three final states are possible: (1) the molecule may have returned to the gas phase, (2) the molecule may be trapped in the shallow minimum of Fig. 2.5, i.e. at z_p, (3) the molecule may be trapped in the deep minimum of Fig. 2.5, i.e. at z_c. Possibility (3) does not of course exist for all molecules, but there is evidence, cited below, that (2) precedes (3) for all molecules. The chance that (2) is the result of a collision is defined as the condensation coefficient c and the chance that (3) is the result of a collision is defined as the sticking probability s. Desorption following (2) is a relatively rapid process while desorption following (3) is a relatively slow process. Both will be considered in Section 2.4.4 below.

Measurements of c and s span a great variety of molecules at various temperatures impinging on many surfaces at different temperatures. We restrict our presentation here to examples which are of interest in uhv systems.

Table 2.23 shows values of c obtained from measurements of pumping speed for He on a panel of molecular sieve at 4·2°K. Note that initially the value approaches unity and has not greatly declined at the end of the experiment when $\theta \sim 0.05$. The surface in this case was highly porous and the high values of c were probably partly caused by this. Values of c for a smooth surface are shown in Table 2.24. In obtaining these data Schafer and Teggers (1953) measured the time dependence of the pressure at the exit of a capillary, at whose input a pulse of gas was applied. These measurements yielded an effective diffusion coefficient for the capillary, which in turn was converted to the fraction of molecules scattered diffusely by the surface (Maxwell's f-factor). Transition probability theory permitted the decomposition of f into the portion caused by true condensation and by surface roughness. The results of Table 2.24 have therefore been obtained indirectly. Note, however, that c falls as the

TABLE 2.23
Condensation coefficients from crysorption pumping of helium at 4·2°K
(on molecular sieve 5A crysorption panel)

1	2	3	4
Helium influx rate (torr litre sec^{-1})	Steady state helium pressure (torr)	Amount of helium adsorbed at end of test cm^3 (STP/gm)	Condensation coefficient c
1.4×10^{-4}	3.7×10^{-8}	0·127^a	0·91
2.0×10^{-4}	5.8×10^{-8}	0·377	0·82
2.8×10^{-4}	7.5×10^{-8}	0·529	0·89
4.5×10^{-4}	1.4×10^{-7}	0·789	0·77
5.3×10^{-4}	1.7×10^{-7}	0·930	0·74
7.5×10^{-4}	2.5×10^{-7}	1·42	0·72
1.1×10^{-3}	3.8×10^{-7}	3·05	0·70
1.5×10^{-3}	5.4×10^{-7}	4·08	0·67
2.0×10^{-3}	6.0×10^{-7}	6·23	0·80

^{a}Amount adsorbed initially was 0·084 cm^3 (STP/gm). Leak rate was then increased in steps as shown in column 1. Monolayer capacity of the sieve was 120 cm^3 (STP/gm).
(from Grenier and Stern 1966).

TABLE 2.24
Condensation coefficient of gases on a glass surface in the temperature range 0 to 100°C

Gas	0°C	50°C	100°C
He	0·24	0·17	0·13
Ne	0·484	0·408	0·340
H$_2$	0·638	0·567	0·495
N$_2$	0·812	0·761	0·704
O$_2$	0·857	0·816	0·766
Ar	0·890	0·855	0·815

(from Schafer and Teggers 1953).

surface temperature rises and falls as the boiling point of the gas falls. Direct measurements of c at room temperature are rare but McCarrol and Ehrlich (1963) have reported $c = 0.4$ for Xe on W measured in an uhv system.

The condensation coefficients of gases impinging upon their own frozen deposits have received extensive study because this is the basic process in a cryogenic pump. Table 2.25 gives values of c for a variety of surface and gas temperatures. Note that c falls as gas temperature increases and often, but not always, as surface temperature increases. The results of Table 2.25 are in general agreement with others. Stickney and Dayton (1963) found c between 0·975 and unity for N_2 gas at ~300°K impinging on frozen N_2 at 10·2°K. In a similar experiment Bachler et al. (1962) found the condensation coefficient for N_2 gas at room temperature remained near unity and constant for surface

TABLE 2.25
Condensation coefficients of common gases impinging upon their own frozen deposits

Surface temperature	N_2 gas at			Ar gas at		
(°K)	77°K	300°K	400°K	77°K	300°K	400°K
10	1·0	0·65	0·49	1·0	0·68	0·50
12·5	0·99	0·63	0·49	1·0	0·68	0·50
15	0·96	0·62	0·49	0·90	0·67	0·50
17·5	0·90	0·61	0·49	0·81	0·66	0·50
20	0·84	0·60	0·49	0·80	0·66	0·50
22·5	0·80	0·60	0·49	0·79	0·66	0·50
25	0·79	0·60	0·49	0·79	0·66	0·50

	O_2 gas at			CO gas at		
	77°K	300°K	400°K	77°K	300°K	400°K
10				1·0	0·90	0·73
12·5				1·0	0·85	0·73
15				1·0	0·85	0·73
17·5				1·0	0·85	0·73
20	1·0	0·86		1·0	0·85	0·73
22·5				1·0	0·85	0·73
25				1·0	0·85	0·73

	CO_2 gas at			N_2O gas at		
	195°K	300°K	400°K	195°K	300°K	400°K
10	1·0	0·75		1·0	0·63	0·50
12·5	0·98	0·70			0·62	0·50
15	0·96	0·67	0·50	0·94	0·62	0·50
17·5	0·92	p·65	0·49		0·61	0·50
20	0·90	0·63	0·49	0·86	0·61	0·50
22·5	0·87	0·63			0·61	0·50
25	0·85	0·63	0·49	0·85	0·61	0·50
77	0·85	0·63	0·49		0·61	0·50

(from Dawson and Haygood 1965).

temperatures below 25°K. Foner et al (1959) found $c = 0.60 \pm 0.06$ for Ar at room temperature impinging on frozen Ar at 4·2°K. Chubb (1966) found for H_2 and D_2 at room temperature impinging on the respective frozen gases at temperatures between 2·2 and 3·9°K that $c \approx 0.9$ for gas impingement rates from 4.6×10^{12} to 1.4×10^{18} molecules cm^{-2} sec^{-1} and total condensed amounts from 5×10^{15} to 7×10^{21} molecules cm^{-2}. More recent values of c have been reviewed by Hobson and Redhead (1968).

Molecular beam techniques have been used extensively for measurement of the condensation coefficient of metal atoms on a variety of surfaces (Wexler 1958; Hirth and Pound 1963). We have classified the quantity measured as a condensation coefficient rather than a sticking probability although this assignment is open to question. Wexler (1958) lists values of c for Ag, Au, Be, Cd, Cu, Cs, Fe, Hg, K, Mo, Na, Pt, Rb, Sb, W, and Zn impinging on a variety of surfaces. Most of the values of c are in the range $0.1 < c \leq 1$, although some values are very low. Hirth and Pound (1963) review data obtained in uhv systems. Generally speaking condensation coefficients approach unity. Rapp et al. (1960, 1961) found c for Ag, Cd and Zn on substrates of the same metal in an uhv system to be very near unity for high beam densities (0·8 to 30 monolayers sec^{-1}). In contrast with the work of others they also found that a background pressure as high as 10^{-4} torr did not reduce c below unity. Ptushinksii (1958), using uhv techniques and much lower beam densities for silver atoms bombarding molybdenum found $c = 1$ for coverages from 2×10^{13} to 3×10^{15} atoms/cm^2 of Ag on Mo. This value of c was independent of target temperature from 300 to 800°K but was reduced to $c = 0.9$ by deliberate residual gas contamination. The value of c for Ag on surfaces of germanium, mica, and glass was 0·96, 0·92 and 0·89 respectively. In contrast with the above results, Sears and Hudson (1963) have reported a very low value of c for zinc atoms (oven temperature 450°C) on Pyrex at room temperature.

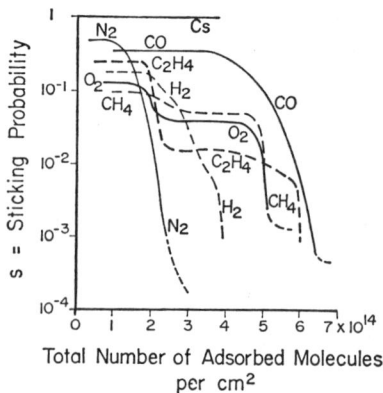

Fig. 2.22 Sticking probability as a function of surface coverage of various gases on the 411 plane of tungsten at room temperature. (Becker 1958).

The initial sticking probability of most gases on clean metal surfaces at 300°K lies between 0·1 and unity. In most cases the sticking probability remains sensibly constant until $\theta \approx 0 \cdot 5$ when it declines rapidly toward zero at monolayer coverage. This constancy of s over a wide range of θ has been interpreted to indicate that the incident gas must become physically adsorbed on the surface before either desorbing or becoming adsorbed in the chemisorbed state. Thus $c > s$ and since s is frequently between 0·1 and 1, c should be in the same range, as appears to be confirmed by the results given in Tables 2.23, 2.24 and 2.25. Figure 2.22 shows the sticking probability of several gases on the 411 plane of tungsten at 300°K (Becker 1958). Considerable variation occurs between the values of sticking probabilities observed by different experimenters as is evidenced by the data of Table 2.26, which shows the initial sticking probabilities of CO, N_2, O_2 and H_2 on W at 300°K. Some of the variability in these data result from experimental difficulties, particularly with the highly chemically active gases such as O_2 and H_2. A further cause of variability is the difference in the crystallographic state of the samples. Sticking probabilities of unity have been observed for CO on a tungsten point at $T = 20-100°K$ (Gomer 1966); this high value of s has been confirmed by King (1966) on evaporated tungsten films at 78°K. The sticking probability of gases on films has been the subject of much study because of its application in gettering (Wagener 1958; della Porta *et al.* 1963). A typical result is shown in Fig. 2.23.

Initial sticking probabilities on semiconductors are usually less than 0·1 (Law 1962).

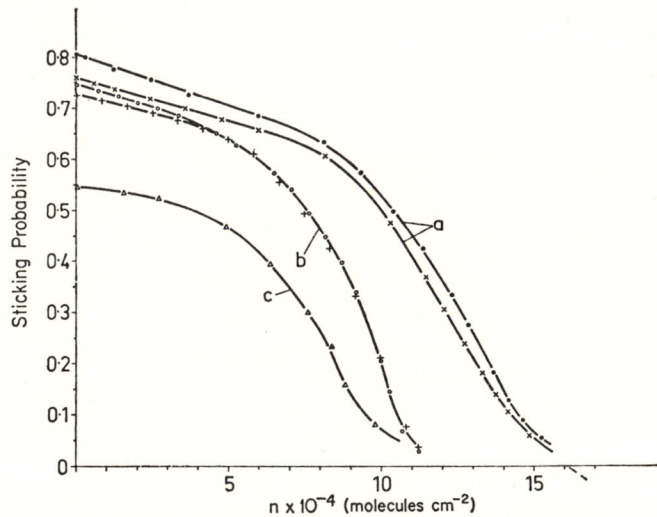

Fig. 2.23 Sticking probability of CO on zirconium film at 300°K. (*a*) Film as deposited, (*b*) film heated to ~400°K for 10 min before measurement, (*c*) film heated to ~460°K for 10 min before measurement. (Hansen and Littman 1965.)

Physical Processes

The variation of sticking probability with temperature of the surface depends on the gas-surface combination, the particular adsorbed phase and the surface structure. Details for particular combinations can be found in the references cited in Table 2.26.

TABLE 2.26
Initial sticking probability of various gases on W at 300°K

Gas	s_0	$\sigma_{max} \times 10^{-14}$	θ_c	Surface	Reference
CO	0·36	6·5	0·54	Ribbon, (411) plane	Becker (1958)
,,	0·18	5·3	0·66	Ribbon, (311) plane	Eisinger (1957)
,,	0·62	5·0	0·30	Ribbon	Schlier (1958)
,,	0·3–0·5	4·5	0·49	Wire	Ehrlich (1961b)
,,	0·5	9·5	0·40	Wire	Redhead (1961)
,,	0·97	—	0·5	Field-emission point	Gomer (1966)
N_2	0·55	3·0	0·33	Ribbon, (411) plane	Becker (1958)
,,	0·30	5·5	0·33	Ribbon, (311) plane	Eisinger (1958a)
,,	0·42	1·8	0·28	Ribbon	Schlier (1958)
,,	0·3	3·0	0·50	Ribbon, (311) plane	Kisliuk (1959b)
,,	0·11 0·13 0·28	2·8	0·27–0·14	Wire	Ehrlich (1961a)
,,	0·2	1·5	0·74	Wire	Oguri (1963)
O_2	0·14	5·2	0·4	Wire	Ageev and Ionov (1966)
,,	0·15	—	0·7	Ribbon	Singleton (1967b)
H_2	0·2	4·0	0·5	Ribbon, (411) plane	Becker (1958)
,,	0·3	7·0	0·43	Ribbon, (311) plane	Eisinger (1958b)
,,	0·08	7·6	0·26	Sheet	Ricca et al. (1965)
,,	0·11	4·5	0·5	Wire	Ageev et al. (1965)

s_0 initial sticking probability.
θ_c relative coverage when s starts to fall.
σ_{max} maximum number of molecules/cm² adsorbed at 300°K.

An updated version of this table may be found in Hobson and Redhead (1968).

2.4.3.4 Theory of the impact between a molecule and a surface. All theories of the impact between a molecule and a surface have considered ideal crystal surfaces, and hence represent only a starting point for theories of most real surfaces. However, even in the case of ideal surfaces each theory has used restrictive assumptions which limit its applicability to certain special situations. The need for restrictive assumptions arises because the general theoretical problem contains many variables, few of which are quantitatively negligible.

Two general theoretical approaches have yielded results of interest in physisorption, and therefore also in chemisorption, since physisorption is apparently a precursor to chemisorption. The first approach was originated by Lennard-Jones and his colleagues in a series of ten papers of which we make reference to three (Lennard-Jones and Devonshire 1936a, 1936b; Devonshire 1937) in which accommodation coefficients of He and H_2 on W were calculated. These authors considered that only single quantum energy exchanges between gas molecule and solid were permitted. The highest energy quantum of the solid is given by $hv_m = kT_D$, T_D being the Debye temperature of the solid. For most solids T_D is several hundred degrees absolute and hence $kT_D \lesssim 1000$ cal/mole. Lennard-Jones and Devonshire (L.J.D.) considered this the maximum amount of energy that could be transferred in either direction in a gas-solid impact. This would limit the validity of their considerations to systems having heats of adsorption less than this, i.e. to systems in which the gas molecule was H_2, He or Ne (Table 2.6). L.J.D. assumed a flat surface for which the potential energy in the normal direction was similar to that given by eqn. (2.18), and was periodic in the two directions parallel to the surface. No vibrations of the crystal atoms were included and hence the calculations apply only to elastic scattering. They calculated the relative intensities of the specular beam and the diffracted beams of first and second orders and obtained excellent agreement with the experimental data of Frisch and Stern (1933) for the impact of He on LiF (similar to the results of Fig. 2.18). They were also able to calculate the locations of anomalous minima in the diffraction results of Frisch and Stern. These were attributed to molecules having particular components of momenta relative to the crystal, which enabled them to skid along the surface of the crystal without loss of energy before being desorbed. This phenomenon was termed selective adsorption and its successful explanation appears to establish the general validity of the model of L.J.D. for the impact of light gases on single crystals. A modern variant of the single phonon theory has been developed by Gadzuk (1967) and Fig. 2.24 shows the result of Gadzuk's calculation of the accommodation coefficient of He on W as a function of T. The experimental points are the same as those of Fig. 2.21 in which it was demonstrated that a formula of Goodman and Wachman (1966), discussed below, gave a good fit to experiment.

Goodman and Wachman's theory is characteristic of the second general theoretical approach to the gas-solid impact which is a classical lattice theory. Gilbey (1962) argues that a classical normal mode formulation has precisely the same form as the L.J.D. formulation and that therefore the L.J.D. results are applicable to multiphonon as well as single phonon processes.

The classical lattice theory was initiated by Zwanzig (1960) who criticized the single quantum approximation, and solved the classical problem of a gas molecule impinging upon a one dimensional array of solid atoms joined by springs. This approach has been developed by others, notably Goodman (1965) in a series of papers. For an atom impinging normally on an atom in a 3-dimensional array of atoms representing the solid Goodman (1965) finds that, in general, the form of α (accommodation coefficient) is independent of the gas-solid system and the typical result is shown in Fig. 2.25. Below a critical gas temperature (T_c) there is complete accommodation. Above $T = T_c$

α decreases as T^{-1} for values of T just above T_c. α then passes through a minimum and finally reaches a high temperature limit

$$\alpha(\infty) = \frac{4\mu}{(1+\mu)^2} \qquad \mu < 0.84. \qquad (2.51)$$

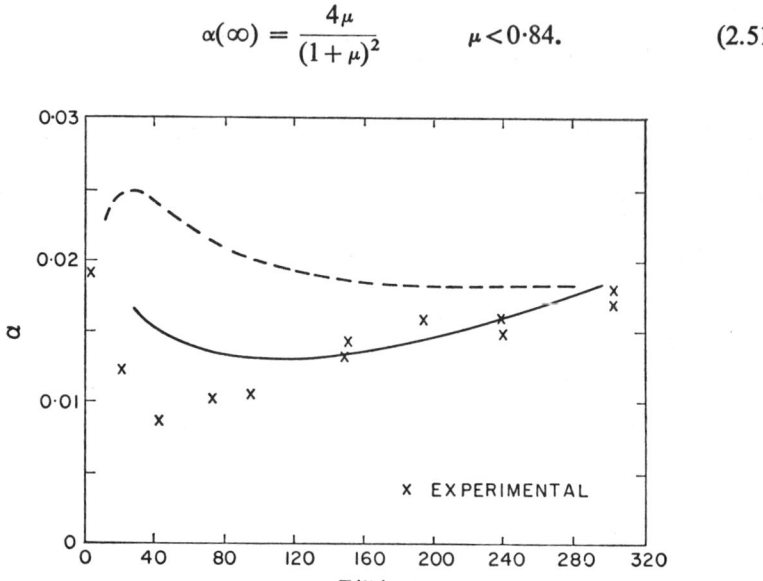

Fig. 2.24 Thermal accommodation coefficient of helium on tungsten (from Gadzuk 1967).
Solid line: single phonon calculation of Gadzuk (1967).
Dashed line: single phonon calculation of Gilbey (1962) based on L.J.D. model.
Points: experimental data of Thomas (1965).
Compare with Fig. 2.21.

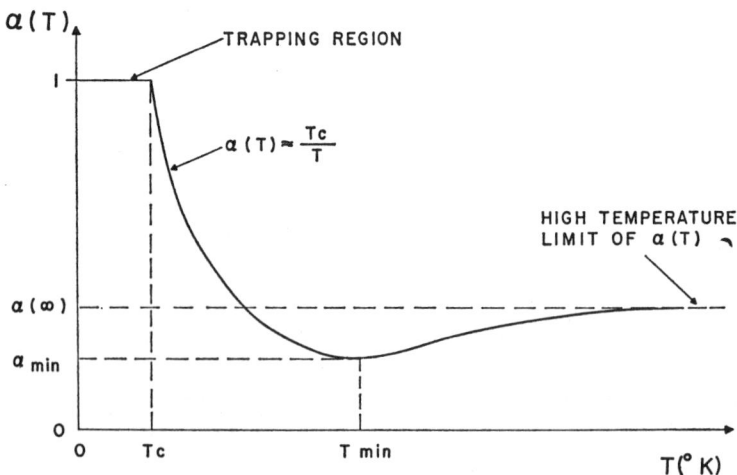

Fig. 2.25 General form of the accommodation coefficient for an atom impinging normally on a 3-dimensional array. Solid atoms initially at rest. T represents gas temperature ($\frac{1}{2}mv^2 = kT$). (from Goodman 1964).

μ is the ratio of mass of gas atom to mass of solid atom. This limit is just the result that would be obtained for the energy transfer in a head-on collision between the two atoms. The parameters T_c, T_{min}, α_{min} depend upon the assumed details of the molecule-solid potential energy (Section 2.3.1).

As might be expected, the value of T_c increases with the value of E_p given in Table 2.6. Specific calculations by Goodman (1964) for the accommodation of the inert gases on tungsten are shown in Fig. 2.26 and compared with experiment. The agreement is seen to be excellent and the general similarity to Fig. 2.20 is clear. The lines of Fig. 2.26 are drawn for gas incident normally on the surface. Goodman and Wachman (1966) have shown that for random incidence the results are not changed greatly and can be expressed by a closed formula:

$$\alpha(T) = 1 - \exp\left(-\frac{T_c}{T}\right) + \frac{2\cdot 4\mu}{(1+\mu)^2}\left[\tanh\frac{ba}{\omega_0}\left(\frac{kT}{M}\right)^{\frac{1}{2}}\right]\exp\left(-\frac{T_c}{T}\right), \quad (2.52)$$

where b is a constant of order unity; $\omega_0 = kT_D/\hbar$ (Debye Temperature); a is an inverse distance with a value of $1\cdot 3$ Å^{-1}; M is the mass of the gas atom.

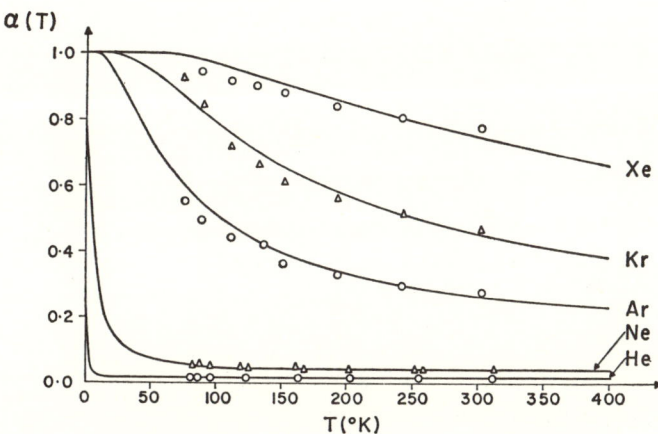

Fig. 2.26 Calculated accommodation coefficients for beams of He, Ne, Ar, Kr, Xe incident normally on W (lines) from Goodman (1964). Experimental points are those of Thomas and Schofield (1955) and Thomas (1961).

The application of eqn. (2.52) is shown in Fig. 2.21 for the case He-W. Goodman's (1964) calculations also yield explicit results for the condensation coefficient $c(T)$ and for the energy loss of the molecules which are reflected. The values of $c(T)$ follow closely the results for $\alpha(T)$ given in Fig. 2.26.

No calculations of α and c have been carried out for heterogeneous surfaces but the outlines of such a theory are already present in the calculations of Goodman to which we have referred. Leonas (1963) has considered the impinging molecule to be diatomic in a 1-dimensional model with the line of centres of the molecule along the direction of incidence. The general conclusion for the practical range of thermal energies is that the influence of the vibrational mode of a diatomic molecule is small.

The angular distribution of the scattered flux (see 2.4.3.1) has been examined theoretically by Logan and Stickney (1966) who regard the solid as an array of surface atoms moving perpendicular to the plane of the surface. Classical collisions take place only between impinging gas atoms and individual solid atoms, which are considered cubes, thus eliminating all forces tangential to the surface. The analysis is successful in predicting the qualitative variations of the angle of maximum reflection (see Fig. 2(19)) as a function of surface temperature, gas temperature, and angle of incidence, and is partially successful in predicting its behaviour with mass of gas atom. Logan and Stickney (1966) also conclude that surface temperature is a significant parameter.

Theories of the sticking probability in chemisorption (Becker 1955; Ehrlich 1956; Kisliuk 1957, 1958; Gavrilyuk 1961) postulate that the incoming molecule first is trapped in the physisorbed state, and then a competition takes place between desorption and capture in the chemisorbed state. By referring to Fig. 2.5 we can derive the formula of Gavrilyuk (1961). Assume a rate of impingement vp molecules cm^{-2} sec^{-1}. The rate of arrival at the minima at z_p will be cvp. There will be two loss rates from this state: (1) a return to the gas phase given by

$$\frac{\sigma_p}{\tau_0} \exp\left[-\frac{E_p}{RT}\right]$$

(see Section 2.4.4 below) where σ_p is the number of molecules cm^{-2} in the physisorbed state, (2) a transmission over the barrier to the chemisorbed state given by

$$\frac{\sigma_p}{\tau_0} \exp\left[-\frac{(E_p+E_a)}{RT}\right].$$

Assume that the number σ_p is independent of time:

$$cvp = \frac{\sigma_p}{\tau_0}\left\{\exp\left[-\frac{E_p}{RT}\right] + \exp\left[-\left(\frac{E_p+E_a}{RT}\right)\right]\right\} \text{molecules cm}^{-2} \text{ sec}^{-1} \quad (2.53)$$

The rate of increase of molecules in the chemisorbed state will be:

$$\frac{d\sigma_c}{dt} = \frac{\sigma_p}{\tau_0} \exp\left[-\left(\frac{E_p+E_a}{RT}\right)\right] - \frac{\sigma_c}{\tau_0} \exp\left[-\frac{E_d}{RT}\right] \text{molecules cm}^{-2} \text{ sec}^{-1} \quad (2.54)$$

If the sticking probability is defined as $s = (1/vp)(d\sigma_c/dt)$, combination of eqns. (2.53) and (2.54) yields

$$s = \frac{c}{1+\exp(E_a/RT)} - \frac{\sigma_c}{\tau_0} \exp\left[-\frac{E_d}{RT}\right]. \quad (2.55)$$

Figure 2.27 shows eqn. (2.55) plotted by Gavrilyuk and Medvedev (1963) using the values $c = 0.7$, $E_a = -235$ cal/mole, $\tau_0 = h/kT$. The variation of E_d with σ_c was determined experimentally from desorption measurements. By assuming a variation of E_a with σ_c Gavrilyuk and Medvedev were able to generate lines very close to those shown dashed through the experimental points. Some of these procedures are arbitrary but the theory does have three important features: (1) the value of s at low coverage is $\leq c$. Thus for small values of E_a, magnitudes of s commensurate with those of c are probable, as seems to be confirmed by the general results of Tables 2.25 and 2.26; (2) the temperature variation of s at low coverage is likely to be difficult to predict

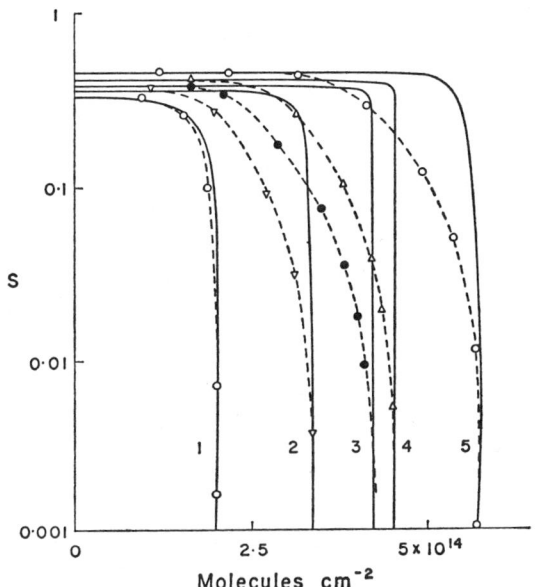

Fig. 2.27 Sticking probability s for CO on W(113) as function of coverage. Surface temperatures °K: (1) 1100, (2) 900, (3) 700, (4) 500, (5) 300. Gas temperature ~ 300°K. Solid lines eqn. 2.55. Points experiment. (Gavrilyuk and Medvedev 1963).

depending upon the magnitude and sign of E_a. As noted in discussion of Table 2.26 this result is found experimentally; (3) the value of s remains constant with coverage until desorption directly from the chemisorbed state causes a decline as monolayer coverage is approached. This result is general in Fig. 2.22.

2.4.4 Desorption of a molecule from a surface

2.4.4.1 Mean time of sojourn. After an impinging molecule has fully accommodated to a surface there is a finite probability that it will desorb at a later time, its time of sojourn being τ. There is also a finite probability that it will migrate over the surface, possibly making contact with other adsorbed molecules, thereby altering its probability of desorption.

If an adsorbed phase is in equilibrium with the gas, then a simple equilibrium equation between the number of molecules arriving and desorbing may be written:

$$vcp = \frac{\sigma}{\tau} \text{ molecules cm}^{-2} \text{ sec}^{-1}. \tag{2.56}$$

where τ is the average time of sojourn of an accommodated molecule, and ν is the specific rate of arrival from the gas phase and is given by

$$\nu = \frac{3 \cdot 5 \times 10^{22}}{(MT)^{\frac{1}{2}}} \text{ molecules cm}^{-2} \text{ sec}^{-1} \text{ torr}^{-1}, \tag{2.57}$$

p being in torr. σ is the number of adsorbed molecules cm^{-2}, and c the condensation coefficient.

Since the pressure and σ are the measured quantities for every adsorption isotherm the product $c\tau$ may be obtained for every isotherm point. The ratio $\tau' = \sigma/\nu p = c\tau$ actually gives the average sojourn time of all impinging molecules, whereas the quantity of main interest here is τ the average sojourn time of all accommodating molecules. The distinction is rarely made in the literature. However c has been found, in Section 2.4.3.3, to be a quantity whose magnitude is frequently near unity and which can be obtained either from theory or measurement and is moreover a quantity which varies relatively slowly with the isotherm variables of pressure, coverage, and temperature. Thus values τ' of obtained from isotherm data are of interest.

The number of values of τ' that can be obtained from the literature in this way is virtually limitless, and we restrict ourselves here to two examples, which are given in Tables 2.27 and 2.28. The first example from the work of Halsey and colleagues (Sams *et al.* 1960) was selected because the adsorbent (graphitized carbon black P33(2700)) was particularly homogeneous and the heat of

TABLE 2.27

Mean sojourn time (τ') of inert gases on graphitized carbon

1	2	3	4	5	6
Gas	T(°K)	Adsorbed quantity from experiment (Molecules gm^{-1} torr^{-1})	$\tau' = \sigma/\nu p$ sec	E_p cal/mole	$\tau_0 = \tau' \exp\left(\frac{-E_p}{RT}\right)$ sec
Ne	78·611	$1 \cdot 14 \times 10^{16}$	$1 \cdot 0 \times 10^{-10}$	758	$7 \cdot 9 \times 10^{-13}$
Ar	166·135	$2 \cdot 87 \times 10^{16}$	$5 \cdot 3 \times 10^{-10}$	2200	$6 \cdot 8 \times 10^{-13}$
Kr	285·119	$4 \cdot 29 \times 10^{15}$	$1 \cdot 5 \times 10^{-10}$	2900	$9 \cdot 2 \times 10^{-13}$
Xe	295·052	$1 \cdot 54 \times 10^{16}$	$6 \cdot 9 \times 10^{-10}$	3806	$10 \cdot 4 \times 10^{-13}$

Adsorbent P33(2700). Calculated from experimental data of Sams *et al.* (1960). All data in Henry's Law range, P ~ 200 torr. Homogeneous surface. Energies used are for 3-9 Potential and Area used in B.E.T. area ($1 \cdot 25 \times 10^5$ cm^2 gm^{-1}).

adsorption likely to be single valued. The data are all in the Henry's Law range where τ' is independent of pressure. Sams et al. (1960) do not analyse their data in this way, and their values of E_p and the area of the adsorbent were obtained without the concept of sojourn time. The final column of Table 2.27 will be discussed in Section 2.4.4.2 below. The second example is taken from the data of Hobson (1961c) for the adsorption of nitrogen on a heterogeneous surface. Note how long τ' can become in the uhv range for typical physical adsorption of a vapour. τ' is also variable reflecting the results that the isotherm does not follow Henry's Law at 77°K. The final column of Table 2.28 will also be discussed in Section 2.4.4.2 below.

TABLE 2.28

Mean sojourn time (τ') of N_2 on Pyrex at 77°K calculated from experimental data of Hobson (1961c)

1	2	3	4	5
σ molecules cm^{-2}	p (torr)	$\tau' = \sigma/\nu p$ sec	Isosteric heat q_{st} cal/mole	$\tau_0 = \tau' \exp\left(\dfrac{-q_{st}}{RT}\right)$ sec
$7\cdot4 \times 10^{11}$	$3\cdot8 \times 10^{-10}$	$2\cdot6$	5860	$6\cdot5 \times 10^{-17}$
$1\cdot3 \times 10^{12}$	$1\cdot1 \times 10^{-9}$	$1\cdot5$	5700	$1\cdot1 \times 10^{-16}$
$3\cdot7 \times 10^{12}$	$1\cdot7 \times 10^{-8}$	$2\cdot9 \times 10^{-1}$	5330	$2\cdot4 \times 10^{-16}$
$8\cdot3 \times 10^{12}$	$1\cdot4 \times 10^{-7}$	$8\cdot0 \times 10^{-2}$	5090	$3\cdot3 \times 10^{-16}$
$2\cdot0 \times 10^{13}$	$1\cdot5 \times 10^{-6}$	$1\cdot8 \times 10^{-2}$	4800	$4\cdot8 \times 10^{-16}$
$5\cdot7 \times 10^{13}$	$3\cdot2 \times 10^{-5}$	$2\cdot4 \times 10^{-3}$	4280	$1\cdot9 \times 10^{-15}$
$1\cdot0 \times 10^{14}$	$4\cdot3 \times 10^{-4}$	$3\cdot2 \times 10^{-4}$	3860	$4\cdot1 \times 10^{-15}$

Heterogeneous surface. Isosteric heats obtained from isotherms at different temperatures. Area used is geometric surface area.

Another method for finding τ' developed by Clausing (1930a, 1930b) and summarized by de Boer (1953) was the measurement of the time delay of a step increase of pressure traversing a capillary. This method has been developed and is still in use (Barrer 1954a, b; Eschbach et al. 1961; Ash et al. 1963; Muller 1965). It may be shown that the capillary in Clausing's method is formally equivalent to a medium of thickness l and diffusion coefficient D where

$$D = \frac{l^2}{6\bar{t}} \qquad (2.58)$$

$\bar{t}$ being the average time which a molecule needs to proceed through the capillary. If it is assumed that every molecule striking the capillary wall is condensed and re-emitted from the same point with a cosine distribution in angle a time τ later then

$$\bar{t} = \frac{l}{2d\bar{v}} + \frac{l^2 \tau}{2d^2} \qquad (2.59)$$

d being the diameter of the capillary and $\bar{v}$ the mean molecular velocity. Once $\bar{t}$

has been measured τ may be evaluated from eqn. (2.59). Table 2.29 shows some values of τ obtained by this time delay method. We have included only results where the full temperature dependence has been given by the authors. It should be noted that eqn. (2.59) should only be used where τ is independent of the number of adsorbed molecules, i.e. where the adsorption isotherm follows Henry's Law. Where the isotherm does not follow Henry's Law more complex relations are found (Clausing 1962; Ash et al. 1963).

TABLE 2.29
Times of sojourn or desorption rates of molecules on surfaces measured by time delay method

Gas	Surface	Temperature range of measurement	τ' (sec)	Reference
Ar	Glass	78—90°K	$\tau = 1\cdot7 \times 10^{-14} \exp[3800/RT]$ sec	Clausing (1930b)
He	Glass	13·8—20·4°K	$\tau_1 = 10^{-9} \exp[229/RT]$ $\tau_2 = 10^{-9} \exp[530/RT]$ } 2 states	Müller (1965)
NH_3	Analcite	278—308°K	$\tau = 3\cdot1 \times 10^{-11} \exp[4300/RT]$ sec	Barrer & Grove (1951)
SO_2	Analcite	278—308°K	$\tau = 6\cdot6 \times 10^{-11} \exp[3010/RT]$ sec	Barrer & Grove (1951)

Table 2.30 is reproduced from de Boer (1953) and shows results obtained from eqn. (2.59) of three values of $\tau = 10^{-12}$, 10^{-10}, 10^{-4} sec respectively. These times have been selected because they correspond approximately to sojourn times of H_2, N_2 and an organic molecule at room temperature. From the table it may be seen that as the capillary becomes smaller the second term in eqn. (2.59), i.e. the adsorption term, becomes relatively more important. This term is always more important for the organic molecule (diffusion pump vapour). Tominaga (1965) has reported experimental values for the adsorption time of oil molecules on borosilicate glass.

The assumption that an accommodated molecule will be desorbed from the same spot as it originally struck will not in general be true, since it is more likely to hop over the potential barrier between surface sites than it is to desorb completely.

A mean time of sojourn between surface hops τ_s may be defined. In general τ_s is much smaller than τ; hence the molecule executes on the average many surface hops for every desorption hop. This surface migration can lead to several important consequences. The most obvious is that the molecules can move along a surface and be desorbed at some distant point. This surface migration alters eqn. (2.59) for the time delay method to the following

$$\bar{t} = \frac{(l^2/2d\bar{v}) + (l^2\tau/2d^2)}{1 + \tfrac{3}{4}(\tau/\tau_s)(a^2/d^2)} \qquad (2.60)$$

TABLE 2.30
Delay times ($\bar{t}$ sec) for transmission of pressure pulse through a capillary of length l and diameter d. Mean sourjourn time is τ and mean molecular velocity $\bar{v}$

	$\tau = 10^{-12}$ sec $\bar{v} = 15 \times 10^4$ cm sec^{-1}	$\tau = 10^{-10}$ $\bar{v} = 5 \times 10^4$ cm sec^{-1}	$\tau = 10^{-4}$ sec $\bar{v} = 10^4$ cm sec^{-1}
$l = 10$ cm $d = 10^{-1}$ cm	(1) $3 \times 10^{-3} + 5 \times 10^{-9}$	(2) $10^{-2} + 5 \times 10^{-7}$	(3) $5 \times 10^{-2} + 5 \times 10^{-1}$
$l = 10$ cm $d = 10^{-4}$ cm	(4) $3 \times 10^{-4} + 5 \times 10^{-7}$	(5) $10^{-3} + 5 \times 10^{-5}$	(6) $5 \times 10^{-3} + 50$
$l = 10^{-2}$ cm $d = 10^{-6}$ cm	(7) $3 \times 10^{-4} + 5 \times 10^{-5}$	(8) $10^{-3} + 5 \times 10^{-3}$	(9) $5 \times 10^{-3} + 5000$
$l = 10^{-3}$ cm $d = 10^{-7}$ cm	(10) $3 \times 10^{-5} + 2 \times 10^{-5}$	(11) $10^{-4} + 5 \times 10^{-3}$	(12) $5 \times 10^{-4} + 5000$

where τ_s is the mean time between surface hops and a is the sideways hop distance and may be set equal to the interatomic distance in the adsorbing surface. It may be noted that the numerator of eqn. (2.60) is the same as eqn. (2.59) and surface diffusion does not contribute quantitatively to $\bar{t}$ until the second term of the denominator becomes commensurate with unity. Surface diffusion is discussed further in Section 2.4.4.4 below.

2.4.4.2 First-order desorption. Instead of viewing adsorption in equilibrium as in Section 2.4.4.1 we can alternatively ask the question: What is the rate of desorption of σ molecules cm^{-2} from a homogeneous surface in the event that none is returning from the gas phase. This question has been answered by the theory of rate processes (Glasstone et al. 1941) as follows:

$$\frac{d\sigma}{dt} = -\sigma \frac{kT}{h} \frac{f_{\ddagger}^+}{f_a} \exp\left[-\frac{E_d}{RT}\right] \text{ molecules cm}^{-2} \text{ sec}^{-1} \quad (2.61)$$

where E_d is the activation energy for desorption and $f_{\ddagger}^+$ and f_a are the partition functions for the so-called activated complex and the adsorbed state respectively. Equation (2.61) is written for first-order desorption in which the desorption rate depends linearly upon σ, i.e. there is no interaction between adsorbed molecules in the desorption process. First-order desorption will apply for physisorbed molecules and non-dissociated chemisorbed molecules.

If $f_{\ddagger}^+/f_a$ is near unity as is often the case (see below) the first order rate constant is given by

$$v_1 = \frac{kT}{h} \text{ sec}^{-1}. \quad (2.62)$$

This has the value 6.2×10^{12} sec^{-1} at room temperature. For this case $1/\tau$ of eqn. (2.56) can be identified with $v_1 \exp[-E_d/RT]$ of eqn. (2.62)

$$\frac{h}{kT} = \tau \exp\left[-\frac{E_d}{RT}\right]. \tag{2.63}$$

Thus from (2.63) the expression $\tau \exp[-E_d/RT]$ may be expected to have a value (at room temperature) of 1.6×10^{-13} sec and to be independent of the specific gas-solid combination. These conclusions are confirmed experimentally in column 6 of Table 2.27 for physisorption, although the magnitude of $\tau \exp[-E_d/RT]$ is found to be nearer 10^{-12} sec than 10^{-13} sec. Similar conclusions may be drawn from the rate constants of Table 2.31 for chemisorption.

An equation similar to eqn. (2.63) was derived many years ago by Frenkel (1924) who concluded that

$$\tau_0 = \tau \exp\left[-\frac{E_d}{RT}\right] \tag{2.64}$$

where τ_0 was the period of vibration of the adsorbed molecule normal to the surface which Frenkel (1924) found to be about 10^{-13} sec. Kiselev and Poshkus (1963) have given a detailed calculation of τ_0 for argon on the basal plane of graphite and find $\tau_0 = 7.0 \times 10^{-13}$ sec, a value in approximate agreement with Frenkel and in excellent agreement with the results given in Table 2.27.

The ratio f_+^+/f_a which was assumed to be unity above is essentially a measure of the difference in entropy between the activated complex and the adsorbed phase. For physical adsorption where the activated complex can be identified with the gas phase a ratio of f_+^+/f_a nearly unity is expected for a highly mobile adsorbed phase (Kruyer 1955) as appears to be confirmed by the results of column 6 of Table 2.27. Results similar to these were found by de Boer and Kruyer (1955) for N_2, H_2, O_2, CO, CO_2, CS_2 and the aliphatic hydrocarbons on charcoal. For water which is polar, however, they found $\tau_0 \approx 10^{-16}$ sec. This was interpreted as an indication of localized adsorption for water, which would give a value of f_+^+/f_a greater than unity. The magnitudes of τ_0 given in column 5 of Table 2.28 suggest localized adsorption in this case also. Baker (1965) and Degras (1965), however, give desorption data for CO from nickel leading to values of τ_0 many orders of magnitude larger than 10^{-13} sec suggesting f_+^+/f_a much smaller than unity. Further discussion of the latter results may be found in Hobson and Redhead (1968).

For comparison with experiment, eqn. (2.61) is more often written simply as

$$\frac{d\sigma}{dt} = -\frac{\sigma}{\tau} = -\sigma v_1 \exp\left[-\frac{E_d}{RT}\right]. \tag{2.65}$$

At fixed temperature this has the simple solution

$$\frac{d\sigma}{dt} = -\frac{\sigma_0}{\tau} \exp[-t/\tau] \tag{2.66}$$

from which τ may be determined from the decay time. Experimentally it is usually more convenient to vary the temperature in a known way (usually linearly with time or inversely with time) and methods have been worked out for measuring E_d and v_1. Ehrlich (1961a) has given a comprehensive account of these methods together with a summary of various experimental checks and precautions required in the interpretation of data. Redhead (1962b) has, in addition, examined this flash filament technique in the presence of high pumping speeds for the gas under study.

Using this flash filament method Ehrlich (1961b) gives for the measured rate of desorption of β-CO from W

$$\frac{d\sigma}{dt} = -3 \times 10^{13} \, \sigma \exp -\frac{75000}{RT} \tag{2.67}$$

the activation energy of desorption being in cal/mole. This result is in good agreement with the magnitude of v_1 (or τ_0) discussed above. Table 2.31 gives other values of v_1 compiled by Ehrlich (1959).

TABLE 2.31
First-order rate constants

	Surface	q (kcal/mole)	v_1 (sec^{-1})
Na	W	62·9	$5·13 \times 10^{12}$
K$^+$	W	58·8	$3·31 \times 10^{13}$
Rb	W	60·0	$4·0 \times 10^{12}$
Rb$^+$	W	43·8	$2·9 \times 10^{11}$
Cs	W	64·4	$1·75 \times 10^{12}$
Cs$^+$	W	47·1	$0·88 \times 10^{12}$
Ba	W	87·2	$3·85 \times 10^{10}$
H	W	67	$2·2 \times 10^{13}$
O	W	147	$1·25 \times 10^{13}$
A	Glass	3·8	$5·88 \times 10^{13}$

(from Ehrlich 1959).

From eqn. (2.67) it would be predicted that a steady increase in the temperature of a tungsten wire would give rise to a single peak in the desorption rate since desorption reduces σ. In practice, more than one peak is usually found. This is clearly shown in the desorption spectra measured by Redhead (1961) which are illustrated in Fig. 2.9 and show that CO may be bound to W with at least four desorption energies. The β-phase of eqn. (2.67) refers to a major phase desorbing at about 1100°K. The surface having several discrete energies of desorption is an intermediate case between a truly homogeneous surface and a heterogeneous one.

Desorption theory for a heterogeneous surface has been developed by Erents *et al.* (1965). These authors consider the isothermal desorption rate from a surface with variable E_d where the initial population of molecules over the site distribution is given by the polynomial:

$$f(E_d) = a_0 + a_1 E_d + a_2 E_d^2 + a_3 E_d^3 \ldots \ldots \quad (2.68)$$

For a band of energies they obtain the solution

$$\frac{d\sigma}{dt} = \frac{RT}{t} \sum_{n=0} a_n \left[RT \ln \frac{t}{\tau_0} \right]^n \quad (2.69)$$

where a_n are the coefficients of eqn. (2.68).

They examine this solution for various values of the coefficients and find, in particular, that if the desorption rate goes as t^{-m} (which is quite a common form in practice), that this behaviour is associated with a population distribution varying as $\exp[E_d/mRT]$.

2.4.4.3 Second-order desorption. Most of the chemically active gases in uhv systems are diatomic (H_2, N_2, O_2, CO) and all (with the exception of CO) are dissociatively adsorbed on metals. This arises (Ehrlich 1959) because the heat of dissociation is less than twice the heat of atomic adsorption (Section 2.3.2). For such a system the process of desorption involves the surface collision of two atoms before desorption as a molecule occurs. Ehrlich (1959) gives the equation

$$\frac{d\sigma_2}{dt} = -2^{\frac{1}{2}} \sigma_1^2 d \left(\frac{\pi RT}{2M} \right)^{\frac{1}{2}} \exp\left[-\left(\frac{2q_a - E^*}{RT} \right) \right] \text{ molecules cm}^2 \text{ sec}^{-1} \quad (2.70)$$

where σ_2 is the number of molecules cm^{-2}, σ_1 is the number of atoms cm^{-2}, d is the collision diameter of an adsorbed atom, and $(\pi RT/2M)^{\frac{1}{2}}$ is the average velocity of atoms migrating on the surface, q_a is the energy of atomic adsorption, and E^* the energy of molecular dissociation.

Equation (2.70) combines the collision probability in a 2-dimensional atomic gas with the desorption probability of the molecule. For comparison with experiment eqn. (2.70) is expressed as

$$\frac{d\sigma_2}{dt} = -\sigma_1^2 v_2 \exp\left[-\frac{E_d}{RT} \right] \text{ molecules cm}^{-2} \text{ sec}^{-1}. \quad (2.71)$$

v_2 is the second-order desorption rate constant. It is not a strong function of atomic mass or temperature and its order of magnitude for nitrogen atoms at 1000°K can be calculated to be $v_2 = 1 \cdot 9 \times 10^{-3}$. Table 2.32 gives values of E_d and v_2 determined experimentally in uhv systems. Values of v_2 are seen to be in general agreement with the magnitude predicted by eqn. (2.70).

In principle, higher order reactions can occur but they are so much less common in ultrahigh vacuum systems than first- and second-order reactions that we restrict the discussion to the latter only. The articles by Ehrlich (1961c) and Redhead (1962b) treat the measurement of the pre-exponential factor and E_d for second-order kinetics as well as for first-order kinetics.

Obviously the solution for eqn. (2.65) will be different in many respects to the solution of eqn. (2.71). Ehrlich points out the precautions which must be taken to ensure that the order of a reaction has not been incorrectly assigned perhaps due to the near proximity of several phases. It should also be noted that Moore and Unterwald (1964a) question deductions about the adsorbed phase on the basis of the order of the desorption kinetics alone; and that Ptushinksii and Chuikov (1967) question the entire basis of the measurements of Becker *et al.* (1961) leading to the results of Table 2.32 on the grounds that the desorbing species may have been incorrectly identified.

TABLE 2.32
Second-order rate constants and activation energies of desorption

Gas	Phase	Surface	v_2 cm^2 sec^{-1}	E_d kcal/mole	Reference
N_2	β	W	$1\cdot 4 \times 10^{-2}$	81	Hickmott and Ehrlich (1958)
H_2	β	W	5×10^{-3}	31	Hickmott (1960a) $\sigma < 3 \times 10^{13}$ molecule cm^{-2}
O_2	First layer	W	3×10^{-3}	106	Becker *et al.* (1961)
O_2	Second layer	W	$2\cdot 5 \times 10^{-3}$	53	Becker *et al.* (1961)

2.4.4.4 Surface diffusion. In the previous section it has been shown that surface diffusion plays a central role in the evaporation of many diatomic molecules. Implied in the discussion is the result that surface diffusion also controls the location of desorption and in some uhv applications the location of desorption may be important. A useful criterion for deciding this point is a consideration of the second term in the denominator of eqn. (2.60). As noted in the brief discussion of this equation, if the quantity $\tau a^2/\tau_s d^2$ becomes larger than unity diffusion relative to desorption plays a significant macroscopic role in the system vacuum conditions. Now 'a' can be taken as a typical interatomic dimensions in a solid say 3 Å. d will of course vary widely with system design but let us consider a nominal value of 1 cm. This means that for surface diffusion to be important $\tau/\tau_s > 10^{15}$. τ_s is determined by an equation analogous to eqn. (2.64)

$$\tau_s = \tau_0 \exp\frac{E_s}{RT}. \tag{2.72}$$

As a first approximation we may take τ_0 here the same as in eqn. (2.64) and hence write

$$\frac{\tau}{\tau_s} = \exp\left[\frac{E_d - E_s}{RT}\right]. \tag{2.73}$$

This equation yields the number of sideways hops an adsorbed molecule makes on the average before a desorption hop. For $\tau/\tau_s < 10^{15}$ eqn. (2.73) yields the result $(E_d - E_s)/RT < 35$. Table 2.11 indicates that this criterion is not difficult to fulfill in practice. Many adsorbed molecules fulfil this criterion simultaneously with the criterion that τ is in the range where desorption, whether accompanied by surface diffusion or not, is of interest. It is clear that if the characteristic diameter d above had been chosen to be the size of a pore in a porous material (say $d = 10^{-6}$ cm) that surface diffusion might well become the dominant mechanism for mass transport. This is so in much of the work on the permeation of gases through porous beds (Barrer 1954). The quantity obtained in these studies is the surface diffusion coefficient D_s (cm^2 sec^{-1}) which may be converted to τ_s with the formula

$$D_s = \frac{a^2}{\tau_s} \frac{\text{cm}^2}{\text{sec}} \qquad (2.74)$$

where a is the interatomic hop distance and may be taken as 3×10^{-8} cm. Table 2.33 shows values of D_s obtained by Barrer and colleagues (Barrer 1954).

TABLE 2.33
Surface diffusion coefficients for gases on porous glass in the Henry's law range

1 Gas	2 T (°K)	3 $D_s \times 10^3$ cm^2 sec^{-1}
N$_2$	273	1·53
	290	2·00
	323	2·26
O$_2$	273	1·30
	290	1·73
	323	2·05
Ar	273	1·31
	290	1·65
	323	1·91
Kr	273	0·50
	290	0·63
	323	0·78
CH$_4$	273	1·24
	294	1·57
	323	1·73
	343	
C$_2$H$_6$	294	0·262
	323	0·466
	343	0·596

(from Barrer 1954b).

Gomer (1959) and his colleagues have tested these considerations of surface diffusion directly in the field emission microscope, particularly for cases where the adsorbate had a sharp boundary which could be observed. The boundary travels a measurable distance $x = (D_s t)^{\frac{1}{2}}$ in a time t yielding D_s immediately. Use of eqn. (2.74) yields τ_s, from which E_s could be obtained from eqn. (2.72). Some of the values of E_s in Table 2.9 were obtained in this way. Further details of surface diffusion may be found in Hayward and Trapnell (1964).

2.4.5 Adsorption of mixtures, cryotrapping and replacement

Work on the physical adsorption of two gases simultaneously on a surface has concentrated to date on the phenomenon of 'cryotrapping'. This is potentially useful as a pumping mechanism in uhv systems (Hengevoss 1965). A surface is held at a certain temperature and two gases strike it simultaneously. The vapour pressure of one gas called 'condensable' is very low at the surface temperature. This gas entrains molecules of the second gas called 'non-condensable' thereby creating a pumping mechanism for the non-condensable gas, lowering its pressure.

The phenomenon is illustrated in Fig. 2.28. The vacuum system contained a surface at 4·2°K, and a steady flow of H_2 was established which came to equilibrium with pumping by all available pumps including the cold surface. Curves a, b, c are for increasing magnitudes of hydrogen flow respectively. A steady flow of Ar was next introduced to the system. This also came to equilibrium with the pumps and had no appreciable effect upon the H_2 partial pressure until a certain critical flow rate was reached, when the H_2 partial pressure fell by about an order of magnitude. The critical argon flow rate was found to correspond to one Ar atom striking the cold surface for every two H_2 molecules. Since it was estimated that the condensation coefficient for H_2 was about 0·4 this meant that the ratio of H_2 molecules to Ar atoms on the surface was about 1:1. Table 2.34 shows combinations of gases studied by various workers and a few main results. There is not complete agreement throughout Table 2.34 but several points appear established: (1) For trapping with water at 77°K the number of condensable to non-condensable molecules is approximately 100, even for vapours of O_2, N_2 and Ar, which are adsorbed strongly at this temperature; (2) there is only a minor cryotrapping effect for He and H_2 at 77°K; (3) for H_2 cryotrapping at 4·2°K the ratio of non-condensable to condensable molecules is of the order unity or less.

Hengevoss (1965) discussing the data of Fig. 2.28 suggests four possible mechanisms: (a) the hydrogen is simply buried by the Ar, but the low condensable to non-condensable ratio of Table 2.34 argues against this; (b) H_2 is physically adsorbed on the continuously recreated surface; (c) a 'chemical' bond is formed between H_2 and Ar; (d) H_2 dissolves in the Ar condensate. No final conclusion is drawn on these possibilities.

Simple kinetic theories of cryotrapping have been given by Schmidlin *et al.* (1962) and by Haygood (1963) applicable at low pressures only. The theories are similar and the nomenclature of Schmidlin *et al.* (1962) is used here. There are two central assumptions: (1) the non-condensable gas is in adsorption equilibrium with the surface (which is continuously being recreated by the condensable gas). For this equilibrium the Langmuir equation is written

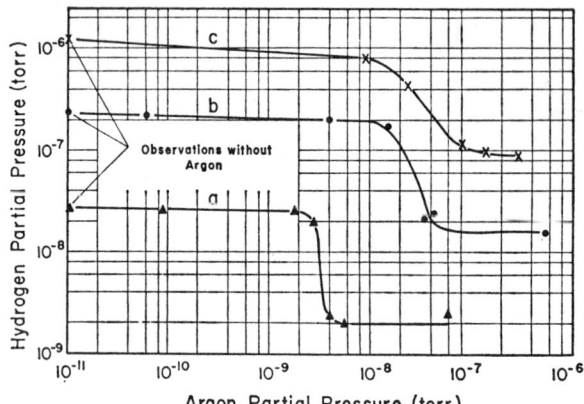

Fig. 2.28 Cryotrapping of hydrogen by argon at 4.2°K. Curves a, b and c are for increasing magnitudes of hydrogen flow (from Hengevoss and Trendelenburg 1963).

TABLE 2.34
Summary of cryotrapping results

1 Non-condensable molecule	2 Condensable molecule	3 Surface temp. (°K)	4 Min. ratio of condensable to non-condensable molecules	5 Lowest pressure of non-condensable reported (torr)	6 Ref.
H_2	H_2O	~20			
He	H_2O	7-14			
O_2	H_2O	60-90	50	2×10^{-5}	Brackman & Fite (1961)
H	H_2O	4-10			
N	H_2O	77	50	2×10^{-5}	
N_2	H_2O	77	125	$2 \cdot 5 \times 10^{-5}$	
Ar	H_2O	77	77	$2 \cdot 5 \times 10^{-5}$	Schmidlin *et al.* (1962)
H_2	H_2O	77	$\gtrsim 3 \times 10^5$		
He	H_2O	77	$\gtrsim 10^5$		
N_2	Nitro-propane	77	$\gtrsim 2 \times 10^4$		
Air	H_2O	77	100	3×10^{-6}	Haygood (1963)
H_2	Ar	4·2	1	2×10^{-9}	Hengevoss & Trendelenburg (1963)
He	Ar	4·2	30		
H_2	N_2	4·2	<0·06	10^{-8}	Degras (1963)
H_2	N_2	4·2	<0·1	10^{-8}	

(from Hobson 1967).

$$\theta_2 = (p'/p_2')/(1+p'/p_2') \tag{2.75}$$

θ_2 being the relative coverage of non-condensable, p' being its partial pressure and p_2' being the temperature dependent constant for the isotherm; (2) non-condensable molecules adsorbed on the surface may be trapped by a direct or near-direct hit by a non-condensable: trapping rate of non-condensables

$$\rho = \theta_2 \omega R \tag{2.76}$$

where ω is the effective fraction of sites active in the trapping process, and R is the rate at which condensables arrive at the surface. Combination of eqns. (2.75) and (2.76) yields

$$f_R = \frac{\dot{Q}_t}{\dot{m}} = \frac{\beta p}{[1+(\beta/\omega)p]} \tag{2.77}$$

where $\beta = \omega/p_2'$, $\dot{Q}_t$ is the in-flow rate of non-condensables and $\dot{m}$ the corresponding rate for condensables (both at steady state equilibrium); p is the value of p' after correction for thermal transpiration.

Verification of the form of eqn. (2.77) is shown in Fig. 2.29 and Haygood (1963) finds an equation similar to eqn. (2.77) verified experimentally. Both authors draw the conclusion that the Langmuir isotherm and, in particular, Henry's Law of adsorption at very low pressures are valid. This conclusion is at variance with the data of Fig. 2.11, but it should be remembered that the experimental conditions are rather different. In the cryotrapping process the equilibrium is dynamic, there being a continuous transfer of matter to the adsorbed phase. The temperature of the gas is above that of the surface. The rate of deposition of the condensable surface is fast, being between 1 and 100 monolayers sec^{-1}. It is possible to explain the cryotrapping effect as physical adsorption on the newly formed substrate. Schmidlin et al. (1962) measure this

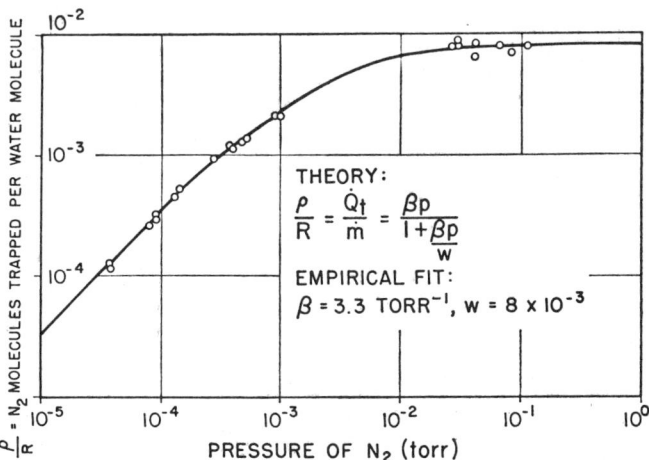

Fig. 2.29 Cryotrapping of nitrogen by water at 77°K. (from Schmidlin et al. 1962).

to represent only 7% of the full cryotrapping effect for the case of trapping of H_2 by H_2O at 77°K.

Replacement is defined as the process wherein gas A adsorbs on a surface and releases gas B which was previously adsorbed. Replacement may occur when an adsorbed layer of B is exposed to gas A or during the simultaneous adsorption of binary gas mixtures, e.g. gases A and B may adsorb simultaneously and independently at first and then, as saturation is approached, gas B may be displaced from the surface by gas A. Replacement is of considerable importance in uhv systems since it can cause changes in the composition of gases admitted to the system.

Replacement is possible when the change in free energy is negative. Thus if the energy of the bond of gas A to the metal is higher than gas B to the metal then replacement of B by A is possible if the change in entropy is small. The rate of replacement of any gas is dependent on (*a*) the replacing gas, (*b*) the nature of the surface (varying on different crystal faces of the same metal), (*c*) the coverage and (*d*) the temperature.

Replacement can, in principle, occur for physisorption or chemisorption. Most of the studies of the adsorption of mixtures and replacement have been in chemisorption, partly because of the considerable relevance of such studies to problems of catalysis. The physisorption of mixtures of gases does not usually result in replacement because one gas can physisorb readily on top of another, i.e. the strength of the physisorption bond is very similar whether the bond is between a gas molecule and a substrate molecule or between two gas molecules. We shall concern ourselves only with chemisorption of mixtures of gases on metals in the following.

Replacement of H_2 by CO has been observed by several experimenters. Robins (1962) found the replacement of adsorbed H_2 by CO very rapid on a tungsten filament at 300°K. When the H_2 coverage was greater than half a monolayer the CO was able to replace 60% of the adsorbed H_2. The replacement was almost complete after an exposure to 10^{-6} torr minutes of CO. Siddiqi and Tompkins (1962) have studied the interaction of H_2, CO and O_2 on evaporated films of nickel and copper at 300°K by means of surface potential measurements. Moore and Unterwald (1964a) have reported a qualitative investigation of the effects of CO on an adsorbed layer of H_2 on tungsten and molybdenum.

Rigby (1965) has made a detailed study of the replacement of H_2, adsorbed on a tungsten filament, by CO. Replacement of the β_2-phase of H_2 can go to completion whereas the β_1 and α phases are only slightly affected.

Ageev *et al.* (1965) have adsorbed CO and H_2 simultaneously on a polycrystalline wire at 300°K. Initially the two gases adsorb independently until about 4×10^{14} mole/cm² of H_2 have been adsorbed, thereafter the CO replaces H_2 on the surface and the hydrogen coverage decreases. It was also found that only H_2 in the β_2-phase ($E_d \approx 34$ kcal/mole) was replaced by CO while the β_1-phase of H_2 ($E_d = 14$ kcal/mole) was unaffected. It is estimated that one

CO molecule replaces two H_2 molecules. The same general behaviour has been reported by Ustinov and Ionov (1966) who also observed that N_2, CO and H_2 are adsorbed independently, at low coverage, in amounts proportional to their respective partial pressures.

Nitrogen has been observed to replace hydrogen from tungsten and molybdenum surfaces. Little et al. (1964) have observed the evolution of H_2 from a saturated layer on an evaporated molybdenum film when exposed to nitrogen at fairly high pressures ($\sim 10^{-2}$) in the temperature range 250 to 300°K. Rigby (1965) found that H_2 would not replace N_2 from a tungsten surface at 300°K. Exposing a saturated layer of N_2 to 10^{-5} torr of H_2 for 5 minutes caused no change in the nitrogen desorption spectrum. The hydrogen desorption spectrum showed that the H_2 had adsorbed in the sites not occupied by N_2 (the saturated layer of N_2 only covers about half the tungsten surface). Nitrogen was observed to replace the weakly bound H_2 as it was thermally desorbed from the surface. When a monolayer of H_2, in equilibrium with 10^{-5} torr of H_2 gas, was exposed to 2×10^{-7} torr of N_2 only 5% of the N_2 monolayer was adsorbed in 30 minutes. However, when the H_2 gas was pumped away the rate of replacement of H_2 by N_2 increased greatly.

Ricca et al. (1962) have studied the sequential adsorption of N_2 and CO on a tungsten sheet at 300°K. The prior adsorption of N_2 produced an exactly equivalent reduction in the number of molecules of CO that could be subsequently adsorbed at saturation. Prior adsorption of CO at less than 3×10^{14} molecules/cm^2 reduced the number of molecules of N_2 adsorbed at saturation by an almost equal amount. When more than 3×10^{14} molecules/cm^2 of CO were preadsorbed the total amount of adsorbed gas (CO+N_2) increased. Rigby (1964) has studied the adsorption of CO and N_2 on a tungsten wire at 300°K. The behaviour of CO adsorbed onto a monolayer of N_2 was similar to that observed by Ricca. When equal pressures of CO and N_2 were maintained over a preadsorbed monolayer of CO, a slow replacement of CO by N_2 was observed. The CO displaced by N_2 came principally from the α and β_3 phases. At all stages of replacement the total amount of gas on the surface was found to be the same.

Oxygen is so strongly bound to most metals that it will replace almost all other chemisorbed gases. The replacement of adsorbed H_2 by O_2 was first observed by Roberts (1935). Siddiqi and Tompkins (1962) have studied the adsorption of mixtures of (CO+O_2) and (H_2+O_2) on copper and nickel films by means of surface potential measurements. Singleton (1967b) has observed that when CO and O_2 are adsorbed simultaneously onto a tungsten surface at 300°K the two gases adsorbed independently at first. As the coverage increases the CO is replaced on the surface by O_2 until at saturation coverage only a small amount of CO was left on the surface.

Replacement of H_2, adsorbed on nickel films by mercury has been studied by Campbell and Thomson (1959). The replacement is extremely rapid but incomplete.

Rigby (1965) distinguishes three types of replacement processes when gas A replaces gas B:

(i) The adsorption of A on identical sites vacated by B with no strong interaction between A and B on the surface (e.g. N_2 replacing H_2).

(ii) The adsorption of A on identical sites vacated by B with a strong interaction between A and B on the surface (e.g. N_2 replacing CO).

(iii) The adsorption of A on non-identical sites which cannot be occupied by B with the subsequent desorption of B from adjacent sites (e.g. CO replacing H_2, Hg replacing H_2).

If the adsorption sites for A and B are identical, sites for A are only produced when B thermally desorbs from a weakly bound phase. Thus type (i) and (ii) replacement takes place via a weakly bound state which is reversibly adsorbed at the operating temperature. In these cases, rates of replacement are governed by (*a*) the rate of desorption from the weakly bound state of B and (*b*) the relative concentration of A and B in the gas phase.

Type (iii) replacement can be considered a direct replacement. In this case the rate of replacement is larger than the rate of thermal desorption of gas B. The rate of replacement is governed principally by the rate at which gas A impinges on the surface (proportional to the pressure of A). The systems with which replacement have been observed are summarized briefly in Table 2.35.

TABLE 2.35
Replacement of chemisorbed gases
$(M + B_{ad}) + A_g \rightarrow (M + A_{ad} + B_{ad}) + B_g$

1 Replacing Gas (*A*)	2 Replaced Gas (*B*)	3 Surface	4 Temperature (°K)	5 Reference
CO	H_2	W wire	300	Robins (1962)
CO	H_2	W, Mo ribbon	300	Moore & Unterwald (1964a)
CO	H_2	W wire	300	Rigby (1965)
CO	H_2	W wire	300	Ageev et al. (1965)
CO	H_2	W wire	300	Ustinov & Ionov (1966)
N_2	H_2	Mo film	250-330	Little et al. (1964)
N_2	H_2	W wire	300	Rigby (1965)
N_2	CO	W sheet	300	Ricca et al. (1962)
N_2	CO	W wire	300	Rigby (1964)
O_2	H_2			Roberts (1935)
O_2	CO	W ribbon	300	Singleton (1967b)
Hg	H_2	Ni	300	Campbell & Thomson (1959)

2.5 DIFFUSION, SOLUTION AND PERMEATION OF GASES IN SOLIDS

In our previous discussions of the interaction of gases with solids we have regarded the solid as impenetrable to the adsorbed gas. However, under particular combinations of gas, solid and temperature, the adsorbed gas may be able to penetrate between surface atoms and so gain access to the bulk interior of the solid. Gas may also penetrate the solid without any identifiable adsorbed phase. If this penetration should happen, solution of the gas in the solid is said to have taken place and there will be an equilibrium established between the concentration (C) of gas in solution, the external pressure, and the temperature, i.e. $f(C,p,T) = 0$ which is analogous to the adsorption isotherm, $f(\theta,p,T)$, which is simultaneously operative at the surface. Of importance in uhv are the relative amounts of gas in solution and adsorbed on the surface. This will vary with the particular situation. Another important point is the rate of change of each following a change of pressure and/or temperature. The answer to this question determines the rate at which a system can be evacuated, and much of the discussion below is directed toward this question of outgassing. General references on diffusion, solution and permeation are Barrer (1951) and Jost (1952).

The most fundamental place to start a discussion of solution and its consequences is in the interior of the solid and we consider a dissolved gas atom undergoing a random walk between sites in the solid. This picture leads to Fick's first law of diffusion

$$\vec{J} = -D\,\vec{\nabla}\,C, \tag{2.78}$$

where $\vec{J}$ is the rate of flow of dissolved matter across unit area (atoms cm^{-2} sec^{-1}) C is the concentration (density) of dissolved matter (atoms cm^{-3}) and D is the diffusion coefficient with units cm^2 sec^{-1}. Figures 2.30a and 2.30b give values of D for typical gas-solid combinations. Often the temperature dependence of D is expressed by

$$D = D_0 \exp -\frac{E_D}{RT}. \tag{2.79}$$

D_0 and E_D being constants. For a homogeneous solid D is constant only if one dissolved atom does not interfere with the motion of another. This is likely to be true in the general range of concentrations found in uhv studies. Constant D greatly simplifies the mathematics of diffusion and we will assume it constant and isotropic at first. Conventional considerations of the conservation of mass lead from eqn. (2.78) to the basic time dependent differential equation of diffusion (Fick's second law of diffusion)

$$\frac{\partial C}{\partial t} = D\,\nabla^2\,C. \tag{2.80}$$

Solutions to this equation for certain standard boundary conditions of interest in uhv are next presented.

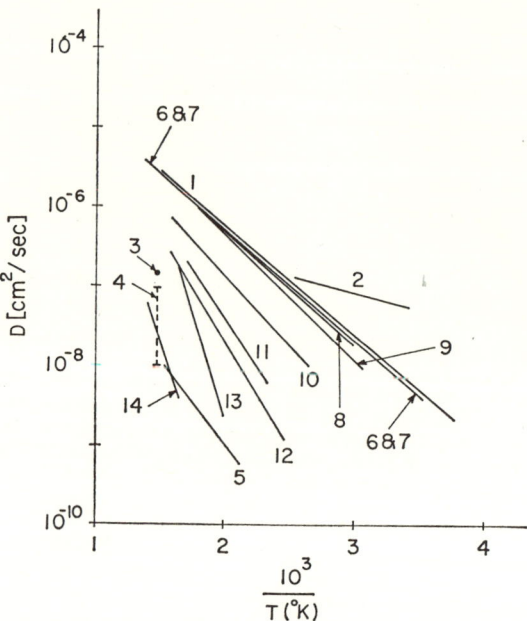

Fig. 2.30a Diffusion constants for various gas non-metal combinations. Units are cm^2 sec^{-1}.

Curve #	Gas non-metal combination	Reference
1	He—Pyrex 7740	Rogers et al. (1954)
2	He—Vycor	Leiby and Chen (1960)
3	H$_2$—Vycor	Leiby and Chen (1960)
4	N$_2$—Vycor	Leiby and Chen (1960)
5	H$_2$—SiO$_2$	Barrer (1951)
6	He—Duran Glass	Eschbach (1960)
7–12	He—Glass (composition given below)	Eschbach (1960)
13–14	H$_2$—Glass (composition given below)	Eschbach (1960)

Glass Composition in %

	SiO$_2$	B$_2$O$_3$	Al$_2$O$_3$	Na$_2$O	K$_2$O	PbO
7	76·1	16·0	1·75	5·4	0·6	—
8	75·9	16·0	0·4	4·9	0·8	—
9	64·7	23·2	4·0	4·0	4·1	—
10	75·3	7·6	6·2	5·7	0·8	—
11	56·2	—	1·2	7·6	4·5	30·0
12	69·1	—	3·3	13·2	1·7	—
13	76·1	16·0	1·75	5·4	0·6	—
14	75·9	16·0	0·4	4·9	0·8	—

Other diffusion constants not shown in the figure are available: He—fused quartz, Swets et al. (1961); Ne—fused quartz, Frank et al. (1961); He—various glasses, Altemose (1961).

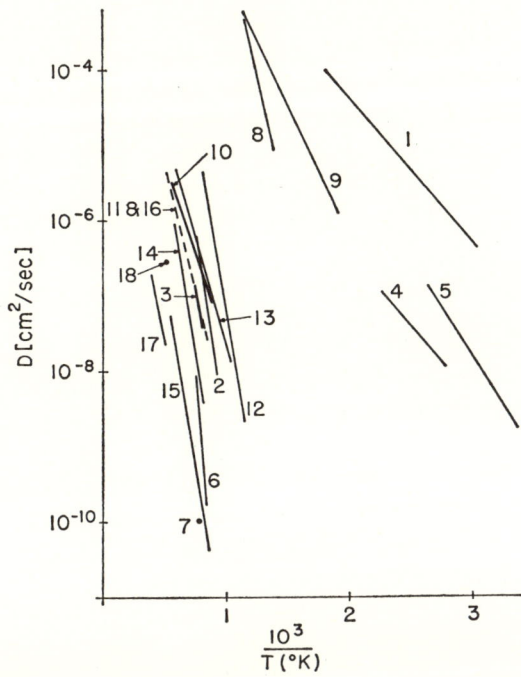

Fig. 2.30*b* Diffusion constants for various gas-metal combinations. Units are $cm^2\ sec^{-1}$.

Curve #	Gas-metal combination	Reference
1	H_2—Pd	Barrer (1951)
2	N_2—Fe	Barrer (1951)
3	CO—Ni	Dushman & Lafferty (1962)
4	H_2—Ni	Barrer (1951)
5	H_2—Fe	Barrer (1951)
6	O_2—Ni	Dushman & Lafferty (1962)
7	O_2—Fe	Barrer (1951)
8	H_2—Al	Smithells (1962)
9	H_2—Cu	Smithells (1962)
10	N_2—Th	Smithells (1962)
11	N_2—Zr (β)	Smithells (1962)
12	O_2—Cu	Smithells (1962)
13	O_2—Ta	Smithells (1962)
14	O_2—Ti	Smithells (1962)
15	N_2—Ti (α)	Smithells (1962)
16	N_2—Ti (β)	Smithells (1962)
17	H_2—W	Moore and Unterwald (1964b)
18	H_2—Mo	Moore and Unterwald (1964b)

Other diffusion constants now shown in the figure are available: H_2—Pd, Hurlbert and Konecny (1961); H_2—Pd, Tiedema *et al.* (1960); H_2—Ge, Frank and Thomas (1960); H_2–hardened steel, Nechai *et al.* (1960); H_2—mild steel, Foster (1960); He—Ti, Tucker and Norton (1960); O_2—Si, O_2—Ge, Haas (1960); Ar—Ag, Ar—Au, Ar—Al, Ar—Pb, Tucker and Norton (1960); H_2—Ag, N_2—Nb, N_2—Ta, O_2—Nb, O_2—Si, O_2—V, O_2—Zr, Smithells (1962).

2.5.1 Solution and de-solution of gas from a semi-infinite solid, a slab, a cylinder and a sphere

Consider a semi-infinite solid occupying the half space $x > 0$ and containing a uniform initial concentration of gas C_i. Consider the half space $x < 0$ to be pumped to a steady gas pressure which would maintain a surface concentration C_f on the solid in equilibrium. Assuming that the rate of desorption from the solid is given by

$$F = -D\left(\frac{\partial C}{\partial x}\right)_{x=0} = \alpha(C_f - C_s), \qquad (2.81)$$

where C_s is the surface concentration, then the solution of eqn. (2.80) in conjunction with the desorption eqn. (2.81) is (Crank 1956)

$$\frac{C - C_i}{C_f - C_i} = \text{erfc}\frac{x}{2(Dt)^{\frac{1}{2}}} - \exp(hx + h^2 Dt)\,\text{erfc}\left\{\frac{x}{2(Dt)^{\frac{1}{2}}} + h(Dt)^{\frac{1}{2}}\right\}, \qquad (2.82)$$

where $h = \alpha/D$ (cm^{-1}).

The value of $C_s(t)$ may be obtained by putting $x = 0$ into eqn. (2.82) to yield the rate of evolution of gas from the solid. It is readily shown that

$$F/F_0 = [\exp h^2 Dt]\,[1 - \text{erf}\,h(Dt)^{\frac{1}{2}}], \qquad (2.83)$$

where F_0 is the initial rate of desorption at zero time. The right-hand side of eqn. (2.83) is a function of $h^2 Dt$ only and is plotted in Fig. 2.31. Equation (2.82) also describes solution of gas in the solid if $C_i < C_f$ and in this case Fig. 2.31 represents the rate of pumping relative to the initial rate of pumping.

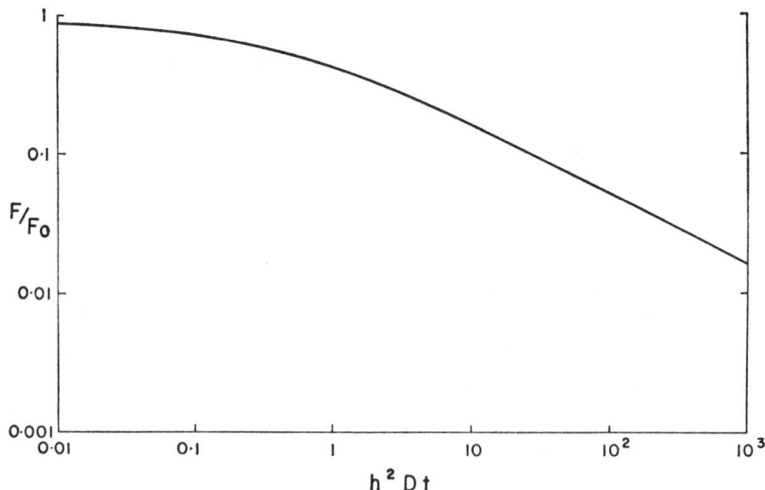

Fig. 2.31 Relative outgassing rate of semi-infinite slab versus time. F_0 is initial outgassing rate. Curve is plot of eqn. 2.83.

Figure 2.31 may be regarded as two zones approximately divided by $h^2 Dt = 1$. For $h^2 Dt > 1$, $F/F_0 \simeq 1/t^{\frac{1}{2}}$ and in this region the absolute value of F (for given C_f, C_i, D) is independent of α the surface desorption rate, i.e. the outgassing rate is controlled by bulk diffusion. For $h^2 Dt < 1$ F/F_0 is given by $\alpha(C_f - C_i)$ and the outgassing rate is controlled by surface desorption and is independent of bulk diffusion. Thus the division $h^2 Dt = 1$ is an important physical condition. Where surface desorption dominates, the results of Section 2.4.4 apply. The constant $h = \alpha/D$ is thus a measure of the relative importance of the desorption and diffusion steps. Large α/D means easy desorption and hence quick establishment of diffusion as the rate limiting step. Small α/D means easy diffusion and hence slow establishment of diffusion as the rate limiting step. A brief discussion of magnitudes will be useful. αC_s is the rate of desorption of molecules cm^{-2} sec^{-1} if C_s is in molecules cm^{-3}. Now the surface concentration σ in molecules cm^{-2} may be written as $C_s d$, where d is the thickness of a monatomic layer which we take as $d = 3 \times 10^{-8}$ cm.

$$\left(\frac{d\sigma}{dt}\right)_{des} = \frac{\alpha\sigma}{d} = \frac{\alpha\sigma}{3 \times 10^{-8}}. \tag{2.84}$$

Thus we may identify

$$\tau = d/\alpha \tag{2.85}$$

where τ is the mean time of adsorption (Section 2.4.4), i.e.

$$\alpha = \frac{3 \times 10^{-8}}{\tau} \text{ cm sec}^{-1} \tag{2.86}$$

and hence

$$h = 3 \times 10^{-8}/\tau D \text{ cm sec}^{-1}. \tag{2.87}$$

As an example let us say we would like to know the values of D and τ which would cause a semi-infinite solid to change from surface limited outgassing to diffusion limited outgassing after $t = 1000$ sec, i.e. $h^2 D = 10^{-3}$. Thus from eqn. (2.87) $\tau^2 D = 9 \times 10^{-13}$.

Typically D will lie in the range $10^{-4} < D < 10^{-10}$ cm^2 sec^{-1}. Thus $10^{-4} < \tau < 10^{-1}$ sec. For adsorption times below this range diffusion limited outgassing will be established quickly; for adsorption times above this range desorption limited outgassing will obtain for long times. Some adsorption times are given in Tables 2.27 and 2.28.

The solution for the semi-infinite solid is useful because it defines most of the qualitative outlines of the solution for finite geometries and is an exact solution for the latter near zero time where the solutions for finite geometries may be tedious because of slowly converging series. A first example of this is the outgassing of an infinite plane sheet occupying the region $-l < x < l$ (thickness of sheet $= 2l$) with boundary conditions the same as those for the semi-infinite solid. The solution for this case is

D*

$$\frac{C-C_i}{C_f-C_i} = 1 - \sum_{n=1}^{\infty} \frac{2L \cos[\beta_n(x/l)] \exp[-\beta_n^2(Dt/l^2)]}{(\beta_n^2+L^2+L)\cos\beta_n} \quad (2.88)$$

where the β_n's are the positive roots of $\beta \tan \beta = L$ (to be found in Crank 1956), where $L = l\alpha/D$ (dimensionless).

The ratio F/F_0 may readily be calculated for this case but a more compact form of the results is the ratio f of the amount remaining in solution in the solid to the total which will have left the solid in infinite time. For the infinite plane sheet

$$f = \sum_{n=1}^{\infty} \frac{2L^2 \exp[-\beta_n^2(Dt/l^2)]}{\beta_n^2(\beta_n^2+L^2+L)}. \quad (2.89)$$

The advantage of this formulation of the outgassing result is that f is a function only of Dt/l^2 and of L. Equation (2.89) is plotted in Fig. 2.32. When the amount in solution is e^{-1} of the total amount which will leave after infinite time $\beta_1^2(Dt/l^2) = 1$, the terms for higher values of n being small. At small values of $L \lesssim 1$, $\beta_1^2 \simeq L$ and this condition becomes $(\alpha t/l) = 1$ which is independent of D and depends only upon the desorption rate parameter α. At large values of $L \to \infty$, $\beta_1^2 = 2.46$ and the condition becomes $2.46(Dt/l^2) = 1$

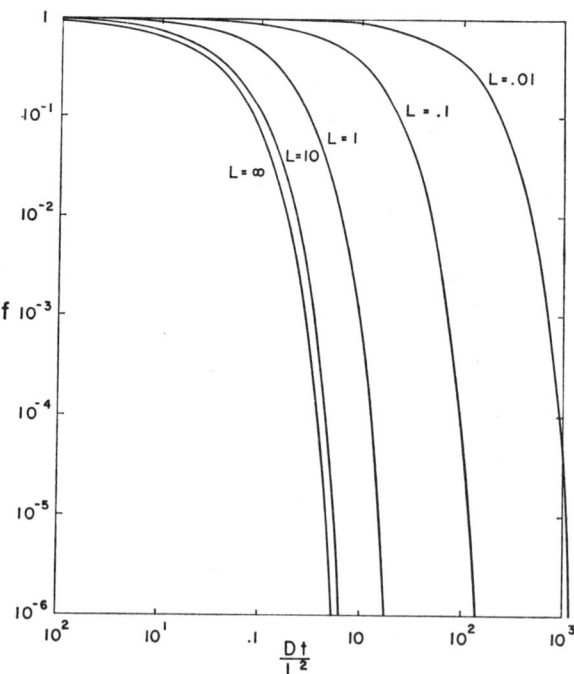

Fig. 2.32 Fraction f of initial amount dissolved in a slab vs time after start of pumping. Slab thickness is $2l$. Curves are plots of eqn. 2.89.

which depends only upon the diffusion coefficient. This relation defines a characteristic time for diffusion

$$\tau_D = \frac{l^2}{2\cdot 46 D} \quad (2.90)$$

for the infinite plane sheet of thickness $2l$. At this time e^{-1} of the material which will eventually leave the sheet by diffusion remains in solution. When the geometry of the sample is further restricted to an infinite cylinder and finally to a sphere, the general form of the results given above is maintained, but the time required to reach a given stage of outgassing is reduced as the surface to volume ratio of the sample is reduced. The relations between the outgassing of a slab, a cylinder, and a sphere are shown in Fig. 2.33.

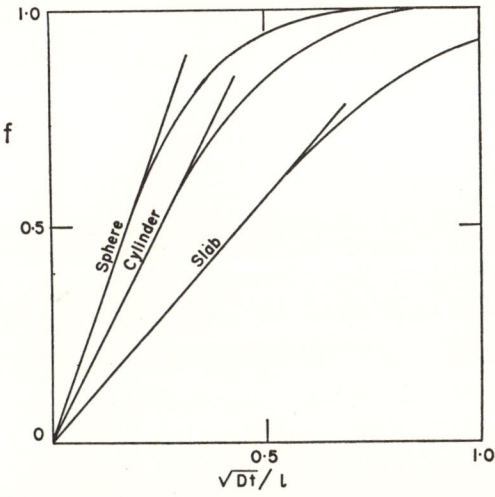

Fig. 2.33 Ratio f of the amount lost by diffusion to the initial amount of diffusing material plotted against $(Dt)^{\frac{1}{2}}/l$ for a slab of thickness $2l$, an infinite cylinder of diameter $2l$, and a sphere of diameter $2l$. (from Todd 1955.)

An important parameter in all the above equations has been the concentration C, representing the density of dissolved gas in the bulk solid. In equilibrium with the gas phase this concentration is given by

$$C = C_0 p^n. \quad (2.91)$$

n is the inverse of the number of atoms (or radicals) into which the molecule splits when it is adsorbed. For gases in non-metals $n = 1$. For most of the simple diatomic gases found in uhv systems (N_2, O_2, H_2, etc.) $n = \frac{1}{2}$. C_0 is a temperature dependent constant known as the solubility. In the above units it is dimensionless and is the quantity of gas (cm^3 at S.T.P.) in 1 cm^3 of solid at 1 atmosphere external pressure. Values for C_0 for typical gas-solid combinations are given in Figs. 2.34a, 2.34b, 2.34c as a function of temperature.

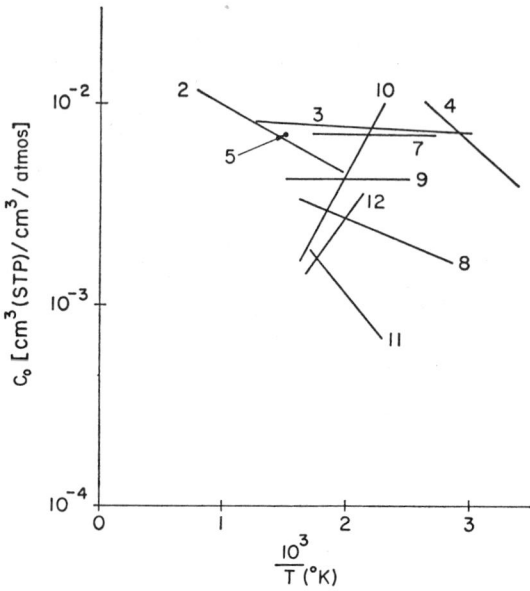

Fig. 2.34a Solubilities for various gas non-metal combinations as a function of temperature. Units of C_0 are dimensionless (quantity of gas in cm³ (S.T.P.) in 1 cm³ of material at 1 atm pressure).

Curve #	Gas non-metal combination				Reference	
1 (not shown on graph)	H_2O—0800 glass, $C_0 \geqq 0.6$ at 300°K				Todd (1955)	
2	H_2—SiO_2				Barrer (1951)	
3	He—Pyrex 7740				Rogers et al. (1954)	
4	He—Vycor				Leiby & Chen (1960)	
5	H_2—Vycor				Leiby & Chen (1960)	
6 (not shown on graph)	N_2—Vycor $10^{-5} \leqq C_0 \leqq 10^{-4}$ at 673°K				Leiby & Chen (1960)	
7–12	He-glass (composition given below)				Eschbach (1960)	
	Glass Composition in %					
	SiO_2	B_2O_3	Al_2O_3	Na_2O	K_2O	PbO
7	76.1	16.0	1.75	5.4	0.6	—
8	75.9	16.0	0.4	4.9	0.8	—
9	64.7	23.2	4.0	4.0	4.1	—
10	75.3	7.6	6.2	5.7	0.8	—
11	56.2	—	1.2	7.6	4.5	30.0
12	69.1	—	3.3	13.2	1.7	—

Other solubilities not shown in the figure are available: He—various glasses (Altemose 1961).

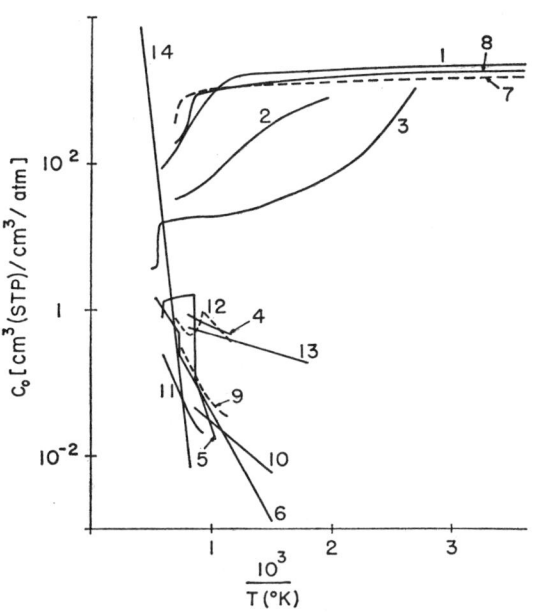

Fig. 2.34b Solubilities for various gas-metal combinations as a function of temperature (units as in Fig. 2.34a).

Curve #	Gas-metal combination	Reference
1	H_2—Ti	Waldschmidt (1954)
2	H_2—Ta	Waldschmidt (1954)
3	H_2—Pd	Waldschmidt (1954)
4	O_2—Cu	Waldschmidt (1954)
5	N_2—Fe	Waldschmidt (1954)
6	H_2—Cu	Waldschmidt (1954)
7	H_2—Ce	Dushman & Lafferty (1962)
8	H_2—Th	Dushman & Lafferty (1962)
9	H_2—Cr	Dushman & Lafferty (1962)
10	H_2—Ag	Dushman & Lafferty (1962)
11	H_2—Pt	Dushman & Lafferty (1962)
12	O_2—Co	Dushman & Lafferty (1962)
13	H_2—Stainless Steel	Dushman & Lafferty (1962)
14	H_2—W	Moore & Unterwald (1964b)

The units of solubility are somewhat inconvenient for uhv purposes but are maintained here because of their widespread use in the literature. 1 cm³ at S.T.P. represents $2 \cdot 7 \times 10^{19}$ molecules.

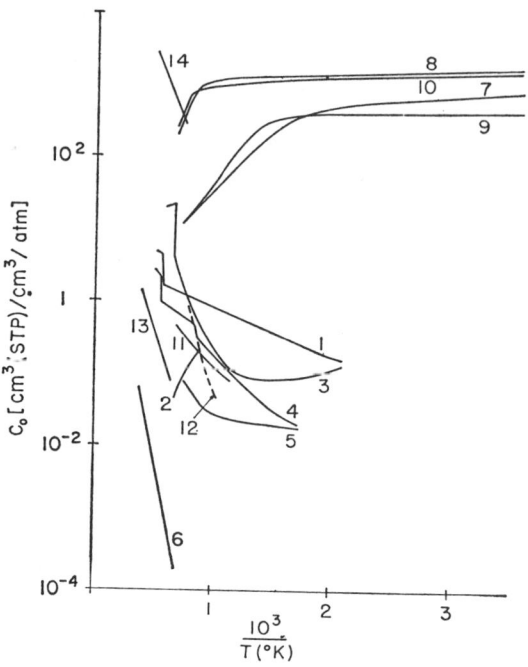

Fig. 2.34c Solubilities for various gas-metal combinations as a function of temperature (units as in Fig. 2.34a).

Curve #	Gas-metal combination	Reference
1	H_2—Ni	Waldschmidt (1954)
2	N_2—Mo	Waldschmidt (1954)
3	O_2—Ag	Waldschmidt (1954)
4	H_2—Fe	Waldschmidt (1954)
5	H_2—Mo	Waldschmidt (1954)
6	N_2—W	Dushman & Lafferty (1962)
7	H_2—V	Dushman & Lafferty (1962)
8	H_2—Zr	Dushman & Lafferty (1962)
9	H_2—Cb	Dushman & Lafferty (1962)
10	H_2—La	Dushman & Lafferty (1962)
11	H_2—Co	Dushman & Lafferty (1962)
12	H_2—Al	Dushman & Lafferty (1962)
13	N_2—Mo	Dushman & Lafferty (1962)
14	H_2—Mo	Dushman & Lafferty (1962)

Other solubilities not given in the figures are available: H_2—FeV alloy (78/82), H_2—PdPt alloy (70/30), H_2—FeNi alloy (47/53), O_2—AgAu alloy (80/20), Espe (1959a); H_2—Mg, H_2—Mn, H_2—Pb, H_2—Sn, H_2—Nb, N_2—various alloys, Smithells (1962); N_2—Co, N_2—Cr, N_2—Al, N_2—various alloys, Smithells (1962); O_2—Cr, O_2—Fe, O_2—Ni, O_2—Pb, O_2—Pd, O_2—V, Smithells (1962).

2.5.2 De-solution of gas from vacuum chamber wall and permeation

A standard vacuum problem is the rate of pump-down to ultimate vacuum. The vacuum walls behave differently from internal components in that the external wall is in contact with the atmosphere which we will assume keeps this wall indefinitely at the initial concentration C_i. We maintain eqn. (2.81) as representative of desorption conditions at the internal wall. This internal condition is not truly realistic except for very fast pumps but it is probably the most useful simple assumption that can be made. Thus we solve the diffusion eqn. (2.80) under the boundary conditions

$$C(x,0) = C_i \qquad 0 < x < l$$
$$C(0,t) = C_i \qquad t > 0 \qquad (2.92)$$

The solution is

$$C = C_i - \frac{L(C_i - C_f)}{l(1+L)} + 2L(C_i - C_f) \sum_{n=1}^{\infty} \frac{\sin \beta_n(x/l) \exp[-\beta_n^2 T]}{(L + L^2 + \beta_n^2) \sin \beta_n} \qquad (2.93)$$

β_n are the roots of $\beta \cot \beta + L = 0$ (see Carslaw & Jaeger 1959). $L'' = lh$ $T = Dt/l^2$.

The flow rate at large times is of major interest

$$F/F_f = 1 + 2L(1+L) \sum_{n=1}^{\infty} \frac{\exp[-\beta_n^2(Dt/l^2)]}{(L + L^2 + \beta_n^2)}. \qquad (2.94)$$

Equation (2.94) is shown in Fig. 2.35. For early times the result is equivalent to that of a slab but at late times the flow rate approaches a constant value,

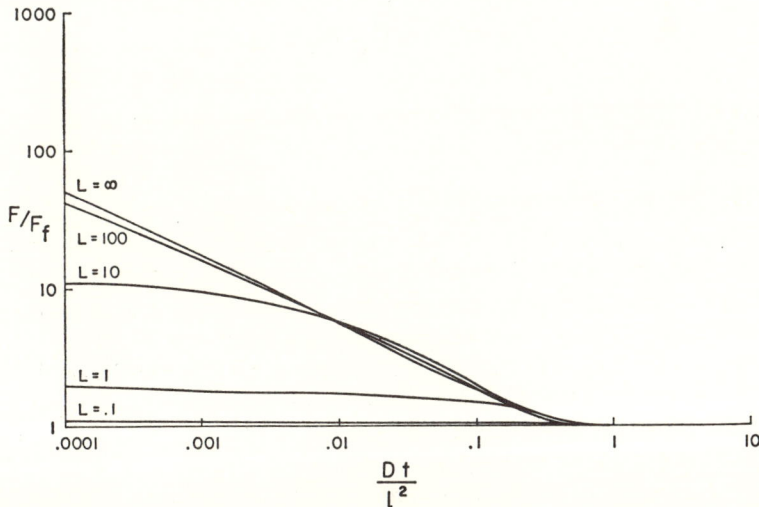

Fig. 2.35 Ratio of outgassing rate to final permeation rate at large times. Curves are plots of eqn. 2.94.

The time at which this limit has been reached is given by the condition

$$\frac{Dt}{l^2} \simeq 1. \tag{2.95}$$

For times greater than this the concentration across the wall is fixed and falls linearly across the wall

$$\frac{C - C_f}{C_i - C_f} = \frac{x}{l}. \tag{2.96}$$

The rate of transfer of diffusing gas is the same at all sections and is given by

$$F = -D\frac{\partial C}{\partial x} = D\frac{(C_i - C_f)}{l}. \tag{2.97}$$

From eqn. (2.91)

$$F = \frac{DC_0(P_i^n - P_f^n)}{l}. \tag{2.98}$$

The product DC_0 is defined as the permeation constant K

$$K = DC_0. \tag{2.99}$$

K is clearly a much less fundamental constant than either D or C_0 but is useful in vacuum studies because it is the constant which describes the passage of gas from the atmosphere to the interior of a vacuum system through the envelope. In the units being used K is the quantity of gas in cm^3 (S.T.P.) passing per second through a wall of area 1 cm^2, thickness 1 cm, when a pressure difference of 1 atmosphere exists across the wall. Values of K are given in Figs. 2.36a and 2.36b. Frequently the temperature dependence of K is given by

$$K = K_0 \exp[-E_K/RT]. \tag{2.100}$$

A useful general result (Norton 1957, 1961) is that the inert gases do not permeate any metal. However, if they are injected into a solid as ions, then their diffusion is not abnormal.

2.5.3 Modifications to idealized solutions for diffusion

The equations that have been presented in Sections 2.5.1 and 2.5.2 are first-order approximations to reality. Next we discuss modifications to these equations which make them more useful in practical situations. Dayton (1959, 1961, 1962 and 1965) in particular has examined these problems.

Standard uhv processing involves a baking period during which the temperature is raised for a period, and then lowered again, with the objective of lowering the time required to reach a given tolerable outgassing rate F. Dayton (1962) provides an illustrative example for the case of the plane sheet. At early times in Fig. 2.37 a plane sheet ($T = T_a$) with a uniform initial concentration of gas is subjected to vacuum on both sides in accord with the assumptions of Section 2.5.2. At $t = 0$ hours an outgassing rate falling as $t^{-\frac{1}{2}}$

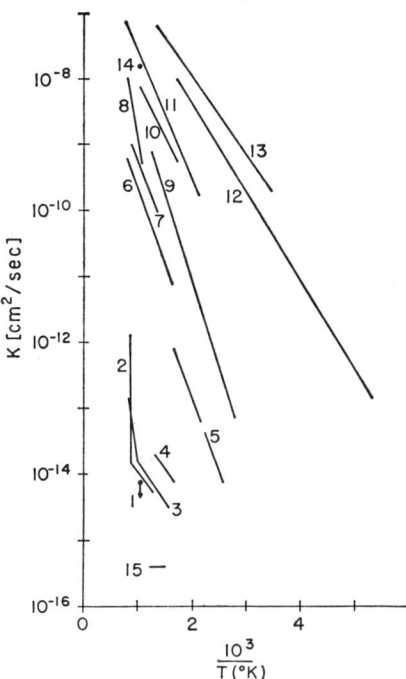

Fig. 2.36a Permeation constants for various gas non-metal combinations as a function of temperature. Units of K are cm²/sec (quantity of gas in cm³ (S.T.P.) passing per second through a wall of area 1 cm², thickness 1 cm, when a pressure difference of 1 atmosphere exists across the wall.

Curve #	Gas-non-metal combination	Reference
1	O_2 or N_2—Pyrex	Norton (1957)
2	Air—Pyroceram	Miller & Shepard (1961)
3	Air—97% alumina ceramic	Miller & Shepard (1961)
4	Air—Pyrex	Miller & Shepard (1961)
5	He—Lead Borate Glass G	Norton (1957)
6	He—97% alumina ceramic	Miller and Shepard (1961)
7	Ne—Vycor	Leiby & Chen (1960)
8	N_2—SiO_2	Barrer (1951)
9	He—1720 glass	Norton (1957)
10	He—Pyroceram 9606	Miller & Shepard (1961)
11	H_2—SiO_2	Barrer (1951)
12	He—Pyrex 7740	Rogers et al. (1954)
13	He—Vycor	Leiby & Chen (1960)
14	H_2—Pyrex	Norton (1957)
15	Air—1720 glass	Miller & Shepard (1961)

Other permeation constants not shown in the figure are available: He—various glasses, Altemose (1961); H_2,Ne,CH_4,Ar,N_2 in various glasses, Altemose (1961). He—various glasses, Espe (1959b); He—various glasses, von Ardenne (1956).

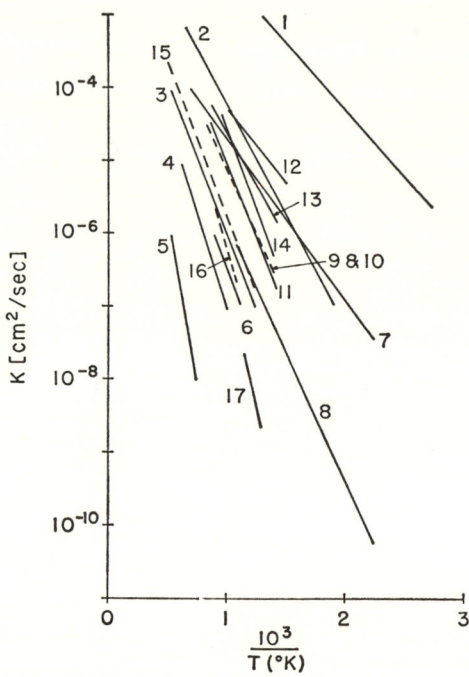

Fig. 2.36b Permeation constants for various gas-metal combinations as a function of temperature (units as in Fig. 2.36a).

Curve #	Gas-metal combination	Reference
1	H_2—Pd	Waldschmidt (1954)
2	H_2—Ni	Waldschmidt (1954)
3	H_2—Mo	Waldschmidt (1954)
4	N_2—Fe	Waldschmidt (1954)
5	N_2—Mo	Waldschmidt (1954)
6	CO—Fe	Waldschmidt (1954)
7	H_2—Fe	Waldschmidt (1954)
8	H_2—Cu	Waldschmidt (1954)
9	H_2—Kovar	Gorman & Nardella (1962)
10	H_2—Stainless steel 303	Gorman & Nardella (1962)
11	H_2—Stainless steel 304	Gorman & Nardella (1962)
12	H_2—Cold drawn steel	Gorman & Nardella (1962)
13	H_2—Monel	Gorman & Nardella (1962)
14	H_2—Inconel	Gorman & Nardella (1962)
15	H_2—Pt	Dushman & Lafferty (1962)
16	H_2—Al	Dushman & Lafferty (1962)
17	O_2—Ag	Dushman & Lafferty (1962)

Other permeation constants not given in the figure are available: H_2—OFHC copper, Gorman and Nardella (1962); H_2, D_2—Ni, Belyakov and Ionov (1961).

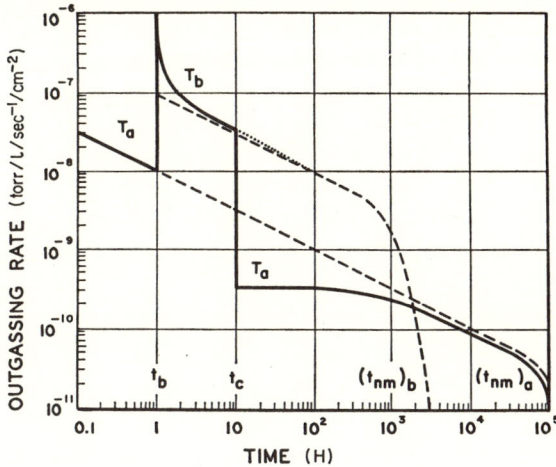

Fig. 2.37 Theoretical curve for effect of bakeout. (Dayton 1962).

has already been achieved and if no change of T occurred the fall in F would have proceeded, along the dashed line shown (c.f. Fig. 2.32). Note at $(t_{nm})_a$ this line bends downwards indicating the start of exhaustion of the gas in the sheet. At $t = t_b$, however, T was raised to T_b. F rises instantaneously by the factor D_b/D_a (chosen to be 100) where D_a and D_b are the diffusion coefficients at T_a and T_b respectively. F now falls assymptotically toward a new $t^{-\frac{1}{2}}$ behaviour (dashed) which is $(D_b/D_a)^{\frac{1}{2}} = 10$ above the original dashed line. Note that the downward bend (exhaustion) in this dashed line at $(t_{nm})_b$ is earlier than $(t_{nm})_a$ and if the temperature T_b is maintained long enough it is clear that a permanent reduction in outgassing rate will be achieved. However, if heating is terminated before $(t_{nm})_b$ is reached as shown at $t = 10$ hours in Fig. 2.37 then the reduction in outgassing rate is not permanent, but may be sufficient for practical purposes. Figure 2.37 has been drawn for $D_a = 10^{-9}$ cm^2 sec^{-1} and a sheet thickness of 1 cm.

After a vacuum system has been pumped, it is frequently opened to the atmosphere for a limited period, and repumped. The time dependence of pump-down on the second pumping has been examined by Rogers (1963). Rogers (1963) examines a semi-infinite solid exposed to a moist atmosphere until the concentration in solution can be considered uniform. The first pump-down exhibits the square-root relationship between outgassing rate and pumping time. If the system is now exposed to a moist atmosphere and again pumped down the resulting outgassing variation will depend on both the previous pump-down time and the following moist atmosphere exposure time. Results are shown in Fig. 2.38 for a range of pump-down times followed by unit exposure time. The upper line for 0 previous pump-down time is in effect a first pump-down falling as $t^{-\frac{1}{2}}$. The second pump-down proceeds more quickly than the first until approximately the final outgassing level of the first pump-

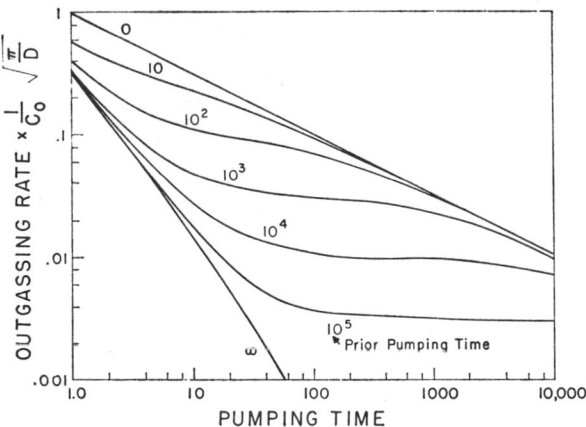

Fig. 2.38 Variation in outgassing rate with prior pumping time for a system with a uniform initial concentration (any consistent time units) (Rogers 1963).

down is reached when the effect of the intervening exposure slowly becomes erased. The limiting second pump-down for infinitely long first pump-down falls as $t^{-3/2}$.

The discussion so far has been based upon straightforward application of the diffusion equations of Section 2.5.1 and 2.5.2. Implied in this application has been the assumption embodied in eqn. (2.81). Qualitatively this means that a negligible fraction of the gas diffusing out of the solid is readsorbed by the solid during the pump-down. Dayton (1959) derives equations from which it may be concluded that for materials and gases commonly encountered in uhv, this condition is satisfied, even for pump speeds of 0·1 litre sec^{-1} for every cm^2 of outgassing area. Of course the assumption of a single diffusion coefficient and a single gas will normally be a gross misinterpretation of reality. Dayton (1959) provides generalized formulae in which the appropriate summations for many gases and many surfaces have been carried out. From an examination of experimental data on the outgassing of untreated metal samples Dayton (1961) concludes that frequently

$$F = \frac{K_1}{t_h\alpha} \tag{2.101}$$

where α may vary from 0·7 to 2 but is frequently in the neighbourhood of 1, K_1 is about 10^{-7} torr litre sec^{-1} cm^2 and t_h is the time in hours since the start of the outgassing. Since the formulae of Sections 2.5.1 and 2.5.2 predict that $\alpha = 0·5$ it is necessary to modify these formulae to give better agreement with experimental results. Dayton (1961) regards the main contribution to outgassing of untreated metal samples to come from a porous surface layer. To this he assigns a distribution function $G(\tau)d\tau$ is the fraction of the layer with a desorption time parameter τ between τ and $\tau+d\tau$. τ is explicitly given by

$$\tau = \frac{\pi l^2}{16D}. \tag{2.102}$$

It is evident that τ so defined is essentially the same as τ in eqn. (2.90). Thus $G(\tau)$ actually defines a heterogeneous distribution of diffusion constants. The main finding is that if $\alpha = 1$ in eqn. (2.101) then

$$G(\tau) = k_1/\tau \tag{2.103}$$

where k_1 is a constant given by

$$k_1 = 1/\ln(\tau_m/\tau_0) \tag{2.104}$$

τ_0 and τ_m being the smallest and largest values of τ present in the physical situation. Since Dayton (1961) is considering a porous layer τ can be related to a mean time of sojourn of a molecule on the surface through any equation similar to eqn. (2.64) of Section 2.4.4.2. He then finds eqn. (2.103) leads to an activation energy of desorption varying linearly with coverage (θ) according to

$$E_d = E_{d0} - b\theta \tag{2.105}$$

E_{d0} being the activation energy of desorption when $\theta = 0$ and b is a constant. This model appears to describe best the outgassing of water vapour from oxide films on metals at temperatures near room temperature.

Dayton (1965) has applied the more generalized model based on eqn. (2.101) to the problem embodied in Fig. 2.38, namely the pump-down behaviour of a vacuum system on the second and subsequent pump-downs. The first pump-down was used, in effect, to provide a description of the heterogeneity of the diffusion coefficients. Subsequent pump-downs were readily calculated from the first pump-down. Results qualitatively similar to Fig. 2.38 were obtained and good agreement with experiments designed to check the theory was found. Viton-A and copper were exposed to moist atmosphere in the measurements. Kraus (1963) provided an alternative theory for the pump-down result (eqn. 2.101) based upon desorption from an adsorption isotherm which does not follow Henry's Law.

2.5.4 Specific examples of diffusion mechanisms in ultrahigh vacuum practice

It is clear from Section 2.5.3 that truly realistic calculations on diffusion mechanisms are complex. Yet these mechanisms are central to the problem of creating vacua in general and uhv in particular. Thus it will be useful to apply the equations of 2.5.1 and 2.5.2 to four representative examples for which parameters are available, in the hope that the calculations will provide at least a first-order representation of reality.

(a) **The outgassing of hydrogen from a nickel slab:** consider a slab of nickel $10 \text{ cm} \times 10 \text{ cm} \times 0.3 \text{ cm}$ whose outgassing of hydrogen will be assumed to follow eqn. (2.88). The total quantity of gas given off during outgassing of a solid varies considerably but an initial concentration near 0·1 cm³ (S.T.P.) cm⁻³ is frequently found (Dushman & Lafferty 1962). We arbitrarily select $C_i = 2 \times 10^{-2}$ from Fig. 2.34c because this is the solubility of hydrogen in nickel at room temperature and 1 atmosphere pressure, and permits us to write a simple equation for the evaluation of α. Equating flow away from the surface to that arriving we write

$$\alpha C_i = \frac{3 \cdot 5 \times 10^{22} p}{(MT)^{\frac{1}{2}}} \text{ molecules cm}^{-2} \text{ sec}^{-1}. \tag{2.106}$$

Substitution of C_i in molecules cm⁻³ and p as 760 torr yields $\alpha = 2 \times 10^6$ cm sec⁻¹. This evaluation of α, the desorption rate constant, for a solid saturated

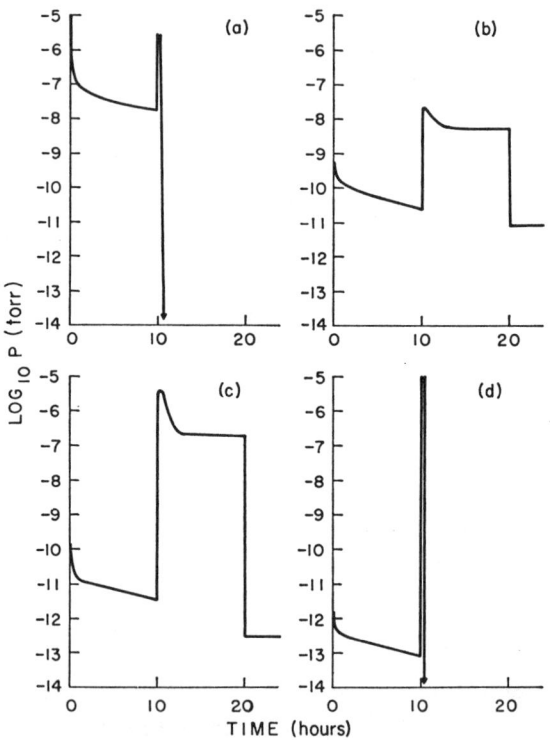

Fig. 2.39a Hydrogen partial pressure vs time for a nickel slab 10 cm × 10 cm × 0·3 cm initially containing 0·1 cm³ (S.T.P.) cm⁻³ of H₂ and placed in a pumping system with speed 10 litres sec⁻¹. Further details in Section 2.5.4a.

Fig. 2.39b Helium partial pressure vs time for a vacuum system with Pyrex walls exposed to a normal atmosphere and pumped with a speed of 0·1 litre sec⁻¹ for helium. Further details in Section 2.5.4b.

Fig. 2.39c Hydrogen partial pressure vs time for a vacuum system with stainless steel walls exposed to a normal atmosphere and pumped with a speed of 10 litres sec⁻¹ for hydrogen. Further details in Section 2.5.4c.

Fig. 2.39d Nitrogen partial pressure vs time for a vacuum system with a Mo wire at 650°K initially containing 1·5 × 10⁻¹ cm³ (S.T.P.)/cm³. System pumping speed for N₂ was 10 litres sec⁻¹. Further details in Section 2.5.4d.

with gas is an uncertain procedure, but possibly less uncertain than the assignment of a heat of desorption and the calculation of α from eqn. (2.86). At time $t = 0$ we assume that the nickel slab is placed in a vacuum system with a speed of 10 litres sec^{-1} for H_2. The initial outgassing rate will be $\alpha C_i = 3 \times 10^4$ torr litres cm^{-2} sec^{-1} and the total initial outgassing rate from 200 cm^2 will be 3×10^6 torr litres sec^{-1}. This rate initially falls very rapidly as will be seen below. Another important parameter is $L = l\alpha/D$. From Fig. 2.30b D for H_2 in Ni at room temperature is $1 \cdot 7 \times 10^{-9}$ cm^2 sec^{-1}, l is the half thickness of the slab ($l = 0 \cdot 15$ cm); from which $L = 1 \cdot 8 \times 10^{14}$ or essentially ∞, which means that the outgassing problem is controlled by diffusion rather than desorption. For simplicity we assume the pump could maintain a zero pressure of hydrogen, i.e. we may assume $C_f = 0$. All quantities are now known for substitution into eqn. (2.88); the calculation has been done and the results are exhibited in Fig. 2.39a as pressure against time for the period zero time to 10 hours. It is clear that outgassing limits the pressure which falls only slowly. From eqn. (2.90) the characteristic time τ_D for diffusion is 1460 hours and only a small fraction of the gas load has left the solid in 10 hours. At $t = 10$ hours we examine the effect of raising the temperature of the slab suddenly to 1400°K. This raises D to an estimated 10^{-4} cm^2/sec and τ_D falls to 90 sec. Calculations similar to those above give a pressure of $1 \cdot 4 \times 10^{-13}$ torr after 37 min, with essentially zero pressure after 6 hours. Cooling of the sample to room temperature after this maintains an extremely low pressure. Despite the crudity of this calculation the main observed features of the outgassing of hydrogen from a nickel slab appear to be reproduced.

(b) The diffusion and permeation of atmospheric helium through Pyrex walls: the helium-Pyrex system has been examined in detail by Rogers *et al.* (1954) and the equations of Section 2.5.2 should be applicable to this problem. We examine the helium partial pressure in a spherical vacuum system of internal volume 1 litre, having walls of Pyrex $l = 1 \cdot 5$ mm thick, with total area 500 cm^2. A normal atmosphere at room temperature is initially both inside and outside this bulb and we assume that the helium partial pressure in this atmosphere is 4×10^{-3} torr (Weast 1966). From Fig. 2.30a D for helium in Pyrex at room temperature is 10^{-8} cm^2 sec^{-1}. Thus from eqn. (2.95) $\tau = l^2/D = 3 \cdot 7$ weeks. While this is a long time, nevertheless most Pyrex before being used as the walls of a vacuum chamber will have come to equilibrium with the helium in the atmosphere. From Fig. 2.34a the solubility of He in Pyrex at room temperature is $6 \cdot 9 \times 10^{-3}$ cm^3 (S.T.P.) cm^{-3} atm^{-1}. Thus at helium pressure of 4×10^{-3} torr the amount of helium in solution in the glass is $2 \cdot 8 \times 10^{-8}$ cm^3 (S.T.P.) cm^{-3}. At time $t = 0$ we start to pump the interior of this system with a speed of $0 \cdot 1$ litre sec^{-1} and calculate the helium pressure as a function of time thereafter. It appears certain that the helium coverage on the surface of the glass at room temperature will be far below a monolayer and we may therefore safely estimate α from eqn. (2.86). From Table 2.10 let us take the heat of desorption for helium on a typical heterogeneous surface as 600 cal/mole. The mean sojourn time is given by eqn. (2.64) to be $\tau = 10^{-12} e^{600/585} \doteqdot 3 \times 10^{-12}$ sec. From eqn. (2.86) $\alpha = 10^4$ cm sec^{-1} and $h = \alpha/D = 10^{12}$. Thus $L = l\alpha/D = 1.5 \times 10^{11}$, i.e. essentially infinite. There seems no doubt that the helium-Pyrex system will be diffusion limited, and we proceed to substitute the above parameters into eqn. (2.94). The result for times up to 10 hours is shown in Fig. 2.39b. After 10 hours the pressure is still about an order of magnitude above its final value of 8.5×10^{-12} torr. This is because the helium originally in solution in the Pyrex has not yet diffused sufficiently to establish a linear gradient of concentration across the wall. Normally a Pyrex uhv system is subjected to a bake of 500°C for perhaps 10 hours and we examine the effect of such a bake on the helium partial pressure in our model. At $T = 773°K$ D becomes 6×10^{-6} cm^2 sec^{-1} and τ falls to about 1 hour. Thus permeation limited emission of helium into the system is established in approximately 1 hour, but since the permeation constant rises by about 1000 times the helium pressure during the bake is calculated to be 6×10^{-9} torr. The solubility of helium in Pyrex however changes very little with temperature and the linear concentration gradient at 500°C is essentially the same as that at room temperature. Thus when the system is cooled to room temperature again at $t = 20$ hours the final equilibrium pressure is achieved immediately. The helium partial pressure is rarely measured during the bake but the other results of this model are in good agreement with observations.

(c) The diffusion and permeation of atmospheric hydrogen through stainless steel: metal walls are frequently used in uhv systems and stainless steel is probably the most common material. Hydrogen is usually found to dominate the residual gases in such systems and it is worth while to examine the problem of permeation of atmospheric hydrogen, in a calculation analogous to that of helium-Pyrex carried out above. The partial pressure of hydrogen in the atmosphere is taken to be 4×10^{-4} torr (Weast 1966) and the gas in the walls is assumed initially to be in equilibrium with this pressure, although this is rather an optimistic assumption since $\tau = l^2/D \doteqdot 100$ years. The system geometry is maintained the same as for the helium-Pyrex problem above but the pump speed is taken to be 10 litres sec^{-1} as more representative for hydrogen. The calculation proceeds as before using solubility and permeation data from Figs. 2.34b and 2.36b and the result is shown in Fig. 2.39c. Low pressures are achieved after 10 hours primarily because of the small value of D (which incidentally is a derived quantity and less reliable than a directly measured quantity). However, the concentration gradient at this time does not even approach the linear gradient which is the steady state condition. Baking at 500°C however lowers $\tau = l^2/D$ to 3 hours and a linear gradient is established in this time. Cooling to room temperature at 20 hours freezes this linear gradient and a pressure of 3.3×10^{-13} torr is immediately achieved. Thus the main effect of the bake has been exhaustion of the hydrogen dissolved in the

Molecule-Molecule and Molecule-Surface Interactions

stainless steel and the relatively rapid achievement of a linear concentration gradient in the wall. However, during the bake the solubility has been raised by roughly an order of magnitude and the final gradient frozen in the wall is ten times as steep as that corresponding to the steady state at room temperature. The actual gradient will fall to the latter value but it will take 100 years for this to be achieved. The result presented in Fig. 2.39c is in reasonable agreement with the experimental results of Das (1962) on the outgassing of a stainless steel bell-jar.

A more detailed examination of this problem has been given by Calder and Lewin (1967) who have drawn the following conclusions:

(i) Permeation of atmospheric hydrogen is 10^{-17} torr litre sec^{-1} cm^{-2} and is small relative to other sources of hydrogen.

(ii) Baking the system under vacuum to 400°C for 12 hours typically gave an outgassing rate of 10^{-12} torr litre sec^{-1} cm^{-2}.

(iii) Baking the system under vacuum to 1000°C for 1 hour reduced the outgassing of H_2 from stainless steel to 10^{-14} torr litre sec^{-1} cm^{-2}.

(d) The outgassing of nitrogen from a molybdenum wire: the assumed molybdenum wire approximates a grid in an ionization gauge. It is 150 cm long, with diam 0·025 cm (0·010″), surface area 11·8 cm^2, volume 0·0735 cm^3. The normal operating temperature is estimated at 650°K and the wire is held at this temperature in a vacuum vessel pumped with a speed of 10 litres per sec for 10 hours. We assume the initial concentration is $1·5 \times 10^{-1}$ cm^3 (S.T.P)/cm^3. From Figs. 2.34c and 2.36b we can obtain C_0 and K at 650°K and from these calculate D to be 5×10^{-10} cm^2 sec^{-1}. The characteristic diffusion time for the wire is about 30 hours. The calculation of the pressure as a function of time proceeds in a similar fashion to the slab with a geometric correction supplied by Fig. 2.33. The resulting pressure is shown in Fig. 2.39d. The pressure falls steadily during the first 10 hours. At $t = 10$ hours the wire is heated to 1750°K (a typical outgas temperature). At this temperature the characteristic diffusion time is about 30 sec. There will therefore be an initial burst of nitrogen which will decay to a negligible value within about 6 minutes. Recooling of the wire to 650°K after this should yield a negligible outgassing of nitrogen. Difficult to estimate and probably not negligible is the outgassing of those portions of the wire which, for various constructional reasons, do not pass through the temperature cycles calculated above and provide a high concentration reservoir of gas which may diffuse down the concentration gradient and escape into the gas phase.

CHAPTER THREE

COLLISION PROCESSES IN GASES

The previous chapter has outlined the interaction of molecules in gases near thermal energies (< 1 eV). The present chapter will review briefly two types of interactions at particle energies far above the thermal range (≥ 1 eV) and below the onset of nuclear reactions:

(i) the collision of electrons with gas molecules,

(ii) the collision of ions and super-thermal atoms with gas molecules.

Knowledge of the processes which occur during such collisions is important for the understanding of uhv phenomena for the following reasons:

(i) almost every measurement of pressure in uhv systems involves the electron impact ionization of molecules;

(ii) the collection of ions in gauges and ionic pumps involves the collision of ions with atoms of a solid. These collisions are similar to those which occur between ions and atoms in the gas phase;

(iii) the excitation products following energetic collisions (photons, secondary electrons, radicals) can cause significant disturbing effects by their interaction with surfaces;

(iv) the collision of ions with atoms is an important process in high density plasma experiments. For reasons of gas purity, such experiments must be performed in systems with uhv capabilities.

Very extensive information, both experimental and theoretical, is available in this field, loosely termed 'atomic collision physics', and it is far beyond the scope of this book to examine it in any detail. Excellent reviews of both theory and experimental data have been given by, among others, Massey and Burhop (1952), Mott and Massey (1965), Hasted (1964), and McDaniel (1964). A good cross-section of the expanding research work in this field is reported in the proceedings of three series of international conferences: those on the Physics of Electronic and Atomic Collisions (McDowell 1963b), those on Ionization Phenomena in Gases (Perovic and Tosic 1965) and those on the Dynamics of Rarefied Gases (de Leeuw 1965). (See also preceding volumes in these series.)

We will limit ourselves to mentioning some of the phenomena which can occur and some theoretical approaches useful in their description. We will consider inelastic collisions, which change the internal energy of one or both of the colliding particles, separately from elastic collisions in which total kinetic energy and momentum are conserved.

Hasted's (1964) notation for atomic collision processes will be used. The most important symbols are: electron = e, photon = ϕ, neutral atom = 0, excited neutral atom = $0'$, singly charged ion = 1, n-charged ion = n.

A selection of collision cross-sections for some of the most common processes is given in Table 3.1.

3.1 ELASTIC COLLISIONS OF ELECTRONS WITH ATOMS AND MOLECULES ($e0/e0$)

According to classical theory, the cross-section for collision of an electron with an atom is infinitely large since arbitrarily small deflections are included. The quantum uncertainty principle, however, sets a lower bound to the theoretically significant deflection during a collision and this gives a corresponding finite limit to the collision cross-section. For an electron with 1 eV kinetic energy, the limiting scatter angle is about 11°; for 100 eV it is about 2·3°.

Many measurements of *total* collision cross-section have been made in the energy range below 100 eV. Except near the upper end of the range, these are usually close approximations to the elastic or 'momentum transfer' cross-sections, since the cross-sections for inelastic collision are relatively small. Two types of experiments have yielded most of the important data. In the first, used originally by Ramsauer (1921a), a beam of mono-energetic electrons is formed in a uniform magnetic field and the ratio of scattered current to primary current in a collision chamber of known geometry is measured. Electron energies in the range 1 eV to 100 eV have been used. The second type of experiment, the 'electron swarm' method, was first used by Townsend and his colleagues (Townsend & Bailey 1922). In this method the steady-state drift velocity of electrons in the direction of a uniform electric field is measured in a gas at known relatively high pressure. Refinements of this technique have been reported (Pack *et al.* 1962) which yield momentum transfer cross-sections covering the energy range 0·003 eV to 2 eV.

Total cross-sections measured by the Ramsauer method for the inert gases are shown in Fig. 3.1. Near 100 eV, the values are a factor of two or three larger than the corresponding ionization cross-section (Smith 1930) (Hanle & Riede 1952). At still higher energies, ionizing collisions become dominant. In the energy range 5 eV to 25 eV, where the inelastic cross-sections tend toward zero, the total cross-sections become large; while in the neighbourhood of 1 eV, very deep minima appear. The latter phenomenon is known as the Ramsauer-Townsend effect. Similar cross-section curves for diatomic gases (see Fig. 3.2) do not show the Ramsauer-Townsend effect. In these gases, inelastic processes are possible at all electron energies via the rotational and vibrational molecular states. Some cross-sections for diatomic gases at very low electron energies, measured by the 'electron swarm' method, appear in Fig. 3.3.

The theoretical description of elastic electron-atom scattering is treated in

TABLE 3.1
Collison cross-sections (cm²)

Process	Total electron atom scattering (a)		Electron induced negative ion formation (total)			Electron induced positive ionization (total)			Photo-ionization			Dissociative recombination			Resonant charge transfer (b)			Proton induced ionization (total)			Elastic atom–atom (viscosity) (c)
Notation	$e0/e0$		$e0/1$			$e0/\Sigma Ne^{n+1}$			$\phi 0/1e$			$e(10)/0'0'$			$10/01$			$10/11e$			$00/00$
Gas	Q_{max}	$E_e(max)$	Q_{max}	$E_e(max)$		Q_{max}	$E_e(max)$		Q_{max}	$EQ(max)$		Q	E_e		Q		E_i	Q_{max}	$E_p(max)$		Q
H	1.7×10^{-15}	2	2.1×10^{-20}	14		7×10^{-17}	65								4×10^{-15}		10	1.7×10^{-16}	7×10^4		
H₂	1.3×10^{-15}	3.0				9.7×10^{-18}	67								1×10^{-15} (max)		10	2.5×10^{-16}	5×10^4		2.4×10^{-15}
He	6×10^{-16}	2.0				3.7×10^{-17}	130		6×10^{-18}	25		1×10^{-15} (He₂⁺)	thermal		2.5×10^{-16}		10	1×10^{-16}	1×10^5		1.35×10^{-15}
N	5×10^{-16}	8				1.5×10^{-16}	110		1.4×10^{-17}	21											
O	6×10^{-16}	25				1.5×10^{-16}	100														
N₂	2.7×10^{-15}	2.5				2.5×10^{-16}	110					3×10^{-14}	,,		3.5×10^{-15}		10				4.3×10^{-15}
CO	3.7×10^{-15}	2.9	2×10^{-19}	10		2.7×10^{-16}	107														
O₂	1.1×10^{-15}	16	2×10^{-18}	6.5		2.7×10^{-16}	122					3×10^{-14}	,,		2.5×10^{-16}		10				4.2×10^{-15}
CO₂			4×10^{-19}	8		3.6×10^{-16}	117														6.8×10^{-15}
Na	4×10^{-14}	1.7							4×10^{-19}	8.9											
K	5.3×10^{-14}	1.25																			
Rb	4.5×10^{-14}	1.25							1.1×10^{-19}	4.2											
Cs	6×10^{-14}	2.5													$\sim 5 \times 10^{-14}$		10				
Li									1.9×10^{-18}	6											
Ne	4×10^{-16}	36				7.8×10^{-17}	187.5		7.5×10^{-18}	30		2×10^{-14} (Ne₂⁺)	,,		2.5×10^{-15}		10	2×10^{-16}	1×10^5		2.1×10^{-15}
Ar	2.5×10^{-15}	12.2				2.9×10^{-16}	92		3.6×10^{-17}	16		7×10^{-14} (Ar₂⁺)	,,		5×10^{-15}		10	6.5×10^{-16}	6×10^4		4.2×10^{-15}
Kr	2.8×10^{-15}	11.5				4.2×10^{-16}	80					$\sim 1 \times 10^{-13}$ (Kr₂⁺)	,,					1.2×10^{-16}	5.5×10^4		5.4×10^{-15}
Xe	4.0×10^{-15}	6.7				5.5×10^{-16}	115					$\sim 2 \times 10^{-13}$ (Xe₂⁺)	,,								7.6×10^{-15}

(a) In most cases predominantly elastic scattering. (b) Positive ions in atoms of the same gas. The cross-section increases toward lower ion energy.
(c) Diameters were taken from Dushman and Lafferty (1962) Table 1.6. $Q = \pi d^2$.

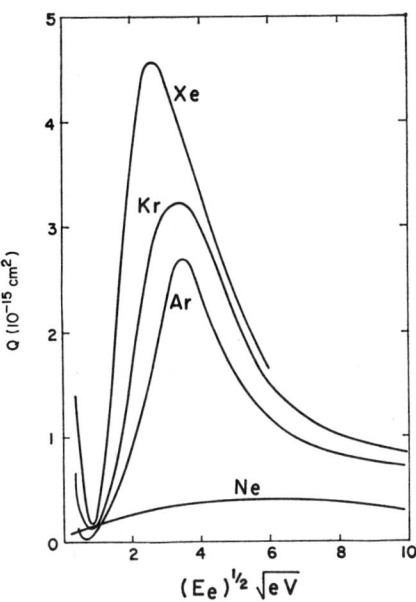

Fig. 3.1 Total collison cross-sections for electrons in rare gases measured by Ramsauer (1921a, 1921b, 1923). (From Hasted 1964).

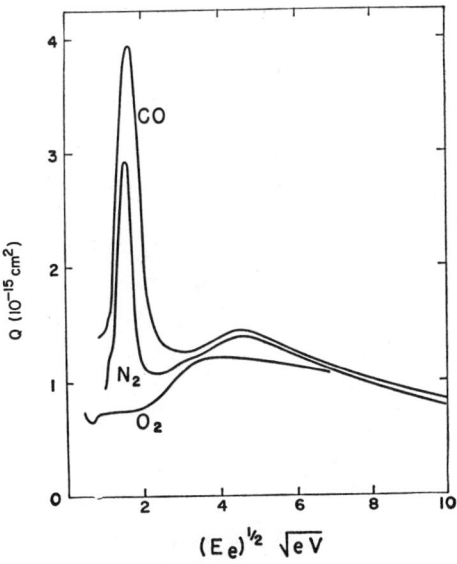

Fig. 3.2 Total collision cross-sections for electrons in diatomic gases. (Hasted 1964).

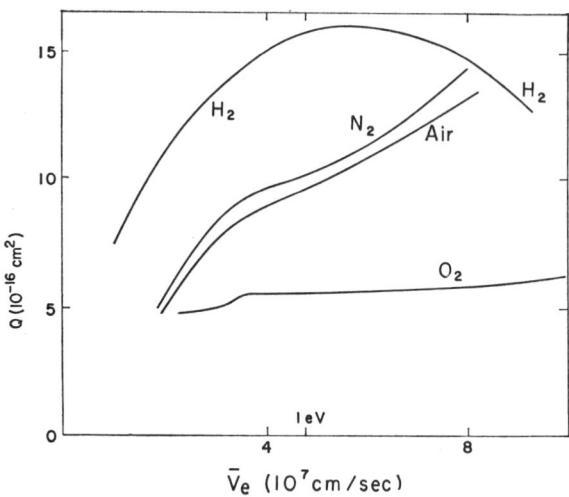

Fig. 3.3 Momentum transfer cross-sections as a function of the mean electron velocity measured by Huxley and Crompton (1962). (From Hasted 1964).

detail by Mott and Massey (1965). The electron is considered a plane wave of the de Broglie wavelength $\lambda_e = h/mv$ incident on the spherically symmetric potential field $V(r)$ of the atom. The scattering is assumed to produce a spherical wave originating at the atomic centre. Solution of the appropriate Schrödinger wave equation yields the total scattering cross-section in terms of the phase shifts experienced in the scattering process. These must in turn be evaluated numerically or by approximation techniques for the chosen form of $V(r)$.

3.2 INELASTIC ELECTRON-ATOM COLLISIONS

3.2.1 Ionization

Fairly accurate measurements of total ionization cross-section of the more common gases were made more than thirty years ago (Compton & van Voorhis 1925; Tate & Smith 1932; Bleakney 1930). The values of Tate and Smith (1932), which are still in common use today, were made with a magnetically collimated electron beam and used a small transverse electric field to extract and collect the ions. Some of their results are shown in Fig. 3.4. The total cross-sections are seen to rise nearly linearly from the thresholds (ionization potentials) to maxima near 100 eV and then to decrease gradually. The maximum values are about one order of magnitude smaller than those for elastic collisions presented in the previous section (see Fig. 3.1). More recently, measurements of the same kind have been repeated and extended by three groups of experimenters (Rapp & Englander-Golden 1965; Schram *et al.*

1965b; Asundi *et al.* 1963). An excellent review of ionization cross-section data is given by Kieffer and Dunn (1966).

Early measurements of cross-sections for ionization to individual charge states were made by Bleakney (1930), Tate and Smith (1934) and Stevenson and Hipple (1942). More recently, results have been reported by, among others, Fox (1959, 1960) and by Morrison and his colleagues (Stuber 1965; Dorman & Morrison 1961). For example, the first six multiple ionization functions for xenon (Fox 1959) are shown normalized in Fig. 3.5 Absolute values of cross-sections for multiple ionization of the noble gases have been published (Schram 1966; Schram *et al.* 1966). In most cases the single ionization process is most probable (see, for example, the data presented by Kieffer and Dunn (1966)). The ejected electrons normally carry away only a few electron volts kinetic energy.

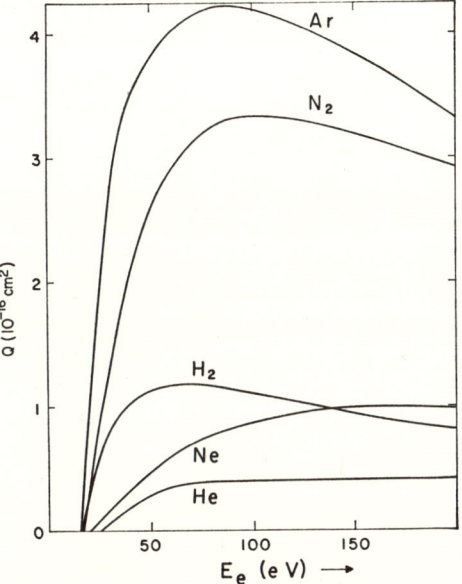

Fig. 3.4 Total electron ionization cross-sections measured by Smith (1930) and Tate and Smith (1932). (From Hasted 1964).

Quantum theoretical calculations have been carried out for the ionization of the hydrogen atom employing the first and second Born approximation and also taking account of electron exchange processes (Oppenheimer 1928). These calculations are discussed in detail by Mott and Massey (1965). A comparison of the first Born approximation prediction with the experimental results of Fite and Brackmann (1958a) is shown in Fig. 3.6. Only above 100 eV, where the required assumption of small scattering phase shifts becomes realistic, are the two in good agreement. Theoretical treatments of the ionization of more complex atoms require more restricting assumptions but give predictions which

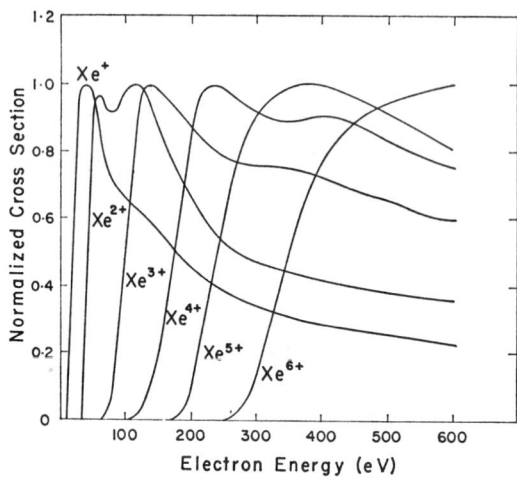

Fig. 3.5 Normalized ionization cross-sections for the first six charge states of xenon (Fox 1959). The relative values of the cross-section maxima are: (1) 1·00; (2) 0·4; (3) 0·17; (4) 0·064; (5) 0·019; (6) 0·0022; (7) 0·00017.

bear qualitatively the same relation to experimental results as that shown in Fig. 3.6. The theoretical complications arise mainly from the existence of many energy levels in the created ions.

Many investigations of the ionization function near threshold have been reported. The data are reviewed by Kieffer and Dunn (1966). Fox and his colleagues, using a 'retarding potential difference method' (Fox 1959) have obtained accurate values of ionization potential for several singly and multiply charged ions. More recently electrostatic energy analysers have been used to obtain similar data (Marmet & Kerwin 1960; Marmet & Morrison 1961). In most cases, more or less distinct breaks are observed in the curves within a few

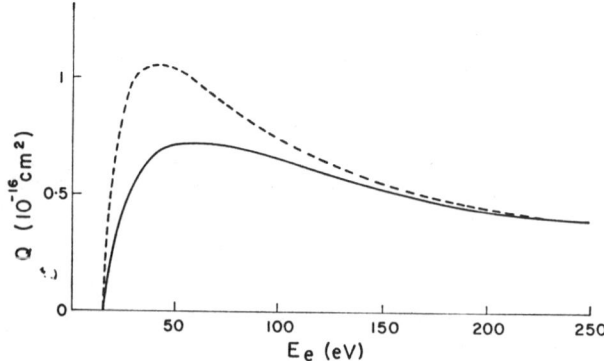

Fig. 3.6 The electron ionization cross-section of atomic hydrogen: Full line, the experimental data of Fite and Brackmann (1958a); dotted line, Born approximation calculation. (From Seaton 1962).

Collision Processes in Gases

volts of threshold (see, for example, Marmet & Morrison (1961)). Although difficult to interpret without ambiguity because of finite electron energy spreads, these are thought to involve transitions to excited levels of the ion. (It should be noted that the thresholds obtained in electron impact ionization are 'vertical' ionization potentials, complying with the Franck-Condon principle, in distinction to the adiabatic values obtained from optical spectroscopy and photo-ionization.)

In the ionization of polyatomic molecules, fragment ions can appear in considerable abundance, and each with its own threshold energy. These have been widely studied in connection with mass spectrometric analysis, particularly of organic compounds. Surveys are given by Field and Franklin (1957) and by Reed (1962). In some cases, the fragment ions carry significant kinetic energies (Hagstrum 1951) and secondary electrons may also carry away unusual amounts of kinetic energy (von Koch & Lindholm 1961).

Some cross-sections for the ionization of ions by electron impact have also been reported, and are reviewed by Kieffer and Dunn (1966). The production of negative ions by electrons is discussed by Rapp and Briglia (1965).

3.2.2 Excitation

The general process of the change of the internal quantum numbers of an atom or molecule by electron impact has been widely studied. Two types of experiments are commonly done: The first involves the bombardment of a gas or molecular beam with a velocity selected electron beam, and the detection of a known part of the optical radiation produced using a prism or grating spectrometer. The excited state which gave rise to the radiation is easily identified in most cases, but detectors adequate for absolute cross-section measurements are difficult to obtain. A thorough review of this type of work is given by Massey and Burhop (1952) and more recently by Fite (1962). Detectors are discussed in some detail by Hasted (1964). When metastable levels are excited, an additional difficulty arises: The excited particles are able to disperse before emission. It then becomes necessary to make measurements of the metastable atom flux (Hasted 1964).

The second type of electron impact excitation experiment involves the measurement of the energy loss suffered by electrons in a mono-energetic beam during collisions (Schulz 1959, 1962). In this case, detection involves the measurement of appropriate electron currents and is relatively simple. Identification of the excited state responsible for the electron energy loss is, however, frequently difficult. An extensive series of papers have been published by Lassettre and co-workers (Lassettre & Francis 1964; Meyer *et al.* 1965; Skerbele & Lassettre 1966) on the electron excitation of simple molecules studied by electron energy loss techniques.

In general the shape of excitation cross-section curves as a function of electron energy are, for optically allowed transitions, similar in shape to those for ionization. For disallowed transitions, they tend to have much sharper

E

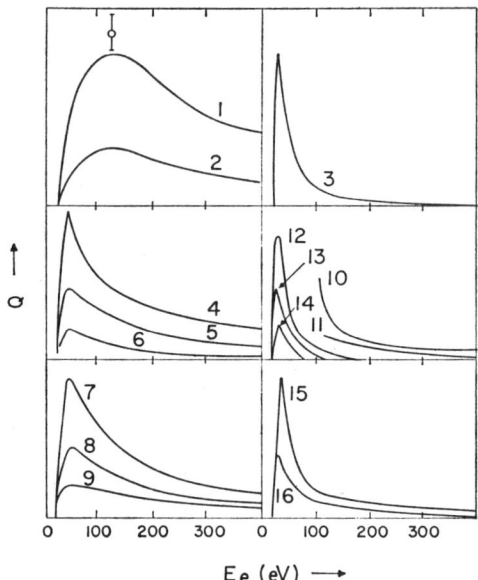

Fig. 3.7 A number of electron excitation functions for helium. Full lines are from the experimental data of Thieme (1932) and are based on optical line intensities. 1. $3^1P \rightarrow 2^1S$; 2. $4^1P \rightarrow 2^1S$; 3. $3^3P \rightarrow 2^3S$; 4. $4^1S \rightarrow 2^1P$; 5. $5^1S \rightarrow 2^1P$; 6. $6^1S \rightarrow 2^1P$; 7. $4^1D \rightarrow 2^1P$; 8. $5^1D \rightarrow 2^1P$; 9. $6^1D \rightarrow 2^1P$; 10. $4^3S \rightarrow 2^3P (\times 10)$; 11. $5^3S \rightarrow 2^3P (\times 10)$; 12. $4^3D \rightarrow 2^3P$; 13. $5^3S \rightarrow 2^3P$; 14. $6^3S \rightarrow 2^3P$; 15. $3^3D \rightarrow 2^3P$; 16. $4^3D \rightarrow 2^3P$. (From Hasted 1964).

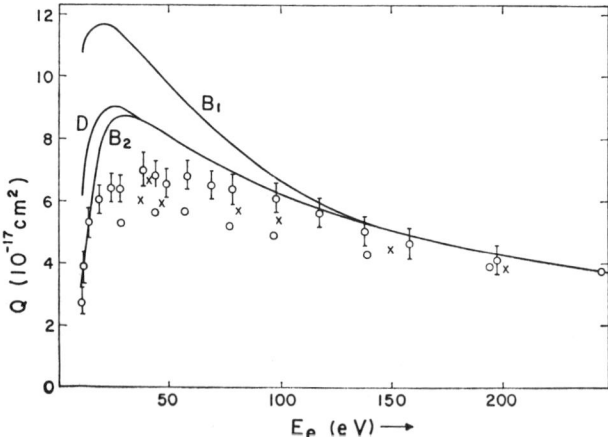

Fig. 3.8 The excitation cross-section for the $1s$–$2p$ transition in atomic hydrogen. Experimental points: open circles with bars, the Lyman α radiation was measured at 90° to the electron beam; crosses, at 54·5° to the electron beam; open circles are points corrected for polarization (Fite and Brackmann 1958b; Fite et al. 1959). Theoretical calculations: B1, the first Born approximation (Massey 1956); B2, the second Born approximation (Burke and Seaton 1961); D, the exchange distorted wave calculation (Khashaba and Massey 1958). (From Hasted 1964).

maxima nearer the threshold energy. For example. in Fig. 3.7 are shown the excitation functions between various levels in the helium atom, and in Fig. 3.8 that for the 1s→2p transition in atomic hydrogen (Fite & Brackmann 1958b; Fite *et al.* 1959). The cross-section maximum in the latter case is approximately equal to that for ionization. Usually the cross-section maxima are at least an order of magnitude smaller than the ionization cross-section for the same atoms.

For heavier atoms, experimental results are reviewed by Massey and Burhop (1952) and by Hasted (1964). The most thoroughly studied atoms are Hg and N. Some prominent excitation functions for the nitrogen atom and molecule appear in Fig. 3.9.

The electron energy loss techniques developed by Schulz (1959, 1962) have been used primarily to observe vibrational excitation functions in molecules. An example of the excitation functions obtained for the nitrogen molecule are shown in Fig. 3.10. In this case it is thought that a temporary negative ion state is formed. Evidence for temporary negative ion states in collisions with monatomic molecules has also been reported (Schulz 1963; Kuyatt *et al.* 1965). These lead to observable anomalies in the electron elastic scattering cross-section over a very narrow range of electron energies.

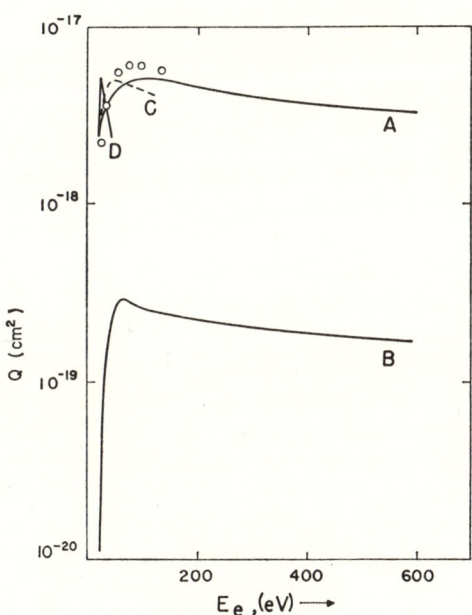

Fig. 3.9 Electron excitation cross-sections for the nitrogen atom and molecule.
A. N_2^+ 0,0 band, first negative system (Sheridan *et al.* 1961). Open circles are the data of Stewart (1956).
B. NI $3s\ ^4P^\circ \rightarrow 3p\ ^4P^\circ$ (Sheridan *et al.* 1961).
C. N_2^+ Meinel bands normalized to A at maximum (Stewart 1956).
D. N_2 second positive bands, normalized to A at maximum (Thieme 1932) (Taken from Hasted 1964).

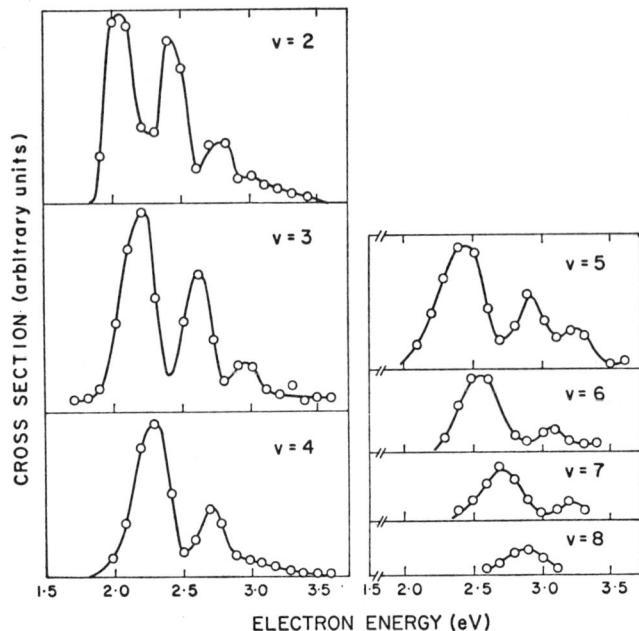

Fig. 3.10 Electron excitation cross-sections to the various vibrational levels of N_2 indicated (2 to 8) in arbitrary units as a function of electron energy in eV. (Schulz 1962).

Quantum theoretical calculations for hydrogen follow lines closely parallel to those used for ionization (see Section 3.2.1). For atomic hydrogen, the results of the first (B1) and second (B2) Born approximation for the $1s \rightarrow 2p$ transition are included in Fig. 3.8 along with that for the exchange distorted wave calculation (D). Again in most cases calculated, the theoretical curves lie *above* the experimental data at electron energies below ~ 100 eV. Quantum theoretical calculations have also been done on a variety of heavier atoms (see, for example, Seaton (1966), Presnyakov et al. (1966) and Peach (1966)).

A classical theory of inelastic electron-atom collisions (both excitation and ionization) has been developed by Gryzinski (1963, 1965). The theory provides relatively simple analytic expressions which predict surprisingly well both the magnitude of the collision cross-section and its variation with electron energy. Account is taken in the theory of the velocity distribution of the atomic electrons and the effects of the coulomb field of the nucleus. Useful empirical expressions for ionization cross-sections have been given by Drawin (1961).

3.3 ELASTIC ATOM-ATOM AND ION-ATOM COLLISIONS (00/00, 10/10)

The study of the scattering of fast atoms and ions by atoms permits the establishment of interatomic potentials at nuclear separation smaller than those which can be deduced from collisions at thermal energies. In contrast to electron-atom scattering, quantum mechanical effects are here usually minor: For scattering angles larger than some critical value ψ_c, where the particle wavelength and the minimum internuclear separation become equal, classical methods can be used in the analysis of the data. For helium-helium collisions, $\psi_c \sim 10°$ at thermal energies, $\sim\frac{1}{2}°$ at 10 eV and $\sim 1/20°$ at 1 keV. For heavier atoms the angles become correspondingly smaller ($\lambda \propto m^{-\frac{1}{2}}$ at constant E).

A variety of atomic systems have been studied in apparatus using fairly small scattering angles and beam energies from a few hundred to a few thousand eV. These are mainly by Amdur (Amdur *et al.* 1961) and Simons (Cramer & Simons 1957) and their co-workers. A comprehensive list of their papers is given by Hasted (1964). Both groups make measurements of total cross-section for scattering into angles greater than some value ψ_0 (determined by the apparatus design) as a function of the beam energy. Amdur's group have used neutral beams obtained by charge exchange from ion beams in their parent gases. Simon's group have used ion beams directly. The scattering results are used to determine the constants of an assumed form (inverse power or Born-Mayer) of the inter-atomic potential. Corrections for non-uniform beam density, finite scattering length and finite beam collimation angle are crucially important in the analysis of the data. A compilation of the potentials derived in this way is given by Mason and Vanderslice (1962). The potentials are valid in the energy region near 1 eV. Theoretical aspects of the scattering process and of the interatomic potential have been treated by, among others, Mason and his collaborators (Mason & Vanderslice 1962), Firsov (1958a, b) and Abrahamson (1963, 1964).

Everhart (Everhart *et al.* 1955) and Fedorenko (Fedorenko *et al.* 1960) and their collaborators have measured large-angle differential cross-sections at higher beam energies (2 to 100 keV). Much more sensitive detectors have in these cases made it possible to approach ideal scattering geometries. Analysis of the data does not require assumptions as to the form of the interatomic potential except that it be monatonic. These experiments have yielded interatomic potential functions in the energy range 1 to 50 keV (separations from ~ 0.01 to 0.4 Å) for various inert gas atom pairs (Lane & Everhart 1960).

In the high energy experiments mentioned above, inelastic processes, particularly charge exchange, electronic excitation and ionization, are actually common (see Section 3.4). The inelastic energy transfers are, however, usually small compared to the kinetic energies of the collision partners in the centre of mass system. This allows the collisions to be described with reasonable accuracy as elastic ones with small perturbations.

In Fig. 3.11 are shown the results obtained for the potential between He^+ and He. The experimental data of Cramer and Simons (1957) were analysed by Mason and Vanderslice (1958) using approximate quantum calculations to determine the forms of the two possible potentials. These results are represented by the full lines in the figure. Careful quantum mechanical calculations gave the results indicated by the open circles. Interaction potential energies

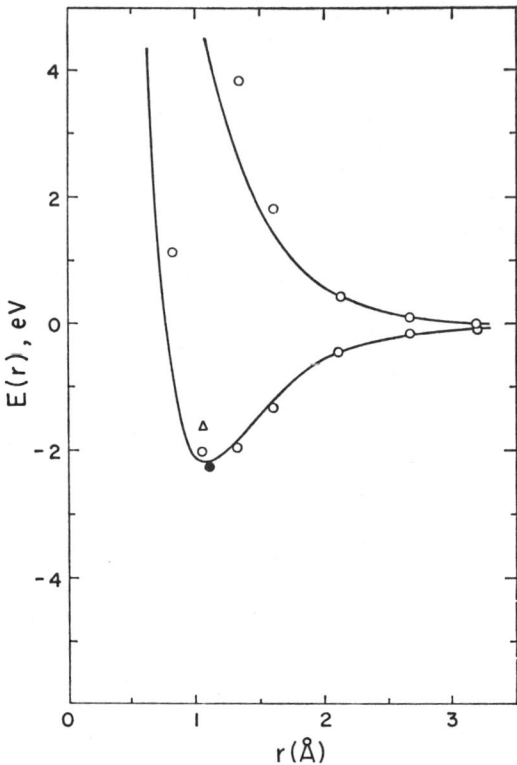

Fig. 3.11 The potential energy of interaction of He$^+$ with He. Full lines give the results of analysis by Mason and Vanderslice (1962) of scattering data (Cramer and Simons 1957). Filled point is the quantum mechanical calculation of Weinbaum (1935); open circles Moiseiwitsch (1956); triangle, Czavinsky (1959). (From Mason and Vanderslice 1962).

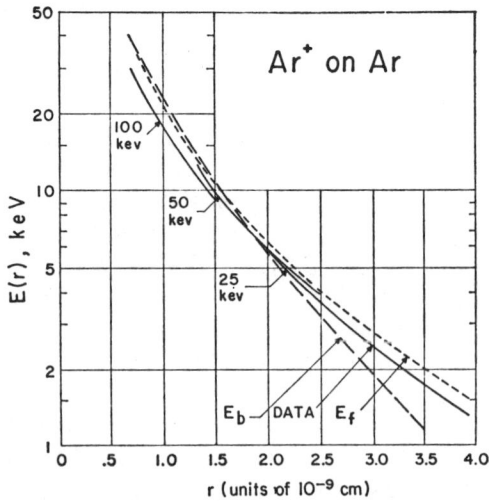

Fig. 3.12 Potential energy of interaction of Ar$^+$ with Ar as a function of internuclear separation. The full lines are derived from large angle scattering data with impact energies 25, 50 and 100 keV. The short-dotted line is a function derived from a statistical model by Firsov (1958a); the long-dotted line is the exponentially screened coulomb potential of Bohr (1948). (From Lane and Everhart 1960).

derived for the Ar^+-Ar system from differential cross-section measurements at high beam energy are reproduced in Fig. 3.12, (Lane & Everhart 1960). Full lines represent experimental data obtained with beam energies 25, 50, and 100 keV. The broken lines marked E_B and E_F are the theoretical potentials using the Bohr function (Bohr 1948) and one due to Firsov (1958a). The separations appearing in Fig. 3.12 are smaller by about a factor 10 than those encountered in thermal energy interaction (see Fig. 2.2.).

3.4 INELASTIC EFFECTS IN ION-ATOM COLLISIONS

The most striking characteristic of the field of inelastic heavy particle collisions is its complexity. The large number of possible final states of the collision partners leads to a large number of possible reactions. The field has been reviewed by McDaniel (1964) and will be discussed only briefly. Recent theoretical work is reviewed by Bates (1962).

3.4.1 Charge Transfer (10/01)

At low kinetic energies (< 1 keV), the predominant inelastic process is that of resonance charge transfer. The cross-sections are very large at low energies, and decrease only slowly with increasing energy. As an example, the resonance charge transfer cross-section for Ar^+ in Ar is shown in Fig. 3.13, where the sources of the data are detailed by Hasted (1964).

The proton-hydrogen atom collision has been carefully studied by Fite and his co-workers (Fite *et al.* 1960, 1962) and compared with theoretical treatments. The results are presented in Fig. 3.14. The most closely agreeing theoretical result (full line) was obtained by a perturbed stationary state calculation

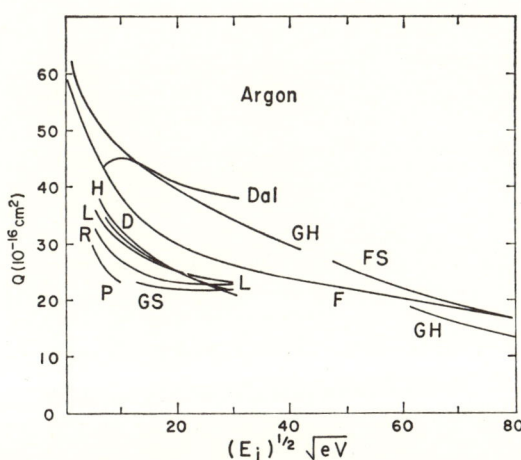

Fig. 3.13 Resonance charge transfer cross-section for Ar+ on Ar. The papers to which the letters refer can be found in Hasted (1962), from whom this figure is taken.

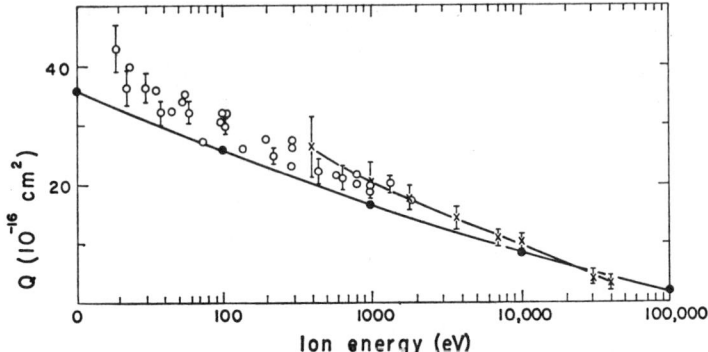

Fig. 3.14 Charge transfer cross-section for protons in atomic hydrogen. The circles are experimental data from Fite *et al.* (1962); the crosses from Fite *et al.* (1960). The filled circles are the results of a perturbed stationary state calculation by Dalgarno and Yadav (1953). (Taken from McDaniel 1964).

(Dalgarno & Yadav 1953). A number of other results are given by McDaniel (1964) with many pertinent references.

The process of non-resonant charge transfer is in general negligibly small until the ion velocity (cm/sec) becomes comparable to $a\Delta E/h$ where ΔE is the difference in the ionization energy (erg) of the colliding particles, a is a length of atomic dimensions (cm) and h is Planck's constant in erg/sec. This velocity is roughly comparable to that of the bound electrons in the atom. Cross-sections for protons in the inert gases appear in Fig. 3.15 (Hasted 1964). The maxima are seen to be 3 to 10×10^{-16} cm^2, and occur in the proton energy range 1 to 20 keV.

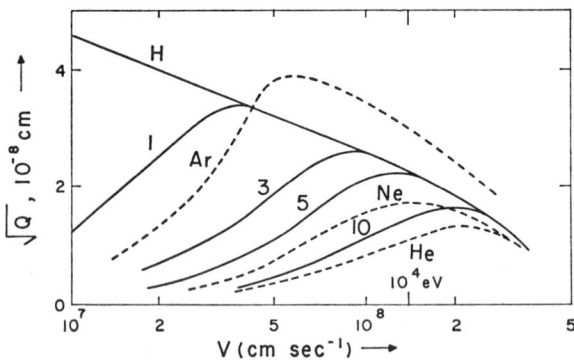

Fig. 3.15 Cross-sections for proton charge transfer in rare gases. The broken lines are the experimental data of Stedeford and Hasted (1955). The full lines are theoretical calculations made by Rapp and Francis (Rapp 1963) for various values of the difference ΔE (eV) between the ionization potentials of H and the gas atoms. Values appropriate for the rare gases are

H^+/Ar ΔE = 2·16
H^+/He   = 7·96
H^+/He   = 10·99   (From Hasted 1964).

Lockwood *et al.* (1963) have measured the resonant electron capture probability of He⁺ in He as a function of energy (0·4 to 25 keV) for a series of scattering angles between 0·4° and 4·4°. These correspond to collisions with impact parameters ~0·1Å. The results exhibit pronounced oscillations with energy (see Fig. 3.16) which are consistent (Everhart 1963) with a quantum mechanical theory of Bates and McCarrol (1958).

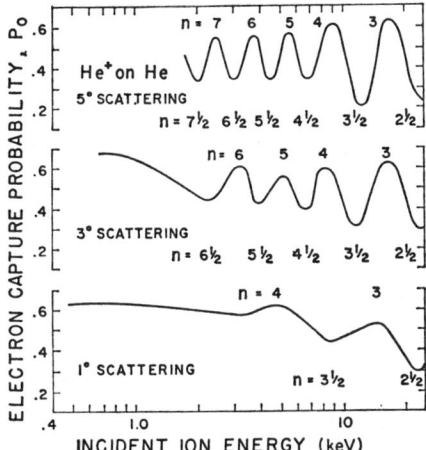

Fig. 3.16 The electron capture probability of He⁺ on He for three scattering angles as a function of ion energy. The n's are index numbers beginning with the first maximum (observed at ~ 250 eV) as $n = 1$. (Lockwood *et al.* 1963).

3.4.2 Excitation (10/10')

Relatively little work has been done in this field, mainly because of the greater difficulty of analysing the photons emitted in de-excitation. Carleton and Lawrence (1958) have studied proton impact on N_2 at energies 1·5 keV to 4·5 keV. They report absolute cross-sections for the production of photons of specific de-exciting transitions. These lie in the range 10^{-17} to 10^{-16} cm². The most extensive work on excitation by fast ions has been done by Kistemaker's group in Holland (Sluyters & Kistemaker 1959). They have studied the excitation of inert gases by Ar^+ ions in the energy range 5 to 25 keV using a vacuum grating spectrograph. Several excitation mechanisms were identified and cross-sections for the production of specific lines were found to be about 10^{-18} cm².

3.4.3 Ionization (10/11)

A review of this subject has been given by Fedorenko (1959). A great deal of experimental data has appeared recently, and only representative samples will be mentioned.

Measurements have been made of the total ion and free electron production in various gases by ion impact. A summary of the results for protons in rare gases, for example, is given in Fig. 3.17 (Hasted 1964). The maximum cross-sections range from 10^{-16} cm^2 to 10^{-15} cm^2 and all occur at a proton energy of about 50 keV. Similar results for He$^+$ appear in Fig. 3.18 (Martin et al. 1963). Experiments of this type are usually done by collecting the slow ions and electrons produced in the collision with transverse electric fields. The slow ions emerge almost normal to the primary beam since most of the collisions involve small beam scattering angles. Experimental techniques have been reviewed by Fite (1962).

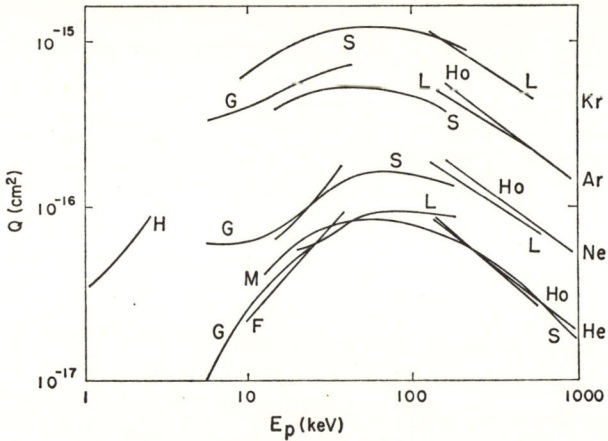

Fig. 3.17 Proton total ionization cross-sections for the rare gases. The original sources are identified in Hasted's book (1964) from which the figure is taken.

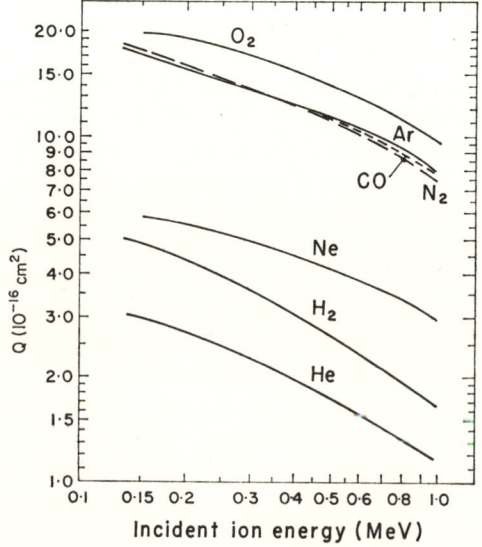

Fig. 3.18 Total positive ionization cross-sections for high energy He$^+$ incident on seven gases measured by Martin et al. (1963). (From McDaniel 1964).

The cross-section for ion production of protons in atomic hydrogen (Fite *et al.* 1960; Ireland & Gilbody 1963) is shown in Fig. 3.19. The full curve is the Born approximation theoretical cross-section (Bates & Griffing 1953). As in the case of electron impact ionization (see Section 3.2.1) the theoretical prediction is too high at low proton energy.

The energy distribution of electrons produced in collisions of K^+ ions with inert gases has been reported by Moe and Petsch (1958). Maxima appear in the spectrum at energies in the range zero to 35 eV. These are thought to be due to Auger processes corresponding to various energy levels of the ion-ion system.

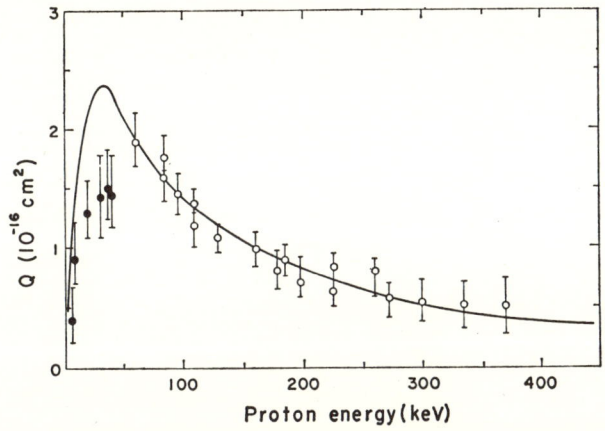

Fig. 3.19 Proton ionization cross-section for atomic hydrogen. The dots are the experimental data of Fite *et al.* (1960); the circles those of Ireland and Gilbody (1964). The solid curve is the result of the Born approximation calculation (Bates & Griffing 1953). (From McDaniel 1964)

Differential cross-sections for the production of ions of specific charge states have been measured by the groups of Everhart (Carbone *et al.* 1956; Fuls *et al.* 1957) and Fedorenko (Fedorenko 1954; Kaminker & Fedorenko 1955). Both studied the relatively rare collisions involving large scattering angle and large momentum transfer. Figure 3.20 shows, for example, the charge state distribution as a function of angle for Ar^+ ions scattered from Ar at 25 keV, 50 keV and 100 keV (Fuls *et al.* 1957). A theoretical interpretation of this data has been given (Russek & Thomas 1958; Russek 1963) which is based on the statistical distribution of the transferred energy among the outer shell atomic electrons. Very good agreement with experiment was obtained.

More detailed measurements of the inelastic energy loss in large angle collisions have also been made (Afrosimov & Fedorenko 1957; Morgan & Everhart 1962). The latter authors examined the charge state and kinetic energy of the recoil products of the collisions with high resolution in both angle and energy. They were able to show that the inelastic loss was simply related to the distance of closest approach in the collision rather than to the kinetic energy of the projectile (see Fig. 3.21).

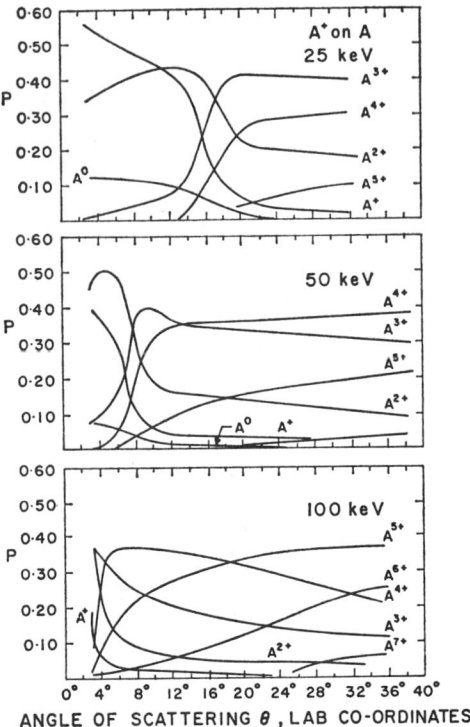

Fig. 3.20 The charge state analysis following single collisions of Ar+ on Ar at 25, 50 and 100 keV as a function of laboratory scattering angle. The probability of the particle being in the specified final charge state is given. (Fuls *et al.* 1957).

Recently both of the groups mentioned above (Afrosimov *et al.* 1965a, b, c; Everhart & Kessel 1965; Kessel *et al.* 1965) have reported coincidence measurements of the recoil and projectile particles in these large-angle collisions with simultaneous measurement of their charge state and kinetic energies. They find that for the collision of Ar^+ with Ar, there appears an excess energy loss (i.e. an inelastic loss in excess of that appearing as ionization) which has characteristic values independent of the final ion charge states.

Fano and Lichten (1965) have suggested that the characteristic losses are caused by the non-adiabatic formation of an excited molecular ion. Upon separation, the excited atoms then emit electrons by auto-ionization. The characteristic electron kinetic energy groups predicted by this theory have recently been found experimentally (Snoek *et al.* 1965).

3.5 PHOTO-ABSORPTION AND PHOTO-IONIZATION

Experimental and theoretical aspects of this field have been reviewed by Weissler (1956) and by Ditchburn and Opik (1962). Shorter reviews are also

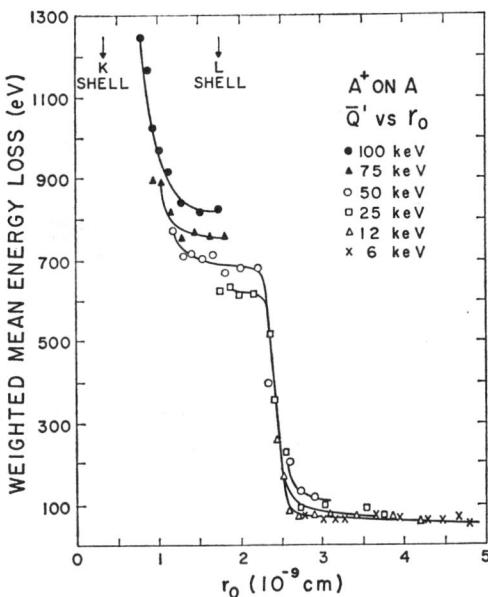

Fig. 3.21 The mean energy loss in large angle Ar⁺ on Ar collisions as a function of the distance of closest approach of the collision partners. The energy loss is nearly independent of the initial ion energy. Argon K- and L-shell radii are indicated in the figure. (Morgan & Everhart 1962).

available (McDaniel 1964; Hasted 1964). Only a very brief summary of the main concepts will be attempted in this section.

For photons with wavelengths long compared to atomic dimensions ($h\nu < 100$ eV), the interaction with atomic systems is via their electric dipole moment. Detailed discussions of this interaction are given in standard textbooks on atomic and molecular spectra (Herzberg 1944, 1950). Magnetic dipole interactions are less probable by $\sim 10^5$ and electric quadrupole by $\sim 10^8$. The lifetime for single-quantum dipole transitions is typically 10^{-8} sec.

Two types of excitation mechanisms can participate in photon absorption:

(i) Resonance transitions to discrete excited states which then decay by either photon re-emission or by dissociation or ionization if the excited levels lie adjacent to the appropriate continua.

(ii) Direct excitation into dissociation or ionization continua in non-resonant transitions.

Resonance transitions have relatively high cross-sections ($\sim 3 \times 10^{-14}$ cm²) over a very narrow wavelength band ($\sim 5 \times 10^{-3}$ Å). They usually account for only a minor part of the total photon absorption.

Cross-sections for photon absorption are usually obtained by measuring the attenuation of a nearly monochromatic photon beam traversing a gas filled absorption cell. Charge collection electrodes in the absorption cell permit the measurement of photo-ionization cross-sections with closely similar apparatus (Weissler 1956). Mass spectrometric analysis of photo-ionization products has led to a better understanding of the processes and energy states involved (Weissler et al. 1959).

Photo-ionization functions for three inert gases are reproduced in Fig. 3.22 (Weissler 1956). The cross-sections are seen to be near their maximum value

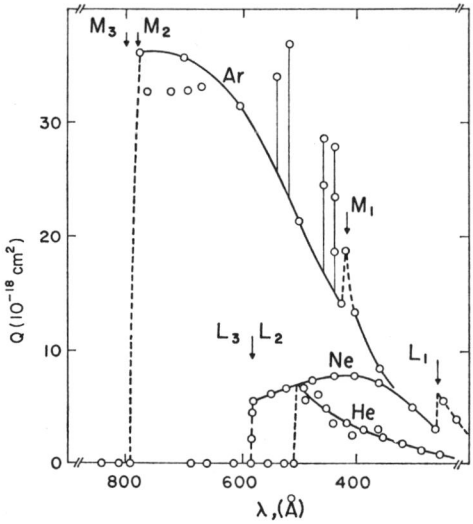

Fig. 3.22 Photo-ionization cross-sections for the rare gases. The experimental data is taken from Weissler (1956). The various absorption edges are indicated by the appropriate letters. (From Hasted 1964).

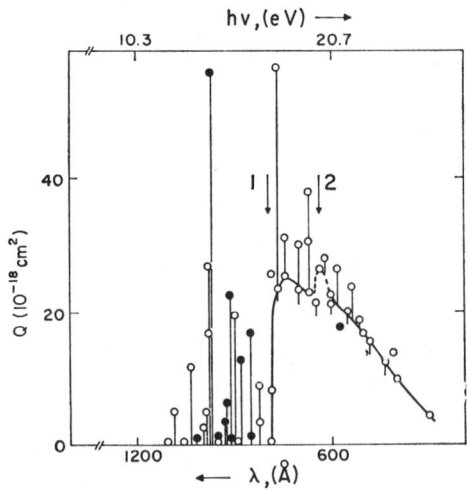

Fig. 3.23 The photo-absorption cross-section of N_2. Open circles give the data of Weissler and his colleagues (1956); the closed circles those of Clarke (1952). (From Hasted 1964).

at the threshold wavelength. Following a plateau of a few hundred ångströms width, the cross-sections then decrease with increasing photon energy. The maximum cross-sections are about one tenth of those obtained in electron impact ionization. Vertical lines indicate resonance absorption to excited states. Arrows mark the various ionization potentials. The photo-absorption function for nitrogen appears in Fig. 3.23, where resonance absorption lines below the ionization threshold are clearly visible.

It should be pointed out that Rayleigh and Raman scattering cross-sections in the same wavelength region are smaller than those for photo-ionization and photo-dissociation by about a factor 10^8 (Heddle 1962).

3.6 POSITIVE ION RECOMBINATION

This subject has been reviewed by Biondi (1963) and Hasted (1964). Experimental results are usually expressed in terms of the 'recombination coefficient' α, where

$$\alpha = \alpha(Q,v) = \int_0^\infty Q(v)vf(v)dv \text{ cm}^3/\text{sec} \qquad (3.1)$$

in which v is the relative velocity of the particles (cm/sec)

$Q(v)$ is the recombination cross-section (cm^2)

and $f(v)$ is the fraction of the encounters having relative velocity between v and $v+dv$.

To a first approximation, the cross-section Q can be obtained by dividing the recombination coefficient by the mean thermal velocity of the electrons. For 300°K

$$Q \approx \frac{\alpha}{v_e} = \frac{\alpha}{10^7} \text{ cm}^2. \qquad (3.2)$$

The only available data on recombination has been obtained from measurements in the afterglow of a plasma following excitation. Two experimental techniques have been used:

(i) Determinations are made of absolute electron or ion density in the afterglow as a function of time using Langmuir probe or microwave conductivity measurements.

(ii) Spectroscopic measurements are made of the afterglow radiation intensities as a function of time. Provided an absolute charge density measurement is made at some time, recombination coefficients can be deduced.

The resulting values of α refer to thermal velocity encounters (i.e. energies from ~0·025 to ~0·25 eV depending on the gas temperature). It is frequently difficult to separate recombination effects from those caused by ambipolar diffusion of the charge carriers. Failure to separate these effects is one of the main sources of error and ambiguity in the data.

Both two-body and three-body interactions can contribute significantly to the recombination. Of the two-body mechanisms, the most commonly observed are:

(i) radiative electron capture $e1/\phi 0$ or $e1/\phi 0'$,

(ii) dielectronic recombination, in which excited levels in one ground configuration of the atom lie in the continuum of a second configuration: $e1/0_1$, $0_1/0_2'\phi$,

(iii) dissociative recombination: $e(10)/0'0'$,

(iv) mutual neutralization: $1\bar{1}/0'0'$.

For the first two mechanisms, experimental work is extremely sparse. Theoretical values of cross-sections derived from the inverse process of (i) (photoionization) yield recombination coefficients at 300°K of $<10^{-11}$ cm^3/sec ($Q \lesssim 10^{-18}$ cm^2). Dissociative recombination coefficients have been much more widely studied, and found to be very much larger. For monatomic gases such as the rare gases, formation of the diatomic ion precedes the dissociation (e.g. $Ne^+ + Ne \rightarrow Ne_2^+$). Typical experimental values of the dissociative recombination coefficient at 300°K are given in Table 3.2, taken from Biondi (1963). The implied cross-section for $\alpha = 10^{-6}$ cm^3/sec is, using eqn. (3.2), $Q \sim 10^{-13}$ cm^2, a very large value. The main reason for the large value is the coulomb attraction between the oppositely charged recombining particles. The complex molecular ions, such as N_4^+ in Table 3.2 quite commonly appear in recombination processes. Mutual neutralization data is very scarce. A few values around 10^{-8} cm^3/sec have been reported for bromine and iodine (Yeung 1958).

TABLE 3.2
Dissociative recombination coefficients of thermal (300°K) electrons with various positive ions

Ion	α(cm^3/sec)
Ne_2^+	$2\cdot 4 \times 10^{-7}$
Ar_2^+	7×10^{-7}
Xe_2^+	$\sim 2 \times 10^{-6}$
N_2^+	3×10^{-7}
N_4^+	$\sim 1 \times 10^{-6}$
O_2^+	$1\cdot 7 \times 10^{-7}$

Biondi (1963).

The two important three-body recombination processes are:

(i) neutral molecule stabilized recombination $e10/0'0$,

(ii) electron stabilized recombination $e1e/0'e$ (the reverse of electron ionization).

In these cases, α has the units cm^6/sec.

The two-body recombination coefficients in general decrease as T^{-n} with increasing gas temperature. The value of n has been found in various investigations to range from 0·9 to 2·2 for energies $\lesssim 0·2$ eV. Choosing $n = 1·5$, the implied value of α at 1 eV would be lower than that given in Table 3.1 by about a factor 100 if the same rate of decrease persisted. Even so, for 1 eV the expected cross-section would be $\sim 10^{-15}$ cm^2, larger than the maximum electron ionization cross-section.

CHAPTER FOUR

INTERACTION OF CHARGED PARTICLES WITH SURFACES

4.1 ELECTRON SCATTERING FROM SURFACES

4.1.1 The energies of scattered electrons

In Chapter 3 the impact of an electron on a gaseous atom or molecule was discussed. As the atoms or molecules are assembled into a solid we may anticipate processes analogous to those occurring in the gas phase but modified by the presence of the ordered structure of the target. Thus elastic reflection of electrons may be expected at low energies (< 20 eV); inelastic processes begin as soon as the incident energy is sufficient to cause processes analogous to excitation and ionization. General reviews are given by McKay (1948), Kollath (1956) and Hachenberg and Brauer (1959). The primary electrons lose energy during this and other inelastic processes and are observed as reflected primaries which have lost energy. The general relationship between the various types of back-scattered electrons is shown in Fig. 4.1 for an

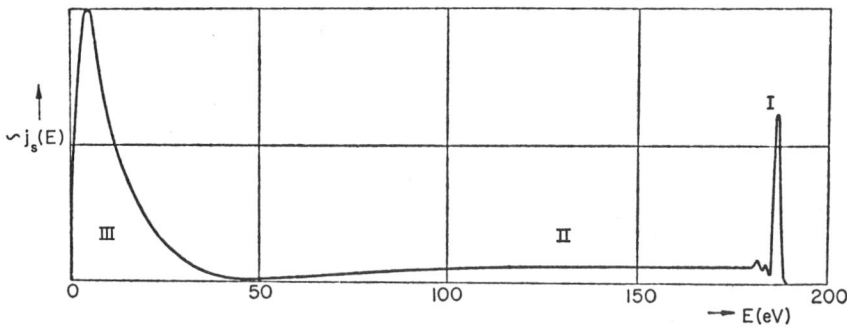

Fig. 4.1 The general shape of the energy distribution of back-scattered electrons. Peak I corresponds to the primary energy of 185 eV. Ordinate is flux of back scattered electrons. (Hachenberg & Brauer 1959).

incident primary energy of about 185 eV, and for a reflected flux integrated over all angles. The elastically reflected primaries are shown as I at the incident energy. The structure to the left of the primary energy represents electrons which have lost certain characteristic amounts of energy (Marton *et al.* 1955; Klemperer & Shepherd 1963; Raether 1965). The magnitude of the characteristic losses is independent of the primary energy, and this group of electrons is considered part of the class II, consisting of electrons which have lost energy in various processes in the solid. III represents true secondary electrons which are arbitrarily defined as electrons with energy below 50 eV. Except at low primary energies ($\lesssim 20$ eV) this group forms the largest fraction of reflected electrons and always has a maximum intensity at a few electron volts. A survey of the magnitudes of the processes exhibited in Fig. 4.1 is the essential content of this section, group I being examined in Section 4.1.2, group II in Section 4.1.3., and group III in Section 4.1.4. The total back-scattering coefficient $\sum$ is defined as the ratio of the total back-scattered current of all energies (i.e. the sum of the currents in I, II and III) to the incident current. The elastic reflection coefficient is defined as the ratio of the current in I to the incident current. It should be noted that the definition of 'elastic' as used in Section 4.1 differs somewhat from that used in Section 4.3.

The parameters to be considered are: the primary energy (0–100 keV), the material of the surface (metal, semiconductor, insulator), the angle of incidence of the primary electrons, the angle of scatter of the reflected electrons, the physical nature of the surface (polycrystalline, single crystal), the cleanliness of the surface (atomically clean or gas-covered) and the surface temperature.

The upper bound of the energy range has been restricted to 100 keV because this is the highest energy at which electrons normally impact on solids in uhv systems. The energy of free electrons in modern accelerators, which have been constructed or are under construction, reaches up to 40 BeV (Neal 1965), but these electrons do not usually strike surfaces. Secondary, but quantitatively important processes, such as the production of radiation, which interacts with solids in the system are treated in Chapter 5. We have thus set the upper bound for discussion at 100 keV, which is the approximate energy range for conventional electron microscopy and electron diffraction, in anticipation that these instruments will, in time, routinely employ uhv techniques (Sewell & Cohen 1965; Grigson et al. 1966; Valdré et al. 1966). The lower bound on electron energies to be discussed will be 0 eV, since many processes such as secondary electron production, photo-emission of electrons, and thermionic emission generate electrons near zero energy. Thermionic emission is of technical importance in uhv systems, but the main features of the interplay between thermionic emission and uhv may be included in a discussion of the reflection of electrons near 0 eV, and we omit any conventional discussion of thermionic emission. The reader is referred to Nottingham (1956) in this regard. The release of molecules and/or ions by electron impact on adsorbed gas layers is considered in Section 4.2 below.

4.1.2 Elastic reflection of electrons

4.1.2.1 General considerations of electron diffraction. A simple account is presented of electron diffraction as used for the study of surfaces. This is a field in which uhv techniques play an important role, and from which many conclusions may be drawn about the behaviour of electrons and surfaces in uhv systems. More extensive accounts of electron diffraction useful for the present purposes may be found in Pinsker (1953), Wood (1964) and Lander (1965). For a non-relativistic electron the electron wavelength is

$$\lambda = \left(\frac{150}{V}\right)^{\frac{1}{2}} \text{Å}, \qquad (4.1)$$

where eV is the kinetic energy in electron volts. Consider a beam of electrons impinging normally on the surface of an ideal crystal Fig. 4.2a which has been assumed to have the simplest cubic symmetry. The electron will be scattered by the atoms of the surface layers, and those that penetrate will also be scattered by the atoms of deeper layers, the magnitude of the scattering

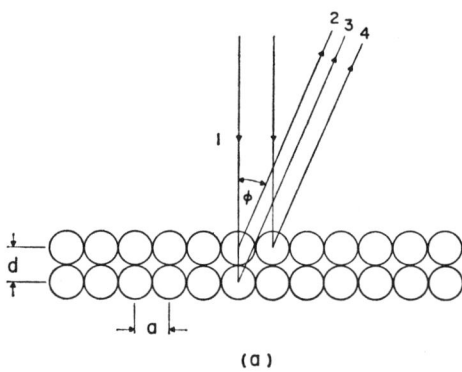

(a)

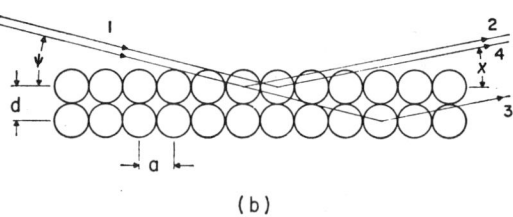

(b)

Fig. 4.2a Scattering of electrons incident normally on the surface of a simple cubic crystal.

Fig. 4.2b Scattering of electrons at glancing incidence on the surface of a simple cubic crystal.

depending upon the identity of the atoms, the depth of penetration, and a number of other factors. In Fig. 4.2a constructive interference between rays 2 and 4 scattered by the row of atoms in the plane of the diagram will take place if

$$a \sin \phi = n_1 \lambda, \qquad (4.2)$$

where $n_1 = 0, 1, 2, 3 \ldots$. Each value of n_1 in eqn. (4.2) defines a cone of revolution about the top row of atoms, the cone for $n_1 = 0$ degenerating into a disc perpendicular to the plane of Fig. 4.2a. For the rows of atoms perpendicular to the plane of the diagram there will be a similar set of cones defined by $n_2 = 0, 1, 2, 3 \ldots$. Constructive interference (diffraction beams) can only occur along lines of intersection of these cones. It is not necessary that the rows defining n_1 and n_2 be at right angles or that the interatomic spacing be the same along both rows. However, only two such rows in the surface are necessary to define the complete set of surface diffraction beams (Wood 1964) which may be derived from a unit mesh, in analogy with a unit cube in three dimensions. If reflection were confined to the surface layer of atoms of the solid, then all these surface beams, each defined by a pair n_1, n_2 would appear simultaneously, but at a given wavelength only some of the surface beams are reinforced constructively by reflections from underlying layers of the solid. If the surface beams in the plane of Fig. 4.2a (i.e. $n_2 = 0$) are to appear, and if the intensity of the reflection from underlying layers is comparable to that from the surface, it is necessary also that rays 2 and 3 reinforce to give

$$d + d \cos \phi = n_3 \lambda \tag{4.3}$$

$n_3 = 1, 2, 3 \ldots$. This third condition does not influence the location of the surface diffraction maxima, but only their intensity. The example of Fig. 4.2a was selected because it represents the geometry used in many modern low energy electron diffraction systems (LEED). Figure 4.3 shows a schematic of such an apparatus. The incident beam travels from right to left; the diffracted beams travel back from the crystal (frequently through field free space), penetrate the two grids which suppress secondary electrons from the crystal, and are accelerated to an energy of about 4 kV before striking the

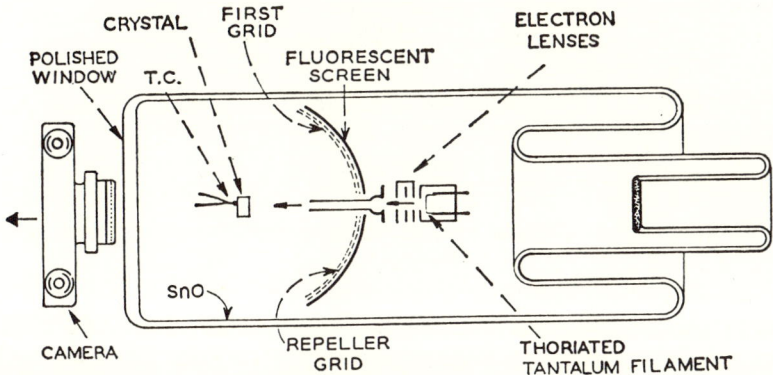

Fig. 4.3 Low energy electron diffraction apparatus. Diameter of screen ~10 cm (MacRae 1963). (*Copyright 1963 by the American Association for the Advancement of Science*).

phosphor screen, where they appear as spots which may be photographed as shown. Typical spot patterns are given later in Figs. 4.8 and 4.9. It is instructive to ask what will be seen on the screen as the incident energy is raised from say 1 eV to 1000 eV, if we assume that the atoms in Fig. 4.2(a) are at the corners of a cube with $a = d = 3$Å. At $V = 1$ volt from eqn. (4.1) $\lambda = 12\cdot3$Å and the only solution of eqn. (4.2) for $\phi < 90°$ is for $n_1 = 0$, $\phi = 0$. Thus the surface diffraction pattern consists only of the 0,0 diffraction beam or the specularly reflected beam. This beam will be modulated in intensity by eqn. (4.3) which becomes $2d = n_3 \lambda$. The 0,0 diffraction beam is not easily observed for normal incidence and may require slight tilting of the crystal (Germer et al. 1966) or an apparatus (e.g. Tucker 1964) of different principle to that of Fig. 4.3. A solution to eqn. (4.2) for $n_1 = 1$ will not be possible until $\lambda = a$, i.e. $V = 16\cdot7$ V. At this voltage the first-order diffraction beam in the plane of Fig. 4.2a (i.e. the beam 1,0) will appear at $\phi = 90°$. The voltage range $\lesssim 20$ eV is characterized by the presence of the 0,0 beam only and this energy region will be examined in Section 4.1.2.2 below. As the energy is raised above 20 eV the beams with $n_1 = 1$ and $n_2 = 1$ move toward the 0,0 beam which remains stationary, and beams with $n_1 = 2$, $3\ldots$, $n_2 = 2, 3\ldots$ appear at $\phi = 90°$ and similarly move toward the 0,0 beam. At $V = 1000$ eV $\lambda = 0\cdot39$Å and diffraction beams with $n < 8$ are present in the plane of Fig. 4.2a. The angular spacing between the 0,0 and the first-order diffraction beam is $\Delta\phi = 7°$. The main crystal property which determines the spatial location of the diffraction beams is the magnitude of 'a' the interatomic distance in the surface of the crystal. If 'a' is doubled in eqn. (4.2) then for given n_1, $\lambda \sin \phi$ is halved. Thus the presence of scattering centres spaced at distances $2a$ on the surface of a crystal will cause twice as many spots in the plane of Fig. 4.2a. A similar multiplication of spots will occur in any plane in which new scattering centres are added. Such additional centres are the frequent result of the adsorption of gas on the crystal surface, and form the basis for many surface structure analyses found in LEED and considered in more detail in Section 4.1.2.3 below.

So far the magnitude of the scattering from each atomic centre of Fig. 4.2(a) has been disregarded. To first-order this follows the elastic scattering of an electron from a gas atom. At electron energies < 20 eV there is usualy a maximum in the total elastic collision cross-section (see Fig. 3.1). In lthis energy range the differential elastic scattering cross-section (angular dependence) is complex showing maxima at various scattering angles (Massey & Burhop 1952). Maxima frequently occur at low scattering angles ($\phi = 180°$) and high scattering angles ($\phi = 0°$) in Fig. 4.2a. These qualitative effects diminish as the electron energy rises and for energies above 1000 eV the scattering is described by the Rutherford formula (Massey & Burhop 1952)

$$I(\phi) \, d\Omega = \frac{Z^2 e^2}{256\pi^2 \varepsilon_0^2 V_p^2} \sec^4 \frac{\phi}{2} \, d\Omega \text{ metre}^2 \qquad (4.4)$$

where $I(\phi)$ is the differential cross-section for scattering into solid angle $d\Omega$, Z, is the atomic number of the scattering atom, e the electronic charge (coulombs), V_p the voltage of the impinging electron, and $\varepsilon_0 = 8\cdot 85 \times 10^{-12}$. Thus as the energy rises the total amount of atomic scattering falls, becoming peaked in the forward direction ($\phi = 0$). Electron diffraction at normal incidence becomes less and less sensitive to scattering by surface atoms as electrons penetrate deeper into the crystal. This tendency may be counteracted by passing over to glancing incidence (Fig. 4.2b) at high energies (so-called Reflection High Energy Electron Diffraction RHEED) (Sewell & Cohen 1965). The set of cones associated with the row of surface atoms in the plane of Fig. 4.2b is defined by

$$a \cos \chi - a \cos \psi = n_1 \lambda \tag{4.5}$$

$n_1 = 0, 1, 2, 3 \ldots$. The set of cones associated with a row of surface atoms perpendicular to the plane of Fig. 4.2b is still defined by an equation like (4.2). At an electron energy of 50 keV $\lambda = 0\cdot 055$Å and the angular displacement between successive orders is about 1°. The spatial resolution of beams with this degree of angular separation requires relatively long flight paths typically of order 1 metre. Angular differences between successive orders in the plane of Fig. 4.2b may in principle be larger than 1° but various effects (Pinsker 1953) tend to obscure clear distinctions between the different orders in this plane.

In this simple account of electron diffraction, refraction at the surface of the crystal (Pinsker 1953), general considerations of the reciprocal lattice (Wood 1964), and phase shifts experienced by the electron when it is scattered (Lander 1965) have been neglected.

4.1.2.2 Elastic reflection at energies < 20 eV. The general relationship between elastic reflection of electrons and secondary emission is shown in Fig. 4.4 which is like Fig. 4.1 but shows the relation between I and III as the incident energy V_p is decreased from 100 V to zero. Note that below $V_p = 5$ V only elastically reflected primaries are present. This result is nearly always found for metals and semiconductors. The depth of penetration of primary electrons into a solid at low energies is uncertain. Shulman and Bazhanova (1967) measure the depth in BaO at $V_p = 5-17$ volts to be 5 atomic layers; Lander (1965) estimates less than 3 atomic layers at 100 volts for most metals; Kanter (1967) measures about 10 atomic layers in gold at $V_p = 1\cdot 1$ volt. The depth is however sufficiently small to make the elastic reflection of slow electron sensitive to surface contamination. It is in this energy range that uhv techniques have made possible for the first time reliable electron impact studies on atomically clean surfaces. The demonstration that a surface is actually atomically clean is a relatively complex task. Hagstrum and D'Amico (1960) concluded that, following thorough outgassing, heating tungsten to 2200°K for a few seconds at $p \sim 10^{-10}$ torr was sufficient to

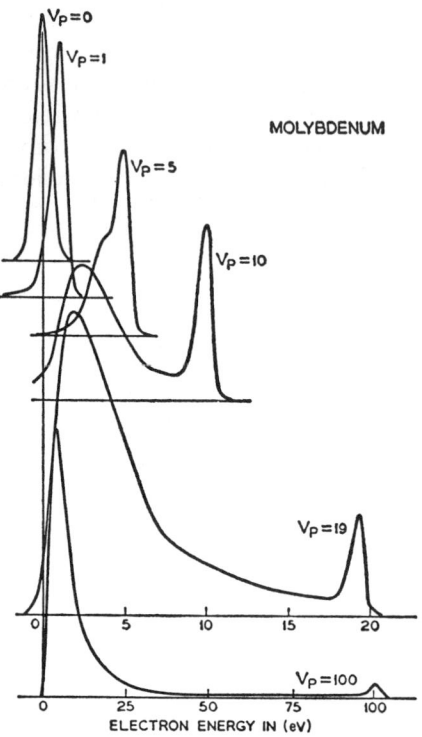

Fig. 4.4 Energy spectra of secondary and reflected primary electrons from a Mo target (polycrystalline) for values of primary voltage V_p as indicated. The ordinate is proportional to the number of electrons per unit energy interval and the abscissa is the energy in electron volts of the electrons as they leave the target surface. Angle of incidence 45°, reflected electrons measured normally to the surface. (Harrower 1956b).

produce an atomically clean surface, but Stern (1964) has shown that carbon impurities in the interior of tungsten diffuse to the (110) surface and create surface structures detectable in LEED. However, the simultaneous study of electron reflection below 20 eV and LEED at somewhat higher energies (Section 4.1.2.3 below) has documented several cases where the reflection of electrons (<20 eV) from atomically clean single crystal surfaces appears to be established. These cases are: W(100), W(112), Khan et al. (1963); Ge (111), Lander and Morrison (1963a); Si (100), Lander and Morrison (1962); Si (111), Lander and Morrison (1963b); Graphite, Lander and Morrison (1964); Pyrographite, Germer et al. (1966); LiF (100), McRae and Caldwell (1964). Reflection at low energies from single crystals of KCl, KBr (Fredericks & Cook 1961) and NaCl (Fredericks & Cook 1963) have yielded results discussed in Section 4.1.3 below but the atomic cleanliness of their surfaces is in doubt.

Figure 4.5 shows the elastic reflection coefficient measured at normal incidence from a tungsten (100) surface both for the atomically clean surface and for the surface covered with at least a monolayer of ambient gas. Zollweg (1964) and Madey and Yates (1967) have confirmed these results for a W(100) crystal obtained from a different source, using a different surface preparation, in apparatus of different principle, and hence the results may be considered to be representative of a clean surface. There are several properties of the result shown in Fig. 4.5 that are general to the results of

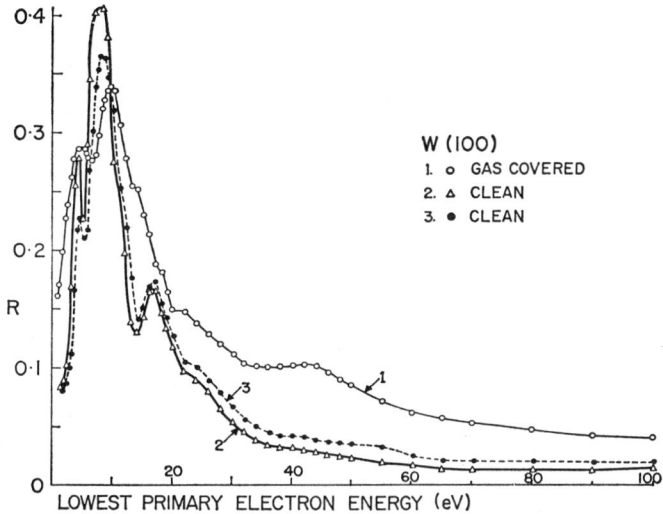

Fig. 4.5 Elastic specular reflection coefficient R for W(100). Normal incidence. (Khan et al. 1963).

all the investigations mentioned above. The elastic reflection coefficient has a magnitude of several tenths below 20 eV and shows a broad maximum in this range, upon which are superimposed a series of subsidiary maxima, associated with the interference of reflections from crystal atoms below the surface layer. These subsidiary maxima depend upon the crystal face being studied and of course upon the chemical identity of the crystal. Figure 4.6 shows Zollweg's (1964) results for elastic reflection from W(100) as the coverage of CO was increased (the discrepancy between the absolute values for the clean surface in Figs. 4.5 and 4.6 probably results from the different definitions of specular reflection used by the respective authors). Note in Fig. 4.6 that the nature of the low energy structure changes as the CO coverage increases but that the general magnitude does not change. The effect of CO on W(110) and W(211) differed in detail only. Results comparable to those of Fig. 4.6 have been published by Armstrong (1966) for H_2 on W(100), W(211), W(111) and W(110) by Kisliuk (1957) for N_2 and O_2 on W(310).

Theoretical treatments of the reflection of slow electrons from a crystal vary, depending upon the model assumed for the surface potential energy of an

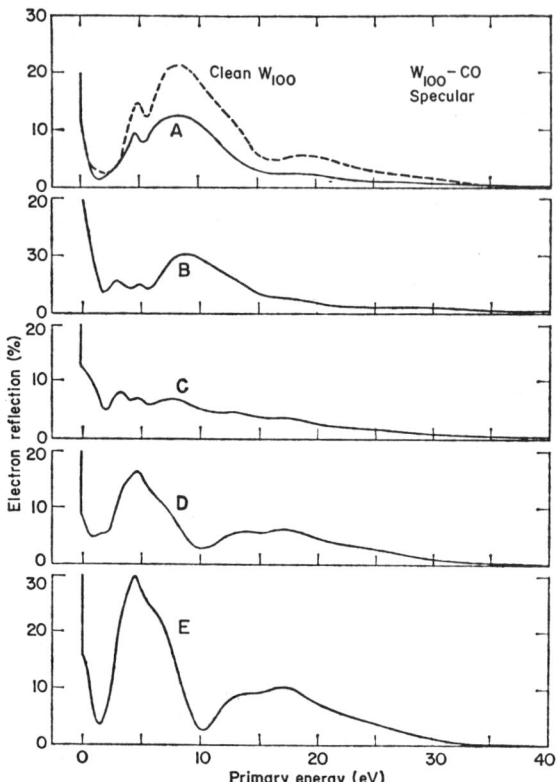

Fig. 4.6 Changes in the specular reflection coefficient for W(100) with CO adsorption at 3×10^{-8} torr. The various curves were obtained at the following times after cleaning by flashing: A, 5 min; B, 15 min; C, 26 min; D, 44 min. Curve E was obtained after raising the CO pressure to near atmospheric and then pumping out to 10^{-5} torr. (Zollweg 1964).

electron. This has ranged from a one-dimensional surface potential (Cutler & Davis 1964) to a completely atomistic potential (Khan *et al.* 1963) (McRae 1966), with various intermediate cases (Lander 1965). A measure of success has been obtained with these treatments but the problem is of considerable complexity (Gafner 1964), and is also in flux at the time of writing, and will not be discussed further here.

The vertical portions of the curves in Fig. 4.6 at 0 eV represent the reflection of electrons by the electrostatic field of the crystal and do not have physical interest in themselves, but the value of the reflection coefficient immediately to the right of the vertical portions, i.e. at $V_p = 0$, is a quantity of physical interest in thermionic emission since it has been shown by Herring and Nichols (1949) that this reflection coefficient is the same as that for electrons leaving the surface. For some years the magnitude of the reflection coefficient at $V_p = 0$ was in some doubt and a summary of the pertinent papers may be found in Kisliuk (1957) but the issue was first conclusively settled experimentally by Shelton (1957). Shelton (1957) measured the reflection by a single crystal of tantalum (211) of an electron beam thermionically emitted by another identical crystal of tantalum, and found that the deviation from an ideally sharp 'knee' in the retarding field plot was only 20 mV wide. This was interpreted as demonstrating that the reflection coefficient at $V_p = 0$ was $\sim 6\%$, a value in agreement with the theoretical predictions of Herring and Nichols (1949). As gas is adsorbed on a single crystal the value of the reflection coefficient at $V_p = 0$ changes but there appears to be no general rule for the direction of change, the result depending upon detailed interference conditions. With adsorption of gas the work function of a crystal face also changes by an amount which depends upon the face, the gas, and the amount adsorbed. In a measurement like that of Fig. 4.6 this change of work function is observed as a shift in the measured location along the energy axis of the vertical portion of the reflection coefficient. Thus most experimental arrangements for measuring the reflection of slow electrons are suitable for measuring the changes in work function which occur when gas is adsorbed on a surface. For example, Shelton's (1957) arrangement is suitable for the measurement of the work function of both emitter and collector (Love & Wilson 1967; Madey & Yates 1967). Table 4.1 is a compilation of the changes in work function of W(110) and W(100) resulting from adsorption of CO, N_2, O_2 measured with uhv apparatus. Note that either sign is possible.

There have been relatively few direct measurements of the elastic reflection

TABLE 4.1
Effect of adsorbed gases on the work function of W(110) and W(100).
Values given are largest changes observed in eV.
First values are those measured by Madey and Yates (1967)

Crystal surface	Gas		
	CO	N_2	O_2
W(110)	+0.20	+0.03	+1.2
	+0.42 [a]	0 [a, b, c]	+1.15 [d]
			+0.62 [a]
			+0.85 [e]
W(100)	+0.50	−0.57	+1.5
	+0.43 [f]	−0.4 [b]	+1.6 [d]
	+0.3− 0.4 [g]	−0.65 [h]	+1.65 [e]
		−0.33 [c]	
		−0.9 [i]	

[a] Stern (1965); [b] Delchar and Ehrlich (1965); [c] Van Oostrom (1966); [d] Hopkins and Pender (1966); [e] Zingerman et al. (1966); [f] Anderson and Estrup (1967); [g] Armstrong (1965); [h] Estrup and Anderson (1967); [i] Holscher (1964).

(From Madey and Yates 1967).

coefficient on atomically clean polycrystalline metal surfaces (e.g. polycrystalline W (Morgulis & Gorodetskii 1956)) partly because the advent of uhv techniques was soon followed by the ready availability of single crystals. However, measurements of the total back-scattering coefficient Σ below the threshold for secondary emission, which is several electron volts, provides a reliable measure of the elastic reflection coefficient (R). Measurements of this type are those of Holzl (1965) for polycrystalline Pt; Fowler and Farnsworth (1958) for polycrystalline Pt; Hobson (1956) for polycrystalline Ta and Cu. The result of measurements of R on polycrystalline metals are essentially like the results of Fig. 4.5 and 4.6 but the detail of the structure is of course different, and cannot be interpreted simply.

The reflection of electrons from gas covered polycrystalline surfaces is the most usual situation in uhv systems where most surfaces are not atomically

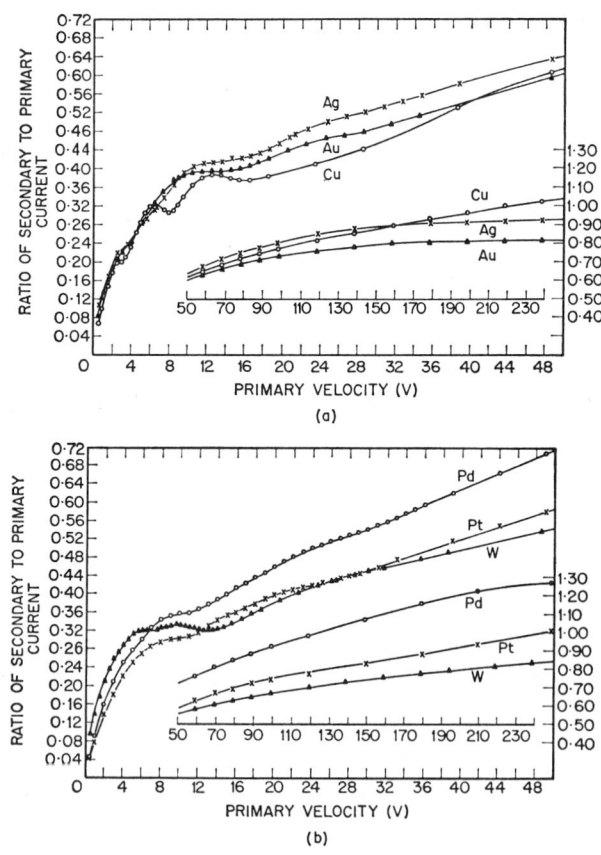

Fig. 4.7a Total back-scattering coefficient for Cu, Ag, and Au after red-heat treatment.
Fig. 4.7b Total back-scattering coefficient for W, Pt, and Pd after red-heat treatment (Farnsworth 1925).
Primary Velocity expressed in volts.

clean. Much early work from the 1920's and 1930's is of value in this regard since prepared surfaces in these experiments were probably comparable to typical component surfaces in modern uhv systems. Figure 4.7a and 4.7b show some early measurements of Σ by Farnsworth (1925) for electron reflection from polycrystalline surfaces of Cu, Ag, Au, W, Pt, Pd. Note the general similarity of all the curves.

4.1.2.3 Low energy electron diffraction (LEED). LEED employs the elastic reflection of electrons mainly in the energy range from 20 to 300 eV with diffraction beams of low order excluding the 0,0 diffraction beam. The latter beam, which has been discussed in some detail in Section 4.1.2.2 above, is excluded because the position of this beam is independent of surface structure, whereas the position of all other beams depends on the spacing of scattering centres in the surface as explained in Section 4.1.2.1 above. Figure 4.8 is based on a diagram presented by Germer (1965) and shows the surface atoms in three surfaces of the face-centred crystal nickel, which has been the subject of many LEED studies (MacRae 1964a; Farnsworth & Tuul 1959; Farnsworth 1963; Park & Farnsworth 1964a; and other papers from these groups). Figure 4.8b shows the unit mesh in each surface. By convention (Wood 1964) the shortest dimension of the unit mesh is defined as having a unit vector denoted by [1,0] pointing down. The three faces illustrate hexagonal (unit vectors at 120°), square, and rectangular unit meshes for Ni(111), Ni(100) and Ni(110) respectively. Wood (1964) shows that these represent three of only five possible surface meshes for 2-dimensional periodic surface structures. When electrons are incident normally on these surfaces oriented as in Fig. 4.8b then the spot patterns showing in Fig. 4.8c arise where the pairs of numbers represent the order of the diffraction beam arising from the [1,0] and [0,1] axes of the unit mesh respectively. Thus all the spots $0,n_2$ will be found at right angles to the [1,0] axis and all spots $n_1,0$ will be at right angles to the [0,1] axis. For a rectangular unit mesh the spacing of spots $0,n_2$ is inversely proportional to the interatomic distance along the [0,1] direction as given by eqn. (4.2). Thus even the symmetry of the spot pattern gives sufficient information for determination of the type of unit mesh while the determination of angle ϕ, along with a measurement of the electron energy (λ), is sufficient for a determination of the lengths of the vectors in the unit mesh. Frequently in modern LEED apparatus the 0,0 beam is obscured more than is shown in Fig. 4.8c but the other spots are usually quite clear for a good clean surface. From patterns like those shown in Fig. 4.8 it has been established that the surface layers of clean Ni, W and graphite are replicas of layers of the same orientations in the crystal while surface layers of Si and Ge form more complex structures than the underlying layers. The measurement of spacing between the surface layer and deeper layers however rests upon an analysis of intensity, to first order through eqn. (4.3), and here we do not find complete agreement between different investigators. MacRae and Germer (1963) conclude that

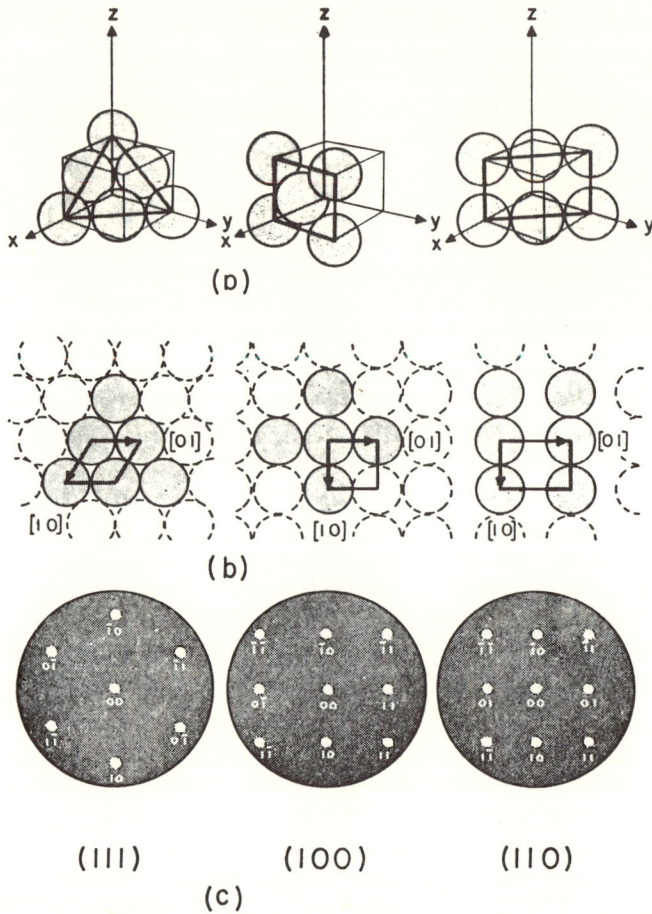

Fig. 4.8(a) Atoms in the surface planes of Ni(111), (100), (110). Ni is a face centred cubic crystal with cube side 3·52 A. (b) Axes of the unit mesh in the plane of the surface define the directions of the rows of closest packed atoms in the surface (all atoms shown are in the surface). (c) Diffraction patterns observed in LEED when an electron beam of energy about 85V is incident normally on the surfaces shown in (b). Typical spot size diam 1–2 mm. (Figure based on a diagram by Germer (1965)).

the normal bulk spacings between (110) and (111) planes of Ni are increased by 5% at the surface, while Park and Farnsworth (1964a) find this spacing to be the same as in the bulk crystal within an experimental error of about 2%.

The preparation of samples to yield spot patterns like those shown in Fig. 4.8 may be complex. Three methods have been used:

(i) *Heating*. This can be successful if surface gas can be desorbed, and gas in the bulk of the sample removed, as a temperature below the melting point of the crystal (e.g. W). Results depend rather critically on the purity of the sample. For example, Khan *et al.* (1963) were able to obtain clean surface spot patterns by heating their W(100), (112),(110) crystals to 2200°K whereas Stern (1964) found that an oxygen exposure of $1 \cdot 4 \times 10^{-2}$ torr/sec at 1550°K and 10^{-1} torr/sec of H_2 exposure at 1550°K was required to remove carbon contamination from his crystal W(110). MacRae (1964a) was unable to clean Ni crystals by heating to 1453°C, the surfaces always being characterized by carbon contamination.

For the lower melting point crystals one of the two methods given below is used.

(ii) *Outgassing, ion bombardment, annealing.* Farnsworth *et al.* (1958) (also other papers) have developed this method and have studied it extensively. For Ti, Ge, Si, Ni (Farnsworth *et al.* 1958) their procedure varies somewhat with each crystal but a typical schedule was as follows: Outgas crystal continuously 700–1000°C for 50 to 100 hours until the system pressure with the crystal hot was below 10^{-8} torr. The crystal was then bombarded with argon ions in a pressure of 10^{-3} torr, at 200–600 V, at a current density of $100 \mu A/cm^2$ for a period of about 5 minutes. It was then annealed at about 500°C for 10 minutes. This cycle of ion bombardment and annealing was repeated until the LEED diffraction pattern was that of a clean surface. This took some 10 cycles. After further repetition of the same cycle 10–15 times the crystal conditions was maintained from 1 day to 1 month. Park (1966) concludes that a LEED study of a Ni(100) damaged by argon bombardment can distinguish between strain resulting from an inbedded ion and that resulting from a vacancy in the lattice.

(iii) *Surface chemical reactions.* Germer and MacRae (1962) developed a chemical method which they have used extensively for cleaning nickel crystals. Ni(110) contaminated with carbon was heated to 500°C in O_2 at a pressure of 1×10^{-6} torr. This produces one of the Ni—O forms summarized in Table 4.2. It is then necessary to heat the crystal in H_2 at a pressure of 10^{-7} torr to produce a clean surface.

General discussions of the preparation of clean surfaces in uhv have been given by Dillon (1961) and Roberts (1963).

When a gas is chemically adsorbed on a surface the spot patterns of the clean surface frequently change. An example is shown in Fig. 4.9a after O_2 is admitted to a LEED system containing Ni(110), whose clean spot pattern is given in the right-hand lower corner of Fig. 4.8. The effect of the O_2 is to double the number of spots $n_1,0$ in such a way that spots $n_1 + \frac{1}{2},0$ appear. These are called half-order spots and are a common characteristic of LEED spot patterns. They are interpreted as arising from a doubling of the dimension of the unit mesh as shown in Fig. 4.9b. Such a surface structure is termed a 2×1 mesh, the [1,0] vector having been doubled in length while the [0,1] remained the same. The interpretation of the physical cause of this doubling is, however, a subject of controversy. Figures 4.9b, 4.9c and 4.9d show surface structures that have been proposed to explain the spot pattern of Fig. 4.9a. Structures represented by Figs. 4.9b and 4.9c are both examples

TABLE 4.2
Surface structure arising from chemisorption of oxygen on nickel as determined by LEED

Structure	Fractional oxygen coverage (θ)	Approximate oxygen sticking probability (s)	Approximate work function increase due to oxygen (V)	Approximate temperature to degrade to structure of lower oxygen content (°C)
Ni(111) clean	—	1		—
Ni(111) (2 × 2)	1/4	0·01	1·2	m.p. [b]
Ni(111) (3 × 3) R(30°) [a]	1/3			400
Ni(001) clean	—	0·04–1		—
Ni(001)p (2 × 2)	1/4	0·02	0·25	m.p. [b]
Ni(001)(c)(2 × 2)	1/2		0·2(0·45) [c]	700
Ni(110) clean	—	1		—
Ni(110) (2 × 1)	1/2	0·2	0·6	m.p. [b]
Ni(110) (5 × 2)	3/5	0·05		800
Ni(110) (3 × 1)	2/3	0·01	0·6(1·2) [c]	350
Ni(110) (5 × 1)	4/5			250

(a) $R(30°)$ means the structure is rotated 30° with respect to the substrate mesh.
(b) Melting point. This temperature is reduced when the crystal is not saturated with O_2 (interpreted as solution of O_2 in Ni).
(c) Figures in brackets are total work function changes starting from clean surface.
(from MacRae 1964a).

of surface 'reconstruction', i.e. the nickel atoms have *moved* from their original positions in the Ni(110) surface. Bauer (1965) summarizes the arguments pro and con for these interpretations. Probably the most important question is: What is the identity of the scattering centres when O_2 interacts with Ni(110)? MacRae (1964a) argues that an oxygen atom ($z = 8$) should scatter much less than a nickel atom ($Z = 28$) since to first-order the scattering will follow a relation like eqn. (4.4). Thus in any mixture of Ni and O atoms in equivalent positions in a surface LEED will detect only the Ni atoms. Bauer (1965), on the other hand, argues that the oxygen is effectively ionized by the surface and the scattering cross-section of an ion is much larger than that of the neutral atom. If one adds to this the assumption that the oxygen atom sits on the surface then Bauer (1965) concludes that it is at least as likely that LEED is sensitive to O atoms as to Ni atoms. The structure proposed by Farnsworth (1963) (Fig. 4.9c) involves many more O atoms than that of Fig. 4.9b although the surface layer is similar. Farnsworth's (1963) conclusion is based in part on the relatively high oxygen exposure 240×10^{-6} torr/sec (Park & Farnsworth 1964b) to which the Ni surface must be subjected to yield the 2 × 1 structure, while Germer and MacRae (1962) find an exposure of 1×10^{-6} torr/sec is sufficient to produce this structure. At the time of writing the controversy concerning reconstruction continues (see Germer 1966; Bauer 1966)

Interaction of Charged Particles with Surfaces 149

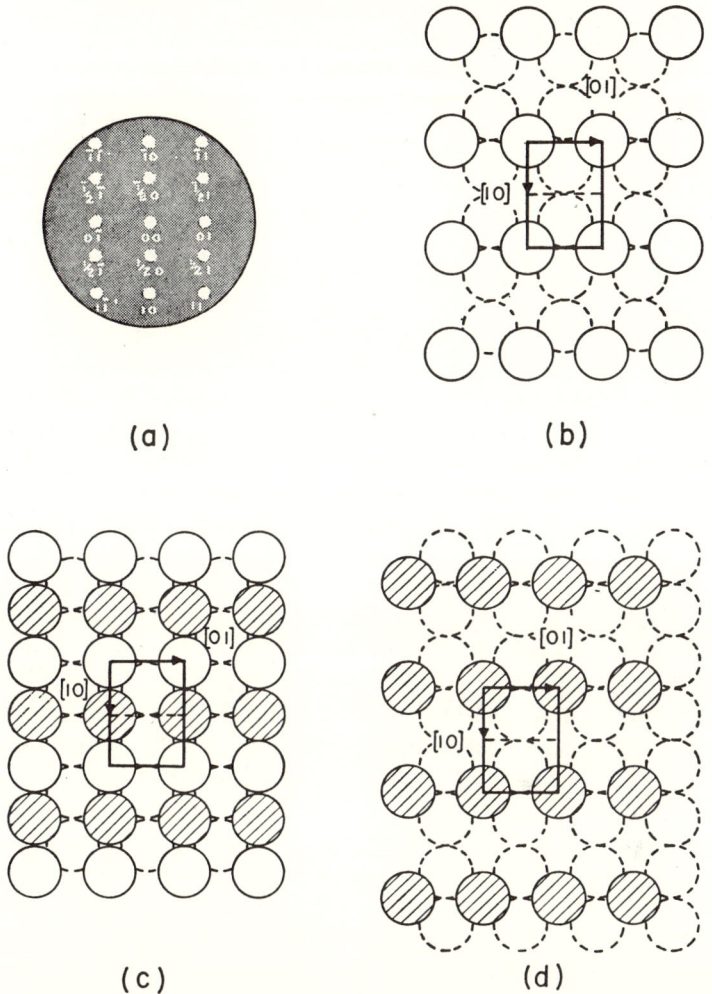

Fig. 4.9(a) Spot pattern from Ni (110) at early stage of oxygen coverage. (b) Surface structure suggested by MacRae (1964a). Circles with solid perimeters are Ni atoms in surface. Circles with dashed perimeters are Ni atoms in second layer. Oxygen atoms are not shown but are assumed located in surface layer between Ni atoms, causing $\theta = 1/2$ for this Ni(110) (2 × 1) surface structure. (c) Surface structure suggested by Farnsworth (1963). Cross-hatched circles are oxygen atoms. This is similar to (b) except that this structure is assumed to be the (110) surface of a crystal of Ni_3O extending some distance into the crystal, i.e. not restricted only to the surface layer. (d) Surface layer of oxygen atoms (cross-hatched) with no nickel atoms in top layer suggested by Bauer (1965). The oxygen atoms are supposed to be scattering elements of the surface. $\theta = 1/2$. (Bauer 1965)

and is not limited to Ni but applies to W as well, and may possibly be extended to other substrates. This uncertainty exists for room temperature for the Ni + O system which has received extensive study. If the substrate temperature is raised, a complex series of other surface structures may be observed. Table 4.2

is taken from MacRae (1964a) and illustrates his conclusions concerning the oxidation of three faces of Ni by oxygen. LEED is sensitive to surface coverages $\theta \gtrsim 0\cdot01$ (Lander 1965), but is normally used in the range $0\cdot1 < \theta < 1$ as indicated in Table 4.2. Sewell and Cohen (1965) have studied the Ni+O system by RHEED and obtain substantial agreement with MacRae (1964a) for the Ni(110)+O system. It may be noted that 50 keV electrons do not destroy the surface structures found in LEED instruments, at least on Ni(110).

The number of gas-crystal combinations that have been studied in LEED apparatus is already large and increasing rapidly. The results are similar in general to those discussed for O_2+Ni(110) above, but the specific details depend upon the gas-crystal combination and the pressure and temperature conditions. For a bibliography to 1965 the reader is referred to Bauer (1965), and for publications since 1965, to many current journals. The largest number of systems that have been studied involve O_2 which is a difficult gas to handle quantitatively in an uhv system because chemical reactions take place readily on various parts of the system (Section 7.2). Most studies to date have been either clean-surface studies or chemisorption studies. LEED has also been applied to physical adsorption on graphite (Lander & Morrison 1967). This work is at an early stage, but the results for the physisorption of Xe on graphite at $T = 90°K$ at $p = 10^{-3}$ torr can be compared with some of the results presented in Chapter 2 for this system, based on quite different considerations. Near a monolayer the xenon was found to form an ordered structure having a unit mesh $\sqrt{3} \times \sqrt{3}$ relative to the graphite mesh and rotated by 30°. The area occupied by the xenon atom is largely determined by the graphite substrate and is $15\cdot7Å^2$. This is somewhat less than the value of $18\cdot8Å^2$ given for Xe in Table 2.17 for packing in a liquid surface, indicating lateral compression of the xenon layer. Raising the substrate temperature 30°C caused the xenon spots to become fuzzy, then to smear into fuzzy rings, to fade, and finally to leave only the undisturbed graphite spots. This behaviour was interpreted as a transition from a two-dimensional (2D) structure (crystal), to a 2D liquid phase, to a 2D gas phase, to desorption. At the low temperatures of the sample holder a background pressure of 10^{-10} torr was essential to prevent contamination of the clean graphite.

Lander and Morrison's (1967) results on the study of physisorption by LEED methods establish that the electron beam does not desorb physisorbed layers, a point on which there had been some discussion (Menzel 1965; Tucker 1965).

Another parameter that can be studied by LEED is the effect of the crystal temperature on the patterns (MacRae & Germer 1962; MacRae 1964b). Figure 4.10 shows the intensity of the 1,0 beam from a Ni(110) surface at two electron energies. The intensities of both maxima fall as the crystal temperature rises but the characteristic temperatures are both below the Debye temperature for Ni which is 390°K and would have yielded the dashed lines. This result is interpreted as indicating that the surface atoms vibrate with greater amplitude than the bulk atoms, an interpretation in agreement with the result shown in Fig. 4.10 that the more highly penetrating electrons (higher voltage) yield a result nearer that expected for bulk atoms. The deviation from linearity of the lower plot from a straight line may be caused by anharmonicity of the vibrations.

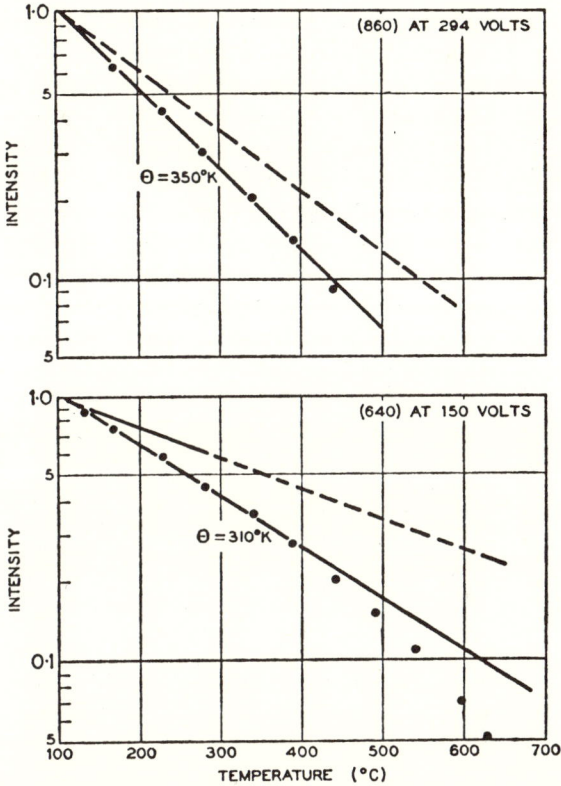

Fig. 4.10 Temperature variation of the intensity of the 1,0 beam from a (110) surface of nickel. The figures (860) and (640) represent Miller indices of maxima being measured. The dashed curves represent the dependence expected for highly penetrating radiation. (MacRae & Germer 1962).

The absence of spots in a LEED pattern indicates a polycrystalline or amorphous surface. A transition from a spot pattern to no pattern at all indicates the build-up of such a layer and such a transition is frequently observed both in chemisorption (Lander 1965) and physisorption (Lander & Morrison 1967).

The majority of LEED studies to date have used normal incidence of electrons, but apparatus for display of the diffraction patterns with angles of incidence from 0 to 90° has been constructed (Fujiwara et al. 1966) and a whole new range of data is thus made available.

4.1.3 Characteristic energy losses and rediffused electrons

The structure to the left of the elastically reflected peak in Fig. 4.1 represents reflected electrons which have lost specific amounts of energy (Marton *et al.* 1955; Klemperer & Shepherd 1963; Raether 1965). The difference $eV_p - eV_s$, where eV_p is the primary energy and eV_s the energy of the reflected electrons is independent of the primary energy and is characteristic of the target material, but the intensity and angular dependence of the loss peaks do depend, in general, upon the incident impact conditions. The energy range $e\Delta V = eV_p - eV_s$ for which characteristic peaks are resolved experimentally is about 0–100 eV. Above this range inelastically scattered electrons are spread out in energy and constitute the remainder of region II in Fig. 4.1. Electrons which have lost characteristic amounts of energy are the result of a process in which a primary electron transfers a specific amount of energy to the solid. A variety of such processes are possible and the cross-section for each depends upon the impact energy. At low impact energies (<10 eV) energy can be transferred to the vibrational state of adsorbed molecules. Figure 4.11 shows

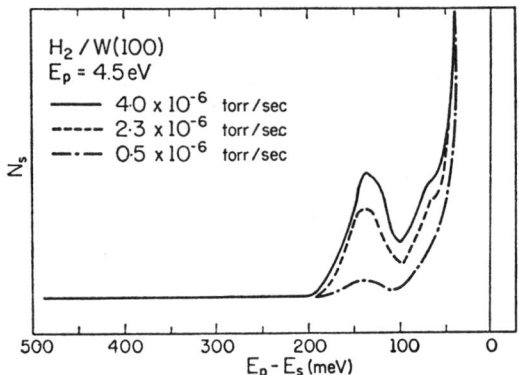

Fig. 4.11 Characteristic energy loss spectra for hydrogen adsorbed on W(100). $E_p = eV_p$; $E_s = eV_s$. (Propst & Piper 1967).

the structure in the reflected beam when electrons with energy 4·5 eV strike at angle of incidence of 45° on a W(100) crystal with hydrogen adsorbed on it. The intensity of the characteristic loss peaks is about 1/100 that of the elastic peak (at $V_p - V_s = 0$). Propst and Piper (1967) interpret their results as indicating that hydrogen is adsorbed as atoms and that each atom is bonded to several W atoms. Similar data were obtained for N_2, CO and H_2O adsorbed on W(100). Schulz (1962) has also reported a characteristic energy loss of 0·25 eV arising from N_2 adsorbed on a slit in his apparatus for measuring vibrational excitation in the gas phase (see Fig. 3.10). The observation of the results shown in Fig. 4.11 requires an electron spectrometer of high resolution and in most apparatus the subsidiary peaks will be lost in the

Interaction of Charged Particles with Surfaces

relatively large elastically reflected peak. The presence of resonant energy transfer in the solid may also be detected as a characteristic change in the absorption coefficient of the solid for electrons with energy. Fredericks and Cook (1961, 1963) have measured the total absorption of electrons from 0·2 to 12 eV impinging normally on the surface of cleaved KCl, KBr and NaCl and have found several prominent characteristic features which they associate with electron interaction with F and V_3 centres in these crystals, since the energies were correlated directly to optical absorption energies. Measurements with insulating crystals of this type require precautions described by Fredericks and Cook (1961, 1963) to prevent charging of the surfaces.

Characteristic energy losses were first discovered in the primary energy range 50–400 volts by Rudberg (1936) for films of the solids Cu, Ag, Au, Ca, Ba, CaO, BaO. Other workers (see Klemperer & Shepherd 1963 for bibliography) have repeated these experiments for other materials, and experimentally the field has divided today into two main streams, the reflection of electrons from solids (750–2500 eV) and the transmission of electrons through suspended thin films (2–100 keV). The results from the former technique are of more direct interest in uhv systems, since the thin films involved in the latter technique require special preparation. Although differences exist between the results of the two methods these are primarily of detail (Klemperer & Shepherd 1963).

Figure 4.12 shows characteristic spectra measured in reflection from an Al surface prepared by evaporation. Al has received more study than any other solid because its characteristic spectra consists of the well defined peaks shown in Fig. 4.12. Note that the energies of the characteristic loss peaks are independent of the primary energy but the amplitudes are not; the intensity of the highest loss peak is some 40 % of the elastically reflected beam at $V_p = 2020$ eV. Two peaks, one at 10·3 eV and one at 15·3 eV, dominate the spectra of Fig. 4.12. The loss at 15·3 eV is identified with a loss to the plasma oscillation of the electron gas in the bulk of the sample. The expected energy of this plasma loss has been derived by Bohm and Pines (1953) to be

$$\hbar\omega_p = \frac{h}{2\pi}\left(\frac{4\pi e^2 n_0}{m}\right)^{\frac{1}{2}} = 3\cdot71 \times 10^{-11} n_0^{1/2} \text{ eV}, \tag{4.6}$$

n_0 is the density of conduction electrons per cm^3 (equal to the density of atoms × valence per atom). The atomic density for Al (6×10^{22} atoms cm^{-3}) and a valence of 3 yields $n_0 = 1\cdot81 \times 10^{23}$ electrons cm^{-3}. Substitution into eqn. (4.6) yields $\hbar\omega_p = 15\cdot8$ eV in good agreement with the observed value of 15·3 eV. If the characteristic loss electrons arise near the surface of a flat solid a second (surface) plasma oscillation is present with an energy given by $\hbar\omega_p\sqrt{2}$ (Ritchie 1957). This would predict a loss in Al of $15\cdot3/\sqrt{2} = 10\cdot8$ eV, also in reasonable agreement with the observed loss of 10·3 eV.

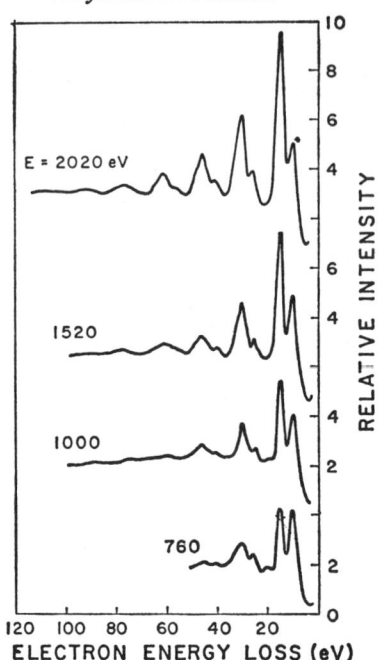

Fig. 4.12 Characteristic energy loss spectra for aluminium for primary electron energies of 760, 1000, 1520 and 2020 eV. The peak of elastically scattered electrons (not shown) has been adjusted to an intensity of 25 units in each case. Primary beam struck the target at angle of 45° and reflected beam was measured at a total deflected angle of 90°. $P \sim 2 \times 10^{-6}$ torr. Target temp. $\sim 200°C$. Primary beam current ~ 1.5 mA/cm^2. (Powell & Swan 1959).

This loss is called a lowered plasma loss. Powell and Swan (1959) show that all the losses of Fig. 4.12 are simple combinations of multiples of the plasma and lowered plasma losses. Klemperer and Thirwell (1966) have shown that volume and surface plasma oscillations in Al can be distinguished by their dependence on the angle of incidence. At glancing incidence the volume plasma loss was found to be greatly reduced, while the surface plasma loss was increased slightly. Powell and Swan (1960) have shown that surface oxidation of Al and Mg eliminates the lowered plasma loss. Oxidation effects may explain some of the discrepancies which exist in the literature on characteristic energy losses (Klemperer & Shepherd 1963). While the Al spectrum is dominated by plasma losses this is not the only mechanism giving rise to characteristic losses. Figure 4.13 shows the losses measured in reflection by Robins and Swan (1960) for the transition metals Ti to Cu. At the side of the figure and in the caption are given the mechanisms assigned to these losses. Only losses b, c, e, are assigned to plasma losses although the main plasma loss at about 20 eV is the major peak in most of the spectra. The losses assigned to shell ionization are those arising from the ejection of an electron from an atom of the solid rather than an interaction with the

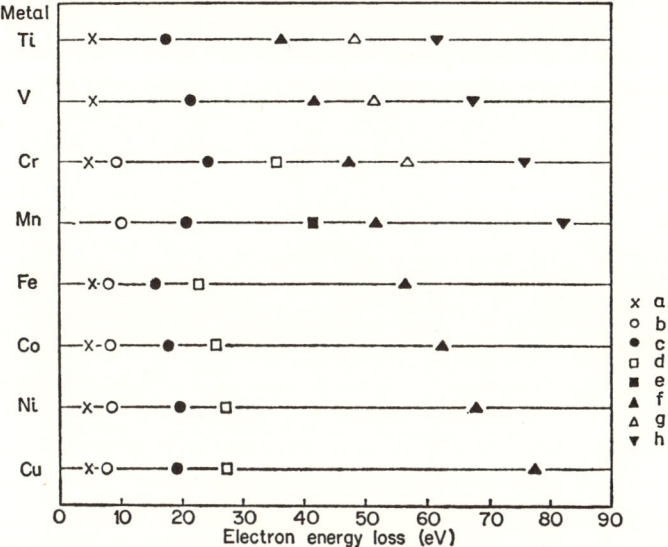

Fig. 4.13 Schematic diagram of characteristic energy losses observed in the transition metals Ti to Cu. Primary energy was 1200 eV. The interpretation of the losses is as follows: (a) $M_{4,5}$ shell ionization, (b) lowered plasma loss, (c) plasma loss, (d) combination or interband transition, (e) multiple of plasma loss, (f) $M_{2,3}$ shell ionization, (g) unidentified, (h) M_1 shell ionization. (Robins & Swan 1960).

solid as a whole. Shell ionization is therefore a potential source of secondary electrons, discussed in Section 4.1.4 below. Interband transitions represent the moving of an electron between two energy levels within the solid without ionization. They correspond to excitation of a gaseous atom. The interpretation of the results of Fig. 4.13 shows that a variety of types of transitions may cause characteristic losses, some of these losses being associated with the collective properties of the solid, others being associated with the atomic properties of the atoms making up the solid. Evidence that the latter can dominate characteristic spectra was provided by Boersch *et al.* (1962) who measured the characteristic losses in Hg as vapour, a liquid, and a solid. The result is shown in Fig. 4.14 and it is clear that the main characteristics of the loss spectrum for the vapour are maintained in the condensed phases. Powell (1965) has found that the characteristic loss spectra for solid and liquid bismuth are similar for total scattering angles of 6° and 70° at the incident energy of 8 keV.

Thus, while the phenomenon of characteristic energy losses is general, the specific results for each material differ. In particular, a large number of materials have been studied in reflection at primary energies 750–2500 eV at a single Australian laboratory (Swan 1964; Hartley & Swan 1966; and preceding papers).

This valuable body of reflection data has two deficiencies. Firstly, back-

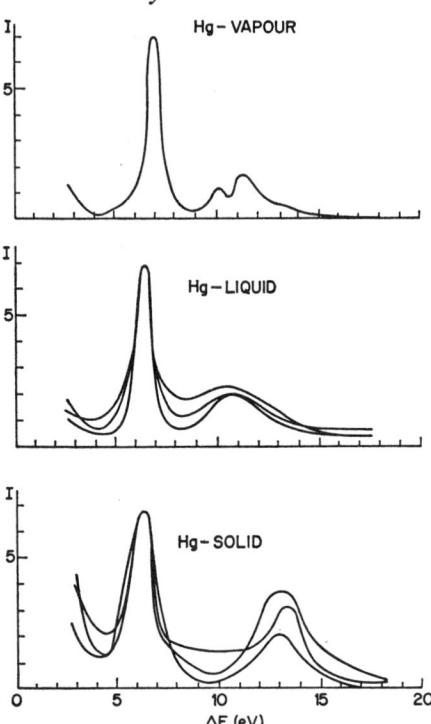

Fig. 4.14 Characteristic energy loss in mercury vapour, liquid and solid measured in transmission at primary energy of 25 keV. The three curves for the condensed phases are for three samples. Ordinate is relative number transmitted with indicated loss. (Boersch *et al.* 1962).

ground pressures were about 10^{-6} torr; the results could be interpreted more decisively if uhv techniques had been used. Secondly, very little information has been given on the magnitude of the reflected beam relative to the incident beam (i.e. the effective reflection coefficient).

In principle the reflection technique suffers from the deficiency that the depth of penetration of the electrons into the solid is not known, hence the cross-sections for the various processes cannot be measured. Also angular dependence of the peaks representing the characteristic losses is not measured. In these respects the transmission technique is superior, and a greater literature exists for transmission measurements than for reflection, although here too emphasis has been concentrated on the values of the characteristic energy losses rather than their cross-sections. However, Swanson and Powell (1966) measure $(1 \cdot 5 \pm 0 \cdot 4) \times 10^{-18}$ cm^2 and $(1 \cdot 5 \pm 0 \cdot 5) \times 10^{-18}$ cm^2 as the cross-sections for the excitation of the 15 eV and 19 eV losses in Al and Be respectively. These authors also give the plasma loss in Al

$$e(V_p - V_s) = 14 \cdot 8 + (1 \cdot 39 \pm 0 \cdot 07) \times 10^4 \phi^2 + (1 \cdot 2 \pm 0 \cdot 2) \times 10^7 \phi^4 \text{ eV} \quad (4.7)$$

where ϕ is the angular displacement from the primary direction measured in radians. There is a sharp decline in intensity as ϕ increases and a complete extinction at a critical value of

$$\phi_c = \frac{k_c}{k} \tag{4.8}$$

where $k_c = 1\cdot46 \pm 0\cdot07\,\text{Å}^{-1}$ and $k = 2\pi/\lambda$ where λ is the wavelength of the incident electron. At an incident energy of 20 keV, $k = 72\cdot5\,\text{Å}^{-1}$ and hence $\phi_c = 0\cdot02$ radians $= 1\cdot15$ degrees. This example shows that the measurement of results expressed by eqn. (4.7) requires very high resolution in energy. Boersch et al. (1966) have reported an energy resolution of 0·07 eV in 30 keV with an analyser aperture of 10^{-4} radians.

Region II in Fig. 4.1 also contains a continuous range of energies associated with 'rediffused' electrons, extending from the characteristic losses down to 50 eV. The latter limit has been arbitrarily adopted as the upper limit for true secondary electrons (see Section 4.1.4 below). The current in regions I and II of Fig. 4.1 as a fraction of the primary current is defined as the rediffusion coefficient η and is the quantity to be discussed next. Gomoyunova and Letunov (1966) use a more refined definition of η, which yields a value 30–40% below that given by the 50 eV cut-off, but the more traditional definition is adopted here.

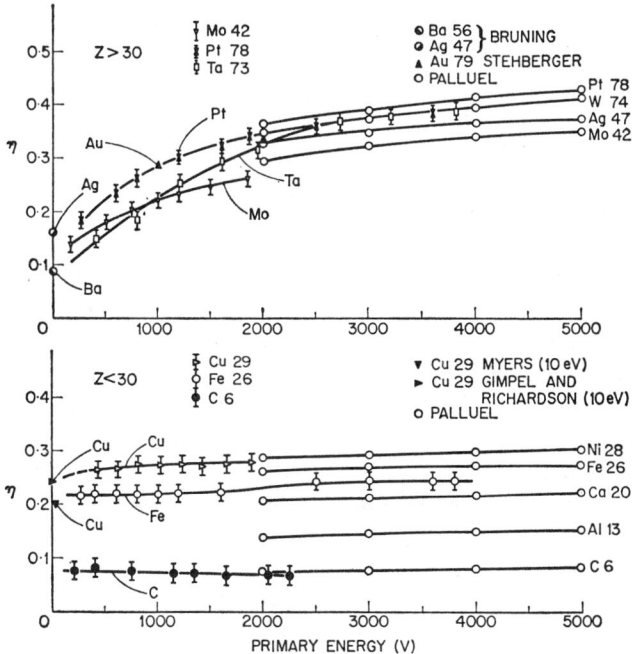

Fig. 4.15 Variation of rediffusion coefficient η with incident energy. Upper diagram – elements of $Z>30$. Lower diagram – elements of $Z<30$. References are: Bruining (1938); Gimpel and Richardson (1943); Myers (1952); Palluel (1947); Stehberger (1928); barred symbols Sternglass (1954). (From Sternglass 1954).

Figure 4.15 shows the variation of η for several metals for incident electron energy up to 5000 eV. Sternglass (1954) divides these metals into two groups with atomic number Z above and below 30 respectively. For $Z \gtrsim 30$, η rises from a value about 0·1 to 0·2 to a limiting value of about 0·5 at an incident energy of several hundred keV. Below 2 keV the change in η is particularly marked. For $Z \lesssim 30$ η is approximately independent of primary energy, showing no decline at 2 keV. Below 2 keV η for $Z \lesssim 30$ can be greater than η for $Z \gtrsim 30$. Limiting values for η at high primary energies are shown in Fig. 4.16 where it is demonstrated that the limiting value is a smooth function of the atomic number, Z, and that this is true also for insulators provided a reasonable value of mean atomic number can be defined. This result is confirmed for 15 alkali-halide compounds by Gomoyunova and Letunov (1966). By definition η contains the elastically reflected peak I of Fig. 4.1 but the fraction of the reflected current in I is small at energies in the kiloelectron volt range (Sternglass 1954). The angular dependence of rediffused electrons from bulk solids is cosine and is insensitive to the angle of incidence (Kanter 1957). The energy of the rediffused electrons is spread over a wide range with some preference for energies near the primary energy (Sternglass 1954; Gomoyunova & Letunov 1966).

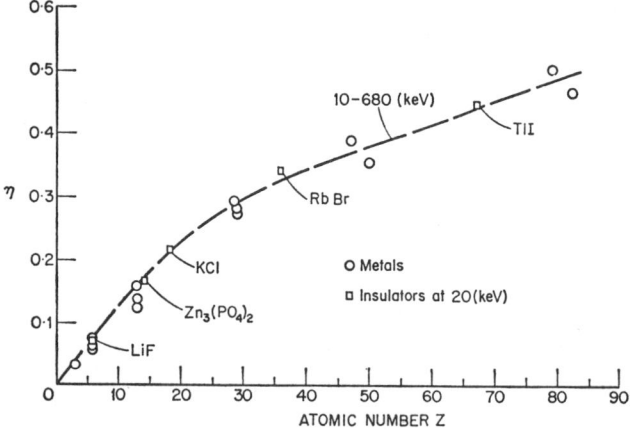

Fig. 4.16 Comparison of limiting rediffusion coefficient, for metals and insulators, as a function of atomic number Z. The insulators are plotted as a function of their average atomic numbers. (Holliday & Sternglass 1957).

Two techniques have been employed in the study of the mechanism of production of rediffused electrons: (i) with evaporated films of a substance with low Z, e.g., Al($Z = 13$), on a substance with high Z, e.g., Au($Z = 79$), it is possible to study the depth of origin of rediffused electrons (Holliday & Sternglass 1959). This technique has also been used to study the relation between the angle of scatter and the depth of origin of rediffused electrons

(Kanter 1964); (ii) the penetration of electrons through self-supporting films (Young 1956b; Kanter 1961) yields more directly information on the origin of rediffused electrons.

A comparison of the results of the two techniques shows that a range R may be defined for the penetration of electrons into a solid, which for the reflection technique is twice the film thickness at which all evidence of the substrate is eliminated, and for the transmission technique is the film thickness at which electrons no longer penetrate the film. It has been found (Holliday & Sternglass 1959) that the range is given by

$$R \approx 1\cdot 1 \times 10^{-6} \, V_p^{1\cdot 4} \text{ mg/cm}^2 \qquad (4.9)$$

$$500 < V_p < 10^6 \, V$$

The advantage of expressing eqn. (4.9) in mg/cm^2 is that it is then approximately true for all solids. The thickness of a layer containing the same mg/cm^2 will vary approximately as Z^{-1}. In order to explain the result of eqn. (4.9) Kanter (1961) suggests a reasonable model in which the number of inelastic collisions per given thickness of solid traversed is approximately independent of Z, but the energy loss per inelastic collision is approximately proportional to Z (76 eV for Al, 680 eV for Au). The latter result is to be expected if, on the average, a high energy electron can excite every possible electronic transition in the atoms of the solid. Such transitions are postulated as the source of the true secondary electrons discussed in Section 4.1.4. Note that the energy transferred to the characteristic ($\sim$20 eV) losses is small relative to that postulated for the rediffused electrons. (A discussion of the range of ions in solids is given in Section 4.3.5 below.)

The view that secondary electrons are generated in processes which degrade the energy of rediffused electrons is supported by the results of Harrower (1956a) who measured fine structure, independent of primary voltage, in region II of Fig. 4.1, and was able to assign this fine structure to the emission of Auger electrons. The latter are ejected when an excited electron in a solid makes a transition to a lower level, and instead of emitting a photon ejects another (Auger) electron of characteristic energy. It appears likely that detailed studies of the energy of Auger electrons in uhv apparatus will develop into an important method for surface studies (Weber and Peria (1967); Harris (1968)).

4.1.4 Secondary electron emission

Secondary electrons are those comprising region III of Fig. 4.1. General discussions of secondary emission have been given by Kollath (1956), Dekker (1958), and Hachenberg and Brauer (1959).

Figure 4.17 shows the total back-scattering coefficient ($\sum$) for two insulators and a metal as a function of the primary voltage. $\sum$ includes all three contributions of Fig. 4.1 but the major contribution (δ) comes from electrons in

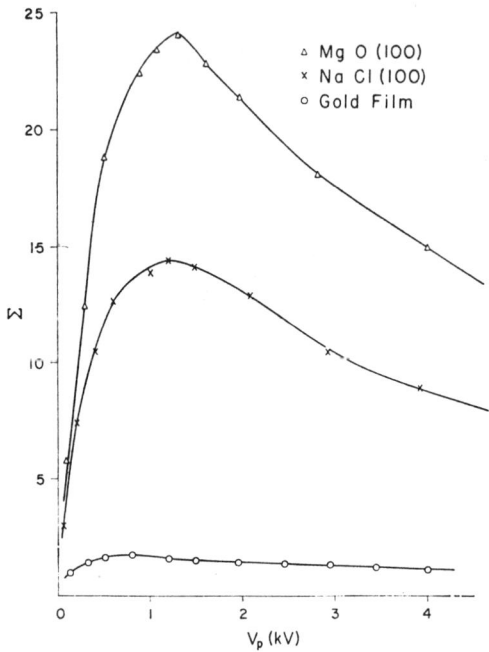

Fig. 4.17 Total back-scattering coefficient Σ as a function of primary electron voltage V_p. MgO cleaved in air before installation in uhv apparatus. (Whetten & Laponsky 1957a). NaCl cleaved in uhv apparatus just prior to measurement at $p = 10^{-10}$ torr. (Whetten 1964).
Au deposited as film in uhv apparatus just prior to measurement at $p = 5 \times 10^{-8}$ torr. (Bronshtein & Denisov 1965).

region III. All total back-scattering coefficients have the general shape of the curves in Fig. 4.17, i.e. they all have a maximum value of $\Sigma = \Sigma_m$ at a value of $V_p = V_{pm}$, but the magnitudes of Σ_m differ greatly as may be seen from Table 4.3. For metals and semiconductors, all values of $\Sigma_m \lesssim 1\cdot 5$ and $V_{pm} \lesssim 800$ V, whereas for insulators both Σ_m and V_{pm} are generally higher than these values, with the value of Σ_m for MgO shown in Fig. 4.17 being the highest.

A theory proposed by Lye and Dekker (1957), although simple, provides a first-order correlation of these results. The theory rests on three assumptions and considers only the x direction normal to the solid surface.

(1) At a depth x in the solid N secondaries are produced per primary according to

$$N(x)dx = -\frac{1}{\beta}\left(\frac{d(eV)}{dx}\right)dx \tag{4.10}$$

where $d(eV)/dx$ is the rate of energy loss of the primary, and β is the energy required to generate one secondary.

TABLE 4.3
Selection of maximum values of the total back-scattering coefficient, Σ_m, and the corresponding primary voltage (V_{pm})

Metals	Σ_m	V_{pm} (volts)
Ti	0·9	280
Fe	1·3	400
Ni	1·35	550
Cu	1·3	600
Zr	1·1	350
Mo	1·25	375
Ag	1·47	800
Ta	1·3	600
W	1·35	650
Pt	1·5	750
Au	1·45	800
Ta	1·1	800
Semiconductors		
Ge (single crystal)	1·2–1·4	400
Si (single crystal)	1·1	250
C (graphite)	1·0	250
Cu_2O	1·19–1·25	400
ZnS	1·8	350
Insulators		
NaCl (single crystal)	14	1200
NaCl (layer)	6–6·8	600
MgO (single crystal)	23	1200
MgO (layer)	4	400
Al_2O_3 (layer)	1·5–9	350–1300
SiO_2 (quartz)	2·4	400
Mica	2·4	300–384
Pyrex	2·3	340–400

Table selected from Kollath (1956) and Hachenberg and Brauer (1959).

(2) Secondaries have an effective mean free path λ so that the probability of a secondary generated at depth x, escaping from the surface is:

$$f(x) = B \exp\left(-\frac{x}{\lambda}\right). \tag{4.11}$$

(3) The range (R) of primaries is expressed by eqn. (4.9) in the form

$$R = A(eV_p)^n \tag{4.12}$$

eV_p being the primary energy. The integral for the secondary emission flux is

$$\delta = \int_0^\infty N(x) f(x) dx. \tag{4.13}$$

d(eV)/dx may be derived from (4.12), inserted in (4.10) which in turn may be inserted into (4.13). The result after reduction is

$$\delta(r) = \frac{B}{\beta}\left(\frac{\lambda}{A}\right)^{1/n} \exp(-r^n) \int_0^r \exp y^n \, dy \equiv \frac{B}{\beta}\left(\frac{\lambda}{A}\right)^{1/n} G_n(r) \quad (4.14)$$

where $r = (R/\lambda)^{1/n}$ and contains the energy dependence of δ. $G_n(r)$ is a function which can readily be calculated and can be shown to yield a maximum value of $\delta = \delta_m$ at a value of $r = r_m$ which is known. It is readily shown that

$$\frac{\delta}{\delta_m} = \frac{G_n(r_m V_p/V_{pm})}{G_n(r_m)} \quad (4.15)$$

where V_{pm} is the primary voltage at which $\delta = \delta_m$. The central result of eqn. (4.15) is that if δ/δ_m is plotted against V_p/V_{pm} all dependence upon the specific parameters β, B, λ, A is removed. Lye and Dekker (1957) have plotted eqn. (4.15) for $n = 1\cdot35$ which is very close to the exponent in eqn. (4.9) and the result is shown in Fig. 4.18 as the theoretical curve. The results of Fig. 4.17 have been replotted in the dimensionless co-ordinates for comparison. It is seen that the agreement is good for such a simple model and

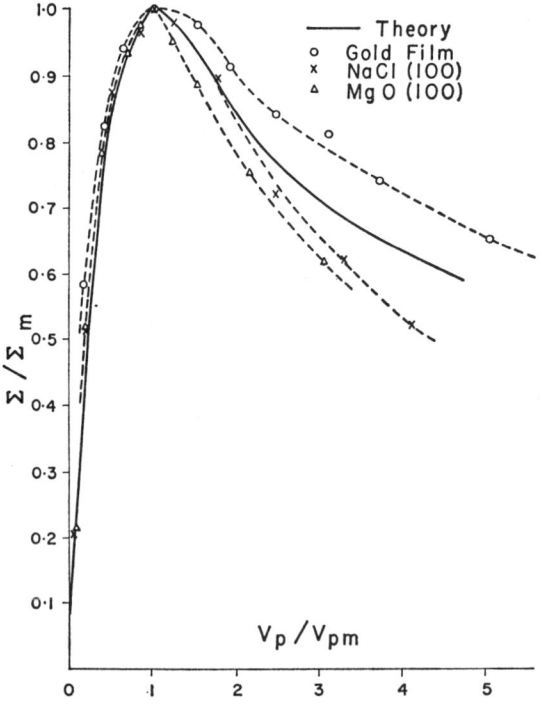

Fig. 4.18 Comparison of eqn. (4.15) with $n = 1\cdot35$ (solid line) with experimental data of Fig. 4.17. Eqn. (4.15) from Lye and Dekker (1957).

appears to confirm the qualitative assumptions of the model. For a given value of n the theory may now be used to yield a value for λ, the mean free path of secondary electrons. Dekker (1958) gets: Pt, 20 Å; Ge, 35Å; MgO, 230Å. Thus secondary electrons in metals have their origin within some 10 atomic layers of the surface while in insulators they are scattered less readily and have their origin within some 100 atomic layers of the surface.

The correlation between δ and the work function ϕ of a solid has been discussed by Hachenberg and Brauer (1959) who conclude that it is difficult to establish because of uncertainties in experimental conditions. Bronshtein and Denisov (1965), however, consider the correlation rests on firm experimental grounds. Gomoyunova and Letunov (1965) show plots correlating δ to the density of both metal and alkali halides.

Dobretsov and Matskevich (1957) have argued that the model described above is over-simplified because it neglects the role of the rediffused electrons (Section 4.1.3) in creating secondaries, and it is known that the rediffusion coefficient, η, is not negligible. Bronshtein and Segal (1960), in particular, have developed a method for separating the effects of true secondaries and rediffused primaries. A brief account of their method will be given since it confirms directly the conclusion that secondaries are generated near the surface of a solid. In all but the simplest apparatus it is possible to make simultaneous measurement of η and δ (the true secondary emission coefficient for electrons with energy < 50 eV) and this can be done at a series of primary voltages. If these measurements are carried out for a number of thicknesses of films of, say, a low Z metal on a high Z metal, then results like those shown in Fig. 4.19 are obtained. The main point illustrated by Fig. 4.19 is that at fixed primary voltage, the δ-η diagram consists of two straight lines, interpreted as region 'a' where layer thickness $d < \lambda$, and region 'b' when $d > \lambda$. It is found experimentally that the film thickness of the break-points is independent of the primary voltage and corresponds to 12 atomic layers of Be. Bronshtein and Segal (1960) however deduce a number of other results from Fig. 4.19. They let δ consist of two parts, (i) δ_0 produced by the incident primaries, and (ii) δ_1 produced by the rediffused primaries. For region 'b' of Fig. 4.19 δ_0 will be constant and the variation of δ will be that of δ_1. δ_0 and δ_1 may therefore be measured. Also

$$\delta = \delta_0 + \delta_1 = \delta_0 + S\eta, \tag{4.16}$$

where S is the number of secondaries produced by one rediffused primary. From 4.16 for a film depth $d > \lambda$

$$\left(\frac{d\delta}{d\eta}\right)_{d>\lambda} = S \tag{4.17}$$

from which S may be obtained as the slope in region 'b' of Fig. 4.19. A table of values of Σ, η, δ, S, δ_1, δ_0, δ_1/δ_0, S/δ_0, λ obtained from this analysis of a η-δ diagram is shown in Table 4.4 for a number of metals and semiconductors.

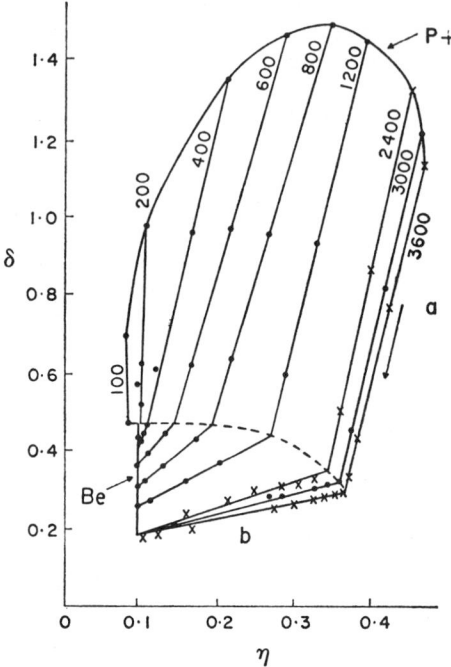

Fig. 4.19 δ vs η diagram for the evaporation of beryllium onto a platinum substrate at $T = 93°K$. The numbers alongside the lines denote the values of V_p. The arrows show the direction of increase of the beryllium layer (thickness d). The envelope obtained by joining all the points at various V_p for $d = 0$ gives the δ vs η characteristic of platinum, while the envelope similarly obtained for d large gives the δ vs η characteristic of beryllium. The dotted line connects the 'break' points which correspond to $d = \lambda = 12$ atomic layers of Be. (Bronshtein & Segal 1960).

The table shows that the role of the rediffused primaries is large and that true secondary electrons do originate close to the surface in metals and semiconductors. Secondary emission therefore is sensitive to surface conditions and uhv techniques are of value for reliable measurements. Table 4.5 has been published by Whetten (1965) and gives measured values for the total backscattering coefficients for MgO as a function of its year of measurement and illustrates the effect of the use of uhv. However, Whetten (1965) also finds good agreement between his values of $\sum$ for alkali halides measured at $p \sim 10^{-10}$ torr and those of others measured at $p = 10^{-6} - 10^{-7}$ torr. Holland and Laurenson (1963a, 1963b) have used changes in $\sum$ for a Ti surface as a measure of contamination by residual vapours in a uhv system.

Since secondary electrons are readily scattered, particularly in metals and semiconductors, it might be expected that the angular distribution of secondary electrons would follow a cosine law and this has been found to be a good first approximation by Jonker (1951, 1957) for polycrystalline and single crystal nickel, independent of the angle of incidence between 0 and 45°, the primary

TABLE 4.4
Parameters of secondary emission of electrons

V_p (kV)	Σ	η	δ	S	δ_1	δ_0	δ_1/δ_0	S/δ_0	λ, atom layers
NICKEL									
0.8	1.33	0.38	0.95	1.7	0.64	0.31	2	5.5	
1.2	1.23	0.38	0.85	1.55	0.59	0.26	2.5	6	
2.0	1.06	0.37	0.69	1.35	0.50	0.19	2.5	7	6
3.0	0.91	0.36	0.55	1.05	0.38	0.17	2	6	
4.0	0.81	0.35	0.46	0.90	0.31	0.15	2	6	
COPPER									
1.2	1.34	0.38	0.96	1.8	0.68	0.26	2.5	7	
2.0	1.14	0.37	0.77	1.5	0.55	0.21	2.5	7	7
3.0	0.96	0.35	0.61	1.2	0.42	0.19	2	6	
4.0	0.83	0.34	0.49	0.91	0.31	0.18	1.5	5	
GALLIUM									
1.2	0.98	0.37	0.61	1.12	0.42	0.20	2	5.5	
2.0	0.87	0.37	0.50	0.94	0.35	0.15	2	6	10
3.0	0.74	0.36	0.38	0.73	0.26	0.12	2	6	
4.0	0.67	0.35	0.32	0.62	0.21	0.11	2	6	
GERMANIUM									
1.2	0.92	0.36	0.56	1.0	0.37	0.19	2	5	
2.0	0.75	0.35	0.40	0.71	0.25	0.15	1.5	5	
3.0	0.65	0.34	0.31	0.56	0.19	0.12	1.5	5	10
4.0	0.59	0.33	0.26	0.45	0.15	0.11	1.5	4	
STRONTIUM									
1.2	0.65	0.34	0.31	0.65	0.22	0.09	2.5	7	
2.0	0.57	0.35	0.22	0.48	0.16	0.06	3	8	
3.0	0.53	0.35	0.18	0.39	0.13	0.05	2.5	8	16
4.0	0.50	0.35	0.15	0.32	0.11	0.04	3	8	
INDIUM									
1.2	1.25	0.39	0.86	1.6	0.63	0.23	3	7	
2.0	1.08	0.4	0.68	1.17	0.47	0.21	2	5.5	11
3.0	0.95	0.4	0.55	0.90	0.36	0.19	2	5	
4.0	0.85	0.41	0.44	0.71	0.29	0.15	2	5	
TIN									
1.2	1.3	0.38	0.92	1.7	0.65	0.27	2.5	6	
2.0	1.12	0.4	0.72	1.34	0.54	0.18	3	7	15
3.0	0.98	0.4	0.58	1.02	0.41	0.17	2.5	6	
4.0	0.89	0.4	0.49	0.85	0.34	0.15	2	6	
THALLIUM									
1.2	1.34	0.42	0.92	1.65	0.64	0.28	2	6	
2.0	1.21	0.45	0.76	1.25	0.51	0.25	2	5	7
3.0	1.10	0.46	0.64	0.96	0.42	0.22	2	4	
4.0	1.00	0.46	0.54	0.80	0.35	0.19	2	4	

(from Bronshtein and Fraiman 1962).

TABLE 4.5
Total back-scattering coefficient of MgO

Author	Σ
Geyer (1942)	2·4
Bruining and de Boer (1939)	4·0
Blankenfeld (1951)	4·5
Johnson & McKay (1953)	7
Wargo et al. (1956)	12
Whetten and Laponsky (1959)	18
Lye (1955)	24
Whetten and Laponsky (1957b)	24

(from Whetten 1965).

energy between 25 and 500 eV, and for all secondary energies between 1·5 and 20 eV.

However, Burns (1960) in a careful measurement under true uhv conditions $p < 10^{-10}$ torr found fine structure in the angle at which secondary electrons were emitted from (100) faces of copper and nickel in all the energy categories 0–10, 10–20, 20–40, 40–90 eV for primary energies 250, 500, and 800 eV. The fine structure was superimposed upon a general cosine background and was assigned to secondaries which escape from the solid without further scattering after their initial creation.

The energy distribution of secondary electrons is shown at the left of Fig. 4.4 and has the same qualitative shape for all solids up to primary energies as high as $1·6 \times 10^6$ eV at least for C, Ni, Al (Schultz & Pomerantz 1963). Bronshtein and Schuchinsky (1964) have given Table 4.6 showing the value of V_{Sm} (most

TABLE 4.6
Values of the most probable energy of secondary electrons V_{Sm} for elements of the 4th period of the periodic table

Z	Element	V_{Sm} (volts)	Work function (volts)
19	K	1·5	2·2
20	Ca	2·1	2·76
22	Ti	2·4	4·09
24	Cr	2·6	4·6
26	Fe	2·8	4·33
28	Ni	3·0	4·74
29	Cu	3·1	4·52
30	Zn	2·8	4·26
32	Ge	2·6	4·76
34	Se	2·6	4·4

(from Bronshtein and Schuchinsky 1964).

probable voltage of secondary emission) for elements of the 4th period of the periodic table. The authors conclude that there is a correlation with the work function also shown in Table 4.6. The energy distribution of true secondaries was found to be insensitive, for MgO, to angles of incidence of 0° and 60° for primary energies of 600, 1000, 4000 eV by Laponsky and Whetten (1960).

The quantum mechanical theory of secondary emission, which involves the directional momenta of the electrons in the periodic potential of the solid has been discussed by Hachenberg and Brauer (1959). Fischbeck (1966) has calculated the energy distribution of secondary electrons from sodium using this theory, and Burns (1960) has applied the theory with some success to experimental results on the angular distribution of secondary electrons from single crystals of Cu(100) and Ni(100).

4.2 ELECTRON-IMPACT DESORPTION

When electrons impinge on the surface of a solid the following particles may be evolved by direct electron excitation:

(i) neutral molecules, atoms or fragments,
(ii) excited neutrals,
(iii) positive or negative ions.

This process is called electron-impact desorption (E.I.D.) or electronic desorption and can best be studied when the power density in the electron beam is sufficiently low that heating of the surface is too small to cause significant thermal desorption. In many practical systems with high power density it is difficult to separate the effects of electronic and thermal desorption. E.I.D. results from electronic excitation or dissociation of surface molecules rather than direct momentum transfer, as can be shown by consideration of the law of momentum conservation. The maximum energy transferred from an electron of mass m_e to a free particle of mass m (when $m/m_e \gg 1$) is given by

$$\Delta E = 4E_i m_e/m = 4E_i/1838M, \qquad (4.18)$$

where E_i is the incident electron energy and M the molecular weight of the bombarded molecule. When the bombarded molecule is not free but bound to a solid then the effective mass of the molecule is modified; this correction is small in the range of electron energies of interest here. Even for very light molecules the maximum energy transferred is small, e.g. for a proton bombarded by 1 keV electrons the maximum energy transferred is only 2·2 eV. Thus in most practical cases the energy transferred by momentum transfer from a bombarding electron to a molecule on the surface of a solid is smaller than typical chemisorption binding energies (1 to 8 eV). Direct momentum transfer will not, in general, break chemisorption bonds but may affect physisorption bonds.

The gaseous and ionic products resulting from E.I.D. can cause serious problems in pressure measurements with ionization gauges and mass spectrometers (see Section 7.4.2). These problems include (*a*) the release of neutrals, which constitute a significant gas source in uhv systems causing a pressure increase, (*b*) the release of excited neutrals or chemically active fragments, resulting in chemical reactions of these species at other surfaces leading to changes in the gas composition, and (*c*) the release of excited neutrals which may be sufficiently highly excited to release electrons from a surface.

E.I.D. from atomically clean surfaces of materials normally used in uhv apparatus is negligible. In the following sections we will review the measurements of E.I.D. from three types of surface:

(i) adsorbed layers of gases and vapours at coverages of less than about one monolayer,

(ii) thick surface layers of material which are readily dissociated on electron impact (e.g. oxides, halides and oils),

(iii) 'technical' surfaces, where the state of the surface and the material adsorbed thereon were not clearly defined.

Some aspects of E.I.D. have been reviewed by Lichtman and McQuistan (1965).

4.2.1 Electron-impact desorption from adsorbed layers

The main features of E.I.D. from adsorbed layers may be summarized as follows:

(i) both neutrals and positive ions may be released. The ratio of positive ions to neutrals evolved is always small (of the order of 10^{-2}),

(ii) the process is highly specific to the adsorbed species; certain molecules are not electronically desorbed at all,

(iii) the cross-section for desorption is strongly dependent on the particular binding state in which the molecule is adsorbed, e.g. E.I.D. from the α-phase of CO on Mo is very probable, whereas the probability of desorption from the β-phase is smaller by a factor of at least 10^4,

(iv) in most cases the cross-sections for E.I.D. are much smaller than corresponding ionization cross-sections in the gas phase,

(v) the positive ions produced by E.I.D. may have considerable kinetic energy (the 0^+ from adsorbed O_2 has a mean kinetic energy of about 7 eV),

(vi) E.I.D. cross-sections vary with electron energy in a similar fashion to gas-phase ionization cross-sections. The cross-section rises to a maximum in the range 50 to 150 eV and then falls slowly with increasing electron energy.

Tables 4.7(*a*, *b*) list some measured values of the *total* cross-section (Q_{TOT}) for E.I.D.; this is the cross-section for desorption without reference to the state of the desorbed entity. *Maximum* cross-sections are given in the table for electron energies in the range 70–150 eV. It can be seen that the desorption cross-sections are considerably smaller than 10^{-16} cm² (typical of gas phase ionization cross-sections) except for three systems: α—CO on Mo (10^{-16} cm²), CO on Ni ($\sim 10^{-17}$ cm²) and H_2 evolution from silicone oil ($\sim 10^{-17}$ cm²).

TABLE 4.7(a)
Electron-impact desorption parameters on tungsten

Molecule	Binding state	Temp. (°K)	Q^+ (cm²)	T_{TOT} (cm²)	V_T (V)	Ion	V_p (V)	Reference
H_2		20		3.5×10^{-20}				Menzel & Gomer (1964a)
H_2		20		5×10^{-21}				Menzel & Gomer (1964a)
N_2		300		N.M*				Ermrich (1965)
N_2	γ	77		Larger than at 300°K	15·5			Ermrich (1965)
O_2		20		4.5×10^{-19}				Menzel & Gomer (1964a)
O_2	2	300	1.5×10^{-20}			O^+	7	Redhead (1965a)
O_2		300	$\sim 10^{-21}$	$\sim 4 \times 10^{-19}$	21·8	O^+	7·5	Yates et al. (1967)
CO	v	20		$2-5 \times 10^{-19}$				Menzel & Gomer (1964b)
CO	α	20		3×10^{-18}				Menzel & Gomer (1964b)
CO	β	20		$5-8 \times 10^{-21}$				Menzel & Gomer (1964b)
CO	$v \to \beta$	20		10^{-19}				Menzel & Gomer (1964b)
CO	α	300		3×10^{-20}				Sugata et al. (1966)
CO	$\alpha + \beta_1$	300	10^{-21}	$\sim 5 \times 10^{-19}$	17·3		0·4 (α)	Yates et al. (1967)
							6·0 (β_1)	Yates et al. (1967)
CO	1	300	$1-2 \times 10^{-20}$	3×10^{-19}	15·1	CO^+		Redhead (1967b)
CO	2	300	$1-2 \times 10^{-20}$	3×10^{-18}	18·7	O^+		Redhead (1967b)
NO	ω	300	$\sim 3 \times 10^{-21}$	$\sim 4 \times 10^{-19}$	16·8		6·8	Yates et al. (1967)
CH_4		300		N.M*				Ermrich (1965)
Ba		20		$<2 \times 10^{-22}$				Menzel & Gomer (1964a)
Cs		77–660		$<6 \times 10^{-22}$				Bennette et al. (1965)
Th		300		4×10^{-26}				Danforth (1961)

* N.M. indicates not measurable.

Measured maximum cross-sections for the desorption of positive ions (Q^+) are also listed in Tables 4.7(a, b), again the cross-sections are much smaller than gas-phase cross-sections except for one measurement of oxygen on molybdenum (Lichtman & Kirst 1966).

The specificity of electron-impact desorption is clearly illustrated in the tabulated cross-sections. In Table 4.7(a) we see that the total desorption cross-section is too small to be measured for CH_4, N_2, Ba, Cs and Th on W. From Table 4.7(b) it can be seen that the ionization cross-section is immeasurably small for N_2 on Mo.

Both the total and ionization desorption cross-sections are strongly dependent on the particular binding state of the adsorbed gas from which E.I.D. occurs as is evident from the data for O_2 on Mo and CO on W (see Tables

TABLE 4.7(b)
Electron-impact desorption parameters on various metals at 300°K

Molecule	Metal	Binding state	Q^+ (cm^2)	Q_{TOT} (cm^2)	V_T (V)	Ion or neutral	V_p (V)	Reference
Cl$_2$	Mo		$\sim 10^{-20}$	10^{-18}				Redhead (1965a)
N$_2$	Mo		N.M*	N.M*				Redhead (1965a)
O$_2$	Mo	1	N.M*	8×10^{-22}				Redhead (1964a)
O$_2$	Mo	2	$2 \cdot 6 \times 10^{-20}$	$1 \cdot 3 \times 10^{-18}$	17·5	O$^+$	7	Redhead (1964a)
O$_2$	Mo	Weakly	2×10^{-17}					Lichtman & Kirst (1966)
CO	Mo		5×10^{-19}			O$^+$		Moore (1961)
CO	Mo	α	$1 \cdot 5 \times 10^{-20}$	1×10^{-16}		O$^+$	2	Redhead (1964b)
CO	Mo	ε	8×10^{-20}	$1 \cdot 8 \times 10^{-16}$		CO$^+$		Degras & Lecante (1965)
CO	Mo	α	$1 \cdot 5 \times 10^{-19}$	4×10^{-17}	15·3	O$^+$		Degras & Lecante (1965)
CO	Mo	α	3×10^{-18}		20			Lichtman et al. (1966)
CO	Mo	β	6×10^{-21}		20			Lichtman et al. (1966)
CO	Mo	α		3×10^{-19}				Sugata et al. (1966)
CO	Ni			$6 \cdot 3 \times 10^{-17}$				Pétermann (1963a, b)
CO	Ni	1	3×10^{-21}	$2 \cdot 3 \times 10^{-18}$	17·0			Hinrichs & Donaldson (1968)
CO	Ni	2	1×10^{-18}	$2 \cdot 6 \times 10^{-17}$	14·6			Hinrichs & Donaldson (1968)
Silicone	Mo		..	$\sim 10^{-17}$		H$_2$		Garbe (1963)
Oil†				$\sim 10^{-18}$		CH$_4$		Garbe (1963)
Oxides	Cu, Ni, Mo, Ta			$\sim 10^{-20}$				Young (1960)

† DC 704.

Columns
Q^+ Maximum ionization cross-section for electrons of 70–150 eV.
Q_{TOT} Maximum total desorption cross-section for electrons of 70–150 eV.
V_T Threshold voltage for desorption of neutral or ion indicated in next column.
Ion Ion released, identified by mass-spectrometry.
V_p eV_p is the most probable energy of the evolved ions.
*N.M. Indicates not measurable.

4.7(a, b). This property allows E.I.D. to be used as a tool to study the kinetics of adsorption of certain gases (see, e.g. Menzel and Gomer 1964b, Redhead 1964a, and Lichtman et al. 1966).

The energy distribution of the positive ions released by electron impact may cover a very wide range. These distributions have been measured by a retarding field method (Redhead 1964a, 1967b). For example, the ion energy distributions for CO on W show two peaks (see Fig. 4.20) at 1 and 7 eV; the O$^+$ ion from O$_2$ adsorbed on W shows a maximum at 7 eV. The most probable energy (eV_p) of the evolved ions, for the few cases that have been measured, are indicated in Tables 4.7(a, b). These energy distributions are discussed in more detail in Section 4.2.4.

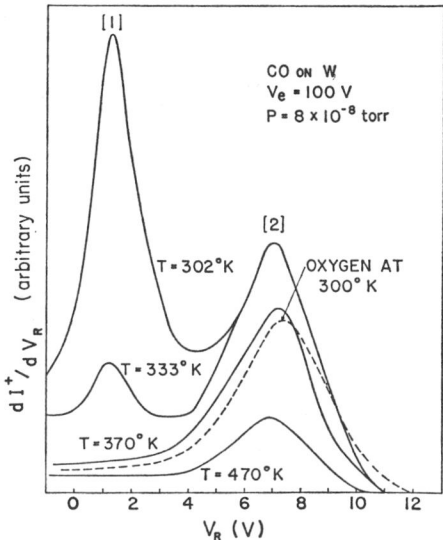

Fig. 4.20 Ion energy distribution of ions desorbed from CO on W and O₂ on W. (Redhead 1967b).

Figure 4.21 shows the normalized ionization cross-section for (a) O^+ from the weakly bound state of O_2, (b) O^+ from α-CO, and (c) Cl^+ from Cl_2 adsorbed on Mo at 300°K. Other experimenters have observed ion currents from electron bombarded adsorbed layers which increase with electron energy up to 500 V or more (Fischer & Mack 1965; Lichtman 1963). These results are thought to be caused by increased electron penetration at high energies of multiple layers or by 'rough surface' effects. In experiments where the metal surface is thoroughly cleaned and multiple layers of gas are avoided the ionization cross-section is always observed to decrease slowly for electron energies above about 100 eV.

The threshold energy required to cause the desorption of an ion or neutral (eV_T) has been measured for a few cases and is indicated in Tables 4.7(a, b), the ion or neutral evolved is also indicated (where known). The threshold quoted for the system O_2 on Mo is the same, within experimental uncertainty, for both the desorption of O^+ ions and neutrals. The equivalence of ion and neutral thresholds has not yet been established for any other adsorbed systems.

The release of excited molecules by electron impact has been proposed by Baker and Pétermann (1966) to explain the anomalously high pumping speed of CO released by electron bombardment from a nickel surface. Redhead (1967b) has shown that CO is desorbed by electron impact as an excited neutral from one of the low-energy binding states (tentatively identified as the virgin state) of tungsten. The excited neutrals were detected by the electron current they released when striking a metal surface. These experimental results could also be explained by assuming that a photon was released when

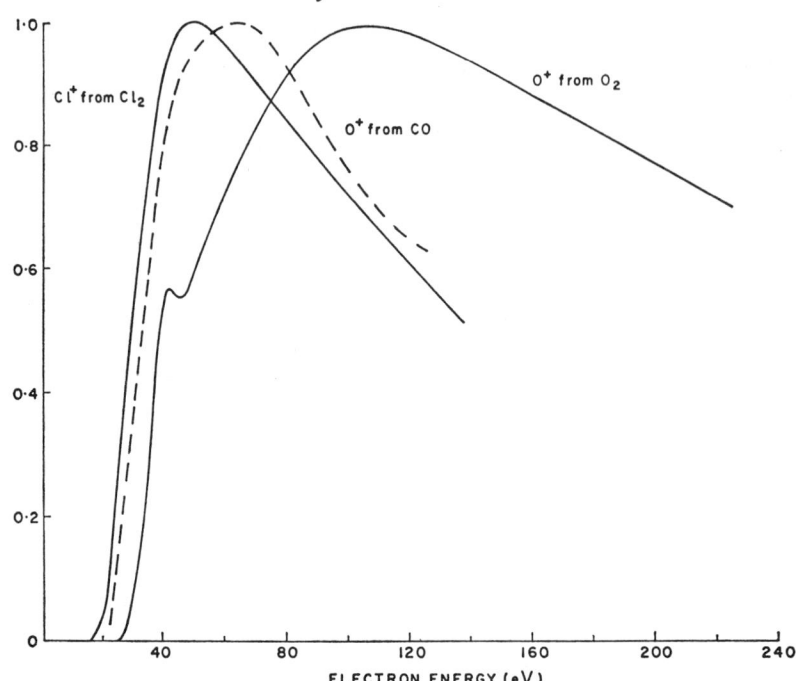

Fig. 4.21 Normalized ionization cross-section for surface ion production from a Mo surface as a function of incident electron energy.

electrons impact on virgin CO; however, the release of a metastable CO molecule seems more probable.

4.2.2 Electron-impact desorption from thick layers

E.I.D. from thick layers of oxides has been observed by several experimenters principally in connection with studies of oxide-coated cathodes. Young (1960) has observed the release of O^+ ions from oxidized surfaces of Cu, Ni, Mo, Ta and Ti. A maximum yield of about 10^{-5} ions/electrons at electron energies of 90 eV was obtained. Chlorine was found to be released from electron bombarded targets which had been exposed to the material evaporated from an oxide cathode. The products released from SrO films by electrons of energy less than 400 eV have been identified (Moore 1959). The major components evolved were Sr^+, Sr^{++}, O^+, Cl^+, Cl^- and CO^+. The yield was of the order of 10^{-7} ions/electron and the results suggested that most of the particles evolved were ionized. The ion yields were independent of the electron energy from slightly above threshold to 200 eV, above 200 eV yields decrease slightly. Table 4.8 shows the measured threshold energies for the production of various ions at the SrO layer compared with the corresponding ionization potential in the gas phase. In all cases the threshold energy for ions from the SrO surface is higher than the ionization potential in the gas phase. The ion yield increased with electron current density in a power-law relationship, the exponent is shown in parentheses: Sr^+ and Sr^{++} (2), O^+ (1·5), Cl^+ and Cl^- (1). A power law relation with exponent greater than unity indicates that the process occurs with more than one step. The ion yields increased slightly with temperature indicating an activation energy of a few tenths of an electron volt.

TABLE 4.8
Thresholds for ion production from SrO layer

Ion	Threshold voltage (surface)	Ionization potential (gas-phase)
CO^+	15·4	14·1
Cl^+	15·0	13·0
Sr^+	14·4	5·7
O^+	26·4	13·6

(from Moore 1959).

Multiply-charged ions, up to 7+, of W and Hg have been observed in a mass-spectrometer (Reynolds 1956). The electron energy (70 eV) in the ion source of the mass spectrometer was too low to produce multiply-charged ions and their production was attributed to secondary electrons, formed in the analyser region of the mass spectrometer, being accelerated backwards down the ion-accelerating potential into the ion source. The behaviour of these mass peaks indicated that the ions came from bombardment of the ion-source walls rather than from the gas phase.

The release of gas by electron bombardment of glasses and ceramics is particularly troublesome and uhv apparatus should always be designed to prevent electrons from impinging on insulators if possible. The outgassing of glass surfaces by the impact of 20 keV electrons has been studied (Todd *et al.* 1960) and it was found that 95 % of the evolved gas was oxygen for five different types of glass.

Electron bombardment of surfaces contaminated with a layer of diffusion pump oil leads to the desorption of hydrogen and various hydrocarbons. Garbe (1963) has investigated the effects of electron bombardment on adsorbed layers of DC 704 silicone oil. The gases released were analysed with a mass-spectrometer and found to consist principally of hydrogen, methane, mass 51 and 77; the latter two being presumably hydrocarbon fragments. Figure 4.22 shows the gas evolution rate as a function of electron energy for the four principal gases released.

Another result of electron bombardment of contaminated surfaces is the production of polymerized organic coatings from adsorbed layers of oils, usually from diffusion pump fluids. The resulting polymerized layers are usually of very low conductivity and cause difficulties when produced on the deflection

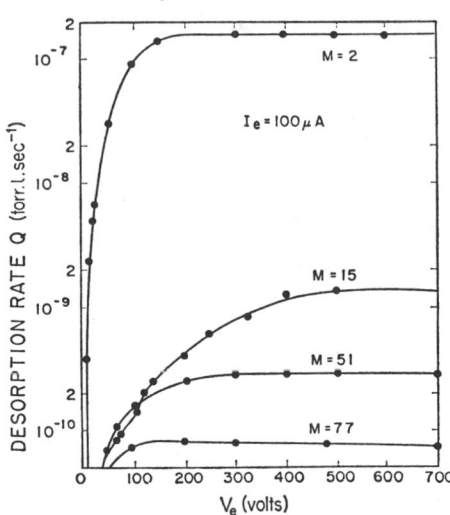

Fig. 4.22 Gas evolution rate from a film of DC704 oil as a function of the bombarding electron energy. M is mass of ion in a.m.u. (Garbe 1963).

plates of mass-spectrometers and other electron optical systems. Holland and Laurenson (1964) have studied the electrical properties of the films produced from silicone diffusion pump fluid (DC 704). The polymerization of silicone and hydrocarbon compounds is discussed by Holland (1965a, p. 161 *et seq.*) which should be referred to for references concerned with these problems.

4.2.3 Electron-impact desorption from surfaces of unknown state

Several measurements have been made of the electron-impact desorption of ions from the electrodes of the ion source of mass-spectrometers. These measurements are all characterized by the unknown state of the surface from which desorption occurs and/or lack of control of the ambient gas. Little quantitative data can be obtained from these experiments; however, the qualitative results may be useful in giving a general indication of what may be expected in similar uhv systems.

Davis (1962) observed the following ions released from the Mo walls of the ion source of his 90° sector mass-spectrometer; O_{16}^+ (from CO), F_{19}^+, Cl_{35}^+ and Cl_{37}^+, and to a lesser extent, Li_6^+, Li_7^+, Na_{23}^+ and K_{39}^+. These surface ion currents were reduced by replacing the Mo with widely spaced Pt wire of small diameter to make the ion source cage. With this change the major ion current, Cl^+, corresponded to about 10^{-11} torr and F^+ and O^+ were in the 10^{-12} range. Robins (1963), using the same type of mass-spectrometer as Davis, showed that the mass peak observed at $16\frac{1}{3}$ was caused by O^+ from adsorbed O_2 being evolved with considerable initial kinetic energy. The O^+ ion energy distribution was measured in the mass-spectrometer and was in good agreement with the results obtained by a retarding field analyser (Redhead 1964a).

Robins estimated that F^+ ions were formed at the surface with less than 1·5 eV kinetic energy and Cl^+ ions with almost no initial energy. Incidental observation of ions produced by electron bombardment of the walls of ion sources have also been made (Marmet & Morrison 1962; Plumlee & Smith 1950).

Qualitative measurements of electron-impact desorption of residual gases from molybdenum and stainless steel surfaces have been reported (Lichtman 1965a). With a Mo target at a total pressure of 5×10^{-10} torr the mass spectrum of the ions produced at the surface is shown in Fig. 4.23 (Lichtman 1965a). The peaks are identified as: 1 – protons from adsorbed H_2, $16 - O^+$ from CO, $17 - HO^+$ from H_2O, $19 - F^+$ from F_2. Figure 4.24 shows the spectrum observed from a stainless steel surface bombarded with 180 V electrons at a pressure of 1×10^{-9} torr (Lichtman & McQuistan 1965).

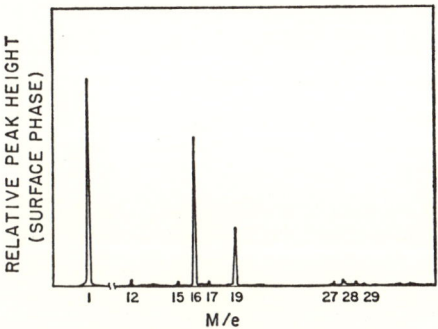

Fig. 4.23 Surface-ion mass spectrum for Mo target with 100 eV incident electrons. (Lichtman 1965a).

Several observations of electron-impact desorption in electron optical apparatus have been reported. Kendall (1965) discusses electron-impact desorption from the dynode surfaces of electron multipliers and concludes that this is likely to be the main source of gas in well designed multipliers. Electron-impact desorption can give rise to pressure increases in charge storage rings via a three-step process: (1) radiation is emitted by the stored beam, (2) photo-emission of electrons occurs from the walls, and (3) the photo-electrons are returned to the walls by the magnetic field causing electron-impact desorption of gas. Bernardini (1965) considers these problems and reports measurements of electron-impact desorption efficiency of stainless steel (304) after various treatments. It is shown that oil covered surfaces have a very high desorption efficiency. Initially, the desorption efficiency was 5×10^{-3} molecules/electron corresponding to a cross-section of 5×10^{-18} cm^2 if we assume 10^{15} molecules/cm^2 of oil on the surface. This compares with Garbe's (1963) figure of 5×10^{-17} cm^2 for the evolution of H_2 from pump oil (DC 704). These high values of desorption efficiency of an oil-covered surface could be reduced to below 3×10^{-6} molecules/electron by a short bake in hydrogen followed by oxygen. Measurements of gas desorption rates from Cu

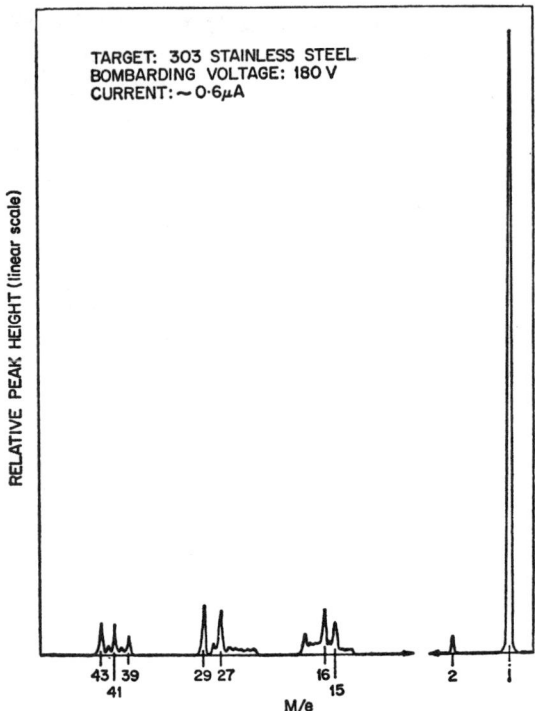

Fig. 4.24 Surface-ion mass spectrum from stainless steel. (Lichtman & McQuistan 1965).

and stainless steel made in connection with the Harvard electron storage ring have been reported (Fischer & Mack 1965).

4.2.4 Theory of electron-impact desorption

The properties of electron-impact desorption outlined above have been explained in general terms by Redhead (1964a) and Menzel and Gomer (1964a), to which papers the reader should refer for a more detailed description. Only a brief outline of the theory will be given here. It is assumed below that the E.I.D. process is confined to one binding state of the adsorbed layer.

The rate of E.I.D. is given by

$$\frac{-d\sigma_d}{dt} = \frac{\sigma_d Q J^-}{e} \text{ molecules/cm}^2 \text{ sec} \tag{4.19}$$

where σ_d is the surface coverage in the binding state which is desorbable by electron impact (this is not necessarily the total coverage),

Q the total desorption cross-section

and J^- the electron current density.

Integration of (4.19) yields

$$\theta_d = \exp -t/\tau \tag{4.20}$$

with θ_d (the relative coverage) $= 1$ at $t = 0$ and $\tau = e/J^- Q$.

Now the ion current from the surface is given by

$$i^+ = I^- Q^+ \sigma_d$$
$$= I^- Q^+ \theta_d \sigma_{d(\max)}, \quad (4.21)$$

where Q^+ is the cross-section for electron-impact desorption as a positive ion. Then from eqns. (4.20) and (4.21)

$$i^+ = i_0^+ \exp -t/\tau. \quad (4.22)$$

Thus, when no readsorption occurs, the ion current from the surface will decay exponentially with time. The time constant of the exponential decay is given by $\tau = e/J^- Q$; measurement of this time constant is the best way of measuring the total desorption cross-section, Q. Note that the decay rate is controlled by the total desorption cross-section Q not the ionization cross-section Q^+.

When readsorption occurs during electron bombardment an equilibrium can be set up between the rate of removal of adsorbed gas by electron-impact desorption and the rate of adsorption. If the desorbing species is adsorbed in a thermally irreversible phase (i.e. thermal desorption from the phase is negligible at the operating temperature), then it can readily be shown that the ion current from the surface at equilibrium is

$$i_{eq}^+ = n p v s(\theta) A e \, Q^+/Q, \quad (4.23)$$

where $s(\theta)$ is the sticking probability,
       A the bombarded area (cm^2),
       v the specific arrival rate of molecules at a surface (molecules cm^{-2} sec^{-1} torr^{-1}),
       $n = 1$ for nondissociative adsorption,
and $n = 2$ for dissociative adsorption (of a homonuclear diatomic gas).
The interesting feature of this equation is that the equilibrium ion current is independent of the electron current at fixed pressure. This independence has been observed experimentally in the case of oxygen adsorbed on molybdenum (Redhead 1964a) at coverages near a monolayer.

When the adsorbed phase is thermally reversible at the operating temperature (e.g. E.I.D. of α–CO at 300°K) the situation is more complex and the equilibrium ion current is given by

$$i_{eq}^+ = \frac{I^- Q^+ v p s(\theta)}{v_1 \exp(-E/RT) + (QJ^-/e)}, \quad (4.24)$$

where it is assumed that thermal desorption occurs by a first-order reaction. It can be seen that in this case the equilibrium ion current is a function of both electron current and pressure.

We next examine the energies involved in the E.I.D. process and for ease of exposition a specific system (O_2 on a molybdenum surface) will be discussed.

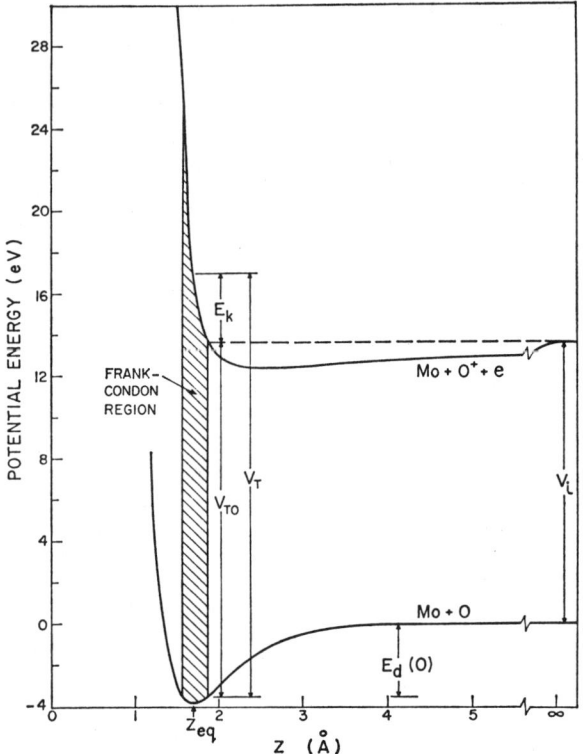

Fig. 4.25 Potential energy vs distance diagram, oxygen on Mo. (Redhead 1964a).

Oxygen is dissociatively adsorbed on Mo or W and thus we consider the impact of an electron on a chemisorbed oxygen atom. Other homonuclear diatomic gases which are dissociatively adsorbed can be expected to behave similarly. Figure 4.25 shows the potential energy of an oxygen atom and an O^+ ion near a Mo surface. The lower curve represents the lowest bound state of the metal-adsorbate system and $E_d(O)$ is the activation energy of desorption of an oxygen atom;

$$E_d(O) = \frac{q(O_2) + E^*(O_2)}{2}, \qquad (4.25)$$

where $q(O_2)$ is the heat of adsorption of the oxygen molecule and $E^*(O_2)$ the heat of dissociation of oxygen. The upper curve represents the antibonding state $M + O^+ + e$, thus at large distances from the surface this curve is V_i (the ionization potential of atomic oxygen) above the ground state. The upper curve has been approximated by a classical image potential for its attractive portion and by a Born-Meyer repulsive portion (see Section 4.3), i.e.

$$V_i^+ = B \exp(-bz) - \frac{3 \cdot 6}{z} \text{ volts}, \qquad (4.26)$$

where z, the spacing, is in ångströms. The limitations of these approximations have been discussed (Redhead 1964a).

Excitation occurs from the ground state to the ionized state within the Frank-Condon region. Excitation to other antibonding states is possible and must be considered for other systems (Menzel & Gomer 1964a); however, for the particular system under discussion (O_2 on Mo) it has been shown experimentally (Redhead 1964a) that at electron energies near threshold the only transition is to the ionized state. Excitation to other states may occur at higher electron energies.

From Fig. 4.25 it can be seen that the threshold energy to produce an O^+ ion of zero kinetic energy at $z = \infty$ is given by

$$V_{T0} \approx V_i + E_d(O), \tag{4.27}$$

then from eqns. (4.25) and (4.27)

$$V_{T0} = V_i + \frac{q(O_2) + E^*(O_2)}{2}. \tag{4.28}$$

The distribution of kinetic energy between the O^+ ion and the metal is governed by conservation of momentum; since the mass of the O^+ ion is much less than the mass of the metal, then the ion acquires essentially all the kinetic energy; this has been demonstrated experimentally (Redhead 1964a). Then the threshold energy (V_T) to produce an ion of kinetic energy E_k is given by

$$V_T = V_{T0} + E_k. \tag{4.29}$$

The transition probability from the ground state to an upper state is governed by the overlap integral which involves the wave functions of both states. In the present case we are only concerned with transition to an upper state which is repulsive or to an attractive state above its dissociation limit, see Fig. 4.25. In these cases the discrete vibrational levels found in attractive states are replaced by a continuum of energy levels. The wave function for the upper state can then be replaced by a Dirac δ-function. The transition probability is then proportional to the square of the wave function in the ground state (the probability density distribution, P.D.D.). The kinetic energy distribution of the atoms in the upper state can then be obtained by reflection of the P.D.D. in the potential energy curve of the upper state.

In the case of an adatom-metal system the P.D.D. is not known. The potential energy curve for the ionized state may be calculated approximately (see eqn. (4.26)) and the measured ion energy distribution curve reflected in it to yield the P.D.D. in the ground state. Calculations of this sort have been made for the O_2/Mo system (Redhead 1964a) and the equilibrium spacing of an adsorbed O atom determined, a value of $1\cdot7\text{Å}$ was found.

Provided that the excitation can be regarded as a one-electron jump the probability should be comparable to similar excitations in diatomic molecules.

Thus the cross-sections should be comparable to those observed in the gas phase, i.e. $\sim 10^{-16}$ cm^2. This argument is justified on quantum mechanical grounds by Menzel and Gomer (1964a). Experimentally, as noted above, the observed cross-sections are orders of magnitude smaller than those in the gas phase. The explanation for this decreased cross-section in terms of bond reformation was proposed independently by Redhead (1964a) and Menzel and Gomer (1964a).

Let us consider a simple case illustrated in Fig. 4.26. Excitation to a repulsive state at z_0 is followed by desorption along this curve or by a transition back to the ground state (rebonding). If the downward transition occurs for $z < z_c$ then the atom has acquired insufficient energy in travelling from z_0 to z on the upper curve to escape over the potential barrier after the transition back

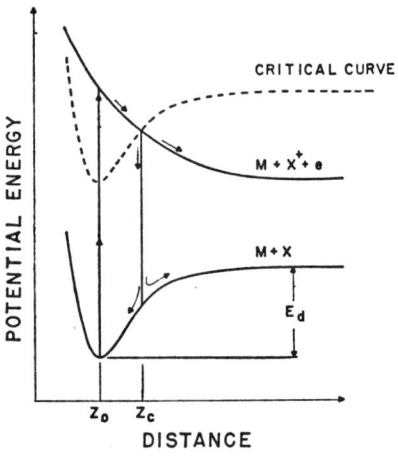

Fig. 4.26 Schematic potential energy diagram to illustrate bond reformation process.

to the ground state. Thus for a downward transition at $z < z_c$ the atom is trapped back on the surface and the kinetic energy acquired on the upper curve is transformed to thermal energy of the bound atom. This retrapping process causes the thermal energy of the adatoms to increase. If the downward transition occurs at $z > z_c$ then the atom returns to the ground state with sufficient kinetic energy to escape over the potential barrier. Specifically, if the upper state is the ionized state, three possibilities exist:

(i) escape along the upper curve as an ion (no rebonding),

(ii) readsorption as an adatom with increased vibrational energy (rebonding at $z < z_c$),

(iii) escape along the lower curve as a neutral (rebonding at $z > z_c$).

The mechanism of the rebonding process depends on the transitions under consideration, for the rebonding of the O$^+$ ion the process is probably an Auger neutralization (Redhead 1964a). Whatever the nature of the rebonding process the probability will be high at the very small spacings appropriate to chemisorption. A detailed quantum mechanical calculation of rebonding and desorption probabilities is not presently possible.

Simple calculations using the above concepts and experimentally measured cross-sections for O_2 on Mo (Redhead 1964a) lead to values of the Auger neutralization rate parameters which are not inconsistent with values obtained by other methods of measurement (Hagstrum 1954b).

The above agreements can explain in general terms the small desorption cross-sections experimentally observed. There are, however, three systems whose measured cross-sections approach 10^{-16} cm^2. These are α-CO on Mo, CO on Ni and silicone oil (see Tables 4.7(a), (b)); in these cases the probability of bond reformation must be very small. The reason for this behaviour is not presently understood. The large cross-section for α-CO on Mo (1×10^{-16} cm^2) is particularly surprising since measurements made in a similar apparatus with α-CO on W gave values of about 4×10^{-18} cm^2 in good agreement with the results obtained in field-emission microscopy (Menzel & Gomer 1964b). Sugata et al. (1966) have also observed a desorption cross-section for CO on Mo ten times larger than for CO on W (see Tables 4.7(a), (b)).

The concept of bond reformation allows some predictions of which adsorbed systems will exhibit electron-impact desorption and which will not. When the binding is by non-localized electrons the probability of tunnelling through the barrier between adsorbent and adsorbate is very large, thus bond reformation is a highly probable process and the desorption cross-section becomes vanishingly small. Non-localized bonding occurs for electropositive adsorbates where $V_i - \phi$ is small (ϕ the work function). This postulate is supported by the cross-section data of Table 4.7(a) where the desorption cross-section for Ba, Th, and Cs, all electropositive on W, is found to be immeasurably small. When $V_i - \phi$ is large and binding is by localized electrons, the probability of tunnelling through the barrier is small and the bond reformation rate becomes finite; in these cases the desorption cross-section may be large.

4.3 IMPACT OF ENERGETIC ATOMS AND IONS ON SURFACES

In this section we will discuss the phenomena which occur during the impact on surfaces of ions or atoms whose kinetic energy is far above the normal thermal range. The energy interval of interest here is from 1 eV to the onset of nuclear reactions, with the main emphasis on energies below 10^5 eV. Except for the phenomena of resonance neutralization (or ionization) and potential ejection of electrons, the effects caused by incident ions and atoms of the same type and energy are in most cases indistinguishable, being determined by the kinetic energy or momentum of the projectile. An extensive review of impact phenomena on metal surfaces has been published (Kaminsky 1965), which discusses sputtering, scattering, ionization and neutralization, and electron emission in some detail, but does not include entrapment and penetration phenomena. A review of sputtering is given by Behrisch (1964) and of ionic entrapment and release by Grant and Carter (1965). Sputtering, entrapment and re-emission play important roles in the production of uhv by ionic pumps (see Section 11.5).

For atoms having unpaired electrons, valence forces (see Section 2.1.1) can have a significant influence on the course of collisions occurring at low kinetic energy (< 100 eV). In the discussion to follow we will not consider these relative complex collisions, restricting our attention to cases in which valence forces are either absent (as in the rare gases) or negligible (as in collisions at energy > 100 eV). The course of these 'simple' collisions is determined almost completely by the repulsive potential energy of interaction at separations less than about 3Å. The form of the interaction potential energy in this region has been discussed both in Chapter 2 (Section 2.3) and Chapter 3 (Section 3.3). For more detailed discussions of the potential energy of interaction, see, for example, Nichols and Van Lint (1966). In addition to the high energy region, they summarize potential energy functions due to Lennard-Jones (see also Section 2.1), Morse, and Buckingham, which are useful at energies < 10 eV but are not suitable at higher repulsive energies. There is considerable evidence, both from differential scattering data and from theoretical work based on various statistical models of the atom, that a repulsive potential energy of the Born-Mayer form

$$E(r) = C \exp\left[\frac{-r}{r_0}\right] \text{ eV} \qquad (4.30)$$

is reasonably satisfactory for free atoms at internuclear separations $3 > r > 1$Å. The characteristic length r is found to vary relatively little from atom to atom, while the constant C increases approximately as $(Z_1 Z_2)^{3/4}$. Constants established for homonuclear pairs of rare gas atoms from calculations published by Abrahamson (1963) are given in Table 4.9 (see also Fig. 2.2). The heteronuclear pair potentials can be obtained to within a few percent by taking the geometric mean of the appropriate homonuclear potentials (Abrahamson 1964).

TABLE 4.9
Born-Mayer constants for rare gases

Atom	C (eV)	r_0 (cm)
He—He	$2 \cdot 26 \times 10^2$	$2 \cdot 42 \times 10^{-9}$
Ne—Ne	$2 \cdot 72 \times 10^3$	$2 \cdot 71 \times 10^{-9}$
Ar—Ar	$7 \cdot 80 \times 10^3$	$2 \cdot 68 \times 10^{-9}$
Kr—Kr	$1 \cdot 77 \times 10^4$	$2 \cdot 82 \times 10^{-9}$
Xe—Xe	$3 \cdot 58 \times 10^4$	$2 \cdot 80 \times 10^{-9}$

(from Abrahamson 1963).

When the collision takes place within a solid, the free-atom potential energy function described above is certainly an overestimate, since the valence force interactions which lead to the formation of the solid have the qualitative effect of reducing the size of the atom. Several potential energy functions for metals have been derived (Gibson et al. 1960; Erginsoy et al. 1964; Johnson 1966) and some non-metallic solids have been discussed for example by Fumi and Tosi (1964). The derived functions are, however, applicable only to pairs of the atoms of the solid near their equilibrium separation (≥ 2Å) and contain important attractive (valence force) terms. No comparable functions have been derived for the case (of primary interest to us here) of a foreign projectile atom interacting with an atom forming part of a solid. Furthermore, it seems likely that surface atoms of a solid will require different potential energy functions from those within the solid, but again theoretical derivations and experimental data are lacking.

For internuclear separations $r \leq 1$Å, an exponentially screened coulomb potential of the type first suggested by Bohr (1948) appears to be fairly accurate:

$$E(r) = \frac{eZ_1Z_2}{4\pi\varepsilon_0 r} \exp\left[-\frac{r(Z_1^{2/3}+Z_2^{2/3})^{\frac{1}{2}}}{a_0}\right] \text{ eV} \qquad (4.31)$$

with r and a_0 in metres, and e in coulombs. A potential energy function of closely similar magnitude based on the Thomas-Fermi screening function χ (see, for example, Gombas 1956) has been suggested by Firsov (1958a)

$$E(r) = \frac{eZ_1Z_2}{4\pi\varepsilon_0 r} \chi\left[(Z_1^{\frac{1}{2}}+Z_2^{\frac{1}{2}})^{2/3}\frac{r}{a_F}\right] \text{ eV} \qquad (4.32)$$

where $a_F = 0.885 a_0 = 0.468$Å.

For very small values of r, both the exponential term of eqn. (4.31) and $\chi(r)$ in eqn. (4.32) approach unity and the unscreened nuclear coulomb potential energy is applicable:

$$E(r) = \frac{eZ_1Z_2}{4\pi\varepsilon_0 r} = (1.44 \times 10^{-9})\frac{Z_1Z_2}{r} \text{ eV} \qquad (4.33)$$

when r is in metres.

To a first approximation, nuclear reactions begin to occur when r becomes equal to the sum of the nuclear radii. Since the nuclear radius is given approximately by

$$r_n = (1.5 \times 10^{-15})A^{1/3} \text{ metre} \qquad (4.34)$$

where A is the mass number (i.e. number of nucleons), then at the onset of nuclear reactions, $r = r_{n1} + r_{n2} = (1.5 \times 10^{-15})[A_1^{1/3} + A_2^{1/3}]$ and the corresponding upper limit in energy to be considered is, using eqn. (4.33)

$$E(r_{n1}+r_{n2}) \approx 10^6 \frac{Z_1Z_2}{[A_1^{1/3}+A_2^{1/3}]} \text{ eV.} \qquad (4.35)$$

The values derived from eqn. (4.35) are plotted as full lines in Fig. 4.27 for various projectile atoms as a function of the mass number of the target atom.

The collision of atoms can, in general, be described in classical terms as long as the de Broglie wavelength of the projectile

$$\lambda = \frac{h}{mv} = \frac{h}{\sqrt{(2mE)}} = \frac{0\cdot 295}{\sqrt{(A_1 E)}} \text{ Å} \quad (4.36)$$

is shorter than the minimum approach distance defined by eqn. (4.33) (Bohr 1948). This leads to a second upper energy limit $E(\lambda)$ above which the scattering of the projectile must be treated in terms of wave diffraction. From eqns. (4.33) and (4.36), $r_{min} = \lambda$ when

$$E(r) = E(\lambda) = (2\cdot 4 \times 10^3) A_1 Z_1^2 Z_2^2 \text{ eV}. \quad (4.37)$$

For hydrogen-hydrogen, helium-helium and neon-neon scattering, the limits $E(\lambda)$ are $(2\cdot 4 \times 10^3)$, $(1\cdot 6 \times 10^5)$ and (5×10^8) eV respectively. The lines $r_{min} = \lambda$ for hydrogen, helium and neon atoms on various targets are drawn dashed in Fig. 4.27.

Thus, in summary, for atoms heavier than neon, no nuclear reactions are to be expected at particle energies below several MeV, and the collisions can

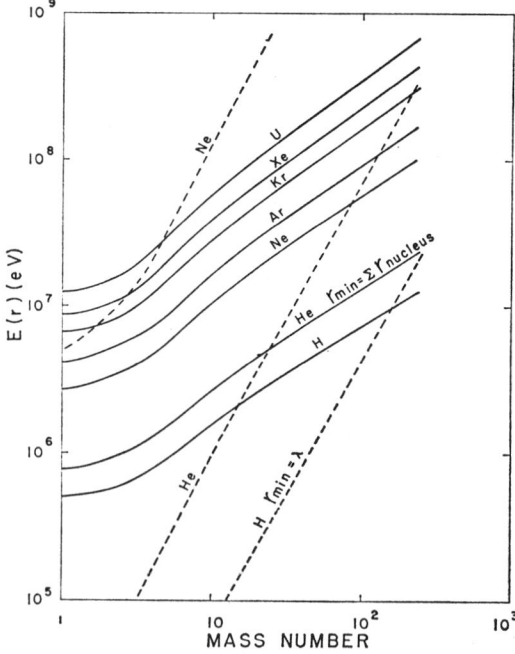

Fig. 4.27 Maximum interaction potential energies for various projectiles as a function of target mass number. Full lines: onset of nuclear reactions; dashed lines: de Broglie wave length $\lambda = r_{min}$.

TABLE 4.10
Internuclear separation (cm) for various repulsive interaction energies (eV)

Gas	Energy (eV)							
	1	10	10^2	10^3	10^4	10^5	10^6	r_K
He—He	$1 \cdot 31 \times 10^{-8}$	$7 \cdot 45 \times 10^{-9}$	$2 \cdot 62 \times 10^{-9}$	$4 \cdot 82 \times 10^{-10}$	$5 \cdot 64 \times 10^{-11}$	$5 \cdot 75 \times 10^{-12}$	$5 \cdot 75 \times 10^{-13}$	$2 \cdot 65 \times 10^{-9}$
Ne—Ne	$2 \cdot 14 \times 10^{-8}$	$1 \cdot 52 \times 10^{-8}$	$8 \cdot 95 \times 10^{-9}$	$3 \cdot 65 \times 10^{-9}$	$9 \cdot 10 \times 10^{-10}$	$1 \cdot 36 \times 10^{-10}$	$1 \cdot 43 \times 10^{-11}$	$5 \cdot 3 \times 10^{-10}$
Ar—Ar	$2 \cdot 40 \times 10^{-8}$	$1 \cdot 78 \times 10^{-8}$	$1 \cdot 17 \times 10^{-8}$	$5 \cdot 66 \times 10^{-9}$	$1 \cdot 88 \times 10^{-9}$	$3 \cdot 68 \times 10^{-10}$	$4 \cdot 51 \times 10^{-11}$	$2 \cdot 94 \times 10^{-10}$
Kr—Kr	$2 \cdot 76 \times 10^{-8}$	$2 \cdot 12 \times 10^{-8}$	$1 \cdot 48 \times 10^{-8}$	$8 \cdot 35 \times 10^{-9}$	$3 \cdot 52 \times 10^{-9}$	$1 \cdot 00 \times 10^{-9}$	$1 \cdot 62 \times 10^{-10}$	$1 \cdot 47 \times 10^{-10}$
Xe—Xe	$2 \cdot 97 \times 10^{-8}$	$2 \cdot 31 \times 10^{-8}$	$1 \cdot 65 \times 10^{-8}$	$1 \cdot 00 \times 10^{-8}$	$4 \cdot 60 \times 10^{-9}$	$1 \cdot 56 \times 10^{-9}$	$3 \cdot 18 \times 10^{-10}$	$9 \cdot 8 \times 10^{-11}$

(from Abrahamson 1963.)

be described classically over the whole energy range of interest. For light projectiles, particularly hydrogen and helium atoms, wave properties can be expected to become significant when collisions occur with light target atoms at energies $> 10^4$ eV, and nuclear reactions can occur above about 10^5 eV.

Returning to more commonly encountered energies, Table 4.10 gives distances of closest approach of various pairs of rare gas atoms for a series of energies in the interval of interest (see also Fig. 2.2). The values are taken from the calculations of Abrahamson (1963) which were based on the Thomas-Fermi-Dirac statistical model of the atoms. In the last column we list the radius of the K-electron shell calculated from the Bohr expression $r_K = a_0/Z$.

The phenomena observed during impact of energetic projectiles on surfaces are the result of two basic processes:

(i) the transfer of momentum from the projectile to the target atom with no change of internal state of the particles (an elastic process),

(ii) changes of internal energy states of the projectile and/or the target atom (an inelastic process).

Among the primarily elastic collision effects are: the scattering of the projectile, sputtering of the surface, radiation damage, projectile entrapment and penetration. The main inelastic processes are neutralization (or ionization) of the projectile, electron ejection from the target and bound electron excitation of the projectile and target atoms with subsequent photon emission or autoionization. The importance of these phenomena in various energy intervals is summarized in a general way in Table 4.11. The main points to be noted are the following:

In energy interval I(< 30 eV) the predominant effect is the scattering of the projectile back into the vacuum space. Sputtering, atomic displacement (radiation damage) in the target, and entrapment are either completely absent or very minor, with thresholds near the top of the interval. The charge of an ionic projectile can be neutralized by Auger or resonance processes if the solid is electrically conducting (metal) and if the ionization potential of the projectile V_i exceeds the target work function ϕ. If $V_i > 2\phi$, electrons may be ejected from the target (potential ejection, see Section 4.3.6). Conversely, if $V_i < \phi$, a neutral projectile may undergo ionization. Projectile velocities are too low to cause bound electron excitation during the collisions.

For projectiles in energy interval II (30 to 10^3 eV), the probability that the energy transferred to target atoms during collisions will exceed their binding energy in the solid increases rapidly. Sputtering and radiation damage then become important effects in addition to the scattering and neutralization or ionization mentioned above (interval I). In interval II, an increasing fraction of the projectiles are able to penetrate a short distance (1–10 lattice constants) into the solid, and as a result, the probability of their entrapment rises rapidly.

For energies 10^3 eV to 3×10^4 (interval III), back-scattering of the projectiles becomes less probable, while the probability of entrapment approaches unity. The distance of closest approach during collisions becomes much smaller than

TABLE 4.11

Impact phenomena at various projectile energies

Energy interval	1 Back-scattering	2 Sputtering	3 Radiation damage	4 Entrapment and re-emission	5 Penetration	6 Electron ejection	7 Excitation effects
(I) 1–30 eV	—$P \approx 1 \cdot 0$ (a) if $V_i > \phi$ —some neutrals ionized if $V_i < \phi$ —collisions occur simultaneously with several target atoms, all in surface —projectile unable to enter target	—threshold energies ~ 15 to 35 eV and related to elastic constants or sublimation energies. Usually yields small. $<10^{-3}$ at 30 eV	—very slight —thresholds (bulk) $>$ sputtering thresholds —most transferred energy to target atom vibration —damage very near surface	—$P \ll 1$ (entrapment) Many thresholds >30 eV —projectile repelled before entering solid	—$P \approx 0$ (see entrapment)	—potential ejection occurs for $V_i > 2\phi$. Potential energy of ionic projectile transferred to conduction electrons —yields for single charged ions $\cong 0.25$ —kinetic ejection does not occur	—none. Projectile velocities too low
(II) 30–1000 eV	—P decreasing to $\sim 1/2$ at 10^3 eV —collisions nearly binary by 10^3 eV —back-scattered atoms in ground state	—occurs for all projectiles and targets —yields rise approx. linearly to ~ 1 atom/ion at 10^3 eV —yields strongly influenced by surface contamination	—up to tens of vacancy-interstitial pairs per projectile —collision chain mechanism important —aggregation of defects (dislocation loops, vacancy clusters, etc) can occur for large numbers of projectiles	—P rises to ~ 0.5 at 10^3 eV —spontaneous re-emission of a small fraction of trapped projectiles —complex thermal desorption spectra dominated by surface atom configurations	—up to ~ 300 eV, average penetration a few lattice constants only —near 10^3 eV, may approach 10 lattice constants —channelling begins to be observable —crystallographic direction of incidence important	—potential ejection still prominent. Yields not greatly dependent on projectile energy —threshold for kinetic ejection generally >500 eV	—thresholds generally near 10^3 eV (see electron emission) —excitation mechanism probably formation of a pseudo-molecule (Russek 1963)
(III) 10^3–3×10^4 eV	—P decreases from $\sim 1/2$ to <0.1 —binary collision a good approximation throughout —some scattered atoms in excited states	—yields rise to broad maximum at $\sim 10^4$ eV (light projectiles), $\sim 10^5$ eV (heavy projectiles) —yields up to a few tens of atoms/ion	—thermal and displacement spikes occur due to high density of energy deposition —aggregation more prominent	—$P \approx 1.0$ —re-emission at 300°K decreases —desorption spectra dominated by bulk diffusion of projectiles from >10 Å inside target	—channelling a prominent influence on average penetration —average depths increase from ~ 10 to $\sim 10^4$ lattice constants in crystals; to $\sim 10^3$ in amorphous material	—potential ejection yield gradually decreases —kinetic yield rises to several electrons/ion for heavy projectiles	—probability of excitation increases monotonically —energy lost in excitation still small compared to that lost in momentum transfer
(IV) 3×10^4–10^6 eV	—$P < 0.1$ —very high excitation and ionization states become the rule —binary approximation practically exact	—yields decrease —some sputtered atoms excited or ionized	—extensive damage and aggregation over depths comparable to the penetration depth —damage reduced by channelling	—$P = 1$ except for those diffusing at bombardment temperature —300°K re-emission very small —desorption occurs by impurity diffusion —trapped atoms may coalesce into bubbles	—penetration depths in a random solid rises from $\sim 10^2$ to 10^3 Å —channelling the most prominent effect in crystalline solids. Maximum penetrations can exceed 10^4 lattice constants (a few microns)	—kinetic ejection yields decrease for lightest projectiles, continue to increase for heavy projectiles	—increasingly high excitation levels become possible —multiply charged projectile and target ions become the rule —decay of excited atoms and ions leads to energetic electrons (>100 eV) and photons

(a) P = probability of occurrence

the interatomic spacing in the solid; distances of penetration become relatively large (>100 lattice constants) and are strongly influenced by the channelling effect in crystalline solids. Sputtering and radiation damage both continue to increase, and projectile velocities become high enough to promote the excitation of bound electrons in both projectile and target atoms particularly in close (large angle) collisions.

Above 3×10^4 (interval IV) inelastic effects begin to play a major role in the energy loss of the projectile. High excitation levels become very common, and channelling continues to be a prominent effect in the penetration into crystalline solids. Sputtering yields tend to decrease.

The crystallographic structure of the target can have a significant effect on all of the phenomena listed in Table 4.11. In particular, the crystallographic face exposed at the surface can affect the back-scattering, sputtering, entrapment and electron ejection at all projectile energies. while the direction of incidence relative to the crystal axes strongly influences penetration depth, radiation damage and excitation effects at relatively high energies.

4.3.1 Back-scattering

A review of the scattering of ions with energies above 10 keV has been given by Snoek and Kistemaker (1965). A general discussion of scattering is also given by Kaminsky (1965). There has been practically no work, either experimental or theoretical on scattering in the energy interval I(1–30 eV). The main experimental reason for this is the difficulty of producing monoenergetic particle beams of such energies. The theoretical problem is complex even for ideal single-crystal surfaces because the collision occur with a number of target atoms simultaneously, and the collisions times are long enough that a partial 'coupling' occurs to the lattice as a whole. Qualitatively it is to be expected that the crystallographic details of the surface, the presence of impurity or adsorbed atoms, and the direction of ion incidence will all play a significant role in the scattering. The attractive dispersion (van der Waals) forces are not usually large enough to exert a significant influence on the collisions even in this energy interval.

In energy interval II ($30-10^3$ eV) a number of experimental scattering studies have been reported. With the exception of Hagstrum (1961) these have employed mainly alkali metal ions which, because of their low ionization potential are not all neutralized during impact (Brunnée 1957) and therefore are easy to detect.

Veksler (1965) has measured the energy distributions of scattered $20-260$ eV Cs^+ ions from the surfaces of a number of polycrystalline metal targets (Fe, Ni, Zr, Nb, Mo, Pd, Ta, W, Re and Pt). The targets were held at high temperatures (>1400°K) to avoid Cs adsorption. The maximum energy of scattered ions was measured for scattering angles 63° to 126°, and the data interpreted in terms of collisions with an 'effective mass' larger than the mass of an individual target atom. Veksler found the effective mass to decrease with increasing ion energy up to ~150 eV and to be constant thereafter. Above 150 eV he found the effective mass (in units of the target atom mass) to decrease with increasing target mass from about 4 for Fe to about 2 for Ta and W.

Veksler has also reported (1966) similar measurements using K^+, Rb^+ and Cs^+ ions scattered from single crystals of Mo and W. He found that the total fraction of ions scattered did not vary noticeably with the direction of ions incident, but that the energy distributions did. In particular, the maximum energy of the scattered ions was found to vary with the angle of the primary beam with respect to the target crystal lattice for a constant scattering angle. An explanation which gives a fair qualitative account of the observed behaviour is given in terms of collision with a group of four lattice atoms.

Brunnée (1957) has measured the total reflection coefficient and energy distribution of normally incident alkali metal ions scattered from clean polycrystalline molybdenum. His values of reflection coefficient are shown in Fig. 4.28 for the five alkali metal ions. The energy spectra of the scattered

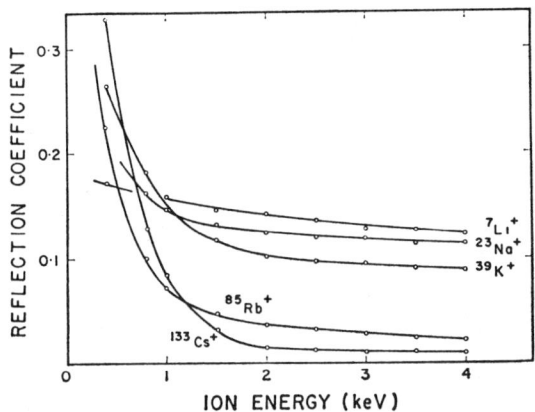

Fig. 4.28 Reflection coefficients for five alkali metal ions from clean polycrystalline molybdenum (Brunnée 1957).

ions were found to contain a low energy component (< 50 eV) most prominent for the heavy ions, and a constant component (dN/dE = constant) extending to the maximum energy allowed by a single binary collision. An interpretation of these results based on multiple scattering of the projectile ions in the target has been published by von Roos (1957).

Hagstrum (1961) has measured the reflection coefficient of rare gas ions as ions (R_{ii}) and as metastable atoms (R_{im}) from W, Mo, Hf, Si and Ge for energies 10^2 to 10^3 eV. The values found for R_{ii} were in all cases small ($< 2 \times 10^{-3}$) and nearly independent of the incident ion energy. Examples for three ions on clean tungsten are given in Fig. 4.29. The reflection coefficient as metastable atoms (R_{im}) was also found to be small but to increase relatively rapidly with increasing ion energy (see, for example, Fig. 4.30). The presence of a chemisorbed gas layer on the surfaces did not greatly change the reflection coefficients for the metals, but led to increases of more than an order of magnitude for Si and Ge, from values 1/10 of those for metals when clean to as much as five times higher than metals when gas covered.

High-speed computer simulations of the impact of copper atoms of energies 25 eV to 10^4 eV on a copper crystallite have been described by Gay and Harrison (1964). They examine the kinetic energy absorbed by the target for

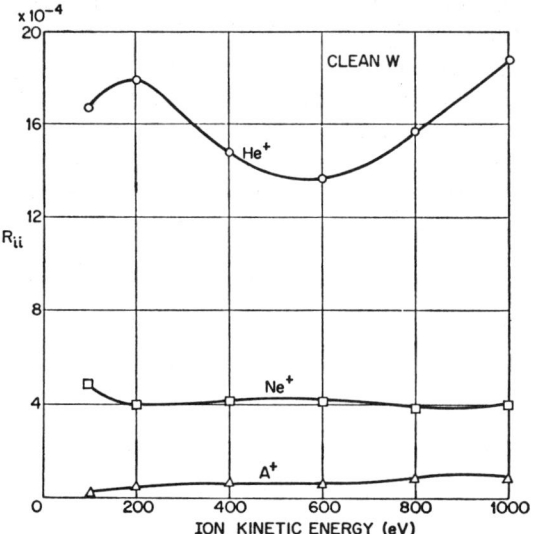

Fig. 4.29 Reflection coefficients for three rare gas ions from clean polycrystalline tungsten (Hagstrum 1961).

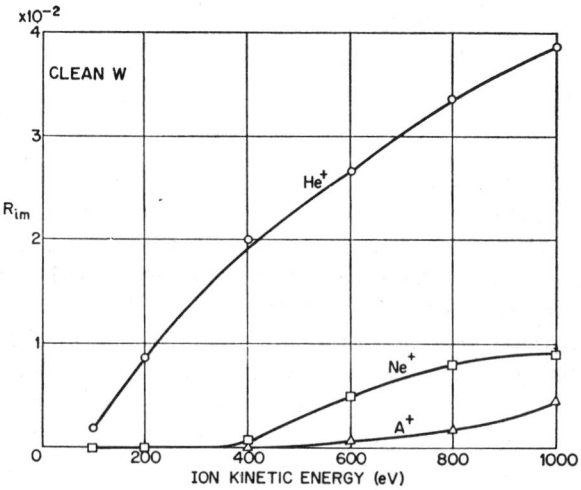

Fig. 4.30 Reflection coefficients of incident ions of the rare gases as metastable atoms from W (Hagstrum 1961).

various points of impact on low index crystal faces using reasonable (Gibson *et al.* 1960) models for the interaction potential energy. They conclude that the 'average effective mass' concept suggested by Veksler (1965, 1966) does not give a good description of the collisions, and that in fact the complex effects dependent on the crystallographic nature of the target cannot be described by any simple average. Binary collision theory was found to be definitely invalid below about 500 eV.

In another simulation, Karpuzov *et al.* (1967) have examined the scattered angle and energy distribution of Cu^+ and Ar^+ ions, of energies 3 and 2·2 keV respectively, normally incident on a Cu(100) surface. They used a Firsov potential function (Firsov 1958a) and assumed sequential binary collisions. Particularly for the Cu^+ ions, well developed 'spot patterns' were found in the scattered flux corresponding to distinct collision sequences at the surface. The energy distributions of the scattered ions contained sharp maxima corresponding to two-, three- and four-collision sequences in addition to the single collision maximum possible for the Ar^+ case. Practically no scattered ions were found within 10° of the plane of the crystal. Total back scattering coefficients were 2×10^{-4} for the Cu^+ ions, 4×10^{-2} for the Ar^+ ions.

Several higher energy scattering experiments have been reported of which a few are summarized below. Datz and Snoek (1964) have examined the charge-to-mass ratios and energy spectra resulting from scattering through large angles (60° to 110°) of 40 to 80 keV Ar^+ ions from Cu, Ag and Au and also from a Cu single crystal. They found multiply charged ions (up to 5+) of both argon and the target lattice atoms with energies corresponding to single binary collisions. In addition there were shoulders on the high energy side of the argon peaks which were thought to result from multiple scattering. The angular resolution was not sufficient to allow measurement of the inelastic energy losses in the collisions (Morgan & Everhart 1962; Afrosimov & Fedorenko 1957). By directing the incident Ar^+ ion beam along a close-packed (110) direction of the copper single crystal, they were able to eliminate multiple collision effects in the scattered beams almost completely. This effect can be understood in terms of the high probability of channelling (Piercy *et al.* 1963; Kornelsen *et al.* 1964) of those ions which do not make a large angle collision in the first atomic layer at the crystal surface. The angular dependence of the relative charge state distributions were similar to those found for gaseous targets (Fuls *et al.* 1957).

Dahl and Magyar (1965) have described similar scattering measurements for 50 keV Ar^+ ions incident on an aluminum surface. They also observed peaks due to various scattered ion charge states superimposed on a continuum. An oxide surface layer on their monocrystalline target prevented the observation of crystallographic effects such as those mentioned above. The resolving power of their apparatus was sufficient for the measurement of inelastic energy losses, which had values of approximately 400 eV for all the observed charge states (Ar^+, Ar^{2+}, Ar^{3+}, Al^+, Al^{2+}). Dahl and Magyar (1965) also made calculations of the number of target atoms contributing to the single collision scattering into various charge states. They conclude that numbers of scattering centres comparable to a single monolayer ($\sim 10^{15}/\mathrm{cm}^2$) are required to account for the observed scattered beam intensities in the most clear-cut cases. This result implies that large angle scattering in the first few atomic layers *alone* gives rise to the scattered charged particles observed.

Mashkova, Molchanov and their colleagues (Mashkova *et al.* 1965, 1966; Molchanov & Soshka 1965; Parilis & Turaev 1965) have published several papers on the energy spectra of 30 keV ions scattered from surfaces. They have observed (Mashkova *et al.* 1965) energies characteristic of two successive binary collisions involving various specific pairs of atoms in a single crystal copper surface. Other qualitative observations include the dissociation of 30 keV N_2^+ ions incident on copper (Molchanov & Soshka 1965) and the scattering of 30 keV Ar^+ and Ne^+ from both components of a silver-copper alloy (Mashkova *et al.* 1966). Parilis and Turaev (1965) have given a theoretical account of the energy spectra observed based on sequential binary collisions assuming a screened-coulomb potential of the form derived by Firsov (1958a) (eqn. (4.32)).

The back-scattering of high energy protons (400 keV) from a Ta single crystal (Bøgh & Uggerhøj 1965) has been used to demonstrate the channelling of the protons in the crystal and to check theoretical channelling predictions made by Lindhard (1965) (see Section 4.3.5). They found the back-scattered fraction at a given scattering angle to decrease dramatically (as much as a factor 80) when the protons entered the crystal within a fraction of a degree of the low index directions (particularly the $<100>$). The angular width of the dip was in good agreement with the theoretically expected critical angle for proton channelling (Lindhard 1965). Tulinov *et al.* (1965) have made similar measurements on a tungsten single crystal using 3 MeV protons. The same authors show a picture of a nuclear emulsion plate which was exposed to the back-scattered protons from a 200 keV bombardment of a tungsten crystal along the $<100>$ direction. Scattering intensity minima corresponding to almost every atomic line and plane (which 'block' the scattered protons travelling in the appropriate directions) are clearly visible. Nelson (1967) has used a similar technique with lower energy protons (20 to 40 keV) and a phosphor screen detector to study the crystallographic properties of surface and epitaxial layers a few tens to hundreds of atomic layers thick on various monocrystalline targets.

tion of atoms of a solid when it is bombarded by fast obably the most widely studied ion impact phenomenon. 1 sputtering have been published for about a century, it -1950's that the strong influence of surface contamination ed (see, for example, Wehner 1955a) and reliable quantitative sputtering measurements were begun. Excellent detailed reviews of the sputtering literature are given by Behrisch (1964) and Kaminsky (1965). We will attempt here to point out only the most prominent effects which are found in various projectile energy intervals.

Sputtering yields (atoms/ions) have been measured for a great many polycrystalline solid materials over a very wide energy range. Yields in the threshold region (15–35 eV) have been measured by Stuart and Wehner (1962) using a spectroscopic method in which the sputtered atoms are excited by the electrons of a high density plasma and the intensity of the resulting characteristic radiation detected. An example of the results obtained appears in Fig. 4.31 which

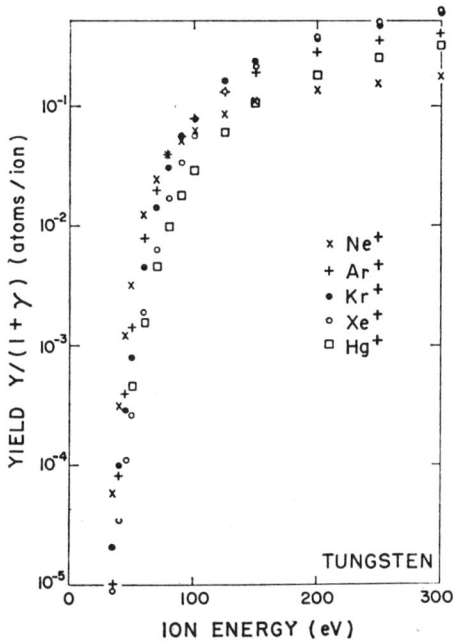

Fig. 4.31 Low-energy sputtering yields of W by four inert gas ions (Stuart & Wehner 1962).

shows the sputtering yield of W bombarded by four inert gases. Tabulations of threshold energies are given by Stuart and Wehner (1962) and by Harrison and Magnuson (1961). The threshold energies are found to be almost independent of the mass ratio of projectile to target atom. There is some evidence of correlation of the threshold energy with the heat of sublimation of the target.

Comprehensive sets of data on the sputtering of metals by inert gas ions and Hg^+ ions of energies 50 eV to 600 eV have been published by Wehner and his co-workers (Wehner 1957; Laegreid & Wehner 1961; Rosenberg & Wehner 1962). For example, Fig. 4.32 shows yields obtained by Laegreid and Wehner (1961) for Ar^+ ions on four metals. In almost all cases, yields were found to rise linearly or somewhat less than linearly with energy between 100 and 600 eV. The magnitude of the yields at a given energy show a good correlation with the reciprocal of the heat of sublimation (Rosenberg & Wehner 1962), see also Table 2.9. No presently available theories have been able to predict in detail the observed variations of yield with either energy or target material.

Sputtering yields continue to rise in the region 1–10 keV and eventually reach broad maxima at energies ranging from $\sim$20 keV (Ne$^+$) to $\sim$120 keV (Xe$^+$). Relatively little yield data has been reported for the energy interval 1 to

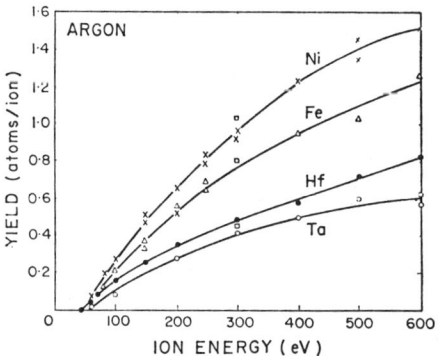

Fig. 4.32 Sputtering yields for Ar$^+$ ions incident on various metals in the energy interval 50 to 600 eV (Laegreid & Wehner 1961).

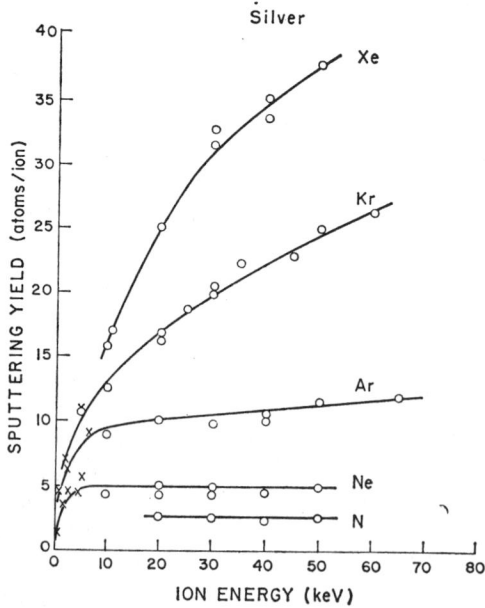

Fig. 4.33 Sputtering yields for silver bombarded by five ions in the energy range 10–60 keV (Almén & Bruce 1961a). Points x are data of Keywell (1955).

5 keV, but the less-than-linear increase appears to continue over this interval (Keywell 1955; Southern *et al.* 1963). More extensive measurements have been made at energies above 5 keV (Almén & Bruce 1961a, 1961b; Rol *et al.* 1960a). Typical values of the sputtering yields observed by Almén and Bruce (1961a) (5 ions on Ag) appear in Fig. 4.33. The plateaus or maxima in the yield curves are about 5 to 10 times those reported at energies of a few hundred electron volts (Laegreid & Wehner 1961). At a given energy, yields are observed to increase with increasing ion mass throughout this interval.

At energies above ~100 keV, data again become scarce, but indicate that yields tend to decrease monotonically with increasing ion energy. Dupp and Scharmann (1966) give yields of copper bombarded with inert gas ions having energies 75 keV to 1 MeV. These are reproduced in Fig. 4.34. The authors make comparisons with other experimental data (lower energies) and with theories of high energy sputtering developed by Martynenko (1965a, 1965b) and others (Goldman & Simon 1958; Thompson 1961).

The presence of foreign atoms on target surfaces, particularly chemisorbed gases, has long been known to exert a strong influence on the sputtering yields at low and moderate ion energies (Wehner 1955a). Such contamination effects make most of the sputtering measurements done before about 1955 of doubtful

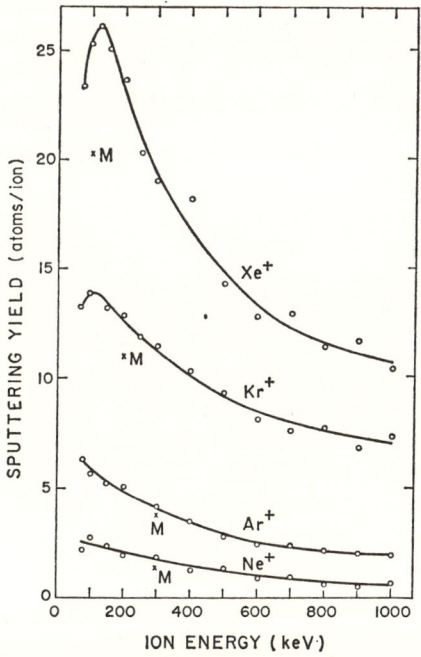

Fig. 4.34 High energy sputtering yields of copper by inert gas ions with energies 75 keV to 1 MeV (Dupp & Scharmann 1966). Points marked M were obtained with mass analysed ion beams.

quantitative value. The difficulty can be appreciated by considering the following simple argument: Let us attempt to sputter a surface which has a sputtering yield Y atoms/ion with an ion beam of current density J amp/cm^2. If the system contains a chemisorbable gas at a partial pressure p torr having a sticking probability s, then the rate of change of coverage can be written

$$\frac{dn}{dt} = vps - \frac{J}{e}\frac{n}{n_m} \cdot Y_g \qquad (4.38)$$

where n_m is the number of gas molecules adsorbed in a complete monolayer, and Y_g is the adsorbed gas sputtering yield (gas atoms/ion for a complete monolayer coverage). Experimental values of Y_g are not available, but based on the previously mentioned correlation of Y for metals with the inverse of their sublimation energy E_s, one might expect an analogous correlation with the gas desorption energy E_d. Since in most cases $E_d \lesssim E_s$ we might expect Y_g to be somewhat larger than Y, but of the same order.

If we require the fractional coverage n/n_m to be <0.05 then at equilibrium the necessary current density is

$$J \gtrsim 1200 \frac{ps}{Y_g} \text{ A/cm}^2. \qquad (4.39)$$

Thus the required current density increases as the yield decreases (low ion energy) and either high current densities or low adsorbable gas pressures are required for quantitative measurements of sputtering yields. Both approaches have been used: Wehner and his co-workers (Wehner 1957; Laegreid & Wehner 1961; Rosenberg & Wehner 1962; Stuart & Wehner 1962) have used a hot-cathode low pressure plasma to obtain high ion current densities ($\sim 10^{-2}$ A/cm^2) in systems with modest adsorbable gas pressure ($\sim 10^{-6}$ torr). Wolsky and Zdanuk (1960) have reported yield data for single crystals of silicon in a system having residual pressure $< 10^{-9}$ torr and an ion current density $\sim 10^{-6}$ to 10^{-5} A/cm^2.

At high bombarding energies (5–50 keV), focussed beams of ions are normally used (Rol et al. 1960a; Almén & Bruce 1961a, 1961b; Magnuson & Carlston 1963a). Beam current densities in the range 10^{-5} to 10^{-4} A/cm^2 are achieved in most of the reported instruments. Since yields in this energy interval are usually in the range 1 to 10 atoms/ion, eqn. (4.39) indicates that the product ps should not exceed about 10^{-7} torr if contamination effects are to be avoided. To establish that contamination errors are absent, it is usually shown experimentally that sputtering yields are independent of ion current density or of system background pressure.

At very high ion energies (>30 keV), and particularly for light ions, a second type of surface contamination can become important in oil diffusion pumped systems: The ions decompose physisorbed oil vapour molecules to form tightly bound fragments which contaminate the surface. Dupp and Scharmann (1966) report such an effect for ions in the energy interval 100 to

1000 keV. They found it necessary to operate their targets at temperatures >250°C to obtain consistent yield measurements. Their system had a base pressure $\sim 1 \times 10^{-5}$ torr and an ion current density $\sim 5 \times 10^{-5}$ A/cm². As an example they found that changing the target temperature from 30°C to 250°C increased the sputtering yield of 300 keV Ne$^+$ ions by about a factor two from 0·7 to 1·34 atoms/ion.

Wehner (1959) has made measurements of the dependence of sputtering yield on the angle of ion incidence. The energy interval studied was 150 eV to 800 eV. The metals studied fell into three groups:

(i) Those with high sputtering yields at normal incidence (Au, Ag, Cu, Pt) which showed only minor variations of yield with angle.

(ii) An intermediate group including Ni and W which showed an increase of a factor ~ 2 to 3 in yield as the angle was increased from 0° (normal) to $\sim 60°$–70°.

(iii) A group with a very large angle variation (up to a factor 10 or more) which included Fe, Mo and Ta.

Reasons for the different types of behaviour have not been suggested. Almén and Bruce (1961a) made some measurements of variation of yield with angle of incidence for 45 keV ions. They found the yields to increase in all cases by about a factor 2 over the range 0° to 60° for both Ne$^+$ and Kr$^+$ ions.

The angular distribution of sputtered atoms from various polycrystalline metals has been reported by Wehner and Rosenberg (1960). Mercury ions of 100 to 1000 eV energy were used, mainly at normal incidence. Maximum sputtered flux densities occurred away from the normal by as much as 60°, especially for the lower ion energies. This effect can be understood from the smaller momentum change required to eject an atom near the grazing angle as compared to the normal.

The effect of target temperature on sputtering yields is complex. Carlston et al. (1965) report cases of both

(i) linearly increasing yield with temperature (about +30% for the body centred cubic metals Mo, W and Ta), and

(ii) no change, or slightly decreasing yields for some single crystal faces (Mo(100), Cu(110), Cu(111))

when temperatures were increased from 350°K to 1000°K. The bombarding ions were 2 to 10 keV Ar$^+$. Almén and Bruce (1961a) also report both increasing (Ag) and decreasing (Pt and Ni) yields over the same temperature interval when bombarding with 45 keV Kr$^+$ ions. The relatively minor variations support momentum transfer as opposed to 'hot-spot evaporation' theories of the sputtering process.

Nelson (1965) has shown, however, that sputtering yields of several metals for 45 keV Ne$^+$, Ar$^+$ and Xe$^+$ ions, while almost contant at lower temperatures, increase very rapidly as the target temperature approaches the melting point. He explains the results in terms of evaporation from a thermal spike of radius about 100Å and excess temperature varying from 150°K (Bi) to 910°K for Au and 1060°K for Ge. The predicted cooling time constants of the spike are between 10^{-12} and 10^{-11} seconds. Supporting evidence based on the energy distribution of gold atoms sputtered by 43 keV Ar^+ and Xe^+ has been reported by Thompson and Nelson (1962).

Anderson (1966) has studied the sputtering yield of (110) and (100) Ge crystal faces bombarded with Ne^+ and Ar^+ ions of 250 to 500 eV energy. For temperatures above a critical value (280°C for Ar^+, 230°C for Ne^+), at which ejection spot patterns could be observed (see discussion of single crystal effects below), differences in yield of up to 50% between the two faces were observed with (100)>(110). Below the critical temperature no yield differences were observed. The annealing of crystal damage in the surface layers is thought to be responsible for the transition.

Velocity distributions of sputtered atoms have been measured by Stuart and Wehner (1964) and Thompson (1963) both using time of flight techniques.

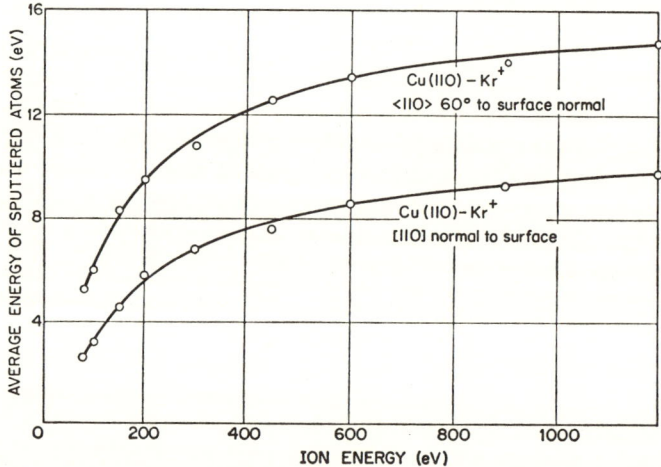

Fig. 4.35 Average ejection energy of Cu atoms sputtered along two <110> directions from a (110) Cu face by normally incident Kr^+ ions. (Stuart & Wehner 1964).

Stuart and Wehner (1964) found, for Cu, average energies $\bar{E}$ of several electron volts and most probable energies of 1 eV to 5 eV, with tails on the distributions extending in all cases to >30 eV when bombarding with Hg and rare gas ions with energies 100–1200 eV. The most probable ejection energies were found to be almost independent of ion energy, while $\bar{E}$ increases slowly with ion energy. Figure 4.35 shows average energies of ejection $\bar{E}$ for two (110) directions for Kr^+ bombardment of a Cu (110) surface. The higher $\bar{E}$ for the ejection 60° away from the normal was also observed for the polycrystalline sample. Both the most probable energy and $\bar{E}$ were found to be lower for the light rare gases than for Kr, Xe and Hg. The authors point out a rather serious disagreement between their work and the average energy measurements of Kopitzki and

Stier (1962, 1963) who used 20 to 60 keV ions. Kopitzki and Stier (1962) report $\bar{E}$ to be almost independent of ion energy for a gold target with values highest for Ne^+ (~100 eV) and lowest for Xe^+ (~20 eV). Their method did not allow examination of particular ejection directions, but they did observe distinct differences in $\bar{E}$ when bombarding different exposed faces of a copper crystal (Kopitzki & Stier 1963).

Thompson (1963) found, in addition to the distributions observed later by Stuart and Wehner (1964), a group of sputtered particles of still higher energy; in his case (40 keV Ar^+ ions on a Au crystal) above 300 eV. Thompson (1963) identified this high energy group with gold atoms being channelled toward the surface along <110> directions in the face centred cubic crystal. He also suggested that the high energy limit of the normal spectrum (280 ± 50 eV) is associated with the maximum focussing energy for the <110> collision sequences in gold.

Measurements on single-crystal targets indicate that atoms are ejected along preferred crystallographic directions during sputtering. The resulting spots of maximum deposition rate, first observed by Wehner (1955b), appear along directions corresponding to close-packed atomic rows in the crystal independent of the orientation of the target surface or the direction of ion incidence.

More detailed observations of atomic ejection patterns have since been made by several groups of workers (Anderson & Wehner 1960; Anderson, G. S., 1962, 1963; Koedam 1959; Koedam & Hoogendoorn 1960; Yurasova et al. 1960). For all bombarding ions and materials, the general results are that

(i) Atomic ejection occurs predominantly along close packed directions in the crystal; but with a significant deflection away from the normal of the surface bombarded.

(ii) Second-nearest-neighbour directions contribute to the patterns at higher energies (> 300 eV) especially in body-centred cubic crystals.

(iii) At low energies, directions requiring minimum momentum change (i.e. nearer the surface plane) were preferred, but at higher energies directions nearer the surface normal became predominant.

(iv) For different metals of the same crystal structure, the patterns are closely similar at the same sputtering yield when different ions are used.

In other experiments, yields have been measured for normally incident ions on various exposed monocrystal faces. Carlston et al. (1965) show yields from Cu (111) and (110) faces and Mo (100) and (110). In both materials, yields were highest for the most densely packed exposed plane, but variations of yield with temperature showed no simple trend; decreasing for Cu (111), increasing for Mo (110), and being independent of temperature for the remaining two. Southern et al. (1963) show yields varying by factors 3 for various Cu faces under Ar^+ ion bombardment from 1 to 5 keV. Their data are summarized in Fig. 4.36. Yield measurements of three Cu planes, the (100), (111) and (110) over the energy interval 100 eV to 200 keV have been made by Snouse and Haughney (1966). They make comparisons with other

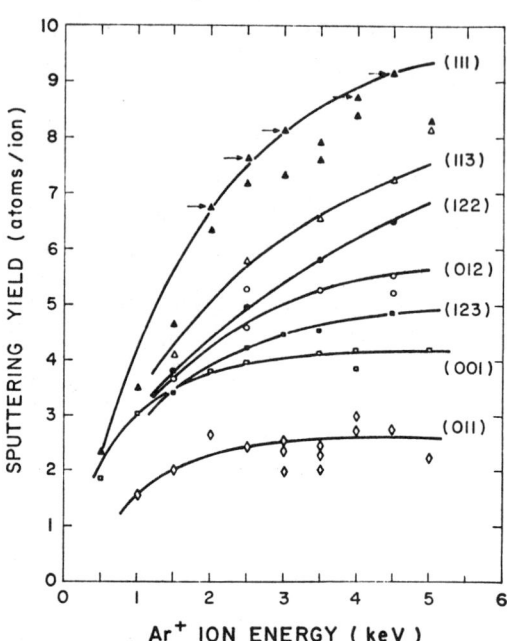

Fig. 4.36 The sputtering yields from monocrystalline Cu faces as a function of the energy of normally incident Ar+ ions. Arrows indicate yields obtained on freshly electropolished (111) surfaces. (Southern et al. 1963).

pertinent observations; scatter of ±10% in yield values between various groups of workers being typical.

Measurements have also been made of the sputtering yield of a given crystal face for varying direction of ion incidence (Almén & Bruce 1961a; Molchanov & Tel'kovskii 1962). Both found the average yield curve to follow that of polycrystalline material, increasing gradually with increasing angle of incidence; but distinct minima were observed at angles corresponding to low index directions (i.e. directions of maximum transparency) (Fluit et al. 1963). This result is thought to reflect the decreased probability of an ion of this energy (45 and 30 keV respectively) making an energetic collision close enough to the surface to provoke a sputtering event.

Theoretical treatments of the sputtering process have been successful only over limited ranges of the important parameters (energy, ion, target material, crystal structure). At the lowest energies, no analytical theories have yet attempted to explain in detail the correlation of threshold energy with the inverse of the heat of sublimation and its lack of dependence on the ion-to-atom mass ratio. It is apparent that two body collisions alone will not be adequate to describe these effects. At intermediate energies (100 to 1000 eV), it has become well accepted that a momentum transfer model of some type must be used to explain the large and diverse body of experimental data. The main supports for this view are the energy distribution of the sputtered atoms, the appearance of ejection spot patterns and the lack of strong dependence of the yields on target temperature. The ejection spot patterns also suggest that focussed collision chains (Silsbee 1957) travelling along close-packed atomic rows are influential in the sputtering. Recent work with computer simulation (Harrison *et al.* 1966) however, suggests that spot patterns might be explained without invoking any collision chains at all for ion energies up to a few kiloelectron volts. In their computed collision sequences Harrison *et al.* (1966), using a fairly realistic potential energy function (Gibson *et al.* 1960), found that even for 10 keV Ar^+ on Cu, four atomic layers contained all the events which led to sputtering, and that collisions in the first two atomic layers were the most effective. Lehmann and Sigmund (1966) give qualitative arguments which lead to similar conclusions. They point out that directions of ejection requiring a minimum of energy to surmount the surface potential barrier will be strongly preferred regardless of the details of energy transfer from atom to atom within the solid. They then present several pieces of experimental evidence which suggest that focussed collision chains, if significant at all in sputtering, must have a length of not more than a few lattice constants at the surface even for high energy bombardment. It is possible, for example, to observe distinct spot patterns for ion energies $\sim$50 eV (Wehner 1955b) at which no penetration of heavy (Xe^+ or Hg^+) ions into the crystal is to be expected (Kornelsen 1964). All primary collisions must in such cases occur in the surface atom layer and it is difficult to see how collision chains could be contributing.

At somewhat higher energies (>5 keV) Rol *et al.* (1960b) have derived the variation of yield with energy and with ion and target mass using the following simple assumptions:

(i) Only the first collision of the ion in the target is significant and the probability of an ion making a collision close enough to the surface to be effective will be inversely proportional to the ion mean free path in the solid.

(ii) The yield is proportional to the maximum energy transfer possible from the ion to the target atom,

$$\text{i.e. } S \propto E_i \frac{4M_1 M_2}{(M_1 + M_2)^2}.$$

Using a simple model of the interaction potential (Bohr 1948) they calculate an effective collision radius and from this the ion mean free path. Their simple model gives a fair qualitative account of the shape of their own experimental yield curves (Rol *et al.* 1960a) and also those of Almén and Bruce (1961a). Crystallographic effects are completely ignored in the model but Thompson (1961) has attempted to include crystallographic effects in a model with similar basic assumptions. Because of other assumptions made, his calculations could be compared only with a relatively few experimental results with light ions (O^+ and N^+) on heavy targets.

Martynenko (1965a, b) has developed a high energy sputtering theory using the concepts of lattice transparency (Fluit *et al.* 1963) and focussing collision chains to derive the dependence of sputtering yield on incidence angle (see also Martynenko 1966) for single crystals (Martynenko 1965a). He then averages over the structure dependent parameters to derive yields for polycrystalline material over the energy range $\sim$10 eV–$\sim$200 keV which is in quite reasonable agreement with the results of Almén and Bruce (1961a) and Dupp and Scharmann (1966).

Other theories for high-energy sputtering (Goldman & Simon 1958; Kinchin & Pease 1955) are able to predict qualitatively the gradual decrease in sputtering yield at energies > 100 keV, but give good approximations only for the lighter ions (Dupp & Scharmann 1966).

Brandt and Laubert (1967) have developed an approximate theory for the sputtering of amorphous targets by heavy projectiles having energies large compared to the displacement energy (see Section 4.3.3) of the target. They derive a single universal yield-energy curve starting with expressions for collision probability of the ion and the collision cascade produced by the struck atom. All the parameters used in their expressions are derivable from the properties of ions and target atoms. They use the reduced energy and reduced range formalism of Lindhard and co-workers (Lindhard *et al.* 1963a, b), taking account of energy lost in electronic excitation was well as elastic

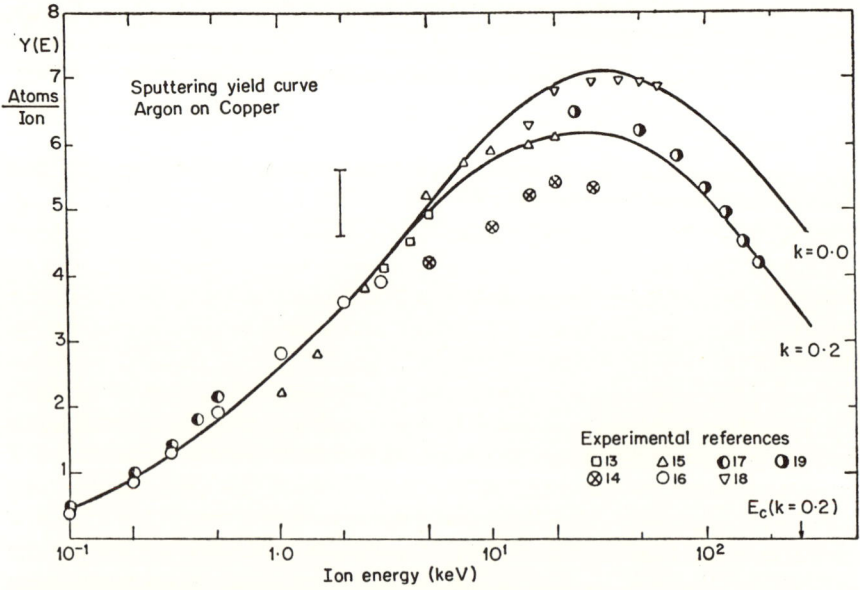

Fig. 4.37 Comparison of the derived sputtering yield curve of Ar$^+$ on Cu (full lines) with various experimental values over the energy range 100 eV to 200 keV. The difference between the curves $k = 0$ and $k = 0 \cdot 2$ is the expected decrease in yield resulting from electronic excitation. See original paper for data references. (Brandt & Laubert 1967).

collisions. Although many approximations are made, the authors claim that no adjustable parameters are used in the derivation. They construct yield energy curves for the case Ar^+ on Cu from 100 eV to 200 keV and show quite satisfactory comparison with a number of sets of experimental results. These results are summarized in Fig. 4.37.

Several experiments have been reported on the release of ionically pumped rare gas by subsequent bombardment of the same surface with ions of a second gas (Carmichael & Trendelenburg 1958; Burtt *et al.* 1961; James *et al.* 1964; Grant & Carter 1967; Kornelsen & Sinha 1966; Erents & Carter 1967). This phenomenon might be considered one of 'gas sputtering' by the action of the second ions. A review of the earlier experiments is given by Grant and Carter (1965).

Two types of experiments have contributed to the, as yet poorly developed, concepts describing this effect:

(i) The release of the primary trapped gas is detected directly during the secondary bombardment by measuring its partial pressure with a mass-spectrometer.

(ii) The changes produced in the desorption spectrum of the primary trapped atoms by the secondary bombardment is detected by thermal desorption following the second bombardment (Kornelsen & Sinha 1966; Erents & Carter 1967).

In the earliest experiments (Carmichael & Trendelenburg 1958) and those on glass surfaces (James & Carter (1962); Grant & Carter (1967)) neither the ion energy, direction of incidence, state of damage nor gas coverage of the target surface were precisely known, making the results difficult to interpret. Even in the recent experiments on clean metals, definitive results are difficult to obtain because

(i) the trapped rare gas atoms occupy sites of varying binding energy and location relative to the surface;

(ii) the second bombardment leads to progressive changes in the state of damage of the surface.

For the case of glass, James & Carter (1962) present fairly clear evidence that the release mechanism is not simply uncovering of the trapped gas by sputtering of the solid. Values of the 'gas release efficiency' deduced by James & Carter (1962) for various sorbed and bombarding ions are given in Table 7.5. The ion energy was ~ 230 eV in every case. The numbers indicate the number of sorbed atoms which would be released per bombarding ion if the total number sorbed were $10^{15}/cm^2$ (i.e. about the equivalent of one monolayer), and may thus be considered gas sputtering yields. From a comparison of Table 7.5 with Figs. 4.31 and 4.32, it can be seen that for the same ion and energy, the gas sputtering yields are much larger than those for the host metals. The evidence favours the operation of a 'thermal spike' (Seitz & Koehler 1956) or similar transient excitation phenomenon which causes the release of trapped gas but does not supply enough energy per particle to sputter the solid. More direct evidence, similar to that presented by Thompson and Nelson (1962) for high temperature, high energy sputtering of metals, is given by Grant and Carter (1967). They show a sharp increase in the gas sputtering yield when the target temperature approached that at which the primary entrapment occurred.

Evidence that a similar effect might occur in gas release from metals was obtained indirectly (Kornelsen 1964) from the approach to saturation in the entrapment of inert gas ions in tungsten. More direct evidence involving the thermal desorption of the primary gas remaining after a 'gas sputtering' bombardment with a second ion has been presented by Kornelsen and Sinha (1966) and Erents and Carter (1967).

4.3.3 Radiation damage

Radiation damage includes all the phenomena which result from the displacement of atoms of solids from their normal position by energetic 'radiation'. The subject has developed mainly out of interest in the severe disorder created in structural materials exposed to intense neutron flux from nuclear reactors. Because neutron mean free paths are long, such damage is created throughout the bulk of the solid. Displacements within a solid can also be created in a controlled way by bombardment with electrons and photons of known direction and high (MeV) energy, and hence the identification of the damage with nuclear radiations. Because of their large mass, ions of even moderate energy (~ 100 eV) have sufficient momentum to cause atomic displacements, but in this case, because of the extremely small penetration, the damage is confined to a relatively thin layer at the surface of the solid. In other respects, the effects produced are similar.

General discussions of radiation damage are given, for example, by Seitz and Koehler (1956), Dienes and Vineyard (1957), Billington and Crawford (1962) and Kelly (1966). A few of the most important concepts and conclusions are summarized briefly below.

The simplest radiation damage event is the creation of a vacancy-interstitial pair, sometimes called a Frenkel pair or defect. The energy which a metal lattice atom must absorb to leave its site and create such a pair has been measured for several materials using monoenergetic MeV electrons. From the electron energy at which changes in the properties (usually electrical resistivity) are first detected, the absorbed energy first leading to displacement can be simply calculated from the maximum momentum transfer relation. A summary of some of these displacement energies appears in Table 4.12 taken from Kelly (1966). The energy required will be a function of the crystallographic direction in which the atom is impelled, so the above numbers should be considered as thresholds for the most favourable directions. Note that the variation is not large, the values lying between 20 and 40 eV for the metals.

The damage created by energetic (MeV) neutrons is more complex: the average energy absorbed by a struck atom from a neutron of energy E_n is of

TABLE 4.12
Some measured values of displacement energies for monatomic solids

Element studied	Displacement energy (eV)
Aluminum	32
Carbon (a) Diamond	80
(b) Graphite	24·7, 60
Copper	22–25
Germanium	31, 14·5, 22·3, 14·5
Gold	28
Iron	24
	37
Nickel	24
Silicon	12·9
Silver	>40
Titanium	29
Tungsten	>35

(from Kelly 1966).

the order $2E_n/A$ where A is the atomic mass number. Since this is of the order of 10's or even 100's of keV, the struck atom (usually called the 'primary knock-on') travels through the lattice creating secondary and tertiary displacements in a collision cascade of very high energy density. Calculations have been made of the total number of displacements created by a single collision event as a function of the neutron energy (see, for example, Kelly (1966)). The results are presented in Fig. 4.38 for materials of various atomic weight. The numbers range from a few hundred to a few thousand displacements. The saturation effect for light materials at high neutron energies is caused by electronic energy loss of the relatively light knock-on nuclei which

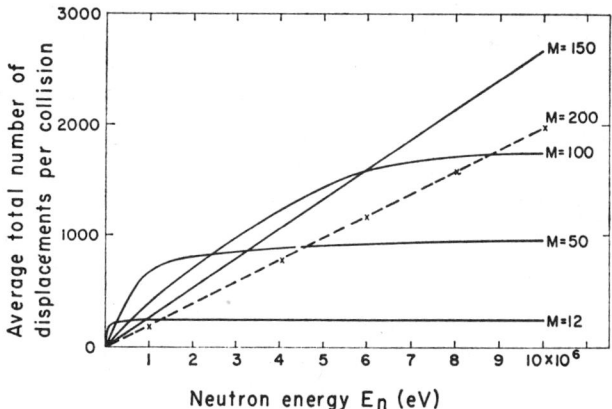

Fig. 4.38 The average number of atomic displacements produced per collision of a neutron of energy E_n on materials of various mass numbers M. (Kelly 1966).

limits the energy available for elastic collisions. Damage can also be created by the recoiling nuclei from nuclear reactions. This damage is not different in character from that described above but can be created by low energy (even thermal) neutrons which are frequently far more numerous.

The damage created by energetic ions will be similar to that created by primary knock-on atoms of the same energy, namely relatively small regions of high defect density. In the ion case, this damage will occur in the vicinity of the target surface, extending to a depth comparable to the ion penetration depth. At lower ion energies (< 10 keV) the transfer of momentum along collision chains in the solid has a strong influence both on the number of displacements actually produced and on the extent of the region in which they appear. The average collision chain length is, for such ion energies, comparable to or larger than the penetration depth.

Attempts have been made to visualize the detailed events which occur during radiation damage at low and moderate energies of the primary knock-on (Gibson *et al.* 1960; Erginsoy *et al.* 1964, 1965). The technique used is to set up a model of a small crystallite of atoms interacting with forces consistent with some known properties of the material, and to have a high-speed digital computer solve the equation of motion of the entire set of atoms when one is given an energy impulse of known magnitude and direction. Such a model of 500 to 1000 atoms representing a Cu crystallite (face-centred cubic structure) was studied by Gibson *et al.* (1960) and another of similar size representing α-Fe (body centred cubic) by Erginsoy *et al.* (1964, 1965). It should be stressed that these visualization techniques are neither experimental nor theoretical in the usual sense. They provide insight into the real phenomena only insofar as the chosen force laws, boundary conditions and limit criteria conform to the actual ones. In the end, only by comparison of the computed effects with experimental observation, and variation of the model to obtain agreement can the input choices be tested, and even then their uniqueness cannot be guaranteed.

Among the effects most clearly evident in the above visualizations are the following:

(i) Even though no thermal vibration energy of the atoms was allowed for, vacancy interstitial pairs were stable only when they were separated by relatively large distances (at least third-nearest neighbours) and the distance required depended on the crystallographic direction, being largest for the close-packed rows.

(ii) Collision chains along low index directions were very prominent when the original impulse was near such a direction. The chains transmitted both energy and mass (i.e. replacement chains were common). As a result vacancies tended to be created near the original event and interstitials were found relatively large distances (tens of atomic spacings) away.

(iii) In both lattices, the interstitials always occurred in a split configuration, two atoms symmetrically sharing one lattice site.

(iv) Displacement thresholds were found to be in fair agreement with the experimental values (see Table 4.12) being 25 eV and 17 eV for the close-packed directions in Cu and Fe respectively.

(v) The rate of energy attenuation along collision chains can be as low as 1 eV per lattice constant in the close-packed directions. Initial kinetic energies for most efficient focussing were generally less than 50 eV.

Although numerical results may be altered by improved input data, it is thought unlikely that qualitative changes will result in the phenomena described above.

The high energy deposition density which occurs when an energetic ion strikes a solid might give rise to a thermal or displacement 'spike' (Brinkman 1954; Seitz and Koehler 1956). A number of atoms around the impact site receive sufficient energy in a very short time interval to essentially melt a small volume of the crystal. The final defect configuration then depends on the way in which the melted region recrystallizes. It is not unanimously agreed whether the time required to dissipate the deposited energy would in fact be long enough for the 'temperature' concept to be meaningful. Seitz and Koehler (1956) make calculations indicating that the spike lifetime might be $\sim 10^{-11}$ sec, or about 10^2 periods of vibration of atoms in the solid. This number depends, however, on both the size and shape of the melted volume as well as the thermal diffusivity of the solid. If spike phenomena occur, their qualitative effect would be to reduce the number of displacements below that calculated on the basis of a cascade of simple binary collisions.

Both focused collision chains (Silsbee 1957; Gibson *et al.* 1960; Nelson *et al.* 1962) and the 'channelling' of energetic atoms along the more open directions in crystal lattices (see Section 4.3.5) also tend to decrease the amount of damage occurring for a given primary energy. Both mechanisms tend to transmit energy rapidly away from the primary impact point without depositing large amounts of energy on individual atoms. Collision chains are most efficient at low kinetic energies, focusing action occurring usually only below 100 eV. Channelling, on the other hand, becomes more important at higher energies, having been observed up into the MeV region. Estimates of the influence of channelling on the number of displacements produced in a cascade (Oen & Robinson 1963) are summarized in Fig. 4.39. Discussions of the effect of focused collision sequences on displacement cascades are given, for example, by Nelson *et al.* (1962).

Studies of point defects indicate that in almost all metals, interstitial atoms become mobile far below room temperature; commonly at $T < 100°K$, while vacancies are mobile in some metals slightly below room temperature and in others somewhat above room temperature. When the damage is created in the bulk of the material, the migration of defects leads to a competition between aggregation of point defects into clusters and mutual annihilation when those of opposite type recombine. The simplest clusters formed are 'loops' of interstitials or vacancies. These are two-dimensional arrays one atom layer thick either inserted between the most densely packed planes (interstitial loop) or missing from a densely packed plane (vacancy loop). The clusters can nucleate either by random association of the defects or at impurity centres, and in general have lower mobility than the individual point defects.

Because of the close proximity of most ion-induced damage to the surface, most of the mobile interstitials and, at higher temperature, vacancies will tend to appear at the surface and, by surface migration, come to rest at the edges of atomic planes. The energy deposition density of the ions, however, coupled

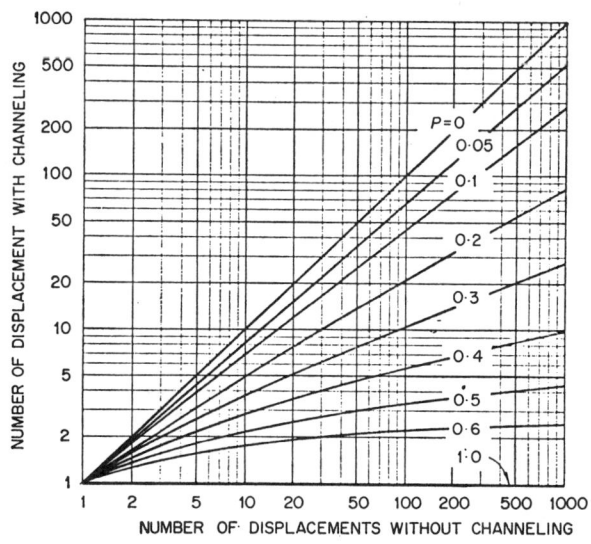

Fig. 4.39 Number of displacements produced in a collision cascade if the probability of both struck and striking atoms becoming channelled is P. (Oen & Robinson 1963).

with the considerable length of replacement collision chains make defect aggregation a definite possibility even for low-energy ions provided bombarding numbers are high ($> 10^{15}$ ions/cm^2). The aggregates do not, in general, have a serious disordering effect on the surface layers of crystals, otherwise the sputtering ejection patterns (see Section 4.3.2) would not be observed.

When target temperatures are sufficiently high to allow impurity diffusion, analogous aggregation of penetrating atoms into 'bubbles' or impurity aggregates can occur (Tucker & Norton 1960; Ruedl and Kelly 1965). Papers reporting bubble formation are listed by Grant and Carter (1965). Such bubbles or aggregates are often very stable against migration and cannot be removed without taking the material well above its self-diffusion temperature (Ruedl & Kelly 1965). The bubbles are thought to migrate, at the high temperature, by a surface diffusion mechanism operating at the bubble surface (Greenwood & Speight 1963).

4.3.4 Entrapment and re-emission

An energetic atom or ion encountering the surface of a solid target finally either passes back through the surface with a fraction of its original energy (see Section 4.3.1) or comes to rest within the solid when its excess kinetic energy is dissipated. The latter phenomenon, known as 'sticking', 'sorption' or 'entrapment' is the subject of this section. For a sufficiently high target temperature, the trapped atom will escape from its trapping site by thermally activated motion, and return to the gas phase. Such a process, which may occur also at room temperature, is termed 're-emission' or 'thermal desorption'.

Grant and Carter (1965) have given a short review of the above phenomena, with emphasis on entrapment. Although the mechanisms involved have been well known qualitatively for several years, relatively few experiments have been done in which all the pertinent parameters were controlled. The importance of the entrapment effect in removing inert gases from an evacuated system was shown by the early work of Alpert and his colleagues (Alpert 1953a; Alpert & Buritz 1954) on uhv production. The ionization gauges used to measure the low pressures created ($\sim 10^{-10}$ torr), themselves provided

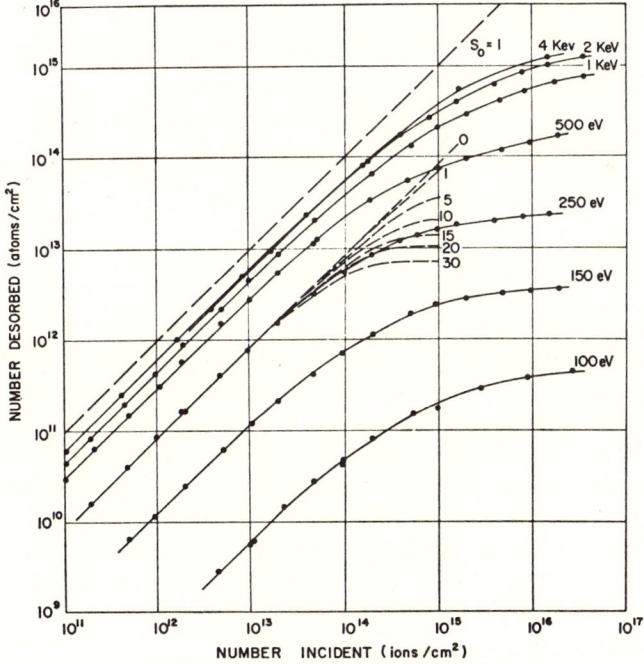

Fig. 4.40 Number of argon atoms desorbed from polycrystalline tungsten by heating to 2400°K, as a function of the number and energy of impinging argon ions. (Kornelsen 1964). Broken lines on 250 eV curve are predictions of a gas sputtering theory; the numbers referring to the ratio of gas-to-metal sputtering yields.

enough ionic entrapment (at their glass envelopes) to maintain the vacuum. Subsequently, several papers were written exploring the first-order behaviour of such pumping systems under variation of the main parameters; average ion energy, temperature, ion type, and number of atoms trapped. See, for example, Varnerin and Carmichael (1957); Hobson (1961b); Cobic et al. (1961a, b); Bills and Carleton (1958) and reviews of early work by Alpert (1958a), and Carter (1959a). Following the development of cold cathode discharge gauges (Hobson & Redhead 1958; Redhead 1959a), some work on their properties as pumps was also published (Kornelsen 1960).

Although useful from a technical point of view, the above measurements gave little quantitative information on the basic phenomenon of entrapment of an ion of known energy impinging on a clean, undisturbed surface. Attempts to obtain data under more clearly defined conditions of ion energy and surface state were made by Burtt et al. (1961), Colligon and Leck (1961) and Kornelsen (1961), using metal targets. The number of argon atoms desorbed from a polycrystalline tungsten wire (n_d) as a function of the number of incident ions (n_i) of various energies is shown in Fig. 4.40 (Kornelsen 1964). The proportionality of n_d to n_i for $n_i < 10^{13}$ ions/cm^2 defines the sticking probability, which is plotted for the four rare gases as a function of ion energy in Fig. 4.41 (Kornelsen 1964).

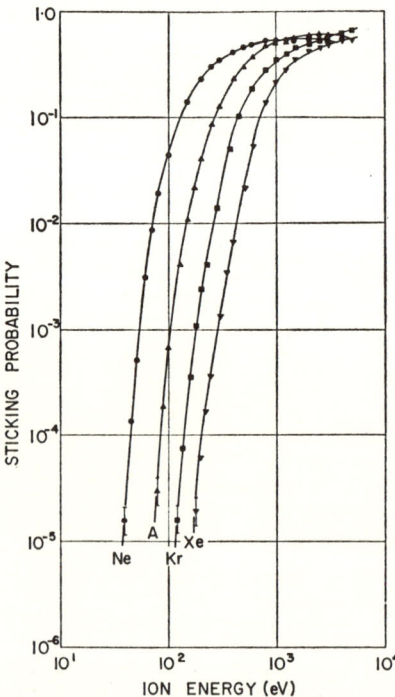

Fig. 4.41 The sticking probabilities of inert gas ions in polycrystalline tungsten as a function of ion energy. (Kornelsen 1964).

Desorption spectra obtained upon heating bombarded targets show a number of fairly distinct features whose temperatures are independent, to a fair approximation, of the type, energy and number of the bombarding ions. A number of spectra obtained for argon (Kornelsen, unpublished) appear in Fig. 4.42, where for each spectrum the enclosed area is proportional to the ion sticking probability (see Fig. 4.41). Arguments have been given (Kornelsen

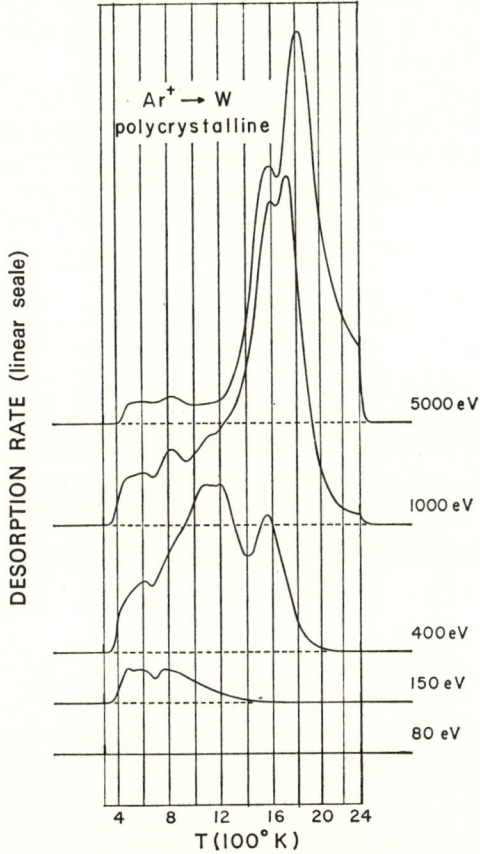

Fig. 4.42 Desorption spectra of argon from polycrystalline tungsten following bombardment at normal incidence with 3×10^{12} ions/cm² of various energies. (Kornelsen, unpublished).

1964) which indicate that (*a*) the peaks observable at temperatures up to 1600°K are associated with desorption from trapping sites not more than a few lattice constants below the target surface, and (*b*) the last peak, at temperatures > 1750°K (dependent on ion energy) involves the diffusion of more deeply penetrating ions as impurities in the bulk tungsten.

Similar data have been published for the entrapment of He^+ ions in tungsten ribbons (Close & Yarwood 1966; Erents & Carter 1966). Close and Yarwood

(1966) give only sticking probabilities, whereas the desorption spectra shown by Erents and Carter are qualitatively similar to those shown by Kornelsen (1964). It should be pointed out that Erents and Carter used much higher numbers of bombarding ions ($\sim 10^{15}/cm^2$) which would tend, for light gases, to enhance the higher energy desorption features (see Kornelsen 1964, Fig. 8).

Erents and Carter (1967) have also examined the effect of prior bombardment of a tungsten surface with Kr^+ on the entrapment of He^+ ions. They observe the sticking probability of the He^+ to rise dramatically for a Kr^+ bombardment of only a few $\times 10^{14}/cm^2$ at 500 eV.

A thermal desorption technique has also been used (Kornelsen & Sinha 1966, and unpublished data) in the bombardment of flat tungsten monocrystals at normal incidence with rare gas ions. The most important result is summarized in Fig. 4.43. For identical ion type, energy, number and incident angle, the desorption spectra are markedly different for the different exposed surface faces. This was found to be true for all ions at all energies and indicates that the atomic configurations surrounding the trapped atom very near to the surface must primarily determine the binding energy. No property of the bulk material alone (such as defect migration energies) can explain such differences.

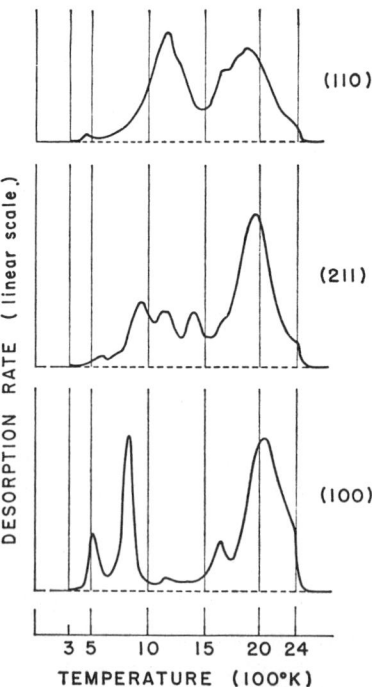

Fig. 4.43 Desorption spectra from three flat tungsten monocrystals heated at 40°K/sec following bombardment with 1500 eV Ne^+ ions at normal incidence. n_i was $\sim 1 \times 10^{13}$ ions/cm² in each case. (Kornelsen & Sinha 1966).

The last peak in each spectrum was found to behave in a similar way with the identity and energy of the ion for all the crystal faces, shifting gradually from a minimum temperature ~1750°K to higher temperatures at higher ion energies. This peak is identified with the highest temperature peak observed in the polycrystalline material (Kornelsen 1964) and is in every way consistent with the desorption of relatively deeply penetrating (> 10 Å) ions by diffusion as impurities in the bulk tungsten.

Contrary to the result reported for the polycrystalline target (Kornelsen 1964) there are easily observable differences in the peak positions and, in fact, even in the *numbers* of peaks, for different gases bombarding the same crystal face. This is additional evidence that surface atomic configurations rather than bulk properties determine the trapping energies, since different configurations might become possible with the varying size of the trapped atom.

Curves of sticking probability vs ion energy on monocrystalline targets were found to have the same qualitative shape as for polycrystals. Surprisingly large variations in the sticking threshold energies were, however, found for the same gas on different crystal faces. The results for Kr^+ ions on the three single crystal faces and a polycrystalline target are shown in Fig. 4.44. The lowest threshold energy for all gases occurs on the (110) crystal surface in which the first atomic plane has the maximum atom density (i.e. minium 'transparency').

In addition to the data described above on sticking probabilities and desorption energies, measurements of the saturation quantity of gas which can be trapped have also been made for some fairly well defined systems. Colligon and Leck (1961) show an approximately linear increase in the

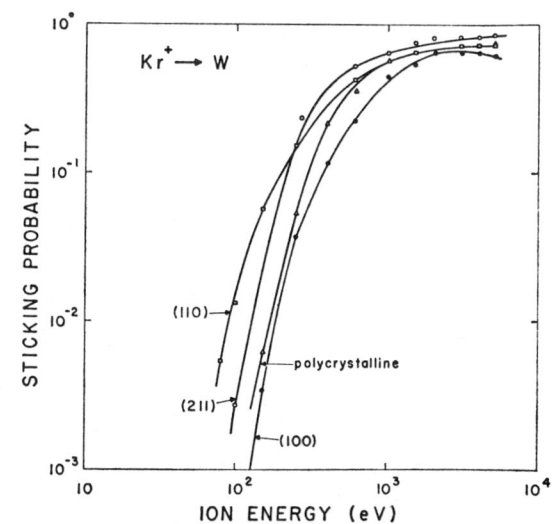

Fig. 4.44 Sticking probabilities of Kr^+ ions incident on three tungsten single crystals and a polycrystalline sample as a function of ion energy. (Kornelsen & Sinha (1968) and unpublished data).

saturation number of neon, argon and krypton atoms absorbed by various targets for energies between 0·5 and 3 keV (see Fig. 4.45). In most cases, ~10^{17} ions/cm² were sufficient to reach the saturation state, and the trapped quantity is about 2 to 10 atomic layers. At much higher energies (45 keV)

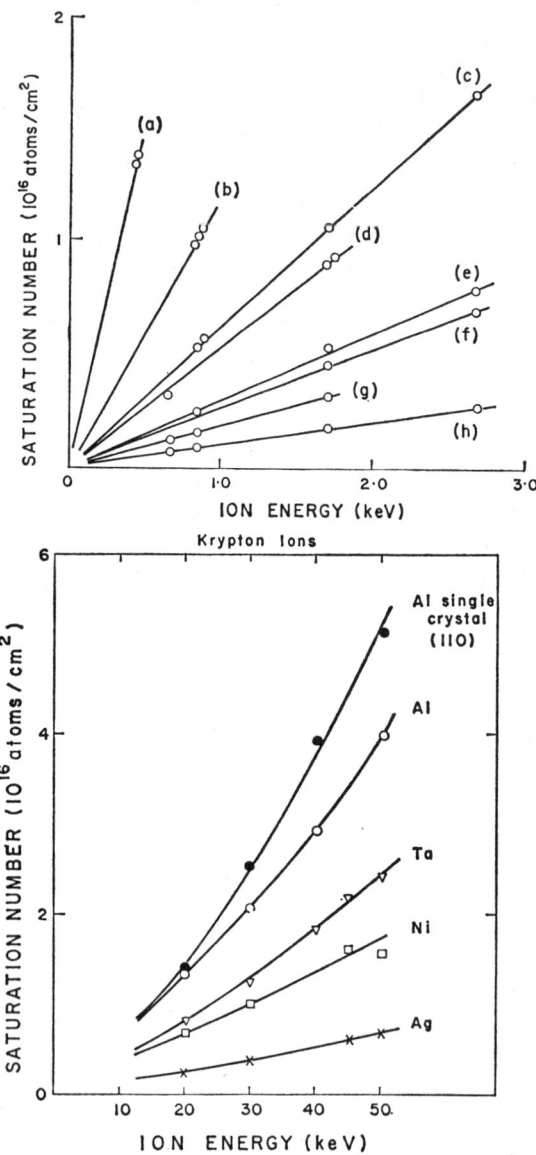

Fig. 4.45 Saturation number of rare gas ions trapped on various targets as a function of the bombarding energy (A) Colligon and Leck (1961); (a) Ne on Mo, (b) Ne on W, (c) Ar on Mo, (d) Ar on Pt, (e) Kr on Mo, (f) Ar on W, (g) Kr on Pt, (h) Kr on W; (B) Almén and Bruce (1961a).

Almén and Bruce (1961a) made detailed studies of the saturation quantity of trapped gas on various targets. These have also been included on Fig. 4.45. The saturation is set up by a dynamic equilibrium between sputtering or 'gas sputtering' and entrapment. Theoretical treatments of the saturation process have been attempted (Kuchai & Rodin 1958; Robinson 1962) but these require assumptions as to the distribution of penetrations of the ions into the solid, a quantity known well for only certain special cases and not, in general, expressible analytically (see Section 4.3.5).

The re-emission of gas at constant temperature following ionic entrapment forms a serious source of gas in uhv systems under certain conditions. Although the effect has been studied in systems where neither the ion energy nor the state of the trapping surface was known, certain interesting parametric variations have been observed. Varnerin and Carmichael (1957) first observed that, for helium ions trapped in Mo, the probability of re-emission varied inversely with the time elapsed since entrapment. They developed a phenomenonological theory to describe the shape of an ionic 'pump-down curve' in an isolated system. They give values of the re-emission constant k, defined by

$$F_r(t) = \frac{nk}{t} \quad (4.40)$$

where $F_r(t)$ is the re-emission rate (molecules/sec) and n the number of molecules trapped at $t = 0$. k was found to decrease with increasing ion energy. Varnerin and Carmichael (1957) also showed that the re-emission constant k was proportional to $(1-s)$ where s is the sticking probability. They concluded that, for He^+ in Mo, the departure of the sticking probability from unity is governed by the same mechanism as produces the long term re-emission. This work was extended by Carmichael and Knoll (1958) who obtain values of k for low energy (≤ 180 eV) helium, neon, argon and krypton ions on molybdenum and nickel targets (see Table 4.13). In all cases, eqn. (4.40) was found to hold with reasonable accuracy. Fox and Knoll (1960) observed the $1/t$ dependence of the re-emission rate to hold for times as short as a few seconds (He^+ on Ni) and also found that the re-emission constant k decreased with decreasing temperature for He^+ on a molybdenum target.

The re-emission of ionically pumped inert gases from Pyrex glass was studied by Smeaton and Carter (1966), following earlier work (Smeaton et al. 1962; Baker and Giorgi 1960; Cavaleru et al. 1964; Robinson and Berz 1959) mainly on ionization gauges. Bombarding with He, Ar and Xe ions having energies distributed from $\sim$50 eV to 200 eV they found the re-emission rate to vary as t^{-m}, t being measured from the cessation of pumping. The re-emission index m was found to be near unity when the pumping time was short ($\sim$1 min), compared to the time over which the re-emission was observed ($\sim 10^2$ min). This is consistent with the results described above (Varnerin & Carmichael 1957; Carmichael & Knoll 1958). When long pumping times t_p were used, the re-emission index m was found to decrease to a value between 0·5 and

TABLE 4.13
Re-emission constants

Gas	Target	Temp.	Ion Energy (max)	k	Reference
He	Mo	330	~200	6.5×10^{-2}	
He	Mo	190	200	3.4×10^{-2}	Fox and Knoll (1960)
He	Mo	77	200	$<10^{-3}$	
He	Mo	~300	150	1.5×10^{-2}	
He	Mo	~300	1100	5.9×10^{-3}	Varnerin and Carmichael
Ar	Mo	~300	2100	4.3×10^{-3}	(1957)
He	Mo	~300	~200	$1.3 \pm 0.05 \times 10^{-2}$	
Ne	Mo	~300	~200	$4.0 \pm 0.3 \times 10^{-2}$	
Ar	Mo	~300	~200	$15 \pm 5 \times 10^{-2}$	
Kr	Mo	~300	~200	$14 \pm 1 \times 10^{-2}$	Carmichael and Knoll
He	Ni	~300	~200	$4.4 \pm 1.8 \times 10^{-2}$	(1958)
Ne	Ni	~300	~200	$5.2 \pm 4.6 \times 10^{-2}$	
Ar	Ni	~300	~200	$5.7 \pm 0.7 \times 10^{-3}$	
Kr	Ni	~300	~200	$3.2 \pm 1.6 \times 10^{-3}$	

0.8, and then to remain constant ($t_p > 250$ min). Smeaton and Carter (1966) give a derivation based on the first-order desorption equation which indicates that the form of the re-emission rate is consistent with an initial population distribution

$$n_0(E) = \exp\left[\frac{E}{RT(1-m)}\right], m < 1. \quad (4.41)$$

The invariance of m with pumping times > 250 min is ascribed to a saturation of the population of all sites for sufficiently high bombarding numbers. The implied trapping time constants lie in the range 10 to 10^5 sec.

The authors also show that lower energy sites are also filled if the glass is cooled to much lower temperatures. Such an effect was also observed earlier (Hobson & Edmonds 1963) in a B–A gauge.

4.3.5 Penetration and channelling

The depth to which atomic projectiles of a given energy penetrate into a solid target has been a subject of considerable recent interest. The most important influence of penetration on pressure in vacuum systems is in the competition between entrapment and release processes occurring during ionic pumping. In addition, uhv plays an important part in the detailed study of heavy ion penetration at low energies (< 40 keV), where target surface conditions become important.

Prior to 1963, no studies of penetration had taken account of the regular arrangement of atoms in crystals. Solid targets were treated as random arrays, i.e. having the properties of a very dense gas, and measured values showed no pronounced effects which required a more sophisticated explanation.

Machine computations of atomic collision sequences in crystals (Robinson & Oen 1963; Beeler & Besco 1963) first suggested that the regular arrangement of the lattice atoms might have a pronounced effect on ion or atom penetration depths. In particular, some ions incident along low index directions on crystals were predicted to undergo a series of small angle collisions which focused them along the open channels between atomic rows characteristic of these directions. About the same time, experimental range distribution measurements (Davies *et al.* 1963; McCargo *et al.* 1963) showed anomalously deep penetration of a fraction of heavy ions incident on polycrystalline aluminum and tungsten. Subsequent experiments on monocrystalline targets verified the channelling effect, predicted by Robinson and Oen (1963) for aluminum (Piercy *et al.* 1963), copper (Lutz & Sizmann 1963), tungsten (Domeij *et al.* 1964a) and silicon (Davies *et al.* 1964). Since that time a large number of papers on channelling have been published and a theoretical treatment of the effect has been developed (Lindhard 1964, 1965).

For ion energies <0.5 keV, penetration depths are usually so small (<10Å) that they cannot be measured by presently available (electro-chemical) techniques (see, for example, McCargo *et al.* 1963). In this interval, entrapment can be said to be the dominant process and penetration, even when assisted by channelling effects, is minor.

Above 1 keV, penetration distributions have been measured for Na, Ar, Kr and Xe ions in 'amorphous' Al_2O_3 and WO_3 (Domeij *et al.* 1964b). The fraction of radioactive ions penetrating through anodically formed oxides of various known thicknesses was obtained from the fraction of the activity remaining after the oxide was dissolved. The results for Kr^{85} ions in Al_2O_3 appear in Fig. 4.46. Below 10 keV ion energy, results were not reproducible

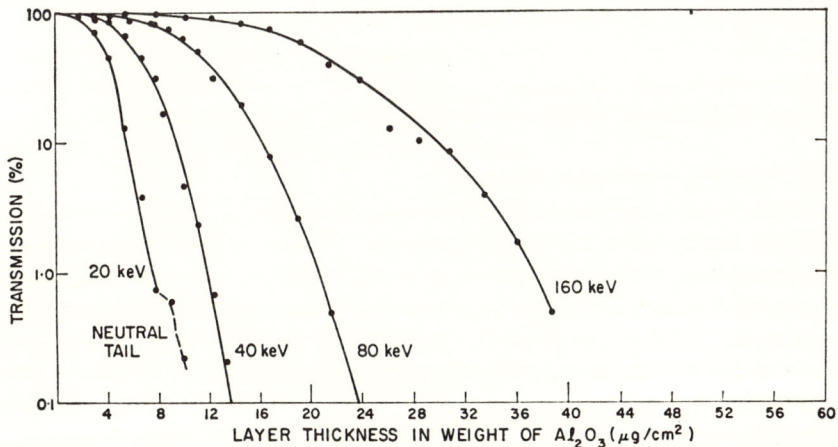

Fig. 4.46 Penetration of Kr^{85} ions through anodic Al_2O_3 (Domeij *et al.* 1964b). The "neutral tail" consisted of 40 keV atoms created by charge exchange in the beam before retardation to 20 keV.

and showed serious scatter. This was attributed to the films being microcrystalline (~50Å crystal size) rather than truly amorphous. The most probable median and mean penetrations were found to be approximately equal, as is to be expected in an amorphous solid (Lindhard et al. 1963b). Values ranged from 3 μg/cm^2 to 80 μg/cm^2 for ion energies 10 to 160 keV.

For comparison with theoretical treatments of penetration in random solids (Lindhard & Scharff 1961; Lindhard et al. 1963b), a correction had to be made for the presence of two types of atoms in the oxide. The correction involves a knowledge of the relative stopping power of the two atomic components. A further complication is that the theory gives the range measured along the path of the projectile rather than the straight-line penetration. The appropriate path length corrections are given by Lindhard et al. (1963b) but only for mass ratios less than 2, which does not include the light ions (Na^{24} and Ar^{41}) in WO$_3$. The agreement between theory and experiment was shown to be quite good ($\pm 10\%$) over the energy range investigated. Powers and Whaling (1962) have used a proton-scattering technique to determine penetration depths of several ions in light materials (Be, B, C, Al) for energies 50 to 500 keV. Agreement with the theory (Lindhard & Scharff 1961) is again quite satisfactory ($\pm 10\%$). They were, however, unable to establish the shape of the penetration distributions. Computer calculations of penetration in amorphous solids (Oen et al. 1963) also give distributions and magnitudes consistent with the theory.

Penetration distributions in single-crystal targets show dramatic evidence of the 'channelling' effect mentioned earlier in this section. Figure 4.47 shows integral penetration distributions for 20 keV Kr^{85} ions in four tungsten targets with the amorphous oxide results included for comparison (Kornelsen et al. 1964). On the depth scale, 1 mg/cm^2 corresponds to $5 \cdot 22 \times 10^3$ Å, indicating that half of the ions penetrate more than 560Å, (175 lattice constants) when incident along the <100> direction. This is longer by about a factor ten than the median penetration in amorphous material, and larger by about a factor two than that obtained from the same bombardment of a crystal not cleaned in uhv. Median penetration of various ions, energies and crystals (all at normal incidence) are given in Table 4.14 (Kornelsen et al 1964). The very much more penetrating 'tails' on the curves of Fig. 4.47, involving a few tenths of a percent of the incident ions, were subsequently found (Davies & Jespersgaard 1966) to involve room temperature interstitial diffusion of atoms brought to rest in undamaged regions at the end of their channel trajectories. Their final location is determined by their chance of encountering a lattice defect capable of trapping them in a site of higher binding energy. The maximum penetration (ignoring this diffusion effect) of Xe ions in three tungsten crystals is shown in Fig. 4.48 for energies 0·5 to ~100 keV. Up to 10 keV, the penetration depth increases roughly as $E^{3/2}$, while the tendency above 100 keV is toward $E^{1/2}$, characteristic of inelastic processes being dominant.

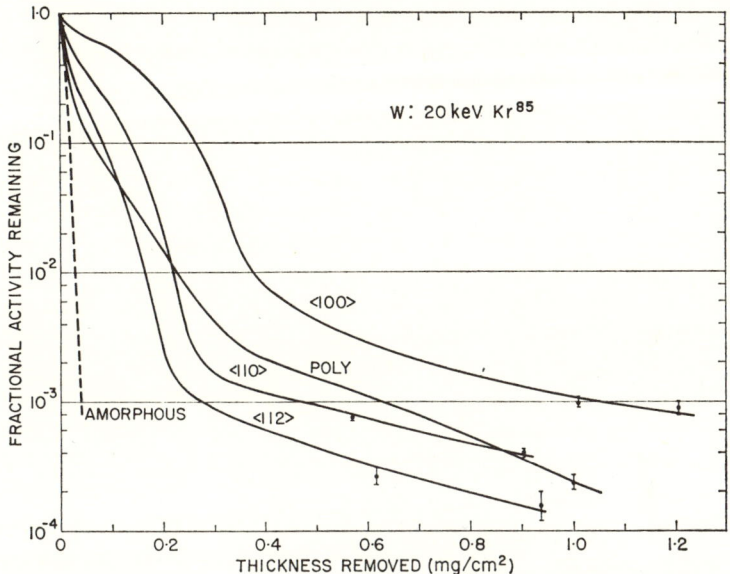

Fig. 4.47 Integral depth distributions for 20 keV Kr$^+$ ions in tungsten. The ions were normally incident on surfaces, along the directions indicated. (Kornelsen *et al.* 1964).

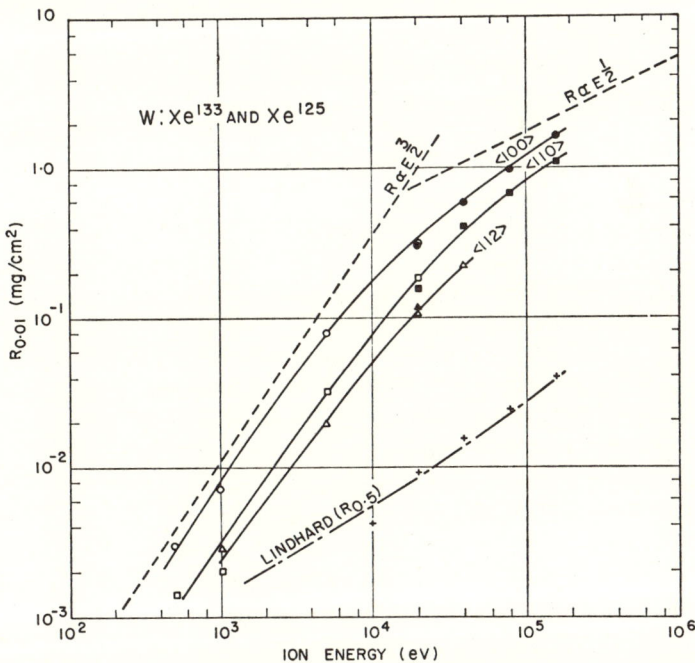

Fig. 4.48 Maximum range of channelled Xe$^+$ ions in tungsten as a function of energy. Lindhard (1965) median range (broken line) and 'amorphous oxide' results (Domeij *et al.* 1964b) (crosses) included. (Kornelsen *et al.* 1964).

TABLE 4.14
Median penetrations [$R_{0.5}$ ($\mu g/cm^2$)] for various bombardment parameters

Ion	Energy (keV)	<111>	<100>	<110>	<112>	Poly	Amorphous (b)
Xe^{133}	1.0		~1.6			~1.0	
	5.0		13	3.4	3.8	3.5	
	20		77	21	9	5.9	
	20(a)	56	33	10.7	10.5		8.0
Xe^{125}	40	135	145	31	12.7	12.0	15.0
	80	195	176	55			24
	160	410	490	89			42
Kr^{85}	0.5						
	1.0		3			~1.0	
	5.0			4.6		4.7	
	20		107	27	13	11.2	
	20(a)		59.4				10.0
	40	176	145	31			19.0
Ar^{41}	40	177	170	70	44		
Na^{24}	40	377	436	95	66		39.1

(a) These bombardments, and all with energies > 20 keV, done in a mass separator at a pressure of ~ 10^{-6} torr.
(b) Domeij et al. (1964b).
(from Kornelsen et al. 1964).

Distribution at higher energies (0.1 to 1.5 MeV) have been obtained (Eriksson et al. 1967; Eriksson 1967) under much more accurate control of the angles of ion incidence ($\pm 0.1°$). They show that the channelling effect is equally prominent at these energies, maximum penetrations at 1 MeV exceeding 2 mg/cm^2 (10^4Å).

Comparisons of channelling experiments with the theoretical treatment (Lindhard 1965) have been made for high energy protons in Al, Si (Bøgh et al. 1964) and Ta (Bøgh and Uggerhøj 1965). In the earlier paper, the intensity of the (p, γ) resonance reaction was measured as a function of the direction of incidence of the protons on the crystals. The critical angle ψ for channelling, defined by Lindhard (1965), was determined from the width of the reaction rate minima observed around the channel axis. In the later paper, the Rutherford scatter yield was used in a similar way. The results are in good agreement with the theory, giving the correct magnitude and variation from one channel to another for the selected energy (400 keV). Andreen and Hines (1967) have obtained critical angles ψ_2 (low ion energies) for H^+, D^+ and He^+ ions in gold by measuring the transmission of 1–17 keV ions through thin monocrystalline foils. Figure 4.49 shows their deduced values of ψ_2 for D^+ ions along the (011) channel. Note the high critical angles (> 10°) for energies below 3 keV. The full line is the theoretical one from an approximate expression of Lindhard (1965). The angles for He^+ were larger at the same energy by a factor 1.13 ± 0.04 whereas the theoretically expected ratio was 1.17.

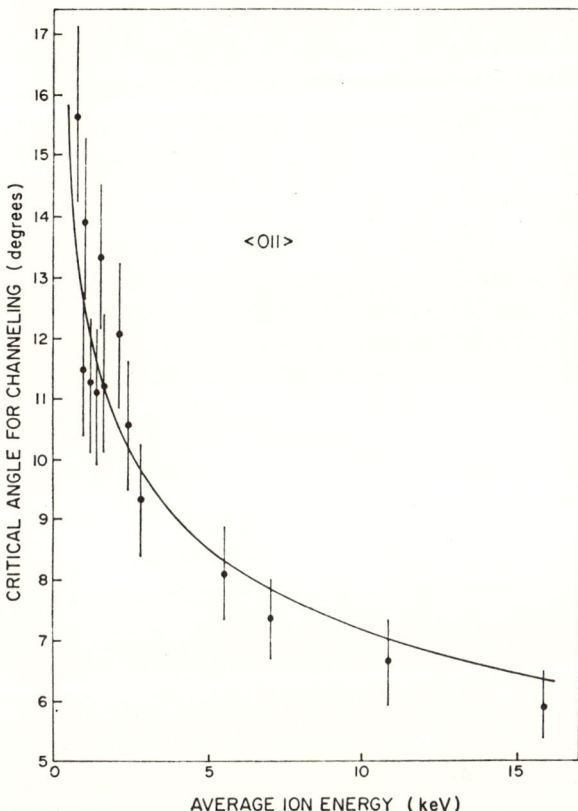

Fig. 4.49 Critical angle for channelling for D^+ ions in Au single crystal films as a function of average ion energy. (Andreen & Hines 1967).

The complementary effect of 'blocking' of the flight of emitted or scattered particles out of a crystal has been observed as a decrease in the intensity of α-particle emission from Rn^{222} entrapped in monocrystalline tungsten, when the emission direction was along close-packed rows or planes (Domeij & Björkqvist 1965). It was concluded that the Rn must be trapped at substitutional sites in the tungsten lattice. The work was extended to other materials (Ta and Au) and to an examination of the influence of surface contamination, lattice damage and temperature (Domeij 1966). A theoretical description of this blocking effect has been published by Oen (1965) based on a shadowing by neighbouring nuclei and including thermal vibration effects. Blocking effects of Rutherford-scattered protons has been observed by Tulinov et al. (1965) and employed as a method of examining crystal layers a few tens to hundreds of atomic layers thick by Nelson (1967). In the former case, the protons were registered on nuclear emulsion; in the latter on a phosphor screen. The shadows of close-packed atom rows and planes can be clearly seen in both cases.

4.3.6 Electron ejection

For ion energies $< 10^3$ eV (intervals I and II) the only significant inelastic process occurring during ion impact on surfaces is the 'potential ejection' of electrons. In this process, an electron tunnelling from the conduction band of the solid to an unfilled state of the ion leads to its neutralization and a second electron is sufficiently excited to allow it to surmount the work function barrier at the surface. The process occurs only for metals and semiconductors. For a detailed discussion of potential emission and of experimental and theoretical work prior to 1964, see Kaminsky (1965). Two mechanisms of of potential ejection are possible:

(i) Resonance neutralization to an excited state followed by Auger de-excitation.

(ii) Direct Auger neutralization to the ground state of the ion.

The mechanisms are energetically equivalent, both requiring that

$$V_i > 2\phi \tag{4.42}$$

where V_i is the ionization potential of the atom and ϕ the work function of the metal or semiconductor. On the basis of the energy distributions of the ejected electrons, Hagstrum (1954b) concludes that direct Auger neutralization is the correct mechanism provided the incident ion energies are low (< 100 eV).

The experimental data on electron ejection which appeared prior to 1950 all applies to surfaces covered to an unknown extent with contaminating layers of adsorbed gas. Results for clean surfaces were first reported by Hagstrum, who studied ejection by rare gas ions from polycrystalline tungsten (Hagstrum 1954a, 1956a) and molybdenum (Hagstrum 1956b) at ion energies from 10 eV to 1000 eV. His total yield (electrons/ion) is given in Fig. 4.50 for singly charged ions in their ground state. It is seen that the yield is approximately independent of ion energy in all cases. The energy distribution of the ejected electrons are shown for the same systems in Fig. 4.51. Almost all of the features of the energy distributions are accounted for by a theory based on the Auger ejection mechanism (Hagstrum 1954b).

Yields for multiply charged ions increase rapidly with charge state. The results observed for tungsten by Hagstrum (1954a) for rare gas ions of several charge states appear in Fig. 4.52; the electron energy distributions for ejection by multiply charged Xe ions are shown in Fig. 4.53. The influence of adsorbed gas on the ejection yield has been discussed in detail (Hagstrum 1956c). The effect of adsorption of a monolayer of nitrogen on tungsten on the yields for inert gas ions is summarized in Fig. 4.54. The yield is reduced for the light ions by about a factor two at low energies, the reduction decreasing as the ion energy is raised. For the heavier ions (Kr^+ and Xe^+) the reduction is even greater and more nearly uniform with energy. The net result is a yield lower throughout but changing more rapidly with ion energy. The reduction is found to be more pronounced for the electrons of higher kinetic energy (Hagstrum 1956c).

More recently, Hagstrum et al. (1965) and Takeishi and Hagstrum (1965) have made careful measurements of the electron energy distributions obtained from single crystal faces (Ni.(111); Ge(111); GaAs(111), (110) and ($\bar{1}\bar{1}\bar{1}$)). They find that the energy broadening in the distributions varies linearly with the ion velocity for energies < 100 eV (He$^+$, Ne$^+$ and Ar$^+$ ions). The techniques have been developed into an 'ion neutralization spectroscopy' (Hagstrum & Becker 1966; Hagstrum 1966) which gives some information concerning the state density functions for electrons in solids.

Some measurements have also been made (MacLennan 1966) of the ejection of electrons by metastable atoms. For helium and neon on W, the yields (at low kinetic energy) were found to be the same, within experimental error, as those of Hagstrum for ions extrapolated to zero energy. This result is consistent with the theory developed by Hagstrum (1954b), since the metastable atom should auto-ionize by a tunnelling of the excited electron into the conduction band of the solid.

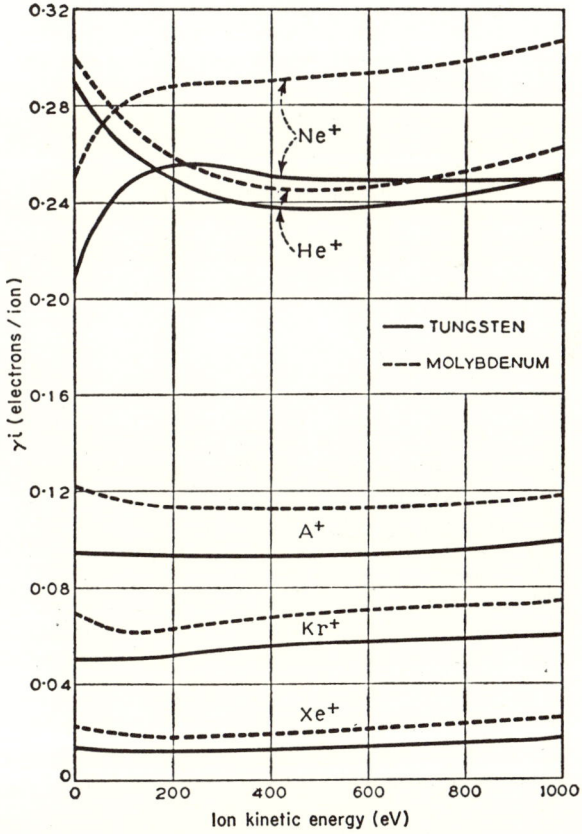

Fig. 4.50 Total electron yield vs ion kinetic energy for singly charged noble gas ions in the ground state on clean polycrystalline tungsten and molybdenum (Hagstrum 1956b).

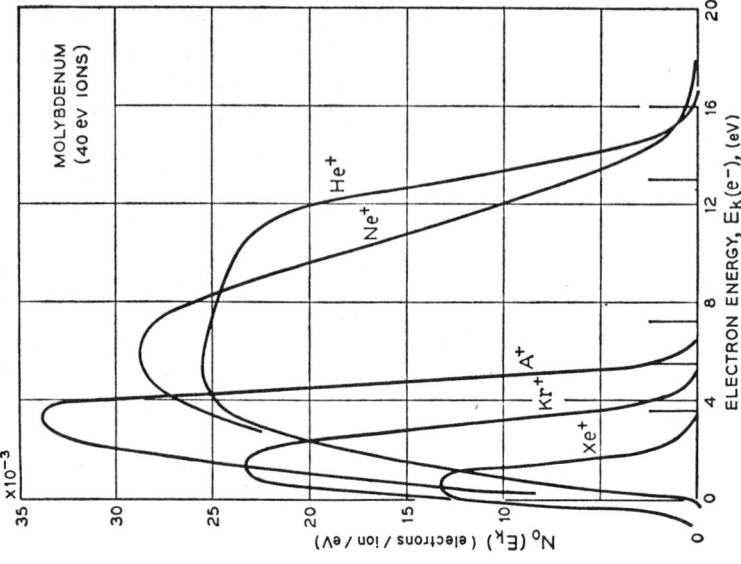
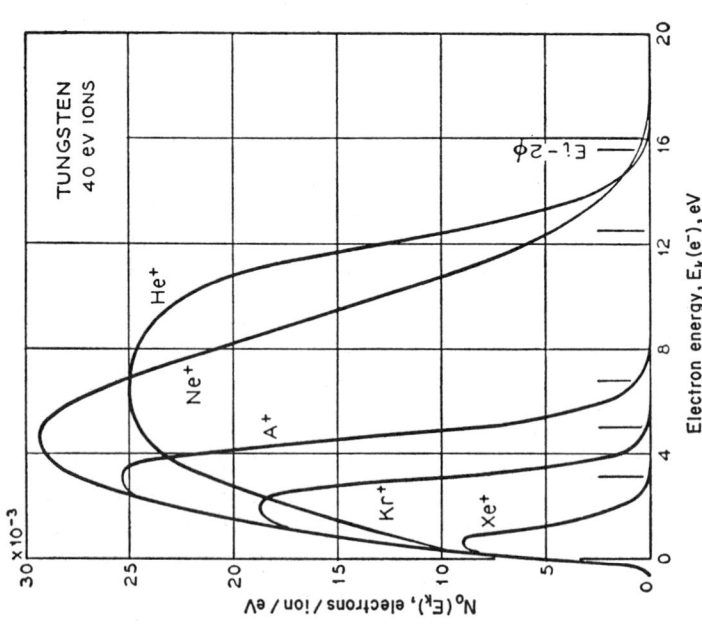

Fig. 4.51 Energy distributions of the electrons ejected from clean polycrystalline tungsten (a), molybdenum (b), by 40 eV noble gas ions. (Hagstrum 1956a, b).

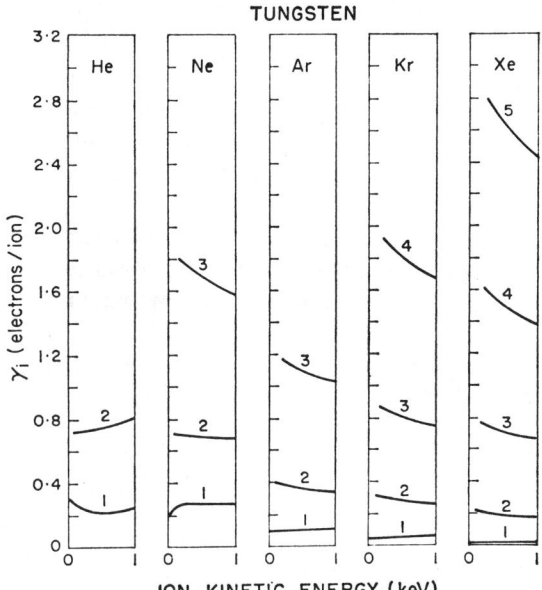

Fig. 4.52 Total electron ejection yield vs kinetic energy for multiply charged ions incident on tungsten. (Hagstrum 1954a).

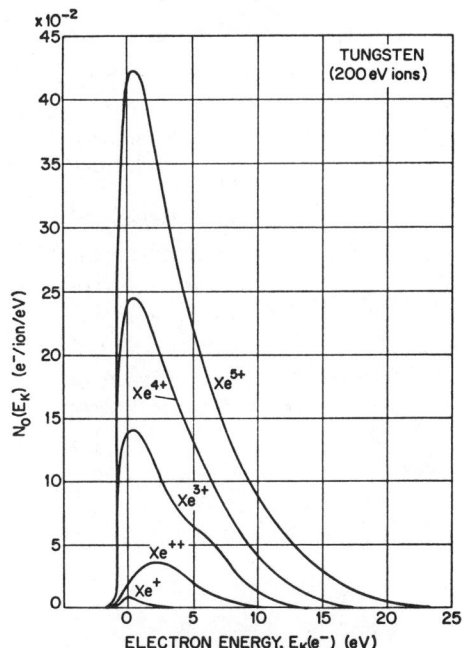

Fig. 4.53 Electron energy distributions for multiply-charged 200 eV Xe ions on tungsten (Hagstrum 1954a).

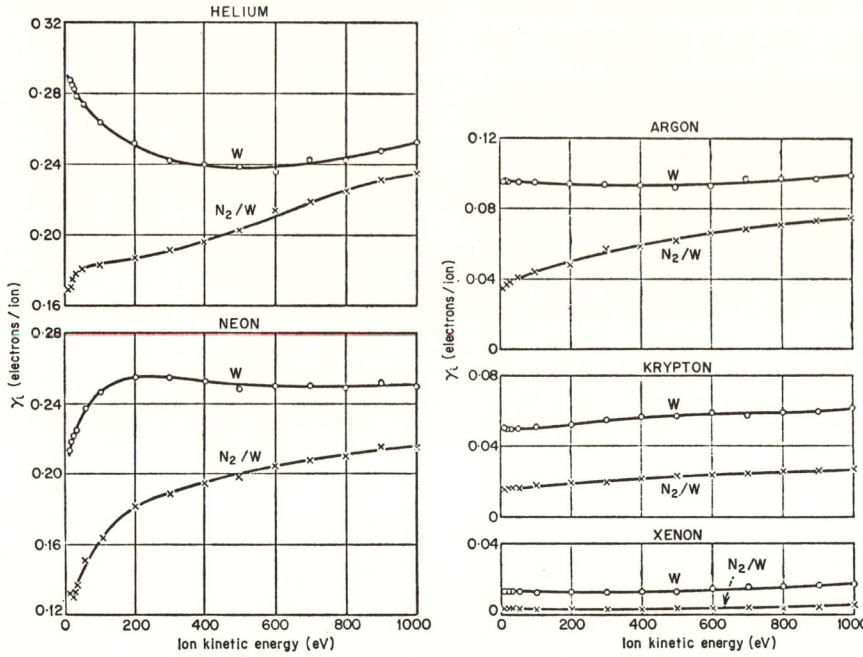

Fig. 4.54 The effect of adsorption of a monolayer of N_2 on W on the electron ejection yield for rare gas ions (Hagstrum 1956c).

The 'kinetic ejection' of electrons by ions, which depends on atomic excitation produced by the rapidly moving ion (see Section 3.4) has been reviewed by Medved and Strausser (1965) and Kaminsky (1965). For ions for which potential ejection does not occur, ($V_i < 2\phi$), yields below ~1 keV are rather small ($\lesssim 10^{-2}$) and in some cases a threshold energy of 1 to 2 keV is indicated. Although the agreement among different sets of experimental data is not as good as in the case of potential ejection, several general points are fairly well established. In the energy interval 1 to 10 keV, the kinetic ejection yield rises approximately linearly with ion energy. Brunnée (1957), for example, gives values for alkali metal ions on clean molybdenum which are summarized in Fig. 4.55. Magnuson and Carlston (1963b) find a linear rise with rare gas ions of 1–10 keV energy on several metals, starting from almost constant values (potential ejection) below 1 keV. At higher energies (10–100 keV) yields continue to rise, but usually less than linearly. Yields for a number of ions on tungsten, measured by Large (1963) are shown in Fig. 4.56. Arifov et al. (1962) and Telkovsky (1957) show that in this energy range, the yield is an almost linear function of the ion velocity. Yields at 100 keV ion energy are in the range 1 to 10 electrons/ion. At higher energies the yields for light ions begin to decrease, those for heavy ions continue to rise. Little experimental data is available for these energies.

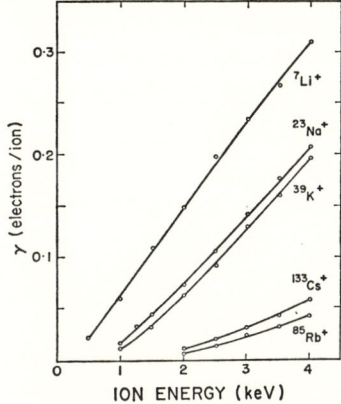

Fig. 4.55 Kinetic ejection of electrons from clean molybdenum by five alkali metal ions at energies up to 4 keV (Brunnée 1957).

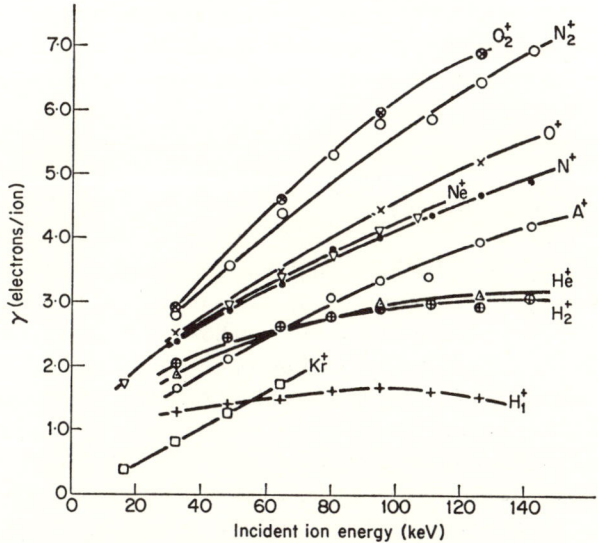

Fig. 4.56 Kinetic ejection yields for several ions of energy 20 to 140 keV on tungsten (Large 1963).

Yields from insulating and semiconducting solids have been measured by Batanov and Abroyan and their colleagues (Batanov 1961; Abroyan 1961; Abroyan & Movnin 1961; Abroyan & Lavrov 1963). The yields are higher by about an order of magnitude than those typical of metals, and thresholds are at a few hundred electron volts ion energy. Damage by the incident ions tends to produce irreversible changes in the yields from the alkali halides. The fact that kinetic ejection occurs for insulators supports the hypothesis that the ejected electrons originate as bound atomic electrons.

The energy distribution of the electrons ejected in the kinetic process almost always has a peak below 3 eV, and a tail extending to about 10 eV (Brunnée 1957; Waters 1958). The most probable energy does not depend strongly on the ion or its energy, whereas the tail extends to higher values at much higher ion energies (up to 1000 eV for 1 MeV protons).

The angular distribution of secondary electron flux ejected by He^+, Ne^+ and Ar^+ from polycrystalline tungsten was found by Klein (1965) to be very nearly 'cosine-law' when kinetic ejection was dominant (energy $\sim$4 keV). For potential ejection (energy 300 eV) the distributions were slightly peaked near the normal for a clean surface, but again cosine law for a gas covered surface. The results were independent of the angle of incidence, angles of 0° and 60° giving the same patterns.

Evdokimov et al. (1967) give extensive measurements of total ejection yield from flat polycrystalline surfaces (Cu, Ag, Zr, Mo, W, Bi and graphite) as a function of the angle of incidence (0 to 85°) of 25 to 30 keV rare gas and nitrogen ions. They find that up to a certain angle (50° to 80° depending on the target material) the yield increased as the inverse cosine of the incident angle, while for larger angles the increase is slower. They give an interpretation on the basis of decreasing 'transparency' of the surface layers with increasing angle, modified at large angles by a shadowing effect of atoms in the surface layers for one another.

The total ejection yield has been measured for monocrystalline targets by several experimenters. Magnuson and Carlston (1963c) found the yields from three flat Cu monocrystals for normally incident Ar^+ ions to differ by more than a factor two, the (111) being highest, (110) lowest. They proposed an explanation based on the opacity of the face to the incoming ions, which gave a fair account of the observed yield ratios. They point out that variations in yield between different polycrystalline samples are hardly surprising in view of their results. Similar results were obtained by Zscheile (1966) using a spherical single crystal of Cu bombarded by 28 keV N_2^+ ions in a beam small compared to the crystal. By rotating the crystal along two axes, he produced a map of relative yield as a function of exposed crystal face in which all low index directions can be easily identified. Again, yield variations spanned about a factor two.

Mashkova and Molchanov (1965a, b) have observed minima in the electron yield from single crystals of Cu, Ni and Zn, when beams of 20 to 40 keV Ar^+ ions were incident along low index directions, as the angle of incidence was varied. These localized minima were superimposed on the generally increasing yield occurring with increasing angle of incidence (see above, Evdokimov et al. 1967).

Theoretical treatments of kinetic electron ejection have been given by von Roos (1957), Parilis and Kishinevskii (1961), Harrison et al. (1965) and Sternglass (1957). In all cases, two separate processes are considered:

(i) The collision between ions and atoms which lead to the production of the electrons.

(ii) The escape of the produced electrons through the surface (see also Section 4.1.4).

The above authors make a variety of assumptions concerning the distribution functions of the ionizing collisions and the probability of escape of the resulting electrons. The ionizing events are predominantly large angle collisions, in which a substantial interpenetration of the atomic electron shells occurs. During such events, according to Russek (1963), tightly bound (inner shell) atomic electrons describing molecular orbits about both nuclei are able to absorb angular momentum due to the relative motion of the nuclei. The resulting excitation energy is distributed statistically among the outer shell electrons leading to ionization. The distribution in depth of such collisions will depend on the energy of the incident ion and the masses of both particles through the collision cross-section. The escape probability of the electrons is determined primarily by the depth at which they are produced and their rate of energy loss in their flight toward the surface. The problem is complicated by the presence of relatively high energy electrons which in turn cause lower energy secondaries. In spite of rather drastic simplifying assumptions, the main features of the experimental results are reproduced reasonably well (see particularly the comparisons made for polycrystalline material by Parilis & Kishinevskii (1961) and for a few monocrystals by Harrison *et al.* (1965)).

4.4 EMISSION OF IONS FROM HOT SURFACES

The emission of positive or negative ions from incandescent surfaces is a common occurrence and may affect the performance of hot-cathode ionization gauges or other uhv devices using heated cathodes.

4.4.1 Thermionic emission of ions

Thermionic emission of positive and negative ions from coated or impregnated cathodes has been observed for many years and ion sources of this type are frequently used in solid source mass spectrometry. When metals are heated close to their melting points, significant ion emission also occurs. The ratio of ions to neutrals evaporated from a metal is very small, e.g., Compton and Langmuir (1930) found that for W at 2800°K this ratio is $2 \cdot 5 \times 10^{-3}$. However, the emission of impurity ions from metals can be quite large, e.g. alkali metal ions from tungsten and other metals.

The current density of positive ions emitted at a temperature T is given by the equation

$$J^+ = \frac{2\pi k^2 eM}{h^3}(1-r)T^2 \exp\left\{-\frac{e\phi^+}{kT} - \frac{1}{k}\int_0^T \frac{dT}{T^2}\int_0^T C_p dT\right\} \quad (4.42)$$

where C_p is the heat capacity at constant pressure of an ion in the condensed state, r is the ion reflection coefficient, ϕ^+ the work function for positive ions, and M the ion mass. This equation is identical in form to that for thermionic electron emission except for a factor 2 and the specific heat term. Wright (1941) has verified this equation for Mo. Provided the same crystal surface is involved then

$$\phi^+ + \phi^- = q_v + V_i \; (\phi^+ \text{ and } \phi^- \text{ are the same sign}) \tag{4.43}$$

where ϕ^- is the electron work function, q_v is the heat of vaporization of the metal atom, and V_i is the ionization potential of the metal atom; Wright has also verified this relation for Mo within experimental error. Reasonable values for ϕ^+ are $8\cdot6$ V (Mo) and $11\cdot9$ V (W). Emission of positive ions from metals at temperature below their melting points has been observed for Cr, Cu, Fe, Mo, Ni, Nb, Rh, Ru, Ta and W.

As with the thermionic emission of electrons, the ions emitted from an incandescent metal surface have a Maxwellian velocity distribution.

The ratio of positive ions to neutral atoms evaporated per unit time can be calculated by thermodynamic equilibrium arguments and is given by the Saha-Langmuir equation

$$\frac{n^+}{n_0} = A^+ \exp\{(\phi^- - V_i)e/kT\} \tag{4.44}$$

where A^+ is a constant involving the ratio of the statical weights of ion and atom. Dobretsov (1952) obtains identical results for alkali and alkali-earth atoms by an argument based on statistical mechanics: for more complex atoms the results differ.

Many metals with alkali metal impurities emit positive ions at quite low temperatures, e.g., Pt emits Na^+, K^+ and Ca^+ at temperatures as low as 1300°K. Tungsten emits alkali ions in pulses above 1300°K (about 10^6 ions per pulse), the major ion being K^+; in the presence of oxygen the rate of emission is greatly increased (Datz et al. 1960; Minturn et al. 1960). The effects of oxygen have been interpreted by Winters et al. (1963) as resulting from the removal of W as an oxide, exposing 'clusters' of alkali atoms. A similar transitory enhancement of the ion pulse emission rate has been observed in the authors' laboratory after oxygen was adsorbed on tungsten at 300°K, the oxygen pumped away and the tungsten then heated. Thus the enhancement effect depends on an adsorbed oxygen layer not an oxygen atmosphere.

The compounds of alkali metals have been found to emit positive alkali metal ions at temperatures as low as 700°K. Copious emission of negative ions is observed from [BaSr]0 cathodes (see Section 7.6). Convenient sources of alkali metal ions (Na^+, K^+, Cs^+) can be constructed using an aluminosilicate molecular sieve material (Weber & Cordes 1966).

4.4.2 Surface ionization

Surface ionization occurs when atoms or molecules, from a molecular beam or vapour, impinge on an incandescent metal surface, are adsorbed for a short time and then evaporate as positive or negative ions. This subject has been reviewed in considerable detail by several authors (Dobretsov 1952; Zandberg & Ionov 1959; Kaminsky 1965) and only a brief account of those aspects relevant to uhv will be given here. The reader is referred to Kaminsky (1965) for an excellent survey of the theory and methods of measurements of surface ionization.

If we assume that the adsorbed particles come to thermal equilibrium with the surface then the Saha-Langmuir equation (4.44) can be used to find the ratio of positive ions to neutrals leaving the surface. The probability that an incident atom will leave the surface as an ion (the ionization efficiency) in the absence of external fields is given by

$$\frac{n^+}{n_i} = (1-C)[1+\exp\{(V_i-\phi^-)e/kT\}/A^+]^{-1} \qquad (4.45)$$

where n_i is the number of incident atoms and C their reflection coefficient. The ionization efficiency will be large for material with low ionization potential on a surface of high work function. For example, Fig. 4.57 shows the ionization

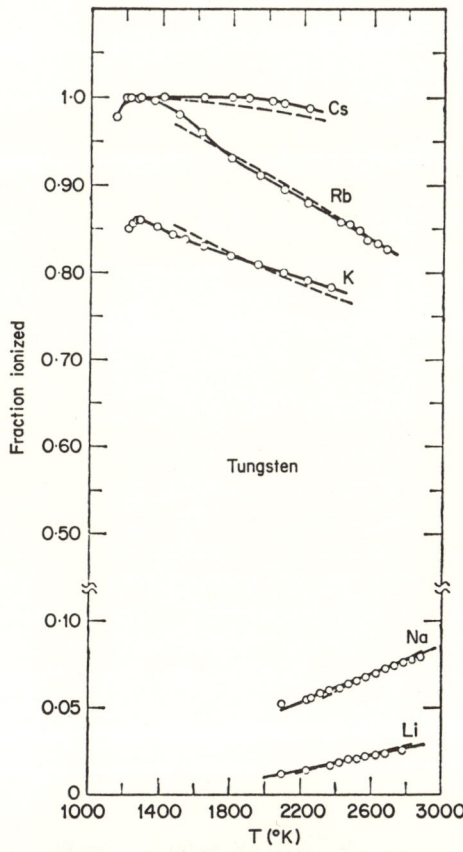

Fig. 4.57 Ionization efficiency of alkali metal atoms incident on tungsten (Datz & Taylor 1956).

efficiency for various alkali metals on tungsten as a function of temperature (Datz & Taylor 1956). The dashed lines show eqn. (4.45) with $C = 0$ and $A^+ = \frac{1}{2}$. On other metals (Pt for example) it is not possible to get agreement between eqn. (4.45) and measurements without assuming that $C \neq 0$ and varies with temperature (Datz & Taylor 1956). For further details of the experimental results on positive surface ionization the reader is referred to Kaminsky (1965).

Fomin et al. (1965) have observed the positive ion emission from an oxidized molybdenum filament exposed only to the residual gases in a mercury-pumped vacuum system (about 7×10^{-6} torr). K^+ and Rb^+ were observed as well as many organic ions with $m/e = 58, 72, 84, 86, 99, 95, 101, 110, 114$. The ion current increased to a maximum at 700°K and decreased rapidly at higher temperatures. The source of these organic ions is not clear since when organic gases were deliberately introduced the ion currents were *decreased*.

Lichtman (1965b) has observed the emission of alkali metal ions (K^+, Na^+, Rb^+, Cs^+, Li^+ in order of prominence) from stainless steel in the temperature range 700–1200°K. Lichtman suggests that impurities migrate through the system during bakeout as alkali halide molecules or clusters and that the original source of the alkali halide contamination is the glass portions of the vacuum system. Such contamination from glass has been observed previously (Bills & Evett 1959; Allen et al. 1960; Donaldson 1962).

TABLE 4.15
Negative Ions from a tungsten surface

Molecule	Filament	Ions observed
O_2	W	WO_3^-
N_2O	W	O^-, WO_3^-
NO_2	W	WO_3^-, O^-, NO_2^-
H_2S	W	HS^-, S^-, S_3^-, S_2^-
SF_6	W	SF_5^-, SF_6^-, F^-
C_6H_6	W, Pt	C_2H^-, C_2^-, CN^-*

* CN^- is a major peak with hydrocarbons and presumably results from an interaction of the gas and adsorbed nitrogen or nitrides on the filament.
(from Herron et al. 1965).

The ratio of negative ions to atoms evaporated in the absence of external fields is described by an equation similar to that for positive ions (eqn. 4.45), namely

$$\frac{n^-}{n_i} = A^- \exp\{(V_a - \phi^-)e/kT\} \qquad (4.46)$$

where A^- is a constant and V_a is the electron affinity. Work functions are usually larger than electron affinities, thus negative ion formation is greatest on low work function surfaces at high temperatures. Most measurements on negative surface ionization have been done for halogens, oxygen and sulphur on tungsten. Herron *et al.* (1965) have observed the negative ions formed by the interaction of some fairly complex molecules with hot tungsten and Pt filaments. Some of these observations with filament temperatures in the range 1200–1900°K are shown in Table 4.15.

CHAPTER FIVE

INTERACTION OF RADIATION WITH SURFACES

When radiation impinges on the surface of a solid, two general processes may occur which are of concern to measurements in uhv, namely (a) photo-electric emission, the release of electrons by photons, and (b) desorption or decomposition of adsorbed gas by the action of photons. The production of photons by electron impact on surfaces is discussed in Section 7.4.1.

5.1 PHOTO-ELECTRIC EMISSION

This subject is described in the classic book by Hughes and DuBridge (1932) and has been reviewed by Weissler (1956). We shall limit ourselves here to a discussion of photo-emission from metal surfaces in the main. Photo-emission from insulators and semiconductors has not been of great importance in most uhv measurements because, for electron optical reasons, insulators are seldom exposed to charged particles or sources of radiation. A detailed discussion of photo-effects in non-metals may be found in the general references listed above.

Photo-emission only occurs when the photon energy ($h\nu$) exceeds the potential barrier to the escape of an electron from the metal. This threshold relationship is expressed by the Einstein photo-electric equation

$$h\nu_0 = \phi \qquad (5.1)$$

where ν_0 is the threshold frequency and ϕ the work function of the metal. The maximum energy of the emitted electron from a metal at a temperature of absolute zero is given by

$$E_{\max} = h\nu - \phi, \qquad (5.2)$$

at non-zero temperature the maximum energy is higher than this value. The photo-electric threshold for most metals occurs at about 4 eV, i.e. at a wavelength, λ, of about 3100 Å. It is convenient to describe photo-electric effects in terms of the quantum energy of the radiation given by

$$h\nu = \frac{12 \cdot 378 \times 10^3}{\lambda} \; (eV), \qquad (5.3)$$

where λ is in ångströms.

The photo-electric yield, defined as the average number of electrons ejected from the surface per incident photon, depends on the frequency, angle of incidence and degree of polarization of the radiation and on the material of the target and the surface conditions (e.g. whether the surface is covered with a layer of adsorbed gas or not). It is the magnitude of the photo-electric yield which is of most interest in uhv measurements and it is on this aspect of photo-emission that we concentrate.

Metals have photo-electric yields of less than 10^{-3} electrons per photon in the visible and ultraviolet regions of the spectrum (2–7 eV). Metals have low yields in the long wavelength region both because of their high reflectivity in this wavelength and because the probability of energy transfer to the conduction electrons (which are relatively free) is very small. Figure 5.1

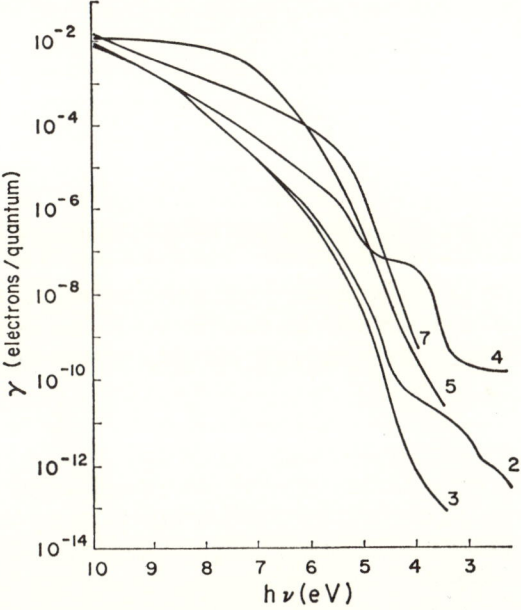

Fig. 5.1 Spectral distribution of photo-electric yield for various materials: (1) Pt, (2) Cu-Be oxidized once, (3) Cu-Be after second oxidation, (4) SrF_2, (5) CuI. (Tyutikov & Shuba 1960).

shows the photo-electric yield for various materials in the range 2–10 eV to indicate the low yields of most metals in the visible and ultraviolet; note the extremely wide variation of photo-electric yield in this limited energy range. The photo-electric yields of most metals in the visible portion of the spectrum do not differ very greatly from that of Pt shown in Fig. 5.1.

The photo-electric yield from semiconductors may be much larger in the visible and ultraviolet. Figure 5.2 shows the spectral distribution of quantum yield for various photosurfaces including both semiconductors and metals. It can be seen that Cs—Sb, for example, has a yield of about 10% over the whole range from 4–11 eV.

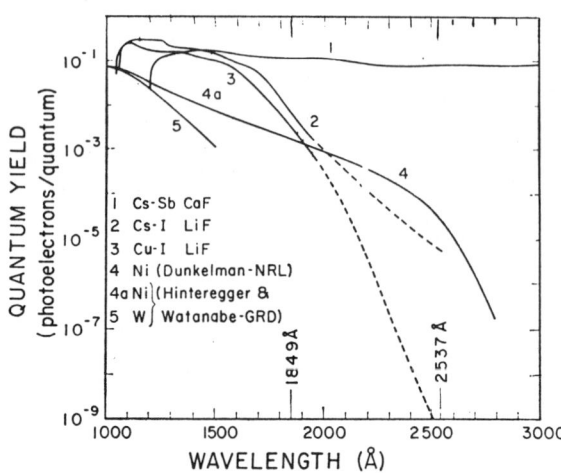

Fig. 5.2 Spectral distribution of photo-electric yield for various photo-surfaces. (Dunkelman et al. 1962).

The shortwavelength region ($\lambda < 1800$Å, $h\nu > 7$ eV) is called the vacuum ultraviolet and soft x-ray region. The wavelength where vacuum ultraviolet becomes soft x-rays is not clear; the differentiation is usually determined by the means used to produce the radiation. It is this short wavelength region extending out to $h\nu \approx 50$ eV which is of prime interest in many uhv problems because it is in this range that much of the radiation lies that causes troublesome effects in uhv measurements (e.g. the x-ray limitation in hot-cathode gauges is one of these problems, see Section 7.4.1). The photo-electric yield in the vacuum ultraviolet is much higher than in the visible. The reasons for this are two-fold: (*a*) the reflectivity of metal surfaces is lower in this wavelength range (see Fig. 7.12, and (*b*) the photons are now sufficiently energetic to be able to excite bound electrons in the solid. Photo-electric yields as high as 20% at $h\nu = 15$ eV have been observed.

Cairns and Samson (1966) have measured the photo-electric yield of W, Ni, Al, Zn, Cu, Be, Fe, Ti, Ta, In, Pt, Sn, Mo, Ag, Au and Pb in the photon energy range 10 to 60 eV. These samples had undergone a routine polishing and cleaning in methanol and were then measured at a pressure of less than 10^{-4} torr without heat treatment. Figure 5.3 shows a compilation of these data for W, Mo, Ta, Pt, Ni and Au. Comes and Wenning (1966) have measured the photo-electric yield of Ni and Au in the photon energy range 13·5 to 18 eV, their values for Ni are in reasonable agreement with Cairns and Samson (1966) but they observe a maximum yield for Au at about 16 eV of 6% compared with the maximum value of 14% observed by Cairns and Samson. All published measurements on the photo-electric yields of metals in the vacuum ultraviolet and soft x-ray region are on 'dirty' surfaces, in the uhv sense. The above data must be used with caution when applying these measured yields to clean metal surfaces in uhv.

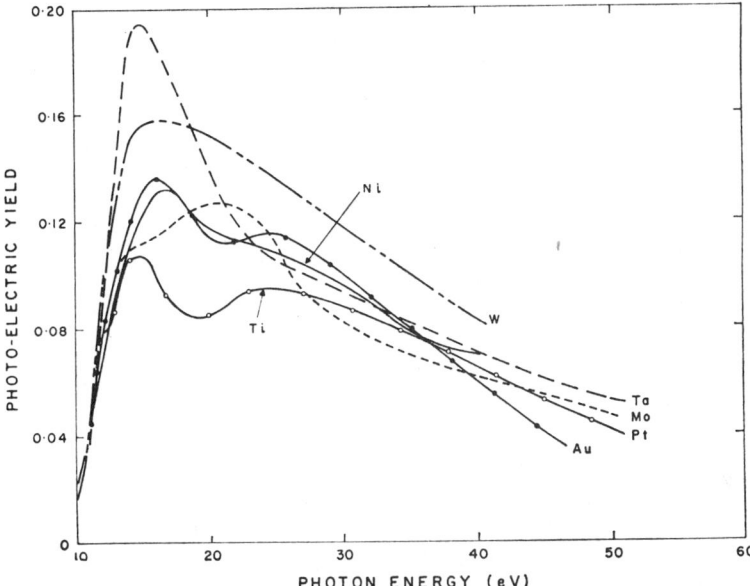

Fig. 5.3 Spectral distribution of photo-electric yield of metals in the vacuum ultraviolet. (Assembled from the data of Cairns and Samson 1966).

The effects of surface cleanliness are very pronounced near the photo-electric threshold. In this region, where the excited electrons have energies only slightly larger than that required for release, small changes of work function radically change the yield. At photon energies greater than 10 eV the electrons are excited to energies sufficiently large that small changes in work function do not seriously affect the yield. The photo-electric yield does not vary as greatly with surface conditions for photon energy greater than 10 eV. For example, Cairns and Samson have measured the photo-electric yield (η) and reflectance (R) of tungsten at 10 eV (*a*) at 25°C and (*b*) at 950°C where surface contamination should be reduced. The results are shown in Table 5.1. Here we see that the measured yield dropped from 28 to 6% while the reflectance increased. Thus the photo-electric yield in terms of *absorbed* photons ($\eta(1-R)$) changed from 38% to 13%. In general, the yield of clean surfaces in the vacuum ultraviolet is less than surfaces with contaminant layers.

The energy distribution of the emitted photo-electrons is too large a subject to be adequately discussed here; the reader is referred to Weissler (1956) for a detailed discussion of this subject. When the photons have sufficient energy to excite electrons from bound states the energy distribution of the emitted photo-electrons shows structure which can be associated with the corresponding energy levels in the metal. Examples of this effect are shown in Fig. 5.4, which shows the energy distributions of photo-electrons from Ni excited by photons with energies from 8·0 to 11·6 eV, and Fig. 5.5, which shows the energy distribution of photo-electrons from gold excited by 1·5 keV photons. It can be seen that even with 1·5 keV photons the vast majority of the photo-electrons have energies less than 50 eV.

TABLE 5.1
Photo-electric yield and reflectance of tungsten at a photon energy of 10 eV and different temperatures

T (°C)	Measured (η)	Reflectance (R)	Yield $\gamma = \eta(1-R)$
25	0·028	0·25	0·038
950	0·006	0·52	0·013

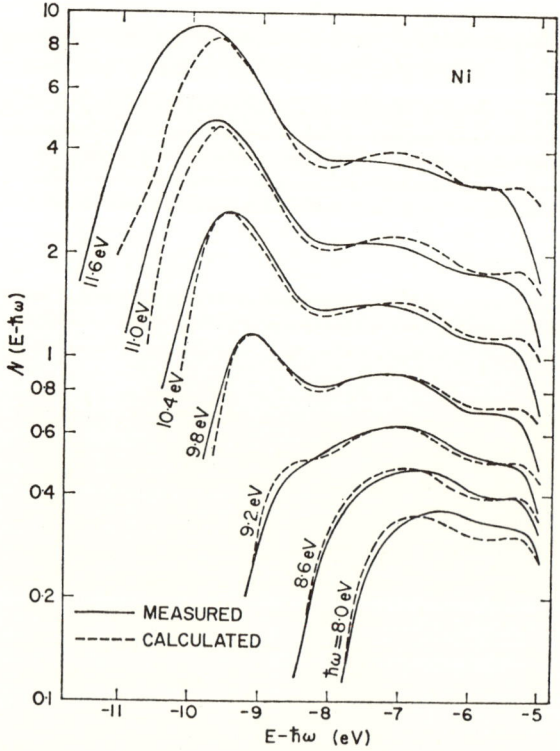

Fig. 5.4 Calculated and measured energy distribution of photo-electrons from Ni. (Blodgett & Spicer 1966).

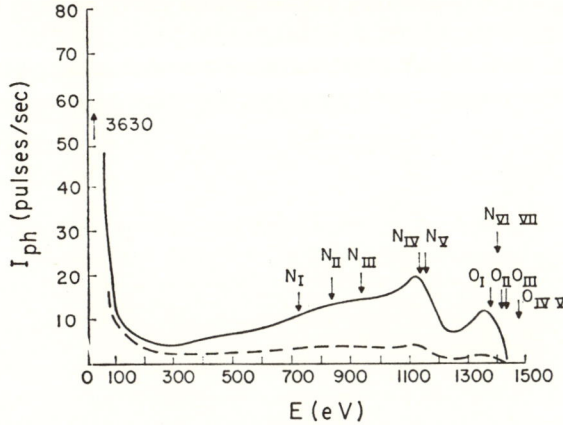

Fig. 5.5 Energy distribution of photo-electrons from gold film excited by 1·5 kV x-ray photons. Full line, uncorrected values; broken line, corrected for variation in effective slit width of electron energy analyser. The count rate at $E = 0$ is 3630 pulses/sec. (Nakhodkin and Mel'nik 1964).

5.2 INTERACTION OF PHOTONS WITH ADSORBED GASES

Four processes involving neutral particles can occur when light impinges on an adsorbed layer, (a) photodesorption of molecules, (b) photosorption of gas under the influence of light absorbed by the solid, (c) photodecomposition of the adsorbed molecules or of the solid, (d) photocatalysis of a reaction between adsorbed molecules or between an adsorbed molecule and the surface. Only process (a) and (c), photodesorption and photodecomposition, are of importance to uhv where they can cause serious technological problems whenever a flux of energetic photons exists.

Only very limited measurements of photodesorption have been made and it is not possible to form firm conclusions about the detailed mechanism of the process. The first observations of photodesorption occurred during work on photoconductors. Work at fairly high pressures (about 10^{-3} torr) in Russia has been reviewed by Terenin and Solonitzin (1959), photodesorption of CO from Ni, H_2O from Cd and Zn, and O_2 from ZnO (with excess Zn) has been observed by various Russian authors referred to by Terenin and Solonitzin. Photosorption of O_2 and CO on ZnO (Zn) and O_2 on silica gel was observed; photodecomposition of H_2O desorbed from Cd and Zn by light of energy greater than 5 eV was also observed. Medved (1958) has observed the photodesorption of O_2 from sintered ZnO. Stone (1964) has summarized much of the early work on photocatalysis (in particular by ZnO suspension) and reports measurement at high pressures ($\sim 10^{-3}$ torr) of photosorption and photodesorption of O_2 on ZnO and TiO_2 and the photo catalysis of CO to CO_2 on ZnO. Haber and Stone (1963) have observed the photodesorption of oxygen from nickel oxide by illumination with red light at 7000Å. Mechanisms to explain the photodesorption from semiconductors are discussed by Stone (1964) and Terenin and Solonitzin (1959).

The photodesorption of gas from metal surfaces at low pressures is of more concern to uhv problems. The parameter of principal technical interest is the photodesorption cross-section as a function of wavelength of the light.

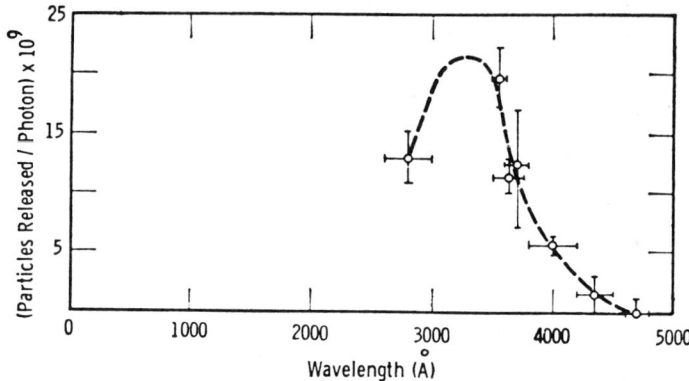

Fig. 5.6 Measured yields for photodesorption of CO from Ni. (Lange 1965).

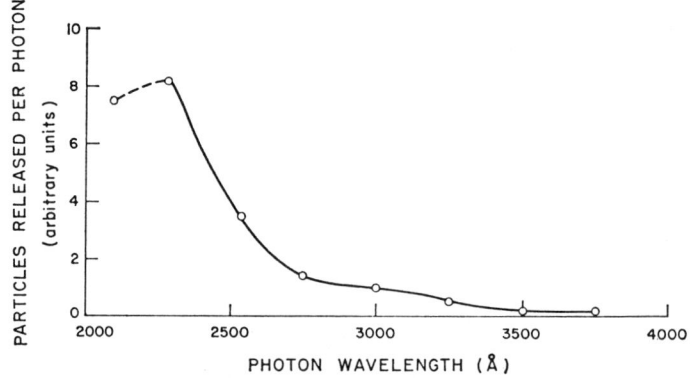

Fig. 5.7 Measured yields (in arbitrary units) for photodesorption of CO from Pyrex glass (Lange 1965).

Lange and Riemersma (1961) and Lange (1965) have studied the photodesorption of CO from Ni and glass, under uhv conditions, as a function of photon energy. Figure 5.6 shows the measured yield (molecules desorbed per incident photon) of CO from a Ni surface as a function of wavelength. A maximum is observed at about 3200Å (3·9 eV). Figure 5.7 shows the measured yields, in arbitrary units, from CO on a Pyrex surface. At an equilibrium pressure of 2×10^{-9} torr the maximum yield, at 2300Å, was 8×10^{-8} molecules per photon with an indeterminate coverage (less than 10^{12} molecules/cm^2). Negligible photodesorption was observed from quartz.

Photodesorption of CO from W was also observed but the data were not repeatable.

To convert the measured yields to a photodesorption cross-section it is necessary to know the surface coverage since the photodesorption cross-section is given by

$$Q_{p-d} = \gamma/\sigma \qquad 5.4$$

where γ is the yield (molecules per photon) and σ the surface coverage (molecules per cm^2). Surface coverage was not directly measured in Lange's experiments so that it is not possible to calculate cross-sections. Lange assumes a coverage of about 10^{14} molecules/cm^2 and, since the maximum yields are about 10^{-8} molecules per photon, he deduces a maximum cross-section of the order of 10^{-22} cm^2. This value may be compared with the maximum cross-section for desorption by electron impact (see Section 4.2.1) of CO on W of about 10^{-18} cm^2.

Adams and Donaldson (1965) have observed the photodesorption of CO from Fe, Ni and Zr in the range of photon energies from 6 to 2 eV. Figure 5.8 shows the measured yield for CO from Ni with a maximum at 2700Å (4·6 eV). The observed yields are two orders of magnitude higher than those found by Lange; the surface coverage was about 5×10^{14}

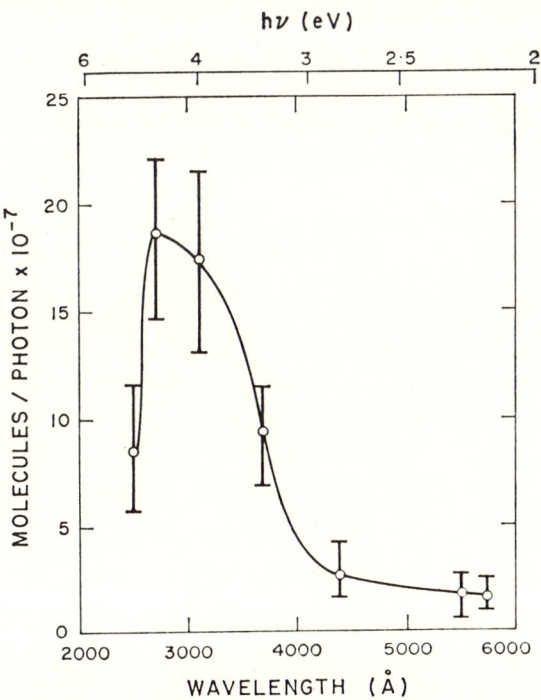

Fig. 5.8 Measured yields for photodesorption of CO from Ni (Adams and Donaldson 1965).

molecules/cm^2, giving a maximum cross-section of $\sim 4 \times 10^{-21}$ cm^2. Photodesorption yields of CO from Zr are still rising at 2500Å (5 eV) suggesting a maximum at higher energies. Photodesorption from Fe was only observed at 250°K and no wavelength dependence could be determined because the signals were too small. No photodesorption of CO from W was observed.

No simple relationship exists between the photon energy for maximum desorption and the activation energy of desorption. It is probable that the incident photon excites the adsorbed molecule to a repulsive state from which it may escape the surface with an energy equal to the difference between the photon energy and the activation energy of desorption. For a discussion of the various energy distributions that may occur see Adams and Donaldson (1965). From the limited amount of data available, photodesorption appears to be highly selective both of the nature of the adsorbed gas and that of the surface. For example, photodesorption was *not* observed for N_2, H_2, and CO_2 on Fe, Ni or Zr in the range 2000 to 6000Å (Adams 1964).

Problems of outgassing induced by synchrotron radiation in particle storage rings have been studied (Fischer & Mack 1965; Bernardini & Malter 1965); these problems are also discussed in Section 4.4. Fischer and Mack (1965) have calculated the photon spectrum of the synchrotron radiation from a 3 GeV electron storage ring of 55 m diameter. Integration under the spectrum shows there are 19×10^{22} photons (with energy greater than 10 eV) per circulating electron. Outgassing of aluminium by gamma radiation (from CO^{60}) has been studied (Muehlhouse *et al* 1965); the magnitude of the outgassing of well cleaned and baked aluminum systems was $(7 \pm 4) \times 10^{-11}$ torr litres/sec cm^2 per watt/gm. Irradiation in a 1 MW reactor showed that the outgassing was primarily due to the gamma radiation rather than the fast neutron flux.

Decomposition of glass by ultraviolet photons can be a serious problem in uhv systems with large ultraviolet fluxes. Kenty (1953) has shown that irradiation of Pyrex glass by photons of about 4 eV caused photodecomposition of H_2O and CO_2 in the body of the glass, the H_2 and CO diffusing out of the glass.

CHAPTER SIX

MECHANICAL PROPERTIES OF MATERIALS AT VERY LOW PRESSURES

The low rate of accumulation of gas on an exposed surface in uhv leads to observable effects in several mechanical properties. In particular, compared to the properties at atmospheric pressure:

(i) large increases can occur in the adhesive forces between surfaces brought into contact;

(ii) friction coefficients and rates of wear of contacting bodies in relative motion can also increase;

(iii) in some cases, observable changes occur in ductility and the resistance to creep, fatigue and fracture.

Careful investigation of these effects has been given impetus by the development of space flight and of uhv systems in which controlled mechanical motion is required. This chapter will give a brief discussion of the above effects.

6.1 ADHESION OR COLD WELDING

The adhesion of thoroughly cleaned metal surfaces brought into contact with only a few grams force in uhv has been observed by Johnson and Keller (1967a, b). The contact areas were very small ($\sim 10^{-7}$ to 10^{-5} cm^2). In their experiments, the metal couples Ag—Ag, Ag—Ni, Cu—Ni (Johnson & Keller 1967b), Ti—Ti and Mo—Mo (Johnson & Keller 1967a) were cleaned by combinations of heating and ion bombardment in a system with a residual pressure of $\sim 5 \times 10^{-10}$ torr. Contact forces were measured with a sensitive torsion beam (precision ± 0.01 gram) and the electrical contact resistance monitored. Deliberate exposure of the surfaces to gas was used to study the effect of adsorption on the adhesion. When adhesion occurred, the tensile force usually required to break the contact was nearly equal to the compressive contacting force applied, and the electrical resistance of the contact remained at a constant low value (10^{-2} to 10^{-3} ohm) during unloading until just before the contact was broken.

Adhesion with no activation energy barrier was observed to occur for all the couples investigated, although sufficiently clean Ti and Mo surfaces were obtained only with extreme difficulty. No evidence was found of the rupturing of the contact bonds by relief of elastic stresses during unloading.

From calculations based on the elastic properties of the metals, the contact load was believed to be carried predominantly by elastic deformation, and the joint strength to approach the bulk value of the weaker member of the couple.

The dependence of adhesion on exposure of the surface to atmospheric gases was much less for the soft metals (Ag, Ni, Cu) than for Ti and Mo. In the former case, even after exposure to 40 torr of dry air for 19 hours, adhesion could be detected, whereas in the latter, 6 hours at 10^{-4} torr caused adhesion to disappear. For the soft metals, water vapour gave a reversible inhibiting effect. In all cases, no adhesion was observed following exposure to atmosphere even after subsequent bakeout and heating in uhv.

The energy per unit area required to cleave lamellar solids such as mica and graphite has been found to be much higher in uhv than in atmosphere (Bryant et al. 1963). The results were interpreted in terms of the coulomb attraction between the surface charges on the lamellae which are very effectively neutralized by the adsorption of polar molecules, in particular, water.

Smith and Gussenhoven (1965) have reported adhesion forces of several kg/cm² when optically flat quartz plates were brought into contact in uhv (no compressive loading). They interpreted the adhesion to be due to van der Waal's dispersion forces at average separations of approximately 100Å. The same force has been observed between non-contacting bodies at larger separations (Sparnaay 1958; Abrikosova & Deryagin 1957) and treated theoretically (Casimir & Polder 1948; Lifshitz 1956).

Studies of adhesion have also been done in uhv using contact forces large enough to cause plastic flow in the asperities of the contacting surfaces. Winslow and McIntyre (1966) measured the bond strength in tension of annealed copper and various steels and titanium alloys after compressive loading at temperatures up to 500°C in uhv (5×10^{-9} torr). No surface preparation other than the vacuum bakeout was used. For static loading, no adhesion was found at room temperature for stresses up to 80% of the compressive yield strength. At temperatures $\geq 300°C$, copper and several of the softer alloys showed adhesion; the strengths of the bonds increasing with the time of compression to values as high as 1·25 kg/mm². The increase is thought to indicate the dispersal of contamination layers at the interface by diffusion into the bulk. The harder alloys showed no adhesion under static loads at any temperature up to 500°C. When the samples were subjected, during compression, to a repeated 2° relative rotation (3 cps for 10 to 300 sec about an axis normal to the contacting surfaces), adhesion was observed in all cases at or below 300°C. Bond strengths in the range 0·1 to 1·0 kg/mm² were common, and in some cases exceeded the applied load. The disruption of surface layers by plastic flow is thought to cause the observed enhancement of adhesion.

Similar static experiments have been reported by Batzer and Bunshah (1967) in an investigation into 'warm' welding techniques. The samples were degassed, argon ion bombarded, and degassed again (400°C, to remove the

Mechanical Properties of Materials at Very Low Pressures

trapped argon) before being compressed to near their yield strength at a temperature 0·1 to 0·5 of the melting point. Welds with nearly 100% of the bulk strength were created with OFHC copper following compression at 400°C. Good welds were also produced in Be and Ti at 400°C as determined by metallographic examination. The latter specimens broke in tension at points of stress concentration rather than at the welds, so that the weld strengths could not be determined.

The sensitivity of adhesion to monolayers of chemisorbed gas has been reported for Al, Cu, Mg, Pb and Ti (Gilbreath & Williams 1966). Notched samples of the metals were fractured in tension, given a known exposure to a selected gas, and the fractured surfaces then brought back into contact with 47·5 kg compressive force. The exposures of various gases required to reduce the resulting bond strength (usually 35 to 40 kg for a minimum exposure) by half are summarized in Table 6.1. Of the pure gases, oxygen and ethylene seem most effective in reducing adhesion.

TABLE 6.1
Exposure necessary to reduce adhesion bond strength by half

Metal (*purity*)	Environment					
	Air	Ar	C_2H_4	H_2	N_2	O_2
Al (99·99%)	−3 (*a*)	4	−2	3	1	−4·2
Cu (OFHC)	−4·9	>1	−2·6	>1	>1	3
Mg (99·95%)	−3	>1	−2	>0	—	−3
Pb (99·99%)	>1	4	—	>4	>2	2
Ti 99%	−4	>0	4·5	−3·7	—	−4

(*a*) Exposure in $\log_{10}$ (torr sec). From Gilbreath and Williams (1966).

Salisbury *et al.* (1964) have examined the adhesive behaviour of silicate powders (3 to 100 micron particle size) in uhv. They sifted the powders at pressures $< 10^{-9}$ torr and observed the geometry of the resulting deposits and the forces necessary to disrupt them. The particles were found to stick predominantly on their first encounter, and to adhere with forces 10 to 100 times their weight. The resulting deposits were relatively loosely packed but exhibited considerable shear and bearing strength. From the reduction of adhesion caused by adsorption of a monolayer of oil vapour or by exposure to atmosphere, the adhesive forces were deduced to be of either the dispersion or chemical bond type rather than gross electrostatic attraction between the particles.

6.2 FRICTION, LUBRICATION AND WEAR

For general discussions of this subject, see the books by Bowden and Tabor (1964) and Rabinowicz (1965). The basic mechanism of friction in the absence of lubricating films is thought to be the adhesion of very small contacting areas and the breaking of the resulting bonds in shear. In the atmosphere, surface films (particularly oxide layers on metals) reduce the number and strength of adhesive bonds, lowering the friction coefficient. The low rate of reformation of surface films in uhv prevents this automatic lubrication, and causes the friction to increase drastically. Thus for example, Brown and Burton (1966) report coefficients of friction for Cu on Cu at $\sim 1 \times 10^{-9}$ torr to vary as shown in Fig. 6.1. It was necessary to reduce the loading force from 100 to 70 g above 200°C to allow sliding to occur at all.

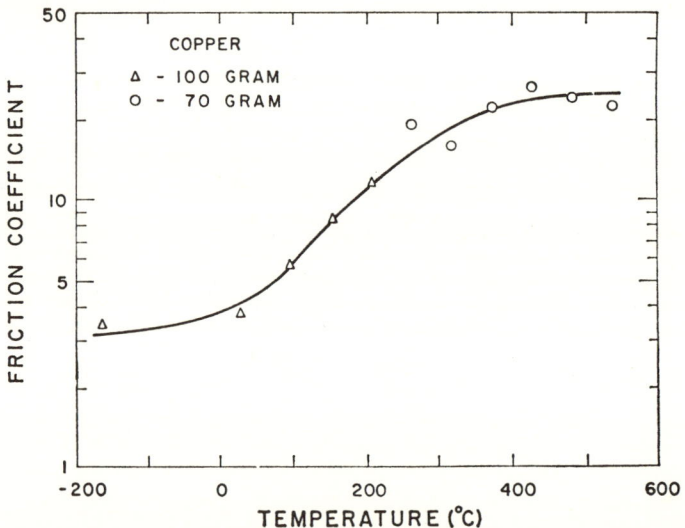

Fig. 6.1 Coefficients of sliding friction of Cu on Cu in uhv as a function of temperature. (Brown & Burton 1966).

Measurements of the friction and wear of diamond on diamond (Bowden & Hanwell 1964) show interesting pressure effects. The equilibrium value of the sliding friction coefficient, determined by a reciprocating slider on a flat surface, is shown as a function of ambient pressure in Fig. 6.2. In each case, the coefficient was found to be 0·05 to 0·1 for the first traverse of the slider, to increase for about 1000 traverses, and thereafter to remain constant at the value shown. It should be noted that the changes occurring between 760 torr and 10^{-6} torr are relatively minor. In the same experiments, the authors were able to show that the rate of wear also increased markedly when the pressure was decreased. They found friction to be lowest (for all exposed faces) when sliding occurred along the most closely packed atomic rows. Most dramatic of all was the effect of exposure of the surfaces to the background gases; an exposure of 6×10^{-7} torr sec (20 min at 5×10^{-10} torr) with the surfaces separated was sufficient to cause the friction coefficient to drop from 0·9 to 0·1. The value returned to 0·9 within 10 to 20 traverses of the slider as the very thin adsorbed layer was removed.

Several types of lubrication studies have been conducted under uhv conditions. The basic objective has been to provide, between the contacting surfaces, a film of very low shear strength which is not removed too quickly under the specified operating conditions of pressure, temperature and bearing load. For applications in which the outgassing rates of the lubricants are not important (e.g. some spacecraft applications) and in which temperatures do not exceed ~150°C, certain types of oils and greases have been shown to

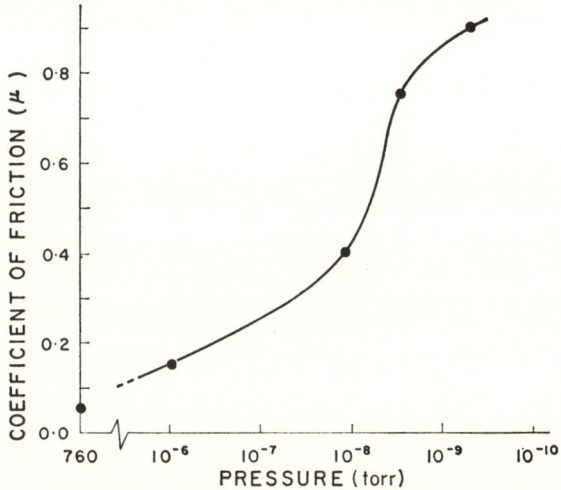

Fig. 6.2 The equilibrium sliding friction coefficient of diamond on diamond as a function of ambient pressure. (Bowden & Hanwell 1964).

perform well over periods of 1 to 2 years (Parcel *et al.* 1963). In their best cases, the lubricants were impregnated into phenolic retainers for ball bearing assemblies which were enclosed in double shieldings to minimize outgassing and evaporation while providing a relatively high local pressure at the bearing surfaces. The assemblies were used to support the rotors of electric motors turning continuously at 8000 rev/min. Chamber pressures were usually between 10^{-7} and 10^{-9} torr.

When outgassing rates must be minimized or high temperatures are encountered, dry lubricants become necessary. Molybdenum disulphide has been fairly widely used in such applications, particularly when high-speed motion is not required. Basic studies of MoS_2 (Johnson & Vaughn 1956) have shown that its lubricating action depends on the existence of an adsorbed layer of amorphous sulphur at the interface, produced by the bearing wear during the initial running. The same authors found exposure to air, and particularly to water vapour, to *increase* the friction coefficient by about a factor 4 over that found in vacuum (a typical value for MoS_2 in vacuum is 0·1). The lifetime of MoS_2 layers depends strongly on the method of bonding and on the applied load, but times up to a few thousand hours have been reported (Parcel *et al.* 1963).

Vapour deposited gold films have also been reported as lubricants in uhv (Spalvins & Buckley 1966). The gold films (1800Å thick), which must be strongly adherent, were evaporated onto surfaces of nickel and nickel alloys

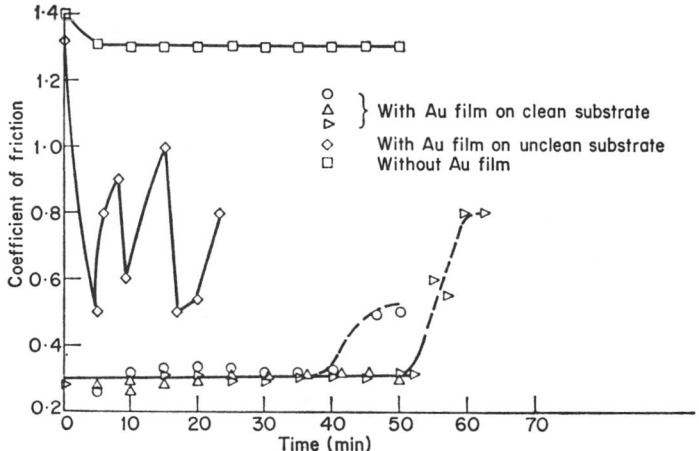

Fig. 6.3 Coefficient of sliding friction of a Nb rider travelling at 2·5 cm/sec on a vapour deposited gold film on Ni/10%Cr. (Spalvins & Buckley 1966).

(Ni, Ni/10%Cr, Ni/5%Re) which had been prepared by electron bombardment heating (~650°C) and which exhibited clear thermal etch patterns. Diffusion type interfaces were produced in all these cases when deposition was carried out in uhv at substrate temperatures ~425°C. The etched grain boundaries are thought to provide reservoirs of the lubricant (gold in this case). The alloys gave longer lifetimes and lower friction coefficients than the pure nickel, a fact attributed to their greater hardness as a base for the gold films. Figure 6.3 shows the coefficient of friction vs time of a 4·8 mm diam Nb slider loaded with 250 g travelling at 2·5 cm/sec at 25°C on gold deposited as described above on a Ni/10%Cr alloy. Note the difference between the 'cleaned' and 'uncleaned' samples, which demonstrates the difference between a diffusion interface bond and a van der Waal's bonding of the gold film.

6.3 CREEP, FATIGUE AND FRACTURE

Creep is the irreversible strain (elongation) of materials at constant stress and temperature. This property has become important in refractory alloys with the

TABLE 6.2
Creep rates of Tantalum alloys in UHV

Alloy	Composition	Duration (hrs)	Stress (kg/mm^2)	Min. creep rate [a] (cm cm^{-1} hr^{-1})	Duration (hrs)	Stress (kg/mm^2)	Min. creep rate (cm cm^{-1} hr^{-1})
			$T = 1095°C$			$T = 1205°C$	
T-222	Ta—9·6%W + 2·4%Hf + 0·01%C	>5000	11·2	2×10^{-7}	>5000	5·64	$1·25 \times 10^{-6}$
Ta-10W	90%Ta/10%W	1400	11·2	$6·4 \times 10^{-6}$	1100	5·64	$9·4 \times 10^{-6}$
T-111	Ta + 8%W + 2%Hf	1400	11·2	$3·8 \times 10^{-3}$	1300	5·36	$1·1 \times 10^{-5}$

(a) Creep rates tend to increase somewhat with time.

development of high-temperature power systems (e.g. turbines and liquid metal heat exchangers) for the aerospace industry. The refractory metals rapidly absorb atmospheric gases which modify (usually increase) the strain rate by promoting grain boundary motion. At atmospheric pressure, the resulting corrosion rate is so rapid as to make these materials unusable and, even at 10^{-6} torr, strain rates are markedly affected. To obtain reliable creep data, an uhv environment is therefore necessary. Equipment for creep testing refractory metal alloys in uhv is described, for example, by Buckman and Hetherington (1966). Their furnace can maintain the test volume (3·8 cm diam, 15 cm long) at 1900°C at a pressure of 1×10^{-8} torr. Typical of the results are those reported by Titran and Hall (1966) who measured the creep rate of various Nb and Ta alloys at 1095 and 1205°C and stresses of ~ 7 kg/mm^2. Relatively minor changes of alloy composition were found to produce large changes in creep rate. Table 6.2, for example, shows the results for three Ta alloys which differ in creep rate under identical conditions by more than two orders of magnitude. The test pressures generally fell below 10^{-8} torr within 20 hours and approached an equilibrium of $\sim 3 \times 10^{-9}$ torr. Chemical analysis of the samples after the completion of the tests indicated that the oxygen content had not increased by more than 10 to 20 ppm, an amount considered unlikely to affect the creep properties.

Kramer and Podlasek (1963) have observed that the stress-strain relation of aluminum single crystals at room temperature depended on the system pressure. The stress required to produce a 6% strain was found to decrease by almost a factor two when the pressure was reduced from atmospheric to the 10^{-8} torr range. They suggest that the surface oxide acts as a barrier preventing the egress of dislocation 'tangles' through the surface. At low pressure, the original oxide film is disrupted, dislocations escape, the internal stresses are relieved and additional strain occurs. A similar mechanism has been postulated to explain the dependence of the fatigue strength of aluminum on pressure (Shen *et al.* 1966) between atmospheric pressure and 10^{-5} torr.

Room temperature tensile tests on molybdenum have shown that its ductility (i.e. strain-at-fracture) is enhanced in uhv (Feuerstein & Rice 1966; Feuerstein *et al.* 1964). The strain-at-fracture is observed to increase from 32% at atmospheric pressure to 38% at 5×10^{-10} torr with significant changes occurring below 10^{-8} torr. The authors suggest that strain induced desorption of gas from grain boundaries strengthens them and inhibits the fracture mechanism (microcracks originating at grain boundaries). The stress-strain relation below the fracture point is independent of pressure.

Part B

PRESSURE MEASUREMENT

CHAPTER SEVEN

GENERAL CONSIDERATIONS OF PRESSURE MEASUREMENT

In this Chapter, attention will be focused primarily on the problems and errors which can arise in the measurement of low pressures with ionization gauges. It should not be forgotten that the dynamic range of pressure over which these gauges are required to operate may be as much as nine orders of magnitude. In this perspective, reliability and accuracy of ionization gauges in a wide variety of systems is really rather impressive. Particular total pressure gauge designs and characteristics are described in Chapter 8, and partial pressure measuring devices in Chapter 9.

7.1 CALIBRATION OF UHV GAUGES

The pressure in a vacuum system is defined as the force exerted by the gas per unit surface area:

$$P = \frac{\text{Force (dynes/cm}^2\text{)}}{(1 \cdot 3332 \times 10^3)} \text{ torr} = (7 \cdot 500 \times 10^{-4}) F_g. \tag{7.1}$$

The calibration of any gauge consists of relating its indication to pressure either by a direct measurement of F_g, or by calculation based on gas kinetic theory, measured values of F_g, and dimensional constants of the system.

We will limit our considerations to measurements which attain accuracies of about $\pm 10\%$ or better.

7.1.1 Absolute pressure measurement

In practice, F_g is measured almost without exception by the difference in levels it creates between two communicating columns of a liquid of known density; one column exposed to the gas and the other to a much lower pressure. When mercury is used as the liquid, both the forces inherent at the mercury-glass interface and the inertial forces produced by mechanical vibrations become troublesome at about 10^{-2} torr for columns of the optimum diameter (2·0–2·5 cm) (Kistemaker 1945; Carr 1964). In addition, the vapour pressure of mercury at room temperature (about 2×10^{-3} torr) requires that a cold trap separate the mercury manometer from the gauge under calibration. The attendant vapour stream effect can lead to errors which will be discussed below (see Section 7.1.2). Using oil as the working fluid and interferometric measurement of the liquid levels, pressure differences as low as 2×10^{-6} torr have been shown to be detectable (Thomas et al. 1962; Aubry and Delbart 1965) if measurements are made to 1/10 of an interference fringe. In this case care must be taken to avoid gas absorption and temperature gradients in the oil, and particularly vibrational disturbances of the liquid surfaces. Absolute accuracies consistent with the above detection limit (i.e. 2% at 10^{-4} torr) have been claimed (Aubry & Delbart 1965).

7.1.2 The McLeod gauge

The most common method of establishing pressures of permanent gases in the range 10^{-5} to 10^{-1} torr from absolute measurements employs the mercury-filled McLeod gauge. In this gauge, a sample of the gas is compressed by an accurately known ratio (usually between 10^4 and 10^6) and the correspondingly increased pressure measured by the difference in mercury levels. Descriptions of various types of McLeod gauges and details of their operation have been given, for example by Leck (1964), and will not be repeated here.

Important assumptions implicit in McLeod gauge measurements are that, during the compression,

(i) the perfect gas law is obeyed,
(ii) the number of molecules in the sample remains constant,
(iii) the gas temperature remains constant.

Corrections for departure from the perfect gas law have been calculated using the van der Waal's constants of the gas (Jansen and Venema 1959). For all the common gases, the corrections are negligible over the pressure range normally used (i.e. compressed sample pressure $< 10^2$ torr). Condensable gases such as H_2O, Hg and C_6H_6 (benzene) cannot be measured in a McLeod gauge except at temperatures such that the vapour pressure exceeds the compressed sample pressure.

The validity of assumption (ii) has been firmly established for some gases by the use of a variable compression technique (Nottingham 1961) in a specially constructed McLeod gauge (Podgurski & Davis 1960). For helium (Podgurski & Davis 1960), and possibly CO and CO_2 (Utterback & Griffith 1966), permeation and/or adsorption may cause the number of molecules in the sample to decrease significantly upon or following compression.

Provided that the ambient temperature is stable and the gas is compressed slowly, assumption (iii) is not difficult to satisfy. From the gas law, the error introduced will be equal to the fractional change of the absolute temperature. Slow compression has the further advantage of avoiding surface electrostatic charging in the capillaries (Leck 1964).

The use of a cold trap to prevent the mercury vapour (vapour pressure of

Hg at 26°C is 2×10^{-3} torr) from the McLeod gauge reaching the gauge being calibrated gives rise to three additional sources of error:

(i) Thermal transpiration effects can cause unpredictable differences in pressure as great as 5% between the two sides of an unsymmetrical cold trap. Smaller errors may occur even across a symmetrical trap (Edmonds & Hobson 1965). The error will vary with pressure and with the depth of immersion. To minimize the effect, cold traps should be, as nearly as possible, symmetrical U-tubes bent from a single piece of glass tubing. (The first bends *outside* the cold zone should also be symmetric.)

(ii) The one-directional flow of mercury vapour from the McLeod gauge to the cold trap (the 'vapour stream' effect) produces a diffusion gradient which is dependent on the gas, the pressure, the temperature and the system geometry (Ishii & Nakayama 1961; Meinke & Reich 1963; de Vries & Rol 1965). This error can be reduced to a small value ($< 1\%$) either by cooling the McLeod gauge (Rothe 1964; deVries & Rol 1965) or by reducing the conductance between the mercury surface and the trap (Meinke & Reich 1963; Edmonds & Hobson 1965).

(iii) Variations in level of the trap coolant can change the number of gas molecules adsorbed on the trap surface. This change can cause pressure differences between the gauges when the connecting conductances are small, as is usually the case in McLeod gauge systems. For many gases, solid CO_2 (196°K) is a more suitable trap coolant than liquid nitrogen (77°K) because it decreases the amount of gas adsorbed on the trap surfaces.

When all the precautions indicated above have been observed, the basic source of error which limits the accuracy of the McLeod gauge is the unpredictable variation of the forces acting at the mercury-glass interface. This produces uncertainties in the capillary depression which are approximately proportional to the depression and therefore increase nearly inversely with the capillary cross-sectional area A. Since the gauge sensitivity is proportional to A, the resulting pressure uncertainty cannot be reduced by decreasing A, but is determined primarily by the size of the compression volume. For the largest practicable volumes (4-5 litres), provided care is taken in preparing the inner surface of the capillaries (Rosenberg 1939; Podgurski & Davis 1960), the resulting uncertainties are about 1% at 10^{-4} torr; 3% at 10^{-5} torr.

Moser and Poltz (1957) have described a major modification of the McLeod gauge capable of extending its lower limit of measurement to about 10^{-8} torr. In their gauge a very small (5×10^{-8} litre) cylindrical chamber with accurately measured dimensions serves to define the compressed volume. Optical reflection is used to establish an accurately reproducible position of the mercury surface which forms the lower circular face of this chamber. The pressure of the compressed gas is read on a 20 mm diameter stand-tube in which the variation of capillary depression with meniscus height can be accurately determined (Kistemaker 1945). To establish

TABLE 7.1
Characteristics of gauge calibration systems

Type (1)	Gauge pressure (2)	Valid time range (3)	Measured pressure range (4)	Claimed accuracy (5)	Reference (6)
a) Static expansion Mode I	$P_a \dfrac{V_1}{V_1+V_2}$	$t \gg \dfrac{V_2}{C_A}$*	10^{-9} to 10^{-3}	$\pm 2\%$	Edmonds and Hobson (1965)
Mode II	$P_1 \dfrac{V_1}{V_1+V_2}$	$t \gg \dfrac{V_2}{C}$	10^{-6} to 10^{-4}	$\pm 10\%$	Schuhmann (1962)
b) Linear rise rate	$\dfrac{P_1 C_1}{V_2}(t-t_L)$	$t_L < t \ll \dfrac{V_2}{C_1}$	10^{-8} to 10^{-2}	$\pm 3\%$	This laboratory
c) Single stage pressure division	$P_1 \dfrac{C_1}{C_1+C_2} \dfrac{1}{1-(C_2/S)}$	$t \gg \dfrac{V_2}{C_2}$	10^{-11} to 10^{-3}	$\sim \pm 5\%$ ($p > 10^{-8}$ torr)	Davis (1963)
d) Quadratic rise rate	$\dfrac{P_1 C_1 C_2}{V_2 V_3}(t-t_L)^2$	$t_L < t \ll \dfrac{V_2}{C_1}, \dfrac{V_3}{C_2}$	10^{-9} to 10^{-3}	$(\sim \pm 10\%)$	Alpert and Buritz (1954)
e) Quadratic rise rate plus pressure division	$\dfrac{P_1 C_1 C_2}{V_2 V_3} \dfrac{C_3(t-t_L)^2}{C_3+C_4(1-C_3/S)}$	$\dfrac{V_3}{C_3} \ll t \ll \dfrac{V_2}{C_1}, \dfrac{V_3}{C_2}$	10^{-10} to 10^{-4}	$\sim \pm 10\%$	Lange and Eriksen (1966)
f) Multistage pressure division	$\dfrac{P_1 C_1 C_2 C_3}{(C_2+C_2')(C_3+C_3')(C_4+C_4')}$	$t \gg \dfrac{V_2}{C_2'}, \dfrac{V_3}{C_3'}, \dfrac{V_4}{C_4'}$	10^{-9} to 10^{-4}	$\pm 10\%$	Roehrig and Simons (1961)

*C_A the conductance of valve A when open.

General Considerations of Pressure Measurement

the zero pressure level in the stand-tube, they used two different initial compression volumes and a constant pressure. Alternatively a very low pressure ($< 10^{-9}$ torr) or two pressures having an accurately known ratio could be used.

7.1.3 Extensions to lower pressure

Only ionization gauges have proved to be practical for uhv pressure measurements (see Chapter 8). Those designs which are suitable for uhv have *upper* limits of measurable pressure, set by electron mean free path and ion space charge effects, of $< 10^{-1}$ torr. Their calibration against absolute pressure gauges (Section 7.1.1) is therefore not possible over any part of their operating range except with the interferometric gauge (Aubry & Choumoff 1965). In the pressure range where accurate comparisons with a McLeod gauge or interferometric gauge can be made ($> 10^{-4}$ torr), uhv ionization gauges are often not exactly linear (see Fig. 8.5). The common practice of assuming the measured sensitivity at 10^{-4} to 10^{-3} torr to apply to all lower pressures can introduce errors of 20% or more. Two general techniques have been used to obtain uhv ionization gauge calibrations:

(i) The sensitivity variation of the gauge with pressure is measured over the range from uhv to $\sim 10^{-3}$ torr. The gauge is then compared with a McLeod gauge or an interferometer gauge above 10^{-4} torr to obtain the absolute sensitivity.

(ii) Gas is allowed to flow from a reservoir at a measured high pressure through two or more constrictions of accurately measured conductance (or conductance ratio) to uhv pumps of known speed, creating calculable pressures in the intermediate volumes.

Both the techniques rely on simple predictions of gas kinetic theory; namely that in a static expansion, the pressure and volume are inversely related, and that the equilibrium rate of flow of gas through a conductance C is given, in the molecular flow range, by $Q = C\Delta P$ where ΔP is the pressure difference and C is a constant.

Brief descriptions of six systems which have been used for calibration are given below. Corresponding schematic diagrams appear in Fig. 7.1 and pertinent characteristics are listed in Table 7.1.

Static expansion. This method requires two volumes whose ratio is accurately known. Two operating modes have been used. In Mode I, the volumes V_1 and V_2 (see Fig. 7.1) are of the same order of magnitude, and the gauge (G_1) is operated in V_1. If an initial pressure P_a is set up in V_1, then upon opening the valve (A) the final pressure P_{b1} is given by

$$\frac{P_{b1}}{P_a} = \frac{V_1}{V_1+V_2}. \tag{7.2}$$

Any change of sensitivity of the gauge can be detected by comparing the

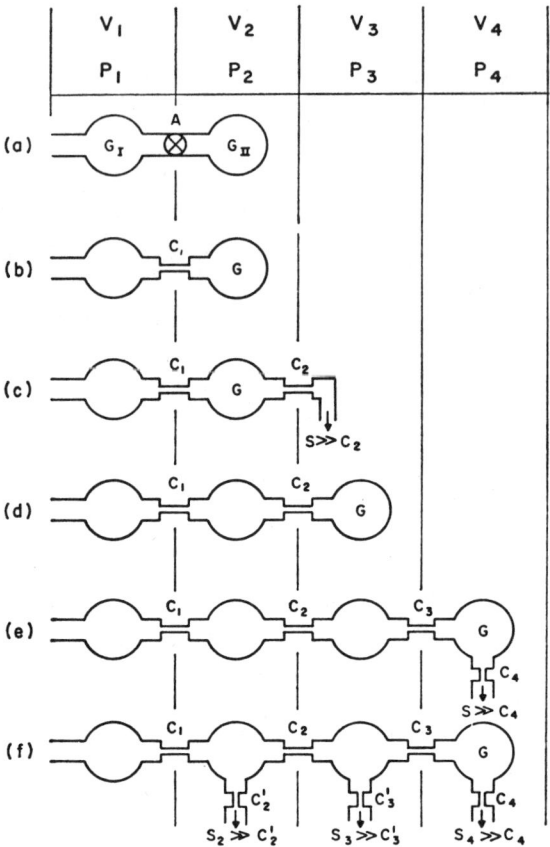

Fig. 7.1 Schematic diagram of various arrangements used for uhv gauge calibration. The letters (a) to (f) correspond to those in Table 7.1.

ratio of its indications before and after opening A with the ratio $V_1/(V_1+V_2)$ If A is reclosed, V_2 evacuated and the cycle repeated n times, then

$$\frac{P_{bn}}{P_a} = \left(\frac{V_1}{V_1+V_2}\right)^n. \tag{7.3}$$

The variation of the gauge sensitivity over the range P_a to P_{bn} can then be established. To complete the calibration the gauge must be compared with a McLeod gauge or interferometric gauge above 10^{-4} torr. A calibration of this type has been reported with $V_1 \approx V_2$ (Edmonds & Hobson 1965).

A second type of static expansion calibration, Mode II, involves $V_2 \gg V_1$, the gauge (G_{II}) being operated in V_2. The initial pressure P_1 is made high enough to be measured directly. A single expansion yields

$$P_2 = P_1 \left(\frac{V_1}{V_1+V_2}\right) \tag{7.4}$$

and the extension to multiple expansions is obvious. A number of variations of this mode have been described (Schuhmann 1962; Smetana & Carley 1966; Reich & Meinke 1966).

The pressure range covered and the accuracy claimed are included in Table 7.1 for good examples of the above two methods. Also included in the Table are the time restrictions which must be observed (column 4) and the expression for the pressure at the gauge (column 3).

It is possible to correct for the initial gas present in the volume and for the background gas evolution by performing a 'blank' experiment. Corrections can also be made for the pumping of the calibration gas if the pumping speeds are not large (see Section 7.2). Replacement phenomena occurring during the pumping, however, give rise to uncertainties. The method becomes impractical for pressures within one order of magnitude of the system background, and for pumping speeds giving time constants less than $\sim 10^3$ seconds.

When multiple expansions are used, volume ratio errors are cumulative, a fractional error f leading to a fractional error nf in the final pressure ratio for n successive expansions.

Linear rise rate. This method assumes that for a constant P_1 (see Fig. 7.1) and $P_2 \ll P_1$, the rate of flow gas through a fixed conductance C_1 is a constant. For high values of P_1, the high pressure end of C_1 may in some cases be in transition flow rather than free molecular flow. Since P_2 is usually in the molecular flow range, a transition region must then exist within C_1. Any variation of this transition region produced by the variation of P_2 would cause the above assumption to be violated. Such violations have not, however, been experimentally observed in the authors' laboratory for values of P_2 up to 10^{-1} torr with $P_1 \approx 10$ torr. Thus in the absence of any pumping or source effects in V_2, we expect

$$P_2 = \frac{P_1 C_1}{V_2} t \qquad (7.5)$$

where t is the time since P_1 was established. Usually $P_1 C_1$ and V_2 are not known absolutely. Only the variation of gauge sensitivity with pressure is determined by dividing the ion current by t. This method is limited in practice to gauges of very low pumping speed and to gases which do not chemisorb.

A second and perhaps more serious source of error is the lag time t_L which occurs between the establishment of P_1 (at $t = 0$) and the start of the linear rise in P_2 (see Section 2.4). Equation 7.5 should actually be

$$P_2 = \frac{P_1 C_1}{V_2} (t - t_L). \qquad (7.6)$$

Provided that the gauge is assumed to be linear in the region immediately above the starting pressure, t_L can be found by extrapolating the straight portion of the ion current trace back to the starting level. It should be noted that this in fact *assumes*, over a small pressure range, the result being examined, namely the constancy of sensitivity with pressure. Using a bakeable uhv valve as the conductance C_1, values of t_L as high as 10 seconds have been observed in the authors' laboratory for the lowest leak rates employed ($\sim 2 \times 10^{-10}$ torr/sec). For higher leak rates ($> 10^{-7}$ torr/sec) values of 1 to 2 seconds were typical. Accurate values of the relative sensitivity (I^+/t) can thus be obtained only for times exceeding about 1 minute. To cover a large range of pressure would then require excessively long times if a single value of P_1 and C_1 are used. It is possible to use a series of values of P_1 and/or C_1 and to normalize the resulting relative sensitivity curves where they overlap. Relative sensitivity curves accurate to about $\pm 1\%$ over the pressure range 10^{-8} to 10^{-2} torr have been obtained in the authors' laboratory for several Bayard-Alpert type gauges (BAG's) measuring inert gas pressures. Because of rapidly increasing values of t_L (P_1 was kept constant at ~ 10 torr) and system background pressures of $1-2 \times 10^{-10}$ torr, extensions to lower pressure were not possible. Since excellent linearity was observed in every case below 10^{-5} torr, this limitation was not considered serious. Calibration against a McLeod gauge above 10^{-4} torr served to complete the absolute calibrations to an accuracy of approximately ± 2 to 3%, over the whole range.

Single-stage pressure division. An accurately known ratio of two greatly different conductances is the basis of this method. A steady-state flow is established with P_1 a directly measurable pressure and the gauge operated in V_2. If the pumping speed S is large compared to C_2, the pressure downstream from C_2 will be small compared to P_2 and need not be known accurately in determining the effective conductance

$$C_2' = C_2 \left[1 - \frac{C_2}{S} \right] \approx C_2. \qquad (7.7)$$

The ratio C_2/S can be determined by measuring the pressure ratio across C_2 when the gas flow is large enough that both are far from background. A gauge in the pumping line is, however, necessary since the correction is usually not negligible. Most frequently the same type of gauge is used for this purpose as that being calibrated and the ratio of their indicated pressures used in the correction.

To provide calibration pressures in the uhv range while retaining P_1 in the direct measurement range, the ratio C_2/C_1 must be very large. A thin-walled aperture of calculable conductance (Bureau *et al.* 1952) is most often used for C_2. Various types of porous media (porcelain, sintered glass or silicon carbide) have been employed for C_1 (Owens 1965; Christian & Leck 1966). These retain their molecular flow characteristics up to easily measurable pressures (> 10

torr) and have conductances in the range 10^{-4} to 10^{-6} l/sec. Accurate values of C_1 can be determined by observing $P_1\,\mathrm{d}V_1/\mathrm{d}t$ or $V_1\mathrm{d}P_1/\mathrm{d}t$ when gas is allowed to flow through the plug into an evacuated chamber ($P_2 \ll P_1$). A variation of the method has been reported in which $P_1\,\mathrm{d}V_1/\mathrm{d}t$ was measured for each calibration point and C_1 never explicitly calculated (Hayward & Jepsen 1962). In another example, lower values of P_1 were measured with a McLeod gauge and maintained constant by flow from a higher pressure reservoir (Davis 1963).

This method has the important advantage that pumping effects in V_2 due to adsorption or ionic pumping in the gauge can be tolerated provided the pumping speeds are small compared to C_2. This is a far less stringent requirement than in the previous two methods, and makes the calibration of most ionization gauges easier for most of the simple gases. It should, however, be pointed out that establishing equilibrium flow at low pressures may require long times due to variation of adsorption pumping speeds. Davis (1963) reports equilibrium times of 30 minutes at 10^{-9} torr and progressively longer times at lower pressures. Near the limits of his system ($2-3 \times 10^{-12}$ torr) the different gauges agreed only within about a factor of two.

Quadratic rise rate. When two conductances are arranged in series, then as long as $P_3 \ll P_2 \ll P_1$, the pressure in the final volume V_3 increases quadratically with time after the establishment of a constant pressure P_1 in V_1. The flow through C_2 must be free molecule flow over the whole range of P_2 encountered in the experiment. The pressure inequality stated above requires $P_1 > 10^5 P_3$, or $P_1 > 100$ torr if P_3 is to extend high enough to allow accurate calibration against a McLeod gauge ($\sim 10^{-3}$ torr).

As is the case with the linear rise rate, this method is highly sensitive to adsorption and gauge pumping in V_3 and to a lesser extent in V_2 also. Likewise the time lag effect can cause errors here. The square root of the gauge indication must be plotted against time, linearity assumed and the intercept with the initial P_3 level used to establish the effective starting time. Any significant non-linearity in the gauge makes this determination difficult.

The advantage of the quadratic leak rate method is that a large pressure range can be covered in a reasonable time with a single set of values of C_1 and C_2. For example, Alpert and Buritz (1954) have reported the calibration of a BAG for helium and argon between 10^{-9} and 10^{-3} torr in experiments lasting about 24 hours each.

Quadratic rise plus single-stage division. By using the final chamber V_3 of the previous method as the initial volume of a pressure division system, some of the advantages of both can be preserved. A system of this type has been described by Lange and Eriksen (1966). Gauge pumping and adsorption phenomena can be made less serious than in the quadratic leak rate method, while a large pressure range can be covered in a reasonable time. In addition to the considerations mentioned under the appropriate two sections above, one further condition must be satisfied: C_3 must be chosen so that only a small fraction of the gas entering V_3 flows through to V_4. Since C_4 must be fairly large to minimize adsorption and gauge pumping errors, P_4 will be much smaller than P_3. To cover the same calibration range, both P_3 and P_1 must then be correspondingly higher, making the requirement that C_2 be in molecular flow more difficult to satisfy.

The time lag in this system will include a term related to the equilibrium time of the volume V_4 through C_4. Establishing t_L can proceed as in the previous method and will contain the same implicit assumptions. Since P_3 is not known absolutely, comparison of the gauge with a McLeod gauge or interferometer gauge above 10^{-4} torr is necessary to complete the calibration.

Multiple-stage pressure division. A number of pressure division stages connected in series have one advantage over a single stage: provided P_1 is sufficiently low ($< 10^{-2}$ torr), all the conductances can be thin-walled orifices whose values are calculable from dimensions (Bureau *et al.* 1952) so that no conductance measurements are necessary. A disadvantage is that a pair of gauges is required to obtain the effective conductance of each pump aperture. If the pumping speeds are constant, this determination needs, in principle, to be made only once. It is also a disadvantage that uncertainties in the values of the conductances have a cumulative effect (see Table 7.1). A four-stage pressure division system has been reported by Roehrig and Simons (1961). They describe calibrations in the range 10^{-9} to 10^{-4} torr for three ionization gauges in nitrogen.

7.1.4 Relative sensitivity for different gases

The relative sensitivity of ionization gauges for different gases is determined primarily by the gas ionization cross-sections under electron impact (see Chapter 3). It is difficult to calculate absolute gauge sensitivities directly from the cross-sections for two reasons:

  (i) the electron paths are not known in detail in most gauges.
  (ii) The fraction of the ions produced which are collected is not known.

The electron paths do not depend on the gas being ionized except at high pressures ($> 10^{-5}$ torr). Since the shape of the cross-section curves is almost the same for all gases except near threshold, the distribution in space of the ions produced is nearly independent of the gas. Provided the ions are formed with zero kinetic energy, the fraction of the ions collected should also be almost independent of the gas being ionized. In this case the relative gauge sensitivities should be nearly proportional to the magnitudes of the ionization cross-sections near their maximum ($\sim$100 eV electron energy.) For many polyatomic gases, some ions are formed with considerable kinetic energy (see Section 3.2.1). For these ions the collection efficiencies may be grossly different in some gauge geometries.

Table 7.2 lists measured relative sensitivities to various gases for a number of ionization gauges, with details of their operating conditions. In most cases the sensitivities are normalized to argon; when Ar data was not available the reference gas is nitrogen. The last column gives the ratios of total ionization cross-sections at an electron energy of 100 eV (Rapp & Englander-Golden 1965). Some of the relative sensitivities for hot-cathode gauges were measured in the pressure range where such gauges tend to be non-linear. The relative sensitivities listed are for the linear region when indicated in the original papers. The variation of *absolute* sensitivity amongst nominally identical gauges may be considerable, e.g. the observed variations of absolute sensitivity among BAG's of the type shown in Fig. 8.6 is about 25%.

7.2 MEASURING SYSTEMS AS SINKS AND SOURCES

Almost all the parts of an uhv system pump and/or release gas to some extent. It is often essential to consider the measuring system as a dynamic combination of sinks and sources of gas (Alpert 1962). At very low pressures it is important (and often difficult) to assess the effect of these sinks and sources on the pressure being measured. As an example of the type of problem that can be encountered, consider two uhv gauges (say hot-cathode gauges) reading the background pressure in a system, and suppose that a molecule travelling from one gauge to the other must make at least one wall encounter. Now suppose that the filament of gauge 1 releases gas for some reason, while that of gauge 2 does not, and suppose that this gas is pumped with unit sticking probability by the walls. Gauge 1 will indicate a higher pressure than gauge 2. With only the indicated pressure available and without the knowledge that the filament of gauge 1 was emitting gas, it is difficult to know which gauge, if either, is measuring correctly. The essential information required if the observations are to be interpreted correctly is the magnitude and location of the sources and sinks in each gauge and in the connecting tubing. Below we subdivide the discussion of this problem into three categories: pumping and re-emission of gas in gauges; gas interactions at hot surfaces; and sinks and sources in connecting tubing.

7.2.1 Pumping and Re-emission in gauges

The simplest situation is the measurement of the pressure of an inert gas, and the reason for the simplicity of this case is that rare gases are not pumped chemically and usually not physically at room temperature. The equation for a single gas is then

$$V\frac{dp}{dt} = -pS_E + F_{int} + F_{ext}, \qquad (7.8)$$

where F_{int} is the rate of emission into the gas phase from the gauge itself (for an inert gas usually the re-emission of gas previously pumped ionically). F_{ext} is the flow of gas into the gauge from external sources, V is the system

TABLE 7.2
Relative sensitivity of ionization gauges to various gases,
normalized to argon and nitrogen

Gas	1 Triode	2	3	4	5	6	7	8	9	10 Hot-cathode magnetron	11 Magnetron	12 Inverted Magnetron (Cold-cathode)	13 Trigger discharge	14 Total ionization cross-section
Ar	1	1	1	1		1	1		1					1
He	0.1283	0.14	0.13	0.17			0.134	0.15			1.76	0.15		0.128
Ne	0.2407	0.22		0.26			0.258		0.22		0.24			0.233
Kr	1.333			1.53			1.34		1.43					1.47
Xe	2.190			2.22	2.5				2.0					1.92
N$_2$	0.2808	0.67	0.75		1	0.53	0.30	0.56		1	1	1	1	0.885
H$_2$		0.28			0.43			0.64		0.43	0.52		0.32	0.324
Cl$_2$														
CO	0.823		0.94		1.1	0.48								0.93
O$_2$						0.48				0.80	0.99		0.69	0.940
NO						0.61								1.10
Hg														
H$_2$O	1.163						0.90							
CO$_2$							1.20				1.29			1.24
HCl							0.91							
NH$_3$		1.7					1.90							
SF$_6$			0.81											2.49
Air														
CH$_4$	1.015													1.28
C$_2$H$_2$	1.992													

Notes on Table 7.2

Column	Gauge designation	V_{gf} (V)	I^- (A)	V_a (kV)	B (gauss)	P (torr)	Reference
1	Leybold IM-1	180	10^{-3}				Moesta and Renn 1957
2	Westinghouse 5966	140	10^{-4}			10^{-5} up	Schulz 1957
3	Veeco RG-75	127.5	10^{-4}			$10^{-5}-10^{-2}$	McGowan and Kerwin 1960
4	Experimental	240	10^{-3}			$10^{-9}-10^{-4}$	Cobic et al. 1961c
5	Westinghouse 5966	145				$< 5 \times 10^{-5}$	Ehrlich (1963a)
6	Veeco RG-75	150	10^{-3}			10^{-4}	Anderson, H.U., 1963
7	Leybold IM-4	180	10^{-4}			$10^{-5}-10^{-3}$	Bennewitz and Dohmann 1965
8	Veeco RG-75	140	10^{-4}			$10^{-6}-10^{-5}$	Shaw 1966
9	As in Fig. 8.6	105	8×10^{-5}			$10^{-8}-10^{-5}$	Authors' laboratory
10	Experimental	300	5×10^{-5} (B=0)			$10^{-10}-10^{-6}$	Lafferty et al. (1967)
11	N.R.C. 552				250	$10^{-7}-10^{-5}$	Barnes et al. 1962
12	Experimental			5	1000	$10^{-8}-10^{-5}$	Hobson and Redhead 1958
13	G.E., T.D.G.			6	2000	$10^{-10}-10^{-7}$	Lafferty et al. 1967
14	—			2	1000	Ratio of total ionization cross-sections at 100 eV	Rapp and Englander-Golden 1965

265

volume and S_E the electronic pumping speed, which is here defined to include all pumping accompanying the emission of electrons from the cathode. A major portion of the electronic pumping speed will be the ionic pumping speed. Electronic pumping speeds for chemically active gases are generally similar in magnitude to those for inert gases.

The most comprehensive single study of the pumping of the rare gases in a Bayard-Alpert gauge has been made by Cobic et al. (1961a) who have studied the dependence of the pumping speed on the number of molecules pumped, the electron current, the grid accelerating potential and the temperature of the gauge walls. All of these variables are important. Ishikawa (1965) has studied the voltage dependence of Bayard-Alpert gauge characteristics for He, Ar and Kr. The general phenomenon of pumping coupled with re-emission is illustrated in Fig. 7.2. Note that immediately following the introduction of gas,

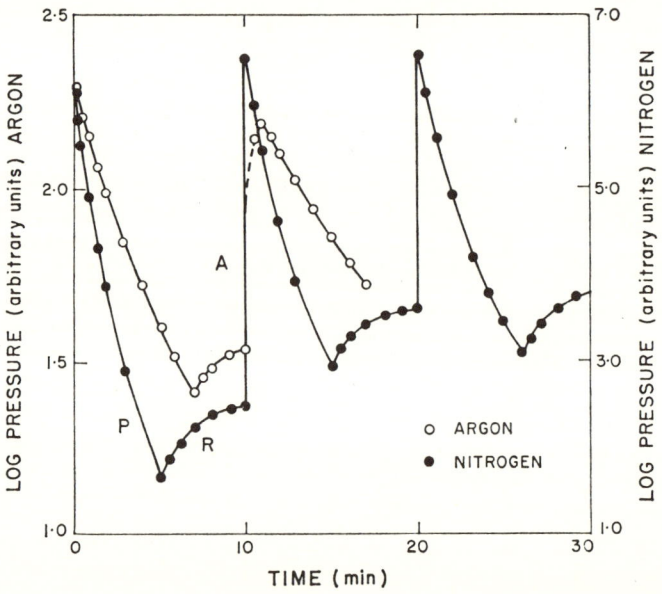

Fig. 7.2 Ionic pumping and re-emission of argon and nitrogen. P, pumping region; R, recovery region; A, region in which fresh gas was admitted. (Cobic et al. 1961b).

the pressure is reduced by gauge pumping and upon removal of the electron beam the pressure rises again as a result of re-emission. Provided F_{int} and F_{ext} are sufficiently small, the pumping speed can be calculated from the data of Fig. 7.2, using the solution to eqn. (7.8)

$$p = p_0 \exp(-tS_E/V) \tag{7.9}$$

p_0 being the pressure at time $t = 0$.

Conversely, the leak rates may be obtained immediately after the electron beam has been turned off at $t = 0$,

General Considerations of Pressure Measurement 267

$$Vp = Vp_a + (F_{\text{int}} + F_{\text{ext}})t, \qquad (7.10)$$

p_a being the pressure at time $t = 0$.

Re-emission probabilities are discussed in detail in Section 7.4.3. They are usually of such a magnitude that a system cannot be pumped ionically more than about four orders of magnitude in pressure without becoming limited by re-emission (Young 1956a; Bills & Carleton 1958 and Redhead et al. 1962b), unless the pumping surfaces are regenerated in some way. A corollary to this general result is that a gauge operated for 15 minutes at 10^{-5} torr cannot subsequently be used below 10^{-10} torr without outgassing (Hobson 1961a).

The electronic pumping speed of a gauge declines as the number of molecules pumped by the collector and the walls approaches approximately a monolayer. Frequently the ion current to the walls is about 5–10 times that at the collector (Young 1956a). Ishikawa (1965) estimates the ionic pumping speed for helium, argon and krypton at the ion collector of a hot-cathode gauge is two to three orders of magnitude lower than the total pumping speed. Table 7.3 contains a representative compilation of pumping speeds obtained by various workers mainly for so-called clean gauges or gauges operating at uhv without having pumped many molecules.

To a first approximation the electronic pumping speed of a hot-cathode gauge is linearly related to its electron current. Thus if $F_{\text{int}} + F_{\text{ext}}$ remain fixed, the equilibrium pressure in the gauge varies inversely as the electron current. However, I^+ at equilibrium is independent of I^-. This may be checked by substituting $S_E = k_1 I^-$ and $I^+ = k_2 p_{\text{eq}} I^-$ into eqn. (7.8) to yield at equilibrium

$$I^+ = \frac{k_2}{k_1}(F_{\text{int}} + F_{\text{ext}}). \qquad (7.11)$$

The electron collector (grid) potential normally changes both the probability of ionization by electrons (Sect. 3.1) and also the energy of impact of the ions on the collecting surfaces (Section 4.3); and the pumping speed changes from both these causes. A typical resulting variation of speed with grid voltage is shown in Fig. 7.3. A variant on the dependence of S_E on grid voltage has been reported by Carter (1959b), who found that under some circumstances with a glass gauge, particularly when it had not been thoroughly outgassed, the glass walls acquired a potential near that of the grid. In this condition the gauge speed (and sensitivity) was reduced by about a factor 5 below the value obtained when the walls were near filament potential. This problem may be avoided with a conducting film on the inside surface of the bulb (Section 8.1.1) or a second grid enclosing the normal grid structure (Nottingham 1961).

The pumping speed increases and the re-emission rate decreases as the wall temperature is lowered. Table 7.4 based on the results of Fox and Knoll (1961) shows this. Fox and Knoll (1960) found the re-emission varied as t^{-1} where t was the time since ionic pumping took place. Thus the re-

TABLE 7.3
Pumping speed of UHV gauges to various gases

Gauge	Pumping speed (litres/sec)								Operating conditions			Reference
	He	Ne	Ar	Kr	Xe	H₂	N₂	O₂	CO	V_{gf} (V)	I^- (mA)	
BAG	10^{-3}									150	10	Alpert 1953a
,,	4×10^{-3}									145	10	Young 1956a
,,	$6 \cdot 2 \times 10^{-3}$	$7 \cdot 0 \times 10^{-3}$	$1 \cdot 7 \times 10^{-2}$	$5 \cdot 2 \times 10^{-2}$	$6 \cdot 05 \times 10^{-2}$					250	10	Cobic et al. 1961a
,,	$1 \cdot 1 \times 10^{-2}$	$1 \cdot 2 \times 10^{-2}$	$2 \cdot 9 \times 10^{-2}$	$6 \cdot 9 \times 10^{-2}$	$9 \cdot 1 \times 10^{-2}$					250	10	Cobic et al. 1961a
,,	$1 \cdot 18 \times 10^{-2}$						1×10^{-1}			105	8	Hobson & Edmonds 1963
,,	$1 \cdot 7 \times 10^{-3}$		$7 \cdot 2 \times 10^{-3}$	$1 \cdot 2 \times 10^{-2}$						105	8	Ishikawa 1965
,,	6×10^{-4}		6×10^{-4}	1×10^{-3}						40	8	Ishikawa 1965
,,							$1 \cdot 8 \times 10^{-1}$			250	10	Cobic et al. 1961b
,,							5×10^{-1}			250	10	Cobic et al. 1961b
,,							3×10^{-1}	$2 \cdot 5$		118	8	Bills & Carleton 1958
,,							2			250	8	Hobson 1961b
,,							$2 \cdot 5 \times 10^{-1}$			250	8	Hobson 1961b
,,							8×10^{-1}		30	~100	10	Alpert 1962
,,						8×10^{-3}				~100	10	Alpert 1958b
,,						$2 \cdot 5 \times 10^{-1}$	20			150	0·1	Singleton 1967a Pétermann & Baker 1965
Trochoidal M.S.	$<2 \times 10^{-6}$										8×10^{-5}	Kornelsen 1959
Magnetron	$1 \cdot 7 \times 10^{-1}$		$1 \cdot 7$			$2 \cdot 0$	$2 \cdot 5$				$V_A = 5$ kV, $B = 1000$ gauss	Barnes et al. 1962
,,	2×10^{-1}						$1 \cdot 4 \times 10^{-1}$	$1 \cdot 5 \times 10^{-1}$			$V_A = 5$ kV, $B = 1000$ gauss	Rhodin & Rovner 1960
,,	1×10^{-1}						1×10^{-1}	$1 \cdot 2 \times 10^{-1}$			After several hours operation in O_2 at 10^{-6} torr	Rhodin & Rovner 1960
Inverted Magnetron	3×10^{-2}		$2 \cdot 5 \times 10^{-1}$				$7 \cdot 5 \times 10^{-1}$				$V_A = 6$ kV	Kornelsen 1960

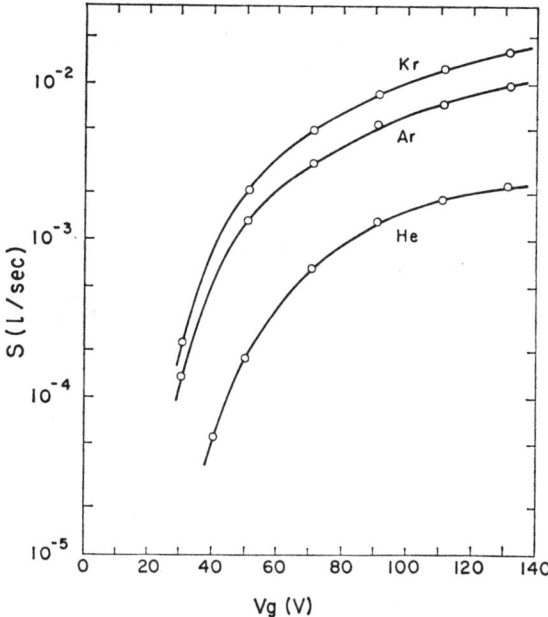

Fig. 7.3 Pumping speed of a BAG to He, Ar, and Kr as a function of grid-cathode voltage (Ishikawa 1965).

emission constant (k) is defined as the re-emission rate (F_{int}) divided by the total quantity pumped ($V\Delta P$) multiplied by t. A relationship between the pumping action and the re-emission of helium was found by Hobson and Edmonds (1963), who noted that the additional gas pumped as a result of a reduced wall temperature was all re-emitted when the wall returned to its original temperature. The material of the walls of an ion gauge influence the pumping speed as shown by Varnerin and Carmichael (1955), Young

TABLE 7.4

Re-emission and pump-down vs wall temperature for helium and molybdenum

Wall Temperature (°K)	Average τ (min)	$V\Delta P =$ Amount of gas pumped (torr-litres)	Re-emission constant (k)
330	2·5	6·9 × 10⁻¹¹	6·5 × 10⁻²
190	1·1	1·26 × 10⁻⁷	3·4 × 10⁻²
77	0·7	—	<10⁻³

$\tau = V/S_E =$ the pumping time constant.

$k = \dfrac{F_{int}\, t}{V\Delta P}$

$t =$ time since pumping occurred.   After Fox and Knoll (1960).

(1956a) and Grant and Carter (1966) who have reported that metal films on the glass walls of a gauge increased the pumping speed for helium. Winters *et al.* (1963) demonstrate that annealing a nickel ion collector greatly alters its collection efficiency for Ar^+. A comparison between ion impact on tungsten (Kornelsen 1961) and on glass surfaces (Carter & Leck 1961) for most of the rare gases has been made by one of us (Hobson 1963), and the pumping effects were qualitatively similar, differing mainly near the threshold of ionic pumping. It was possible that the glass surfaces of Carter and Leck (1961) were covered with thin metal films, but Grant and Carter (1966) argue that thin metal films decisively increase the electronic pumping of helium only.

Ion fluxes in mass-spectrometers are usually about an order of magnitude below those of ionization gauges, with electronic pumping speed also about an order of magnitude lower. There is a dearth of experimental values of

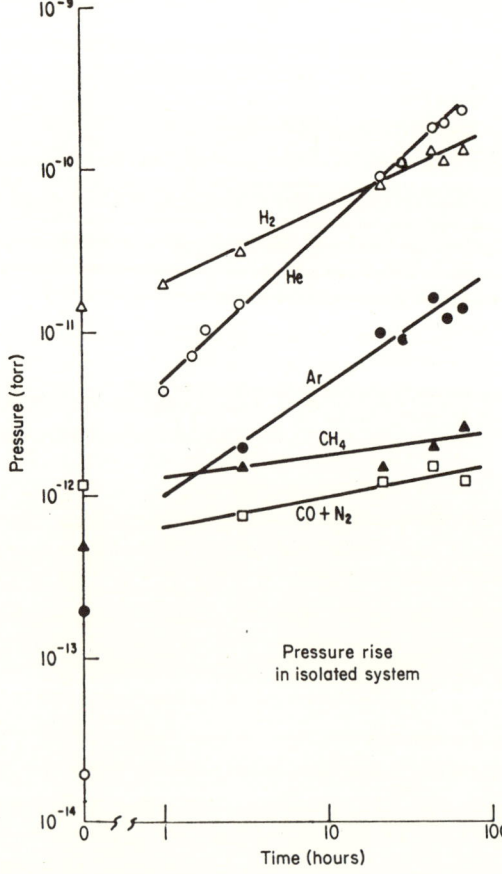

Fig. 7.4 Pressure increase following isolation of mass spectrometer (Davis 1962).

pumping speeds for mass-spectrometers, nor are the re-emission rates for mass-spectrometers well documented. Davis (1962) has published the rates of rise of pressure in an isolated mass-spectrometer having a stainless steel envelope (Fig. 7.4). The total pressure in this system at zero time was about 10^{-11} torr.

Kendall (1965) has examined various sources of gas evolution from electron multipliers. For a beryllium-copper multiplier he identified the following processes:

(i) Electron-bombardment-induced gas desorption from the final dynodes and collector. In a typical uhv system the main gases evolved are likely to be CO and H_2. The pressure rose about 1×10^{-9} torr when the final collected current was 2×10^{-5} A.

(ii) Thermal outgassing and possibly chemical reactions at the surfaces of internal dynode resistors. When carbon film resistors were used which had been baked, a pressure rise of 1×10^{-10} torr was observed when the resistor temperature reached 240°C.

(iii) Electrolysis and momentary discharges in insulators. Pulses of H_2, O_2, and several other gases with total gas contents $< 10^{-7}$ torr litres were observed.

An error in the measurement of pressure of an inert gas arises only if either the pumping term or the evolution term in eqn. (7.8) becomes large relative to corresponding terms for the remainder of the system. It is normally possible to avoid this by keeping conductances to the gauge relatively large or, in the case of hot-cathode gauges, by reducing the electron current or electrode voltages (Ishikawa 1965) to reduce pumping speed. For cold-cathode gauges however (Table 7.3) pumping speeds are not readily controllable and may be undesirably high. In experiments where a static gas pressure must be measured over a long time, the pumping speeds of ion gauges, even though small, may reduce the amount of gas in a system and the gauges should then be operated intermittently.

In addition to thermal re-emission, the impact of one inert gas ion on a surface may release a gas atom previously pumped. This effect is known as ionic replacement and has been studied by Brown and Leck (1955), Carmichael and Trendelenburg (1958) and by James and Carter (1963). A measure of the magnitude of the effect is given by the coefficient

$$Y_G = \frac{\text{Number of atoms of A released}}{\text{Number of ions of B incident}} \times \frac{1}{\theta} \qquad (7.12)$$

where θ is the relative coverage of gas A on the surface. Values of Y_G obtained by James and Carter (1962) for glass are given in Table 7.5, which includes interpolated estimates for the important case when A is the same as B. It is readily seen that near $\theta = 1$, the effect is large, while at low values of the coverage the effect may be negligible. Kornelsen (1964) has deduced values of Y_G for the case A the same as B for a tungsten surface. Kornelsen's values

TABLE 7.5

Gas release efficiency (sputtering coefficient Y_G) for inert gases on glass. Gas sorbed initially A, bombarding gas B. Energies $\sim$ 230 eV. Interpolated results for the case when A is the same as B are shown in brackets

Gas B Gas A	He	Ne	Ar	Kr	Xe
He	(5·5)	12·0	50·0	40·0	5·0
Ne	9·0	(20·5)	35·0	33·0	10·0
Ar	10·0	35·0	(22·0)	18·0	12·5
Kr	7·5	38·0	12·0	(11·5)	6·0
Xe	13·0	24·0	40·0	62·0	(63·0)

After James and Carter (1962).

are smaller than those of Table 7.5 but are nevertheless much greater than unity.

Electronic pumping in a gauge is evidently quite a complex phenomenon; two alternative theoretical viewpoints have been put forward. Baker (1962) postulates that the ions cause damage to some depth at the pumping surfaces and that the subsequent re-emission is controlled by diffusion through, and annealing of, this damage. Carter and Leck (1961), however, postulate a distribution of trapping sites with differing binding energies for the pumped atoms and explain the re-emission as simple desorption from this distribution of sites. It seems likely that elements of both situations exist in practice.

A much more complex situation exists for the measurement of pressure of the chemically active gases (N_2, CO, O_2, H_2, etc.). Chemical adsorption now occurs in addition to the pumping of ions. The rate equation now becomes

$$V\frac{dP}{dT} = -PS_E - PS_C + F_{int} + F_{ext} \tag{7.13}$$

where S_C is the chemical pumping speed and F_{int} now includes any re-emission of chemisorbed species. Exposed metal surfaces in a gauge will typically have an area of at least 20 cm². The pumping speed of these surfaces is some 200 litres/sec for nitrogen with unit sticking probability. This speed is far in excess of normal tubing conductances and should such a gauge be connected to a system through a 2·5 cm diam tube, 5 cm long, the gauge reading will be low by the ratio of the gauge speed to tubing conductance, i.e. by about a factor 20 for N_2. If it is anticipated that a gauge might develop such an error, the usual solution is to use a so-called nude gauge in which the gauge structure is inserted directly into the chamber where the pressure is to be measured. Further discussion of effects of tubulation are given in Section 7.2.3 below. Normally the chemical pumping speed of a clean gauge,

General Considerations of Pressure Measurement

while high, does not approach its maximum possible value (Alpert 1962), which, if the walls are included, is several thousands of litres/sec^{-1}. Chemisorption pumps saturate and eventually reach an equilibrium in which large quantities of adsorbed gas may be held on a surface above which the pressure may be very low (see, for example, the isotherms of H_2 on W measured by Hickmott 1960a). The complications introduced into eqn. 7.13 by such surface phases have been discussed by Comsa and Iosifescu (1961) and Hobson and Earnshaw (1967).

It is possible to measure gauge pumping speed from the time constant of the pump-down when a sudden pulse of gas is introduced to the system, when the gauge is suddenly switched on, or when the flow is suddenly cut off. The shortest possible time-constant for a gauge of volume 0·5 litre is 10^{-4} sec corresponding to a speed of 5000 litres sec^{-1} (H_2 pumping with unity sticking probability on the walls). Amplifier noise limitations restrict the minimum current measurable as a pulse with this time constant to about 5×10^{-10} A corresponding to a pressure pulse of $2·5 \times 10^{-9}$ torr in a typical gauge without a multiplier. This in turn corresponds to a total pulse of 10^6 ions. Thus electronic considerations in practice do not limit this type of measurement which will probably be limited by the rate at which the initiating physical impulse is performed.

Another method of measuring the speed is to control F_{ext} in a known way and to measure the resultant equilibrium pressure

$$S_E + S_C = F_{ext}/P_{eq}. \qquad (7.14)$$

A third method is to introduce another pump to the system of known speed S_K and to evaluate $S_E + S_C$ from the pressures with and without S_K, P_1 and P_2 respectively, then

$$\frac{S_K + S_E + S_C}{S_E + S_C} = \frac{P_2}{P_1}. \qquad (7.15)$$

Such a known pump could lie behind a valve which is suddenly opened or might be a clean filament which is suddenly cooled, or a portion of the system which is suddenly cooled so that it becomes a cryosorption pump. A typical example of a flash filament used in this way is provided by Hickmott's (1960b) studies on the interactions of H_2 on glass. Figure 7.5 illustrates the sudden cooling of a portion of the system in liquid N_2 when a steady leak of nitrogen gas is entering the system. These two types of adsorption pump have an advantage in that the gas pumped by the additional pump can be released suddenly into the system as shown in Fig. 7.5. If the response of the gauge is fast enough, $F_{int} + F_{ext}$ is given directly by the desorption peak height $\Delta P/\Delta t$. The characteristic pumpdown time $\tau_G = V/(S_E + S_C)$ may be measured directly and compared with the value obtained from eqn. 7.14. Even if the response time of the gauge and amplifier is not sufficiently fast, the area under the desorption curve will be preserved. It

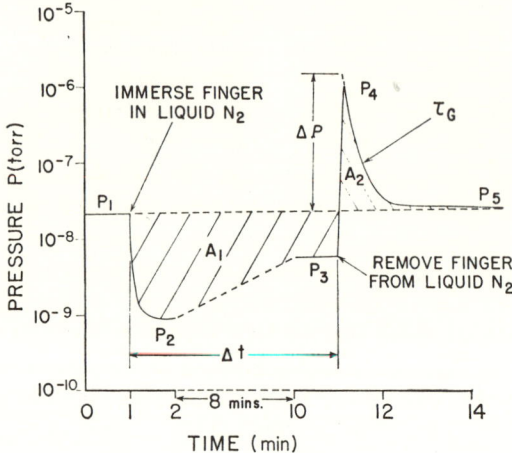

Fig. 7.5 Adsorption and desorption of N_2 by a cold finger in an uhv system (Hobson 1961b).

may be shown that if eqn. 7.8 holds throughout the operation then $A_1 = A_2$ (Fig. 7.5).

Mechanisms other than ionic pumping and simple chemisorption can contribute to gauge pumping speed. Hickmott (1960b) has shown that a tungsten filament of area 2 cm², operated at a temperature greater than 1475°K, produces a speed of 0 1 litre/sec for H_2 simply due to dissociation of H_2 on the filament with subsequent adsorption of the desorbed hydrogen atoms on the walls of the gauge (see Section 7.2.2). Eisinger (1959) estimates a pumping speed for oxygen due to decomposition on a W filament at 1700°K as 0·04 litres/sec^{-1}. Winters *et al.* (1963) show that a pumping speed of 10^{-4} litre/sec^{-1} for nitrogen is introduced by a tungsten filament at 2100°C.

Winters *et al.* (1963) find that 70% of the pumping speed for N_2 at 100 eV electron energy, in a tube similar to an ionization gauge, remains when the ion collector voltage is reduced from 100 volts attractive to zero (i.e. ions are no longer attracted to the collector. Collectors of nickel and nichrome were used). They assign this result to the pumping of the nitrogen atoms created in the dissociation of the N_2 molecule by electrons and find this pumping mechanism is present for electron energies between 10 and 16 eV where no molecular ions of nitrogen can be formed. On the other hand, Jaeckel and Teloy (1961) interpret similar experimental results as pumping caused by the metastable states of molecular nitrogen. Both these studies suffer from the lack of a mass-spectrometer. In contrast, Klopfer (1963), using a mass-spectrometer, does not confirm any significant increase in the pumping of CO and N_2 due to non-ionic pumping in a discharge. However, Klopfer's pumping tube was a magnetron discharge (6 kV, 1000 G) in which electron and ion energies are poorly defined, and Klopfer also assumed that the ion sticking probability was unity.

7.2.2 Gas interactions at hot surfaces

It is not as a result of their relatively high pumping speeds that the active gases present their most serious problem in the uhv gauges, but because chemically active gases may interact in or on heated gauge parts to produce other gases. Some chemically active gases (H_2, H_2O, CO_2, O_2 and hydrocarbons) can dissociate at the surface of the hot cathode of an ionization gauge of mass-spectrometer. The dissociation products are usually highly reactive and may interact with other gases or surfaces to produce new gaseous products. Thus a hot cathode may cause considerable changes in the gas composition of an uhv system. Some of these problems, as they relate to errors in pressure measurement, have been discussed previously (Redhead 1960a) and by Alpert (1962). The possible reactions at the hot tungsten filament listed by Alpert (1962) are:

$$H_2 \xrightarrow{W(2000°C)} H + H \begin{matrix} \nearrow H_2O \\ \searrow C_xH_y \end{matrix}$$

$$C_xH_y + X_aO_b \xrightarrow{W(2000°C)} CO + \ldots \ldots$$

$$O_2 + W + C \begin{matrix} \xrightarrow{2000°C} CO \\ \xrightarrow{2000°C} WO_3 \end{matrix}$$

$$H_2O + W + C \xrightarrow{2000°C} \begin{matrix} CO \\ WO_3 \\ H \end{matrix}$$

Hydrogen. The dissociation of hydrogen by a hot tungsten filament was first observed by Langmuir (1915) and has been studied with uhv techniques by Hickmott (1960a). Atomic hydrogen is formed at an appreciable rate when $T_f > 1000°K$ and the activation energy for evaporation of atomic hydrogen is 67 kcals/mole. Figure 7.6 shows the dependence of the rate of formation of atomic hydrogen on tungsten filament temperature and hydrogen pressure obtained by Hickmott (1960a) with an omegatron.

The atomic hydrogen evolved from a hot cathode is readily adsorbed on most surfaces, thus an anomalously high pumping speed is observed for hydrogen when $T_f > 1000°K$. Contaminant species, in particular CO, are produced by interaction of the atomic hydrogen with the vacuum chamber walls and electrodes (Hickmott 1960b; Mimeault & Hansen 1963; Singleton 1967a). These effects can cause a radical change in gas composition when hydrogen is exposed to an incandescent filament with $T_f > 1000$. Tables 7.6 and 7.7 compiled from the above authors' works show the gas composition of their systems at various filament temperatures and pressures.

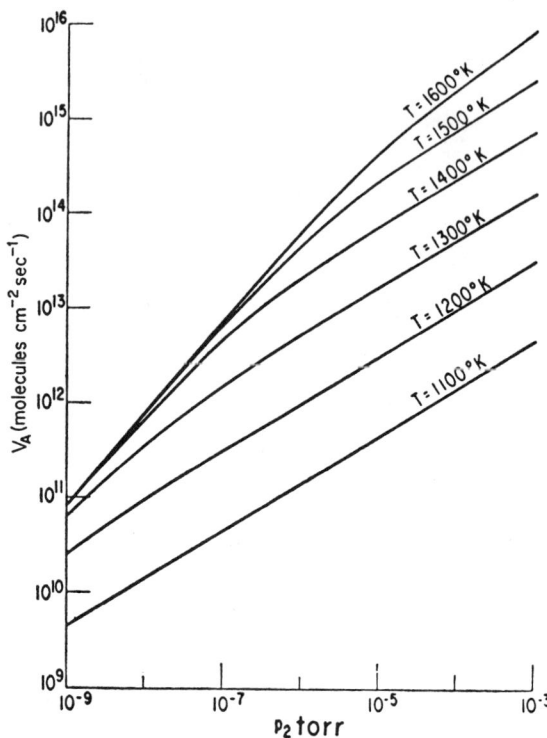

Fig. 7.6 Dependence of the rate of formation of atomic hydrogen (V_A) on tungsten filament temperature and hydrogen pressure (p_2). Gas temperature, 298°K. (Hickmott 1960a).

TABLE 7.6

Relative ion currents $\left[\dfrac{i^+}{i^+ (H_2)} \% \right]$ as a function of tungsten filament temperature in hydrogen

Filament temperature (°K)	CO_2 %	H_2O_2 %	CO %	H_2O %	CH_4 %	Reference
300	—	0·26	0·56	0·33	0·02	Hickmott 1960b
1010	—	0·25	0·65	0·28	0·01	,,
1550	—	0·44	0·82	0·44	0·04	,,
1840	—	0·78	3·1	0·95	0·46	,,
1975	—	0·66	11·3	0·65	2·5	,,
300	0·04	—	0·3	0·6	~0·001	Singleton 1967a
1427	0·03	—	0·3	0·7	~0·001	,,
2365	0·2	—	20	20	~10	,,

H_2 pressure about 10^{-7} torr.

TABLE 7.7

Relative ion currents $\left[\frac{i^+}{i^+(H_2)}\%\right]$ as a function of hydrogen pressure

Hydrogen pressure (torr)	Gauge filament conditions	CO_2 %	CO %	H_2O %	CH_4 %	Reference
7×10^{-12}*	Thoria-coated; $I^- = 10$ mA	570	570	115	14	Singleton 1967a
2×10^{-9}	,,	2	2	7	0·6	,,
4×10^{-9}	,,	0·3	1	2·5	0·1	,,
1×10^{-8}	,,	0·1	0·6	4·0	0·02	,,
2×10^{-7}	,,	0·04	0·3	0·6	0·04	,,
1×10^{-6}	Tungsten; $I^- = 10$ mA	0·4	19	7	2	Mimeault & Hansen 1963
$\sim 3 \times 10^{-6}$	Gauge off	0·18	0·18	—	—	,,
$3·2 \times 10^{-6}$	Tungsten; $I^- = 0·5$ mA	—	3·2	0·4	1·8	Hickmott 1960b
$3·7 \times 10^{-6}$	LaB_6; $I^- = 0·5$ mA	—	0·5	0·7	0·5	,,
$\sim 3·7 \times 10^{-6}$	Gauge off	—	0·4	0·3	0·5	,,

*Background conditions.

The measurement of H_2 pressure in an uhv system has been the subject of controversy (Hickmott 1965) which may not yet be complete. It is clear that many complex reactions may take place with H_2, and that a cool filament (lanthanum boride or thorium oxide) reduces the complexity of these reactions. Pétermann and Baker (1965) have demonstrated that the pumping of H_2 by a Bayard-Alpert gauge is a complex function of the temperatures of the grid and the filament. They suggest that under normal clean operation the filament pumps by dissolving hydrogen, while a molybdenum grid pumps by adsorbing hydrogen. They also point out that simple conclusions drawn from the linear equation 7.13 lead to ambiguous values of $S_E + S_C$.

Singleton (1967a) has observed that operation of a hot-cathode gauge at low emission currents, as suggested by Hickmott (1965), can give rise to large errors in pressure measurement due to electronic desorption of ions from the grid of the gauge (see Section 7.4). Singleton suggests that the gas responsible for the spurious current from the grid is carbon monoxide *not* hydrogen. Operation of the gauge at 10 mA electron current greatly reduces this error but the use of a modulator electrode in the ion gauge, to determine and correct for the residual ion current, is desirable at pressures below 10^{-7} torr. The use of a low temperature emitter (such as thoria) resulted in reasonably low levels of impurity at 10 mA electron current and the pumping speed of the gauge for hydrogen was only 0·08 litre/sec. Progressive improvements in the impurity level could be produced if the system had previously been operated in an oxygen atmosphere with specific processing to reduce carbon impurities in the electron emitter (Becker *et al.* 1961). Singleton found that filament temperatures significantly higher than the 1100°K suggested by Hickmott (1965) did not result in significant impurity problems with thoria filaments.

Moore and Unterwald (1964a and 1964b) have observed the production of CO, H_2O, and CH_4 when H_2 was admitted to an uhv system. The atomic hydrogen produced by a hot cathode can interfere directly with measurements of hydrogen adsorption (Law 1958).

Oxygen. Oxygen reacts with carbon impurities in incandescent filaments to produce CO and CO_2 (Young 1959; Schlier 1958; Rhodin & Rovner 1960). Young has shown that the amounts of CO and CO_2 produced were strongly affected by the amount of carbon in his various filaments (W, Mo, Re and Ta). Figure 7.7 shows results obtained by Schuemann *et al.* (1963) in a system consisting of a Bayard-Alpert gauge, a suppressor gauge and an omegatron. A background pressure of 7×10^{-11} torr (mainly CO) was achieved before pure O_2 was introduced to the system at $t = 0$ and a steady O_2 pressure of 10^{-7} torr as measured by the omegatron was maintained for 35 minutes. Initially the pressure as indicated by both total pressure gauges (operated at different electron currents) was a factor 20 below that at the omegatron. This implies gauge pumping speeds for O_2 of about 30 litres/sec, since the conductance

General Considerations of Pressure Measurement

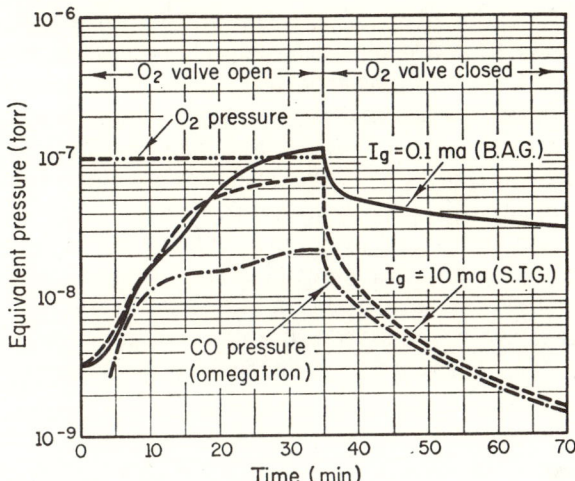

Fig. 7.7 Pressure variations during admission of O_2 to an uhv system. (Schuemann *et al.* 1963).

between gauges was about 1·5 litres/sec. This discrepancy declined slowly. However, despite the fact that only O_2 was being admitted to the system, the CO pressure as measured by the omegatron rose slowly. Upon closure of the O_2 valve at $t = 35$ minutes, the O_2 pressure fell to less than 10^{-10} torr in a few seconds, leaving the resultant system pressure mainly CO, which was pumped away slowly (much more slowly than the true pumping speed would imply). The primary purpose of this experiment was the study of anomalous pressure readings in gauges operating at different electron currents. Conclusions about desorption of surface ions are treated in Section 7.4.

Singleton (1966a) has presented a detailed study of the interaction of oxygen with hot tungsten. After processing the electron emitters at about 2000°K in oxygen pressures of about 10^{-6} to remove carbon, as described by Becker *et al.* (1961), the carbon monoxide level was considerably reduced. After this treatment the only significant impurities at 4×10^{-8} torr of oxygen were: CO, 0·3%; CO_2, 0·2%, H_2, 0·1%. The filaments in both ion gauge and mass-spectrometer were thoria-coated and the gauge was operated at 10 mA and the mass-spectrometer at 20 μA.

Singleton (1966a) shows that removal of adsorbed O_2 from W occurs only by evaporation of the oxide below 1650°K and by evaporation of oxide *and* oxygen above this temperature. The interaction of O_2 with W has also been studied by Schissel and Trulson (1965) who find that above 1400°K the main products are W_3O_9, W_2O_6, WO_3 and WO_2.

Other gases. Schissel (1962) has shown that CO_2 interacts with a heated gauge filament (in this case, thoria-coated iridium) to yield O_2 and CO. Garbe *et al.* (1960) have studied the reactions of water with hot filaments leading to the formation of H_2, CH_4, CO and CO_2. Baronetzky and Klopfer (1960) have studied the decomposition of CH_4 on three different cathodes (oxide, thoriated Pt-Ir, and tungsten).

The contaminating effects caused by the interaction of chemically active gases with an incandescent filament may be minimized by the following steps:

(i) surrounding the hot cathode by a metal surface at which atomic species can readily combine. Far less CO is produced from H_2 in this case than is obtained when the atomic hydrogen can interact with a glass surface;

(ii) the carbon impurity near the surface of a tungsten electrode can be reduced by heat treatment in an oxygen atmosphere. Heating to about 2000°K in 10^{-7} torr of oxygen for more than 20 hours is effective. After this treatment the conversion of O_2 or H_2 to CO, when the electrode is subsequently heated, is greatly reduced;

(iii) the operating temperature of cathodes in gauges or mass-spectrometers should be reduced as much as possible by the use of low work function emitters.

Another possible method of reducing the effects of a hot cathode on pressure measurement is to separate the gauge into two separately pumped portions. In one chamber the cathode is located, electrons are accelerated through a small slit into the second chamber where ions are produced and detected. A gauge of this type has been constructed and calibrated from 10^{-4} to 10^{-10} torr (Blauth & Venus 1965).

7.2.3 Sinks and sources in tubing (Blears effect)

It was first observed by Blears (1947) that an ionization gauge inserted directly into a large chamber gave a pressure reading higher by a factor 10 than an identical gauge enclosed in an envelope and attached to the chamber through a tube. The first type of gauge has been called 'nude' and the second 'tubulated'. Blears (1947) demonstrated that the readings of the gauge arose from the vapour of the diffusion pump, and that the discrepancies in the two gauge readings arose primarily because of vapour adsorption in the tubulation. The pressure range used by Blears was 10^{-8} to 10^{-4} torr, but Haefer and Hengevoss (1960) have repeated Blears result nearer the uhv range (2×10^{-9} to 2×10^{-8} torr). They interpret their results as arising from the difference in conductance of the tubulation for oil vapour (10^{-4} litres/sec^{-1}) and for the gases produced by cracking of the vapour in the tubulated gauge (1 litre/sec^{-1}), and argue that this is a steady-state condition. Reich (1960) on the other hand interprets his similar results as indicative of adsorption of oil vapour in the tubulation but calculates that the discrepancy between the gauges will disappear in 16 days. However he does not demonstrate this experimentally. Reich also shows that the 'Blears effect' exists for water vapour and in this case the discrepancy does disappear in time, the time being shorter the higher the system temperature. The 'Blears effect' would not be of major significance in uhv systems if it were found only for oil and water vapour since it is possible to eliminate these vapours by suitable processing (Chapter 10).

However, Apgar (1963) has found a similar effect in an uhv system free of these vapours. Apgar has observed that a 'tubulated' hot-filament ionization gauge separated from a 'nude' gauge of similar design by a length of glass tubing (12 mm diam, 10 cm long) reads lower by a factor of 10 for CO. Smaller factors were found for H_2 and N_2, and both gauges indicated the same pressure for He. In general the divergence depended upon the pressure, and upon the tubing temperature. Figure 7.8 shows a typical result obtained by Apgar (1963)

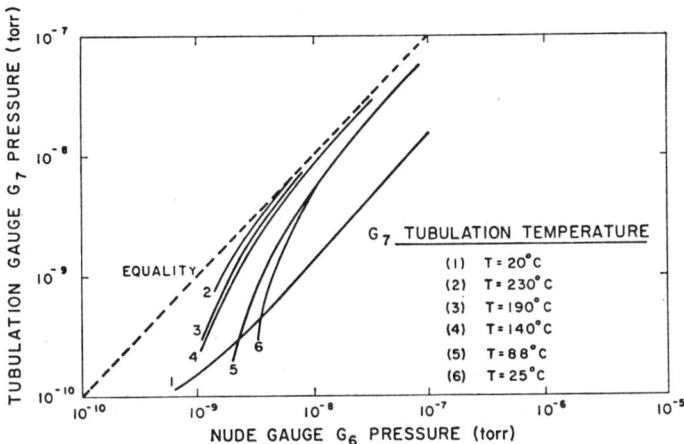

Fig. 7.8 Pressures of nude and tubulated gauges in CO at low pressures. (Apgar 1963).

in a system containing mainly CO. While in principle this result can be assigned to a high pumping speed of the tubulated gauge causing a pressure drop across the tubulation, Apgar (1963) finds that he cannot support this contention on the basis of the magnitude of the conductance of the tubulation and the pumping speed of the gauge, and assigns the result to adsorption and desorption of CO on the walls of the tubulation.

7.3 PRESSURE MEASUREMENTS IN NON-UNIFORM ENVIRONMENTS

So far in the discussion of pressure measurement two conditions have been implicitly assumed which may be violated in many practical situations. These are:

(i) The flux density of gas per unit solid angle in an enclosure is independent of direction and position, i.e. at a given point in space there is a pressure which is independent of the orientation of a pressure gauge at that point.

(ii) All parts of a system are at the same temperature and the pressure quoted is at that temperature.

All ionization gauges measure the density of gas molecules in a given volume and in general are not sensitive to the direction of motion of these molecules or to the magnitude of their molecular velocity. We may illustrate these comments with two idealized examples. Consider first Fig. 7.9 in which gas flows at low pressure from right to left. Suppose that surface X–X has unit sticking probability for this gas. An ideal gauge having no internal sources of gas set in orientation (a) in Fig. 7.9 will read zero, while in orientation (b) it will read an equilibrium pressure determined by equality of the rate of influx and the rate of return flow through the tubulation. Neither of these readings represents the true pressure defined as the force per unit area on a body in the flow.

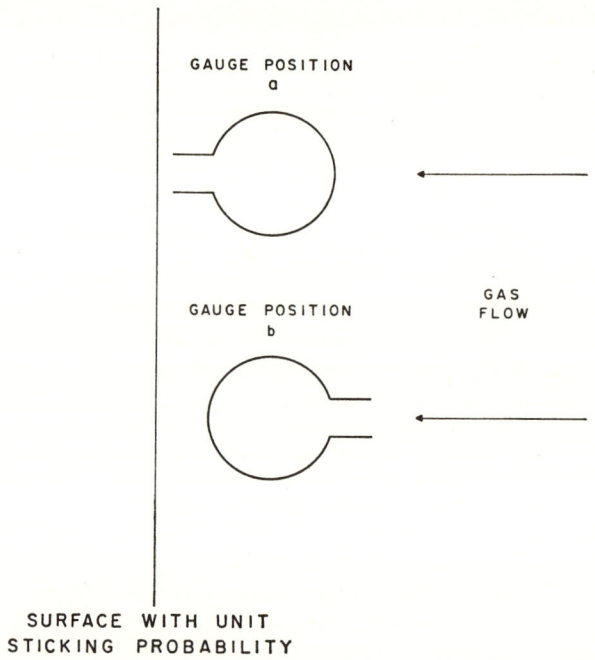

Fig. 7.9 Illustrating the dependence of gauge reading on gauge orientation in the presence of surface XX with unit sticking probability.
Case (a): gauge reads zero.
Case (b): gauge reads positive pressure.
    See also Moore (1961).

Practical problems, similar in general to the above, are described by Stickney and Dayton (1963) who attempt to evaluate, from gauge measurements, the sticking probability of N_2 on a cold surface of copper covered with frozen nitrogen at $10\cdot2°K$, and by Holkeboer (1962) who examined the problem of converting the pressure reading of a gauge mounted on the wall of a space chamber to the reading of a gauge mounted on the vehicle in the chamber.

As a second example, consider a sealed off ionization gauge measuring the static pressure P_1 (of helium, say) $T_1 = 300°K$. Now let this gauge be immersed in liquid nitrogen ($T_2 = 77°K$). The helium will not be physically adsorbed on the walls at this temperature, but the mean molecular velocity will drop to a value corresponding to 77°K. The pressure will drop to a value P_2 given by

$$\frac{P_2}{P_1} = \frac{T_2}{T_1} = \frac{77}{300} = 0 \cdot 257. \tag{7.16}$$

However, the density of gas in the gauge will be unchanged and its reading will be unchanged despite the fact that the true pressure has dropped by a factor 4. Experiments have been done, however, in which the density is deliberately made to depend upon molecular velocity (Morrison & Tuzi 1965).

While the examples given by Fig. 7.9 and eqn. (7.16) are rather extreme, elements of the two situations exist in most uhv systems. Each particular case must be treated individually in detail, and we limit ourselves here to describing only ideal situations from which an analysis of a practical situation may be built. The objective of all such analyses is the deduction of the gas density within the sensitive volume of the ionization gauge, and the starting point of a detailed analysis must be the individual atomic interactions between the gas molecules and the surfaces of the system.

The normal set of assumptions which we will use at first are

(i) A molecule accommodating to a surface simply disappears from the gas phase upon impact.

(ii) The desorbing molecules leave a surface with a cosine distribution in angle.

(iii) The desorbing molecules leave a surface with a velocity corresponding to the temperature of the surface.

(iv) A reflected molecule suffers an elastic collision with the surface.

These assumptions are not without exception (see Section 2.4) but they appear to be useful as a general starting basis.

The simplest example of the use of assumptions is the isothermal enclosure, which we designate volume 1, containing no sinks and sources. This case is the basis of the kinetic theory of gases and has been treated extensively in texts (Kennard 1938; Knudsen 1950) and will not be considered here. If an aperture is created in the thin wall of an isothermal enclosure the molecules effuse from the aperture according to the cosine law, i.e. the flux varies as the cosine of the angle between the direction of effusion and the normal. The effusion pattern is independent of the shape of the volume into which molecules effuse (volume 2) but it is convenient for later discussion to assume that the latter is spherical. Here the flux density of molecules in a solid angle (dω) at an angle θ to the normal to the aperture is given by:

$$\mathrm{d}n = B \cos \mathrm{d}\omega \text{ molecules cm}^{-2} \text{ sec}^{-1}, \tag{7.17}$$

B being a constant.

The integral of eqn. (7.17) over the whole hemisphere yields the well known expression for the number of molecules/sec passing unit area in one second

$$n = 1/4\, \rho\, \bar{v} \text{ molecules cm}^{-2} \text{ sec}^{-1}, \tag{7.18}$$

ρ is density of molecules, $\bar{v}$ is the mean molecular velocity. If the molecules entering volume 2 are pumped by the walls on the first impact then the reading of a pressure gauge in volume 2 will depend critically upon its location and orientation. An example of a practical situation like this has been described recently by Kornelsen and Domeij (1966). It is an interesting and useful property of the spherical receiver (Knudsen 1950; Moore 1965) that the gas flux per unit area of the sphere is the same everywhere. This greatly simplifies calculations of the reflected flux in the case where the sticking probability on the walls is not unity (Moore 1965). For the case that the sticking probability is zero (or any known magnitude) the gas flux arising after the first (and later) impact can be explicitly calculated everywhere. While there is a net flow of gas entering volume 2 the density of this chamber will be determined by the pattern of the incoming flow and the uniform pattern of the gas molecules bouncing back and forth in 2. For other than a spherical receiver the second contribution is not necessarily uniform but may be made approximately so by suitable design (Dayton 1955). For the case that there is no net flow and the temperatures of the two enclosures are the same, then of course the system returns to that of the isothermal enclosure in equilibrium and the gas density is uniform everywhere.

The case where the temperatures of the two enclosures T_1 and T_2 are not the same represents the situation in the presence of a cryogenic pump and in many adsorption and vapour pressure measurements. In the case of no net gas flow across the aperture the fluxes from both sides must be equal

$$1/4\, \rho_1\, \bar{v}_1 = 1/4\, \rho_2\, \bar{v}_2, \tag{7.19}$$

which leads to

$$\frac{P_1}{P_2} = \left(\frac{T_1}{T_2}\right)^{\frac{1}{2}}.$$

This well known expression of the law of thermal transpiration at very low pressures has recently been verified experimentally by Edmonds and Hobson (1965) for an aperture. The ideal aperture between two enclosures is often inconvenient experimentally and is frequently replaced by a tube. The effusion pattern for gas entering volume 2 and volume 1 is now more collimated than the cosine distribution from an aperture. This collimating principle is widely used in molecular beam ovens (Ramsey 1956; Giordmaine & Wang 1960) where the objective is to concentrate the gas effusing from the oven into as small a solid angle as possible. The treatments of this subject emphasize the effusion pattern; however, the spatial pattern of the molecules entering the tube but returning to enclosure 1 before reaching 2 may also be of interest in uhv systems. Dayton (1956) has theoretically considered both these patterns together. Figure 7.10 shows Dayton's patterns for the reflected and transmitted flux densities for a tube with length to radius ratio of 4. Note that the two patterns are complementary, the sum giving cos θ. This complementary result is important in understanding why the presence of a tube in an isothermal enclosure does not lead to pressure gradients. The molecules entering 2 arise from two sources: (a) those coming from 1, (b) those re-entering 2, having failed to reach 1. The effusion patterns of each of these

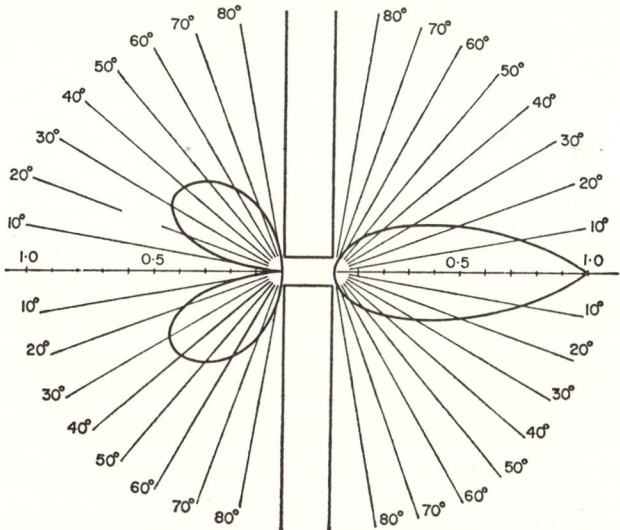

Fig. 7.10 Patterns of molecules reflected and transmitted by a pipe of length to radius ratio 4. Incident beam left to right. (Dayton 1956).

components are complementary and add up to a cosine distribution which will just cancel (for $P_1 = P_2$) the cosine distribution of the molecules leaving enclosure 2 to enter the tube. In the event $P_1 \neq P_2$ there will always be some lack lack of spatial uniformity in both enclosures. An example of this situation has been studied experimentally by Holland and Priestland (1966), who measured the spatial distribution of pressure in a chamber with a directed input. The system of two volumes joined by a tube with $T_1 > T_2$ has been studied by Edmonds and Hobson (1965). It is readily shown that if spatial patterns of the form of Fig. 7.10 are unaffected by the temperature change then the result of eqn. (7.19) will be obtained for this case also. Now if the four assumptions given at the beginning of this section are maintained, the spatial patterns of Fig 7.10 will be preserved. Experimentally Edmonds and Hobson (1965) found for glass tubes separating the two enclosures and with the gases helium and neon, that

$$\frac{P_1}{P_2} = b \left(\frac{T_1}{T_2}\right)^{\frac{1}{2}}, \qquad (7.20)$$

where $b < 1$ for $T_1 > T_2$. This result was assigned to failure of assumption 2 above.

Moore (1964) has examined theoretically the more general problem of the pressures and densities existing in a stationary system (no sinks or sources) in which various surfaces are held at various temperatures, the cosine law holding at all surfaces. He finds first that the molecular flux (i.e. the number of molecules crossing any plane per unit time) is everywhere the same, despite the fact that molecular velocities vary with position. Since the flux is isotropic everywhere the reading of a tubulated gauge is independent of orientation and depends only upon its temperature, all other parameters

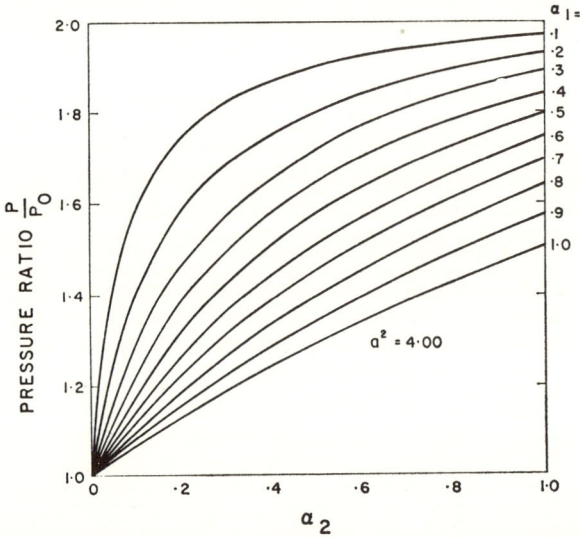

Fig. 7.11 Ratio of wall pressure (P) to tubulated gauge pressure (P_0) for infinite parallel planes at a temperature ratio $a^2 = 4$; $\alpha_1 =$ thermal accommodation coefficient of wall at T; $\alpha_2 =$ thermal accommodation coefficient of wall at a^2T. (Moore 1964).

remaining fixed. Moore (1964) then calculates the relationship between the gauge reading and the pressure exerted on the walls of the system for two special cases: (a) two infinite parallel planes with temperatures T and a^2T, (b) two concentric spheres, the outer sphere having temperature T, and the inner sphere a^2T. The results are given as a function of the accommodation coefficients of the two walls. Typical of Moore's results are those for the parallel planes shown in Fig. 7.11. For concentric spheres the pressures on the inner and outer sphere are unequal. Santeler (1959) has also examined the problem of concentric spheres, assuming constant outgassing of both,

General Considerations of Pressure Measurement

in an attempt to assess the degree to which such a system might be used to simulate pressure conditions on a vehicle in outer space. The important conclusion is that the number of times a molecule originally leaving the inner sphere returns before it is pumped at the outer sphere is

$$N = g(1-c)/c, \qquad (7.21)$$

where c is the ratio of the wall pumping speed to its maximum theoretical value (i.e. the effective condensation coefficient, or sticking coefficient), and g is a geometric factor representing the chance a molecule from the outer sphere will strike the inner sphere on the first attempt. The ideal objective of a space simulator is to reduce N in eqn. (7.21) to zero, i.e. to make $c = 1$. It is clear that this cannot be done simply by connecting a very fast pump to a portion of the outer sphere, rather the whole outer sphere must be made into a pump. Santeler (1959) also shows that the pressure experienced by the inner sphere (vehicle) is only measured directly by a gauge at the inner sphere looking outward toward the outer sphere.

7.4 RESIDUAL CURRENTS

In most ionization gauges there exist various processes which give rise to a current to the ion-collector, the residual current, which is independent to first order of the pressure in the gauge. This residual current establishes a lower limit to the pressure measurable with a particular gauge. The residual current is defined as that portion of the ion-collector current which would remain if the molecular density in the ionization region of the gauge were suddenly reduced to zero. While these effects are most troublesome in hot-cathode gauges they also occur to some extent in mass-spectrometers. The residual current results from two general processes:

(i) photo-electron emission resulting from soft x-rays or other radiation,

(ii) desorption of positive ions by electron bombardment of material adsorbed on the electron collector (electron-impact desorption).

Methods of measuring residual currents are the subject of Section 8.1.7.

7.4.1 Soft x-ray photo-emission

Since cold-cathode ionization gauges do not exhibit an x-ray limit (see Section 8.2) we are only concerned in this section with photo-electron emission resulting from soft x-rays produced by electrons in the energy range used in hot-cathode gauges, i.e. up to 600 eV. This energy range is an extremely difficult one for the production and measurement of x-rays and thus there is a paucity of experimental data. In many experiments, relevant to the problem of soft x-rays in gauges, only the overall effect has been measured (i.e. the relations between the primary electron flux and the ultimate photo-electron current) without breaking down the process into its two stages;

(a) the production of soft x-ray photons by electron impact on a target (grid or anode), and (*b*) the production of photo-electrons by these x-ray photons at a collector.

Schütze and Ehlbeck (1961) have measured the angular distribution of soft x-rays from electrons in the energy range 300–500 eV and found it to follow the cosine law approximately.

No measurements of the variation of soft x-ray flux with electron energy in this low energy range have been reported; however, there have been many studies of the variations of the resulting photocurrent with energy of the initial electrons. Most investigations have found a power law relationship between photocurrent and electron energy ($I_{p.e} = kV_e^n$) but considerable divergence exists amongst various experimenters' values for the exponent; Venema and Bandringa (1958) find n as high as 1·8, while Noble and Jacob (1956) obtained a value as low as $n = 1·15$ for targets of Cu and C. Other experimenters have found values between these two extremes. Unpublished measurements made in the authors' laboratory show that, when the photocurrent-electron energy characteristic is measured in sufficient detail, the relationship is found not to be a simple power law. In the energy range 20–300 eV the photocurrent-electron energy curve appears to consist of a number of linear segments, with the breaks corresponding to energy levels in the target material. These breaks were used in the early days of x-ray spectroscopy to determine energy levels, before the advent of x-ray diffraction gratings (see, e.g. Thomas 1925). In view of the above it is suggested that the exponent in the power law relationship has no fundamental significance but is merely an approximate representation of the condition in a particular system.

The variation of resulting photocurrent with the material of the *target* (the electron collector) appears to be quite small and there is no simple dependence on atomic number as would be observed for high-energy x-rays. Lange (1960) has observed the photo-emission due to soft x-rays produced by electrons (striking target of Mo, Ni, Be, and Cu 55%–Ni 45%) with energies in the range 50–2000 eV. For all the above metals the photocurrents were very similar and the exponent of the power law relationship about 1·6. No change in photocurrents was observed after prolonged electron bombardment. A target of Ag—Mg behaved quite differently showing wide variation in photocurrent depending on surface oxide condition. After electron bombardment the results with the Ag—Mg target approached those of other targets.

Experiments in the authors' laboratory indicate that the presence of adsorbed layers of gas (O_2, N_2 and CO) on the electron collector have little or no effect on the photocurrent-electron curves in the range 20–500 eV. Evaporated layers of some metals do however have a profound effect, e.g. an evaporated layer of Au on a Mo target lowered the photocurrent by a factor 4 at 100 eV electron energy. Material evaporated from a thoria-coated tungsten filament (all tests suggest that the material was thorium) caused pronounced structure to appear in the photocurrent-electron energy curves in the range 80 to 110 eV.

Considerably greater variation of photocurrents is observed when the material of the photo-emitting electrode is changed. Schütze and Ehlbeck (1961) have observed the variation of photocurrents from photo-emitters of Mo, Ni, carbonized Ni, Ni-clad Fe, and Al-clad Fe. Exponents in the power law relation varied from 1·3 (Mo) to 1·8 (Al-clad Fe). Similar measurements in the presence of a [BaSr]O cathode yielded photocurrent curves which were very similar for different photo-emitter materials suggesting that the surfaces were all contaminated with products from the oxide cathode.

The reflection of soft x-rays can be a troublesome problem in certain types of gauge (in particular, the suppressor gauge—see Section 8.1.2). LeBlanc et al. (1964) have measured the reflectance of Mo, Rh, W and Ta in the far ultraviolet and the results are shown in Fig. 7.12. It can be seen that the behaviour of these

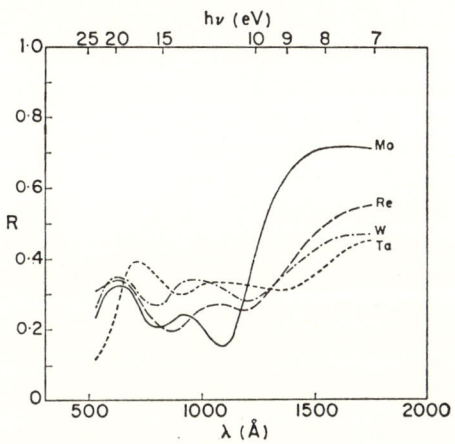

Fig. 7.12 Reflectance of transition metals to radiation in the energy range 7 to 25 eV for an angle of incidence of 15°. (LeBlanc et al. 1964).

transition metals is similar with a reflectance of about 30% in the photon energy range of 10 to 25 eV. The reflectance of Al is about 100% at 10 eV dropping thereafter to about 0·1% at about 45 eV (Philipp & Ehrenreich 1964). The reflectance of Ag is about 8% in the range 6 to 22 eV photon energy (Ehrenreich & Philipp 1962).

Data on quantum yields and the energy distribution of the electrons emitted from metal surfaces by soft x-rays can be found in Section 5.1.

Approximate values of the yield of photo-electrons per initial electron can be estimated from the data of Lange (1960) who measured the ratio of photo-electron current ($I_{p.e}$) to initial electron current (I^-) for electron energies (V_e) from 50 to 2000 eV. The photo-emitting surface, of tantalum, collected essentially all the x-ray photons. The results for Be, Ni, Mo, and Cu 55%–Ni 45% were very similar with $I_{p.e}/I^-$ of about 3×10^{-7} at $V_e = 100$ V, rising to about 10^{-5} at $V_e = 1$ kV in a power law relation, $I_{p.e}/I^- = kV_e^{1.6}$. The photocurrent resulting from soft x-rays in any particular gauge structure can be estimated from the above values of $I_{p.e}/I^-$ and the 'geometric factor', i.e. the fraction of the x-rays which strike the electrode under consideration, which can be readily calculated. Geometric factors for the Bayard-Alpert gauge have been calculated by Schütze and Ehlbeck (1961), Schütze and Stork (1962b) and Groszkowski (1966a). Factors for the extractor gauge have been calculated by Redhead (1966) and Groszkowski (1966a).

7.4.2 Electron-impact desorption

When certain chemically active gases are adsorbed on the electron collector of a gauge or mass-spectrometer and are struck by electrons, positive ions and/or neutrals may be released. This process is called electron-impact desorption and has been discussed in some detail in Section 4.2. Here we are concerned only with the errors in pressure measurement that can occur as a result of the E.I.D. process. The errors are particularly large after exposure of the gauge to relatively high pressures of chemically active gases.

The gases that most commonly cause trouble from E.I.D. in gauges are: O_2, CO, H_2, H_2O, hydrocarbons and halogens. Adsorbed layers of these gases when struck by electrons may evolve positive ions, neutrals or excited neutrals. Some of the positive ions released from the electron collector of a gauge or mass-spectrometer reach the ion-collector and cause an output signal which is not proportional to gas pressure but rather to the gas coverage on the electron collector which is a complex function of prior exposure to electronically desorbable gas. The desorption of neutrals by electron bombardment causes an increase in pressure in the gauge or mass-spectrometer. The pressure increase is frequently very localized because the electronically desorbed neutrals are often highly reactive fragments which readily adsorb on the surfaces of the gauge or mass-spectrometer before they can reach other parts of the system.

The magnitude of the error in pressure measurement caused by E.I.D. is strongly dependent on the electron current used. With large electron currents the adsorbed layer of gas is rapidly removed from the electron collector and thus the error in pressure measurement decreases rapidly. Conversely, the error in pressure measurement due to E.I.D. is usually greatest at low electron currents. Figure 7.13 illustrates both the general magnitude of errors that can result from E.I.D. and the effects of changing electron current. Oxygen was admitted to a system (containing a modulated BAG, an extractor gauge and a small magnetron pump) with the pump switched off and both gauges operated with $I^- = 80$ μA, until the pressure had stabilized at about 5×10^{-7} torr. At time zero the oxygen valve was shut and for the next 20 minutes the pressure dropped very slowly because the system pumping speed for O_2 was small. During this period the pressure indications of the BAG and extractor gauge (EG) were in agreement. At $t = 16$ minutes the getter-ion pump was switched on and the pressure dropped very rapidly. The curve labelled P_E is the pressure indicated by the EG, which is less sensitive to E.I.D., (see Section 8.1.3) and indicated $2 \cdot 5 \times 10^{-10}$ torr after 10 minutes. The BAG reading (P_B) was about

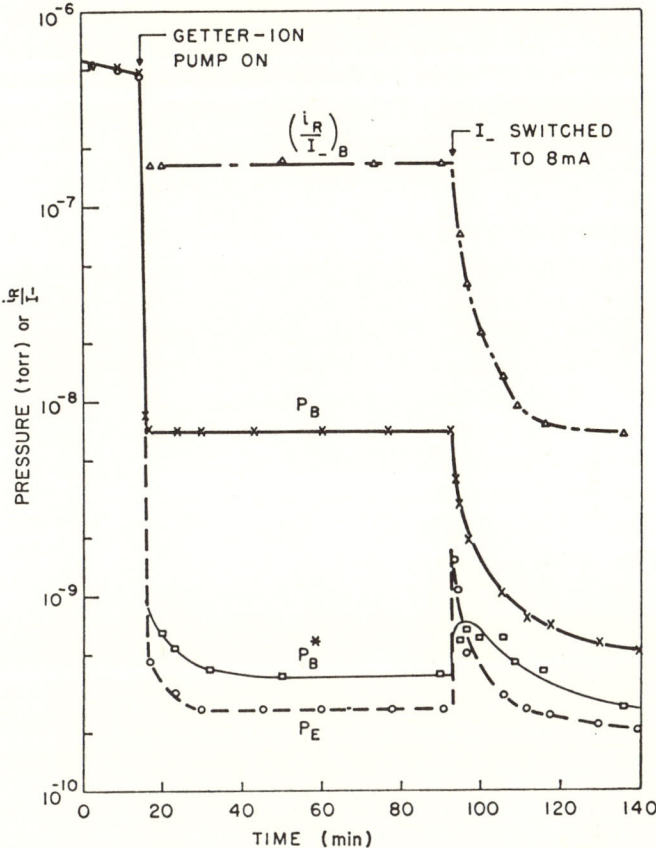

Fig. 7.13 Indicated pressure and normalized residual current (i_R/I^-) as a function of time for an extractor gauge and a BAG after oxygen exposure (Redhead 1966).

7×10^{-9} torr, i.e. about thirty times too high. Pressure measured by modulating the BAG (P_B^*) was in reasonable agreement with P_E. The upper curve shows the residual current of the BAG (i_R), normalized to the electron current, determined from the modulation measurements (see Section 8.1.7). At $t = 93$ minutes the electron current in both gauges was increased to 8 mA. The true pressure, as indicated by P_E, shows a large burst due to gas released from the grids of the two gauges (partly by electron-impact-desorption and partly due to an increase in temperature of the grids). P_B^* also shows this burst but because of the difficulty of following a transient with the modulation method the points are scattered. P_B however shows a marked decrease and eventually agreed with P_E after 20 hours. The residual current also dropped slowly to a value typical of a 'clean' grid in about a day.

A detailed examination has been made of the effects of O_2 on pressure measurement with a BAG (Redhead 1963); pressure indications too high by as much as 75 times were observed. It was also shown that it was possible to obtain reasonably accurate pressure measurement in the presence of large currents of ions from the grid surface by use of the modulation method (Redhead 1960b). Singleton (1966a) has also shown that static oxygen pressure can be measured with reasonable accuracy by the modulation technique. Table 7.8 shows the pressure of oxygen indicated by a BAG, a

TABLE 7.8
Equilibrium pressure indications (torr) in oxygen

BAG	Modulated BAG	Mass Spectrometer
$2 \cdot 22 \times 10^{-8}$	$7 \cdot 6 \times 10^{-9}$	$7 \cdot 7 \times 10^{-9}$
$3 \cdot 25 \times 10^{-8}$	$1 \cdot 30 \times 10^{-8}$	$1 \cdot 28 \times 10^{-8}$
$1 \cdot 77 \times 10^{-8}$	$6 \cdot 0 \times 10^{-9}$	$6 \cdot 0 \times 10^{-9}$
$2 \cdot 04 \times 10^{-8}$	$6 \cdot 9 \times 10^{-9}$	$7 \cdot 0 \times 10^{-9}$

Gauge electron current 10 mA.
After Singleton (1966a).

modulated BAG and a mass-spectrometer, indicating good agreement between the O_2 pressure measured with the mass-spectrometer and the modulated BAG. Hartman (1963) has also observed large residual currents after exposure of a BAG to oxygen. Schuemann et al. (1963) have observed large errors in pressure indication of the suppressor gauge (see Section 8.1.2) with exposure to oxygen, and have studied the way in which the anomalous pressure signal decreases, after oxygen is removed, as a function of electron current. The rate of decrease increases monotonically with electron current, thus at $I^- = 10$ mA the error disappears in a few minutes while at $I^- = 10$ μA the error persists for many hours.

Schuemann et al. (1963) have made similar measurements with CO and find the effect to be smaller than with O_2. For $I^- = 0 \cdot 1$ mA the maximum error in the reading of a BAG was a factor of 4. Similar results were obtained by Redhead (1962a).

Singleton (1967a) has observed errors in the pressure indicated by a BAG in H_2 caused by E.I.D. from the grid. He interprets his results to indicate that the E.I.D. error is caused by the release of 0^+ ions from the grid by electron impact on adsorbed CO formed by reactions of the H_2 with the hot filament (see Section 7.2.2). Operation of the gauge at high electron currents (10 mA) greatly reduced the error but nevertheless it was desirable to use the modulation method to reduce the error to negligible proportions.

Non-linear pressure vs electron current characteristics of a BAG were observed by Mizushima and Oda (1959) which were very probably the result of E.I.D. effects but were misinterpreted at the time. Ackley et al. (1962)

also observed non-linear behaviour of BAGs which was later shown to result from E.I.D. Hoch (1963) has observed E.I.D. effects with the residual gases (CO and H_2) in a system containing a BAG, a Penning gauge and an omegatron.

Alkali metal impurities evaporated from the incandescent tungsten filament of a BAG can adsorb on the grid and Denison *et al.* (1963) concluded that E.I.D. then occurred resulting in an error in pressure measurement. This problem can be easily overcome by prolonged heat treatment to remove the alkali metal impurities from the tungsten or by the use of a low-temperature emitter (ThO_2 or LaB_6).

Electron-impact desorption effects are not usually a problem in cold-cathode gauges because the electron current to the anode, causing release of positive ions, is proportional to pressure. However, the release of ions from the anode by electron impact may affect the value of the Townsend gamma factor and hence change the striking characteristics of the discharge. Reikhrudel and Sheretov (1966a) have observed a large change in the striking characteristic of a cold-cathode magnetron gauge after the anode had been heated to 1900°K to remove gas. Similar outgassing of the cathode caused little change in the striking characteristics.

Mass peaks resulting from ions released by electron bombardment of adsorbed gas on the electrodes of the ion source can be observed in certain types of mass-spectrometer. Mass-spectrometers in which the electron beam is well collimated by a magnetic field (and thus does not strike the inside of the ionization box) do not show peaks due to surface ions. Mass-spectrometers in which the electron beam is not collimated frequently show surface ion peaks. Surface ions can be distinguished from gas phase ions in single-focusing mass-spectrometers (i.e. those instruments which focus for variation in initial angle of the ions but not energy (see Sections 9.1 to 9.3) since the peak caused by surface ions appears in the mass spectra slightly shifted with respect to the gas phase peak. This shift results from a combination of (*a*) the initial kinetic energy with which most electronically desorbed ions are released (see Section 4.2), and (*b*) the potential difference between the wall of the ionization box and the position where most ions are formed. This effect was first reported by Davis (1962), using a 90° sector instrument, who observed surface ion peaks at $16\frac{1}{4}$ (O^+), $19\frac{1}{4}$ (F^+), 35 and 37 (Cl^+). Robins (1963) using the same type of mass-spectrometer as Davis, was able to show conclusively that these peaks were caused by surface ions, and made rough measurements of the distribution of the initial kinetic energy of the ions.

In many mass-spectrometers the mass peaks due to ions produced at surfaces can be readily distinguished and thus do not cause a serious problem in partial pressure measurement. The E.I.D. effect can cause serious errors in measurement with hot-cathode gauges. The gases causing trouble, in order of their nuisance value, are O_2, halogens, CO, H_2 and hydrocarbons.

The errors in pressure measurement with hot-cathode gauges in the presence of the above gases can be minimized by:
 (i) using a high electron current (10 mA in a BAG) to keep the grid clean,
 (ii) use of the modulator technique;
 (iii) use of a gauge type in which the E.I.D. error is small (e.g. the extractor gauge).

7.5 METHODS OF CURRENT MEASUREMENT

With rare exceptions, the only signals available for the interpretation of phenomena occurring in uhv systems are electronic currents. These currents range from 1 A down to the lowest measurable ($\sim 10^{-20}$ A). This section will give a short summary of the problems encountered and the techniques used in the measurement of small currents.

7.5.1 General problems of current collection in UHV

The currents of interest in uhv systems usually involve the arrival or departure of ions or electrons at a conducting surface. Two phenomena make accurate measurements of these currents difficult:
 (i) ion reflection and electron reflection can cause an under-estimate of the rates of arrival of these particles (see, for example, Brunnée (1957) and Khan *et al.* (1963));
 (ii) ejection of secondary electrons can cause an over-estimate of ion arrival rates (Hagstrum 1956a) and an under-estimate of electron arrival rates (Khan *et al.* 1963).

Since the magnitudes of the above effects depend on the surface state of the collector, the secondary particles must either be measured or prevented from escaping the collector if accurate current values are required.

The most common current measurement in uhv systems is that of ion currents produced by electron bombardment of the ambient gas. These currents form the basis of almost all total and partial pressure measurements. The probability of a 100 eV electron creating an ion in argon is about 10 Lp, where L is the length of its flight (cm) and p is in torr. Phenomena occurring at the electron collector with a comparable (small) probability due to electron impact can give rise to extraneous currents. Two such phenomena are photon production and the desorption of positive ions (see Section 7.4). The former gives rise to the well known x-ray limit of ionization gauges when the photons in turn eject photo-electrons from the ion collector. The latter can sometimes cause even more serious effects by collection of the desorbed ions along with those produced in the gas phase (Redhead 1963).

If, as is not uncommon, high frequency oscillations exist in an uhv device, a small fraction of emitted electrons can gain kinetic energies of as much as 10 or 20 eV and so unexpectedly reach ion collectors (see Section 8.1.1).

In addition to the problems mentioned above, a number of phenomena

not related to the uhv space itself can be troublesome in the measurement of small currents:
  (i) resistive leakage across or through insulators;
  (ii) piezoelectric effect in insulators following mechanical stress;
  (iii) electrostatic charging effects on insulators resulting from friction; this effect is particularly troublesome in coaxial cables unless special low noise types are used;
  (iv) capacitive coupling to surrounding electrodes;
  (v) magnetic induction from time varying magnetic fields;
  (vi) radio frequency interference.

All can cause currents competing with the desired signal, but can usually be reduced to $< 10^{-14}$ A by proper choice of materials, mechanical design and electric or magnetic shielding.

7.5.2 Measurements with d.c. amplifiers

Presently available direct current amplifiers are capable of measuring currents of $\sim 10^{-13}$ A while retaining a response time of $\sim 10^{-1}$ sec. Somewhat lower currents are measurable with longer response times. These instruments use electrometer tubes having input grid currents of 1 to 5×10^{-14} A, input resistors of up to 10^{12} Ω and feedback amplifier systems for stability. Zero drifts are ordinarily $< 10^{-13}$ A in a working day, and reliability is excellent. With dynamic capacitor techniques, grid current and zero drift effects can be eliminated and currents of $\sim 10^{-15}$ A measured with response times ~ 1 sec. Limits are set by thermal noise in the sensing resistor. Below 10^{-15} A, dynamic capacitor amplifiers can be used in a 'charge accumulation' mode, but long measuring times and unavoidable stray currents become troublesome at $\sim 10^{-16}$ A.

7.5.3 Electron multipliers

Much lower currents can be measured using secondary-electron multiplier techniques (Allen 1939; 1950). Devices of this type have been made and operated in uhv systems with gains as high as 10^8 and background currents less than 1 ion/sec ($\sim 10^{-19}$ A) (Davis 1962). The gains of these devices often tend to decrease with time, particularly if they are exposed to air, but can be restored by baking either in the presence of $CsNO_3$ at 425°C (Davis 1962) or in 1 atmosphere of O_2 at 300°C (Young 1966b). The latter treatment also tends to reduce the background currents. The highest gains can be utilized only for the very lowest currents, since multiplier output currents greater than 10^{-7} or 10^{-8} A have been observed to cause decreases in multiplier gain with time ('fatigue' effects). Errors due to gain variation can be minimized in high gain multipliers by counting the output pulses individually. Expensive high-speed counting circuits are, however, required for rates above about 10^6/sec (equivalent to about 10^{-13} A).

A parameter of fundamental interest in ion detection by pulse counting is the initial ion-to-electron conversion ratio. If this ratio is much larger than one, all spurious pulses due to the emission of individual electrons, even from the first dynode, will be much smaller than the pulses produced by the ions. The spurious pulses can then be rejected in the counting system by discrimination techniques and a large improvement in the signal to background count ratio obtained. A large conversion ratio also minimizes errors due to the statistical probability of an incident ion producing no electrons. Only when the conversion ratio is greater than five is this error reduced to less than 1% (Bernhard et al. 1961).

When very low currents are being measured, the statistics of random fluctuations sets a fundamental limit to the rate at which data can be accumulated. This limit can be expressed either in terms of the 'counting statistics' of the arriving ions, or as the 'shot noise' content of the current:

$$i_{sn} = \sqrt{(2eI\Delta f)} \text{ A rms} \tag{7.22}$$

where e is the electronic charge
I is the total current being measured
Δf is the measuring bandwidth $\Delta f = 1/(2\pi\tau)$
and τ is the measuring time constant.

When no spurious currents contribute (background counting rate = 0), I can be identified directly with the current of arriving ions i^+. A multiplier of gain G used to detect the ion current will give an output noise current

$$Gi_{sn} = G\sqrt{(2ei^+\Delta f)} \text{ A rms.} \tag{7.23}$$

If this is in turn amplified by an electrometer with an input resistor R, the thermal noise fluctuations in the resistor will contribute an equivalent noise current

$$i_{Rn} = \sqrt{(4kT\Delta f/R)} \text{ A rms} \tag{7.24}$$

a limit which can, in fact, be fairly closely approached in modern electrometer amplifiers. Eqns. (7.23) and (7.24) indicate that the multiplier gain required to make the shot noise current and resistor noise current equal is

$$G_o = \sqrt{(2kT/ei^+R)} \tag{7.25}$$

or, for $T = 300°K$, $G_o = 0.227/\sqrt{(i^+R)}$

Increases in gain beyond this value can improve the signal-to-noise ratio by at most a factor $\sqrt{2}$, and thus give very little increase in available information.

In electrometer amplifiers, noise is minimized by choosing R as large as is possible while retaining the required time constant. For most collector systems, $R = 10^{12}\ \Omega$ is practicable value when the required τ is ≥ 0.1 sec, with proportionally smaller values being required for smaller τ's. The largest value of R generally practicable because of stray leakages is $R = 10^{13}\ \Omega$ and this is normally associated with a response time $\tau \approx 1$ sec. Thus for $\tau \leq 1$ sec

$$R/\tau \approx 10^{13}. \tag{7.26}$$

The fractional probable error set by counting statistics in the measurement of a current i^+ with a time constant τ is given by

$$\text{P.E.} = \frac{0 \cdot 674}{\sqrt{n}} = \frac{0 \cdot 674}{\sqrt{(i^+ \tau/e)}} \tag{7.27}$$

where n is the number of ions arriving in a time interval τ seconds. Thus the minimum measurable value of the product $i^+\tau$ will depend on the accuracy of current measurement required. From eqn. (7.27) we can write

$$i^+\tau(\text{P.E.})^2 = 7 \cdot 3 \times 10^{-20}. \tag{7.28}$$

A product $i^+\tau = 3 \times 10^{-19}$, which yields P.E. $= 0 \cdot 5$, probably represents about the smallest value of any quantitative significance.

Substituting $i^+\tau = 3 \times 10^{-19}$ and $R/\tau = 10^{13}$ into eqn. (7.25) indicates that for *any* value of $\tau \leq 1$ sec, the two noise components are of equal size when $G_0 = 133$, and that increasing the multiplier gain beyond this value yields little further increase in available information. If we retain $\tau \leq 1$ sec, measurements with smaller probable error can be done only on the correspondingly larger currents calculated from eqn. (7.28), and the useful gain G_0 will tend to decrease. More detailed consideration of noise limits in detection systems has been given by Mamyrin (1967) who comes to qualitatively similar conclusions. He includes the effect of background counting rate and statistical gain variation in multipliers in his calculations.

For ion currents less than 10^{-18} A, quantitative measurements require proportionally larger effective time constants and are best made by increasing periods of individual ion pulse counting. Statistical errors in current and background rate determinations are described by eqn. (7.27) with τ replaced by the counting interval. The multiplier gain must be sufficient to allow the output pulses to be increased further in amplitude by pulse amplifiers to a size suitable for counting apparatus. For such low counting rates (low amplifier bandwidth) gains of $\sim 10^4$ are normally satisfactory.

Two types of current multipliers have been reported which utilize continuous resistive secondary emitting surfaces instead of individual dynodes. In the first type, a pair of parallel resistive strips are placed in a transverse magnetic field to form a system of crossed electric and magnetic fields which guide the electrons (Goodrich & Wiley 1961; White *et al.* 1961). In the second type, resistive-walled channels are used, and no magnetic fields are required (Oshchepkov *et al.* 1960; Wiley & Hendee 1962). A constant potential difference of a few kV is applied between the ends of the strip or channel to provide the electron accelerating fields. Gains of 10^6 to 10^7 have been realized in both types in relatively small structures. Ion feedback effects occur in the channel multipliers at gains $> 10^6$, but these can be avoided by curving the channels (Evans 1965). One important feature of channel multipliers is that their gain at a given voltage is determined by the *ratio* of length to diameter independent of the size. Ratios of 50:1 to 100:1 are most commonly used. It is possible to make extremely small multipliers (total length a few millimeters) if the current to be detected is confined to a small area so that it can be made to strike the mouth of the channel.

Output currents $> 10^{-7}$ A have been observed to cause fatigue effects (Goodrich & Wiley 1961) similar to those observed in dynode-type multipliers. A discussion of gain variation effects in channel multipliers is given by Smith (1966).

7.5.4 Conversion-scintillation detectors

A considerable effort has recently gone into the development of 'ion converter' scintillation detectors of the type first described by Schütze and Bernhard (1956). To avoid the damage caused by ion impact on phosphor screens (Richards & Hays 1950), the ions are accelerated into a metal plate held at a potential of -20 to -30 kV, and the resulting secondary electrons accelerated by the same potential to a phosphor screen. The resulting light pulse is detected by a photomultiplier outside the vacuum system. A review and bibliography of recent work on detectors of this type is given by Herold and Schönheit (1965). They describe a system with a Cu—Be conversion dynode operating at a potential of -25 kV, which gave the following performance for argon ions:

Secondary emission coefficient	9 electrons/ion
Transfer factor of secondary electrons	1
Phosphor screen (ZnS(Ag)) yield	2×10^3 quanta/electron
Light transfer to photo cathode	7×10^{-2}
Photo cathode yield	5×10^{-2} electrons/quantum
Photomultiplier gain	8×10^6
Product of above (total gain)	$\sim 5 \times 10^8$

The gain was found to vary only slowly with ion mass (Herold 1965) at this energy. This minimizes the mass discrimination effect common in electron multipliers (Higatsberger *et al.* 1954; Schram *et al.* 1965a), where the ion energies are usually less than about 5 keV. The very high gains and particularly the high ion to electron conversion ratio make counting and discrimination easy. Response times are extremely rapid, allowing a relatively large dynamic range of current measurement; from about 1 ion/sec to $> 10^7$ ions/sec. Duffy and Ruiz (1966) have used two photomultipliers operating in a coincidence-counting mode in a scintillation detector. They report a background counting rate of about 0·1 ion/sec, a factor 10 better than could be obtained with a single photomultiplier.

Detectors of the ion converter type require only a single potential to be introduced into the vacuum system. The electrode must, however, support a negative potential of 20 to 30 kV, and the surrounding structure must be fairly carefully designed to effect an efficient focusing both of the ions onto the conversion electrode and of the resulting electrons onto the phosphor

General Considerations of Pressure Measurement 299

screen. No detectors of this type operating in uhv have yet been reported, but fundamental difficulties are not to be expected.

7.6 CATHODES AND CATHODE EFFECTS

The hot cathodes used in ionization gauges or mass-spectrometers can cause severe problems in the measurement of very low pressures which may be divided into three categories:

(i) thermal effects, resulting from the heating of electrodes by radiation from the hot cathode,

(ii) chemical interactions at the hot cathode resulting in changes in gas composition in the system,

(iii) evolution of positive ions or neutrals from the cathode.

The first two problems can be minimized by lowering the temperature of the cathode as much as possible. The low work-function coatings necessary to achieve adequate emission at low temperatures are sometimes the cause of an increase in the problem listed as (iii) above. We first consider the various types of cathodes used in gauges and mass-spectrometers, with particular emphasis on low work-function emitters, and then examine in more detail some of the problems listed above.

Tungsten is the most widely used cathode material although the power radiated at normal levels of electron emission is sufficient to cause considerable temperature increases of the other gauge electrodes and the envelope. The only other pure metal that has been used to any extent as cathode in gauges and mass-spectrometers is rhenium (Robinson and Sharkey 1958; Kotowski & Tischer 1964).

Lanthanum boride, in the form of a sintered rod or supported on carbon, is an excellent low temperature emitter with a work function of 2·66 V (Lafferty 1951). The rate of evaporation is relatively low, when supported on carbon, corresponding to a heat of evaporation of 169 kcal/mole. When LaB_6 is supported on a refractory metal the boron diffuses into the metal support and the rate of evaporation of lanthanum increases greatly. Buckingham (1965) has measured the rate of evaporation of lanthanum from LaB_6 cataphoretically deposited on rhenium and finds it to be high, corresponding to a heat of evaporation of 75 kcal/mole. This exceedingly high evaporation rate makes the LaB_6 on metal cathode unsuitable for uhv use except in special cases where the cathode is never heated very hot and the emission density is maintained below 10^{-4} A/cm^2.

Thorium oxide can be cataphoretically coated onto tungsten or iridium to produce low work-function cathodes with properties that are very suitable for uhv measurements. The method of preparation and the properties of these cathodes have been described (Hanley 1948; Weinreich 1951). Although the work function of the thoria-coated cathodes is not as low as lanthanum boride cathodes, the rate of evaporation from the thoria cathodes is very much less.

When using thoria-coated cathodes in an ionization gauge and outgassing by electron bombardment it is essential to avoid the use of ac high-voltage supply, as can be used with tungsten cathodes. Bombardment of the thoria-coated cathode by electrons from the hot grid during the reverse half-cycle can damage the thoria coating.

Normal oxide-coated cathodes, [BaSr]O, have not been widely used in uhv instruments. Although these cathodes have the lowest work function available they are easily poisoned by chemically active gases and they emit negative ions in substantial quantities. The principal ions emitted are Cl^-, O^- and H^-; C^-, CH^-, F^-, Ca^-, O_2^-, Br^- are emitted to a lesser extent (Sloane & Watt 1948). Because of these problems [BaSr]O cathodes have only been used at very low temperatures to provide small emission currents (e.g. 10^{-6} A emission in an omegatron).

The properties of the various cathodes mentioned above are compared in Table 7.9. For uhv use a cathode should generally have a very low evaporation

TABLE 7.9
Comparison of cathodes used in uhv instruments

Cathode	ϕ (V)	T_e (°K)	$W_{(T=T_e)}$ (g/cm² sec)	$\eta_{(T=T_e)}$ (A/Watt)	Reference
W	4·53	2180	$6·4 \times 10^{-12}$	$1·3 \times 10^{-4}$	Jones & Langmuir 1927
LaB$_6$/C	2·66	1370	$\sim 10^{-15}$	$\sim 10^{-3}$	Lafferty 1951
LaB$_6$/Rh	2·70	1370	$\sim 10^{-10}$	$\sim 10^{-3}$	Buckingham 1965
ThO$_2$/W	2·96	1500	$\sim 2 \times 10^{-16}$	$\sim 10^{-3}$	Shapiro 1952
					Hanley 1948
[BaSr]O	1·5	750	$\sim 10^{-17}$	$\sim 10^{-2}$	Wagener 1951

ϕ average work function
T_e temperature for emission density of 10^{-2} A/cm²
W rate of evaporation at $T = T_e$
η emission efficiency, amps of emission current per watt of heating power at $T = T_e$.

rate and low operating temperature. If maximum emission efficiency is required, then [BaSr]O is better than any other type of cathode. The figure on rate of evaporation in Table 7.8 must be used with circumspection because the rates for LaB$_6$ and oxides are very dependent on the nature of the substrate, the presence of minute quantities of impurities in the cathode material and the prior history of cathode operation.

Figure 7.14 shows the maximum emission density available from the various cathodes discussed above as a function of temperature. The approximate evaporation rates as a function of emission density are shown in Fig. 7.15. The evaporation from the hot cathode of an ionization gauge causes an increase in the pressure indicated by the gauge. This effect can be roughly estimated on the assumption that the evaporating molecules pass only once through the

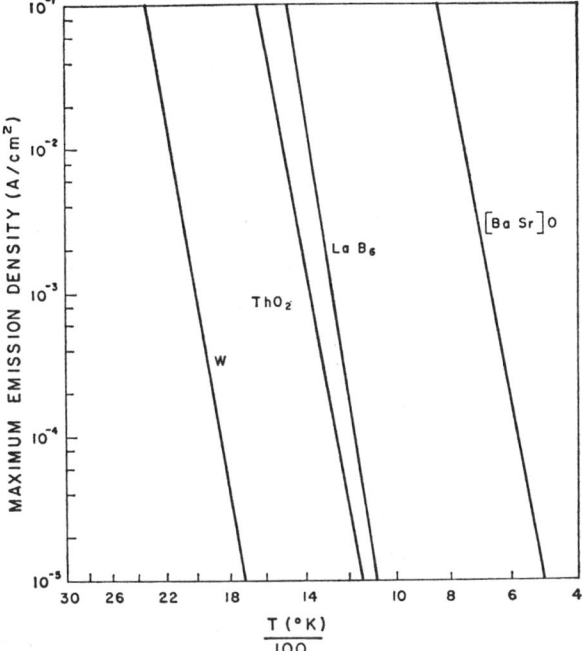

Fig. 7.14 Maximum electron emission density as function of cathode temperature for various cathode materials.

ionization region of the gauge and are then adsorbed on a surface. The approximate pressure indications of a Bayard-Alpert gauge caused by evaporated materials, corresponding to the indicated evaporation rates, are also shown in Fig. 7.15.

Experience in the authors' laboratories indicates that ThO_2 on tungsten or iridium is the best type of cathode for use in gauges and mass-spectrometers. It has reasonable emission efficiency, insignificant evaporation rates (for $J^- < 10^{-2}$ A/cm^2), is reasonably unaffected by chemically active gases and can be repeatedly exposed to air. ThO_2 on iridium has the advantage that it can be exposed to air when hot without damage.

Other types of thermionic cathodes have been extensively studied but none have yet shown significant advantages over the older types. These newer types of cahtodes have been reviewed recently (Tsarev 1965). Sources of electrons, other than thermionic, have not been used to any extent in uhv instruments. The various non-thermionic sources of electrons have been reviewed (Schagen 1965).

Neutral particles evaporated from the hot cathode of an ionization gauge can be ionized by electron impact as they pass through the grid cage and give rise to false pressure indications. This process sets a limit to the lowest pressure

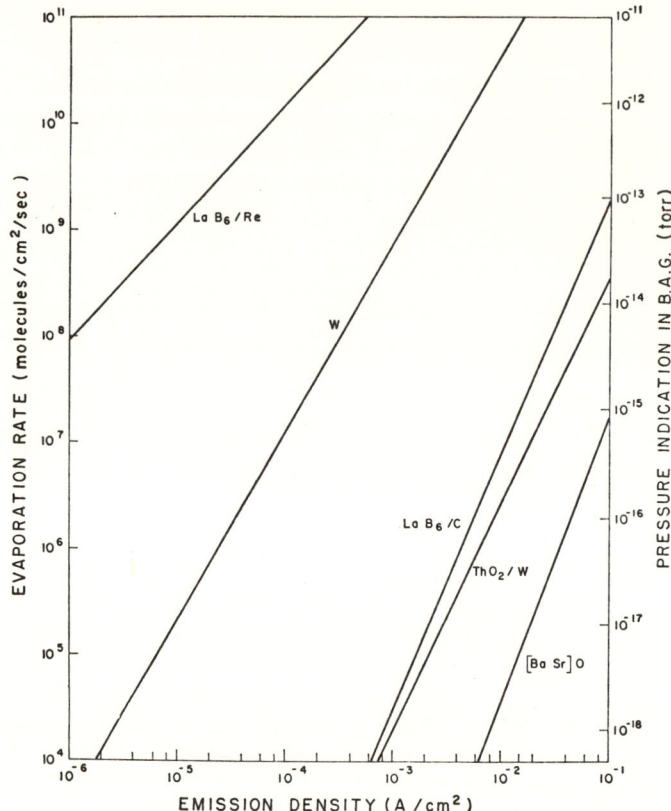

Fig. 7.15 Approximate evaporation rate and corresponding pressure indication in a BAG as a function of electron emission density.

measurable with a hot-cathode gauge. This limitation was first discussed by Alpert and Buritz (1954) who measured the pressure indicated by a BAG when an auxiliary tungsten filament was heated to temperatures between 2600 and 3300°K. From these data one can estimate the relation between apparent pressure (P_{eq}) and rate of evaporation (R', molecules/cm² sec) for a gauge with a tungsten cathode

$$P_{eq} = 1.33 \times 10^{-24} A_f R',$$

where A_f is the cathode surface area (cm²).

Tungsten wire usually contains alkali metal impurities (Na, K, Rb, Cs) which are evaporated as neutrals and positive ions (see Section 4.4). The rate of production of positive ions is increased when the cathode is bombarded with rare gas ions (Riddoch & Leck 1958) or when oxygen interacts with the hot cathode to cause removal of tungsten and exposure of pockets of alkali metal atoms (Winters *et al.* 1963). Denison *et al.* (1963) have shown that alkali metal impurities are evaporated from the cathode onto the grid from whence they can be released as positive ions by electron bombardment. The positive ion current from the grid can result in errors of pressure measurement. Measurements by others indicate that alkali metals cannot be electronically desorbed (see Section 4.2); it is possible that in this case alkali metal compounds are involved. Prolonged heating of tungsten filaments at a high temperature usually suffices to reduce the evaporation of neutrals or positive ions to a negligible level.

The rates of evaporation of the basic cathode material from (*a*) LaB_6 on carbon, (*b*) ThO_2 on W or Ir, and (*c*) [BaSr]O are too low, at the emission levels used in uhv instruments, to cause any problems. However, evaporation of trace impurities from these cathodes can prevent serious problems under certain conditions, particularly in the early life of a cathode. Continued heating always reduces the evolution of impurity ions or neutrals.

The interaction of chemically active gases with hot cathodes resulting in changes in gas composition is a serious cause of contamination. These processes were discussed in Section 7.2.2.

CHAPTER EIGHT

TOTAL PRESSURE GAUGES

We will only be concerned here with gauges suitable for measurements in the ultrahigh vacuum range. For general reviews of gauges and pressure measurement techniques in other pressure ranges the reader is referred to Leck (1964), Steckelmacher (1965) and Schwarz (1951, 1952, 1960, 1961). Review articles concerned with ultrahigh vacuum pressure measurements have been published by Alpert (1958a), Grigorev (1959), Paty (1959), Morgulis and Marchenko (1960), Redhead et al. (1962b), Kornelsen (1965) and Reich (1966). Bibliographies on low pressure measurement have been published by Brombacher (1961, 1967).

Only the ionization gauge, either hot or cold-cathode, has proved really practical for uhv pressure measurements. Only one type of gauge, other than ionization gauges, shows any potential for use in the uhv range. Beams and his colleagues (1962) have demonstrated that measurement of the rate of deceleration of a magnetically suspended rotor, rotating at about 10^6 c/s can be used as a measure of pressure down to 5×10^{-10} torr. The requirements on temperature control ($\pm 10^{-2}$°C) and vibration are so stringent as to make the use of this gauge impractical except in some special cases. The following sections are concerned exclusively with ionization gauges.

8.1 HOT-CATHODE GAUGES

Hot-cathode ionization gauges were used for many years before it was realized that there was a limit to the lowest pressure measurable with such gauges. Nottingham (1947) pointed out that electrons striking the grid would create soft x-rays which were capable of liberating photo-electrons from the ion-collector. The resulting photocurrent would be indistinguishable, in the measuring circuit, from the positive ion current. This so-called 'x-ray limit' has been discussed in Section 7.4. The lowest pressure measurable with conventional ionization gauges was about 10^{-8} torr at the time of Nottingham's comment (1947). Several attempts were then made to design gauges in which the x-ray limit was decreased, either by reducing the solid-angle subtended by the ion-collector at the x-ray source, or by attempting to suppress, by an electrostatic field, the induced photo-electrons. Some early attempts were only partly successful (Lander 1950; Metson 1951) whereas the design due to Bayard and Alpert (1950) was simple and effective and has since been very widely used. It can be argued with considerable justification that the development of the Bayard-Alpert gauge was the most significant step in establishing ultrahigh vacuum technology.

Many problems common to all types of hot-cathode gauges are discussed in Section 8.1.1 below in connection with Bayard-Alpert gauges. Discussions of the basic method of operation and elementary principles of the various gauges can be found elsewhere (e.g. Leck (1964), Dushman & Lafferty (1962)).

8.1.1 The Bayard-Alpert gauge

The original design of Bayard and Alpert is shown in Fig. 8.1. The ion-collector C is a fine wire placed *inside* the grid B. Electrons from the tungsten filaments A are accelerated to the grid, from which they liberate soft x-rays on impact. Because of the geometry of the gauge only a small fraction of the total emitted x-rays are intercepted at the ion-collector. In a conventional

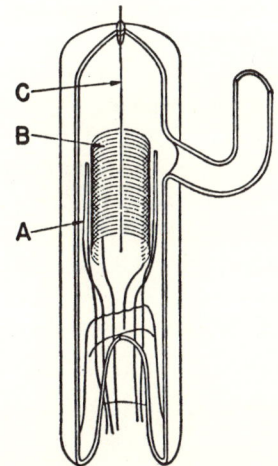

Fig. 8.1 Schematic of first design of Bayard-Alpert gauge, showing filament (A), grid (B), and ion-collector (C). (Bayard & Alpert 1950).

ionization gauge the ion-collector was a large cylinder surrounding the grid which intercepted almost all the soft x-rays. For an ion-collector wire of about 0.2 mm diam the fraction of the total x-rays intercepted at the ion-collector (the geometric factor) is about 1/300. Thus one would expect the Bayard-Alpert gauge to have an x-ray limit 300 times less than a conventional ionization gauge. The sensitivity is however only changed by a small amount.

That the principal limiting process was caused by x-rays, and that this new design was capable of significantly reducing the x-ray limit, was demonstrated by the measurement of Bayard and Alpert (1950) who found the ion-collector current vs grid voltage characteristics at different pressures for a Bayard-Alpert and a conventional gauge to be as shown in Fig. 8.2. At high pressures the characteristics of the BAG and conventional gauge are very similar and resemble an ionization cross-section curve. However, at pressures of about 10^{-8} torr the characteristic for the conventional gauge has changed to a power law behaviour typical of x-ray photocurrents in this energy range. The BAG at 4×10^{-9} torr still shows a 'gas ionization' characteristic superimposed on a monotonically rising curve caused by photo-electrons. At pressures of about 5×10^{-11} torr the BAG shows a characteristic similar in shape to that for the conventional gauge at $\sim 10^{-8}$ but reduced by a factor of more than 100. These experiments clearly demonstrate that the new design had reduced the x-ray limit by two orders of magnitude or more and that pressures of about 10^{-10} torr were now measurable. Limiting processes other than x-ray photo-emission can occur and may have been present in these original experiments.

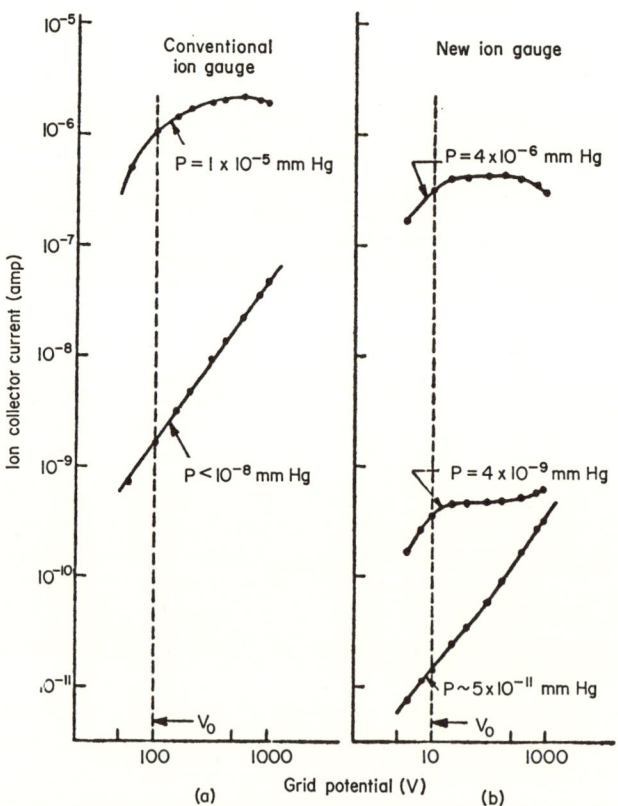

Fig. 8.2 Ion-collector current vs grid voltage for (*a*) RCA type 1949 gauge, (*b*) Bayard-Alpert gauge. (Bayard & Alpert 1950).

Various minor modifications have been made to the original design of Bayard and Alpert but the basic feature of the gauge, an ion-collector consisting of a fine wire, remains unaltered. A typical modern design of commercial BAG is shown in Fig. 8.3. This gauge has an ion-collector of 0·125 mm diam and an x-ray limit of about 2×10^{-11} torr with a grid voltage of $+130$ V and a filament voltage of $+45$ V. This gauge is normally operated at an electron current of 4 mA and has a sensitivity factor for nitrogen

$$K = \frac{I^+}{I^-} \cdot \frac{1}{P} = 25 \text{ torr}^{-1}.$$

To minimize errors in pressure measurement, the grid of a BAG must be very rigorously outgassed. Some designs of gauge can be outgassed by passing current through the grid, which is wound as an unsupported helix. Because of the lack of grid support the material of the grid helix must be very carefully prepared to minimize sagging and distortion when the grid is at a high

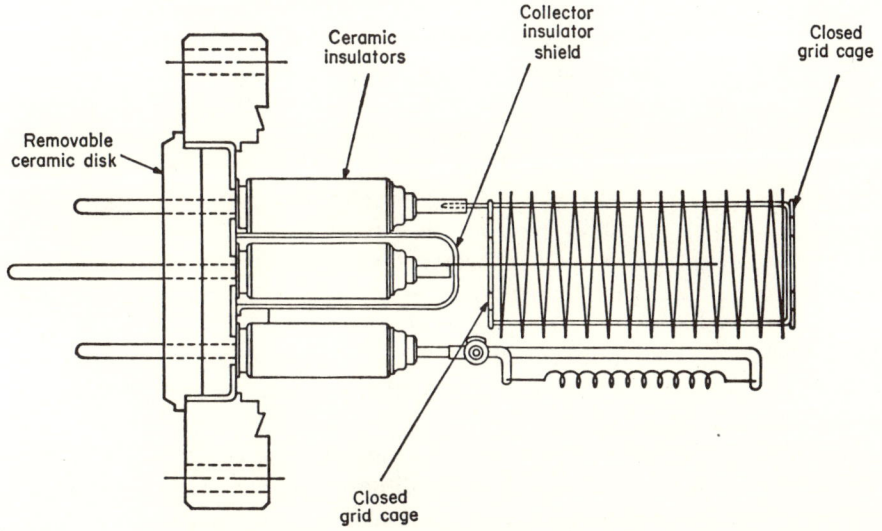

Fig. 8.3 Design of modern Bayard-Alpert gauge for demountable use. (*Courtesy of Varian Associates*).

temperature. Several commercial gauges are of this type and the grids are made of a specially treated, non-sag tungsten or molybdenum. In some cases the special treatment consists of carburizing the grids. To prevent excessive sagging of these grids the outgassing power is quite low, typically about 7·5 watts/cm².

Alternatively, the grid and ion-collector of the gauge may be outgassed by electron bombardment. This method permits the use of a grid of fine wire welded to heavy support rods which will not sag under high temperature outgassing. Molybdenum grids in the gauges used in the authors' laboratory are outgassed with 1 kV electrons at a power density of 15 watts/cm^2, yielding an outgassing temperature of about 1800°K. Outgassing by electron bombardment is preferable for accurate pressure measurements for two reasons: (a) sufficiently high temperature for rigorous outgassing cannot be obtained by resistance heating without grid sagging, and (b) the electron bombardment itself, rather than the concommitant heating, is a very efficient method of removing certain contaminating materials from the grid (see Section 4.2). Visible deposits, which may be formed on the grid during the baking period if water vapour is present in large amounts, can be rapidly removed by electron bombardment. The above comments concerning outgassing of the grid are equally applicable to other types of hot filament gauges which use a grid similar to the BAG (the suppressor gauge and the extractor gauge).

The original design of the BAG and many succeeding designs use uncoated glass bulbs as envelopes. The potential that the glass surface assumes depends on many factors and may exhibit bistability (Carter & Leck 1959). The potential of the glass will stabilize either: (a) at about cathode potential, where the electron current to the glass is equal to the positive ion current (in this state the potential of the glass increases slightly, approaching cathode potential more closely, as the pressure increases) or (b) at about grid potential, where the electron current to the glass equals the secondary electron current from the glass. Condition (a) is the normal state and condition, (b) only occurs infrequently in normal operation but may occur during outgassing by electron bombardment. Condition (b) should be avoided because it causes changes in gauge sensitivity and pumping speed and because considerable outgassing of the glass may occur. Although no net current flows to the glass surface in condition (b) there is nevertheless a transfer of power to the surface because the energy of the incident electrons is much higher than that of the secondaries leaving the surface.

High-frequency electronic oscillations occur in Bayard-Alpert and other positive grid gauges under almost all conditions. The potential of the inner surface of the gauge envelope has a profound effect on these oscillations. If these oscillations build up to large amplitude it is possible for some of the electrons to gain sufficient energy from the rf field to cause three deleterious effects:

(i) Excess energy electrons can reach the ion-collector.

(ii) Excess energy electrons can reach the envelope. If the envelope is insulated it will go negative, causing a change in gauge sensitivity.

(iii) The rf signal may be rectified in the input stage of the ion-current amplifier.

All of the above effects can cause erroneous pressure indications and in some cases may produce a negative collector current. These oscillations are purely electronic in nature (Barkhausen-Kurtz) and the frequency is unaffected by changes in the external circuits (Redhead 1960a). In typical BAGs with electron currents about 1 mA the oscillation frequency lies in the range 40–80 Mc/s. Methods of minimizing the effects of B—K oscillations on pressure measurements have been reported (Muller 1960; Pierre 1961).

Variation in the potential of the envelope of a BAG can also change the x-ray limit of the gauge. This effect, called the 'reverse x-ray effect', is discussed in Section 8.1.7.

To avoid or reduce the problems noted above that can be caused by an insulated envelope surface, it is best to have a conducting layer on the inner surface of a glass envelope whose potential can be controlled from outside. A transparent conducting film of tin oxide (Gomer 1953) is very suitable. During outgassing by electron bombardment the conducting film must be connected to the filament. An alternative method to reduce the influence that the potential of the glass bulb has on the gauge characteristics is to surround the electrode structure with an additional grid (Nottingham 1954).

The above comments concerning the effects of the potential of the glass bulb are also applicable to other positive-grid gauges.

Several useful improvements have been made in the design of Bayard-Alpert gauges. The sensitivity can be increased, over the original design, by closing the ends of the grids to prevent escape of ions axially (Nottingham 1954). The design shown in Fig. 8.3 has this feature. The use of low temperature filaments, such as thoria-coated tungsten or rhenium, reduces the operating temperature of the gauge and, more importantly, reduces chemical effects of residual gases at the hot filament (see Section 7.2.2).

The effects of variation of electrode spacing and dimensions on the sensitivity of the BAG have been investigated (Nottingham 1961; Schütze & Stork 1962a, b; Freytag & Schram 1962; Groszkowski 1965a, b). The sensitivity of a BAG varies only slightly when the position of the collector wire (collector length equal to grid length) is changed (Groszkowski 1965a). The sensitivity is also independent of position or length of the filament and of filament-to-grid spacing which was varied from 2–15 mm (Groszkowski 1965b). Sensitivity is a fairly strong function of the collector diameter and length (Groszkowski 1965c) using a grid with open ends.

The variation of sensitivity and pumping speed to rare gases of a modulated BAG as a function of electrode voltage has been reported by Ishikawa (1965).

Reduction of the x-ray limit of the BAG can be achieved by reducing the diameter and/or length of the ion-collector still further. Venema and Bandringa (1958) used a 25-micron diam ion-collector supported at both ends and reduced the x-ray limit to about 10^{-12} torr. The x-ray limit is here defined as the pressure at which the positive ion current equals the x-ray induced photocurrent. Van Oostrom (1961) used a 4-micron diam collector and achieved an x-ray limit of about 10^{-12} torr. Schütze and Stork (1962b) have examined gauges with various ion-collector diameters and

L

lengths. By extrapolation of their data they expect an x-ray limit of below 5×10^{-12} torr for a gauge with collector diameter 40 microns and length of 10 mm, while operating with a grid-filament voltage of 50 V.

Reduction of the collector diameter brings attendant problems, the principal one being a loss in sensitivity which can be reclaimed (to a degree) by increasing the grid-collector voltage. The decrease in sensitivity results from a decreased ion-collection efficiency for small diameter collector wires. Ions formed inside the grid experience a radially inward force; however, since angular momentum must be conserved, an ion formed with initial kinetic energy may not strike the collector wire, but rather go into orbit around the wire. This type of orbit is called a Kingdon orbit (Kingdon 1923) and will be discussed in more detail in the section on the orbitron gauge (Section 8.1.4). Mott-Smith and Langmuir (1926) have calculated the fraction of the total number of ions emitted from a cylinder of radius r_1 which are collected on a coaxial cylinder of radius r_2, where $r_1 \gg r_2$. This fraction is given by

$$f = \frac{2r_2}{\pi r_1}\left\{\left(\frac{V_1}{V_0}\right)^{\frac{1}{2}} + \left(1+\frac{V_1}{V_0}\right)\sin^{-1}\left(1+\frac{V_1}{V_0}\right)^{-\frac{1}{2}}\right\}, \qquad (8.1)$$

where V_1 is the potential at r_1, the potential at r_2 being zero, and ϵV_0 is the initial energy of the ions (assumed uniform in magnitude but randomly directed).

When

$$\left(1+\frac{V_1}{V_0}\right)^{\frac{1}{2}} \gg 1,$$

then

$$f = \frac{2r_2}{\pi r_1}\left(1+\frac{V_1}{V_0}\right)^{\frac{1}{2}}. \qquad (8.2)$$

Figure 8.4 shows how the ion collection efficiency (f) varies with the radius at which the ion is formed (r_1) for two different gauges, calculated from eqn. 8.2; here r_g is the grid radius. The 'normal' BAG (solid curves) has a collector diameter of 200 microns and a grid diameter of 25 mm. The fine collector BAG (dashed curves) has dimensions similar to Van Oostrom's gauge; collector diameter of 4 microns and grid diameter of 20 mm. The two initial energies chosen are (a) 0·025 eV, thermal energy at 300°K and (b) the most extreme case $e + H_2 \rightarrow H + H^+ + 2e$, where the H^+ has a mean initial energy of 9 V (Dunn & Kieffer 1963). When the ions are formed with a Maxwellian velocity distribution describable by a temperature T then Comsa (1966) has shown that

$$f = erf\left[\left(\frac{V_1}{kT}\right)^{\frac{1}{2}}\frac{r_1}{r_2}\right] \qquad (8.3)$$

The curves of Fig. 8.4 show that the collection efficiency of ions in a fine

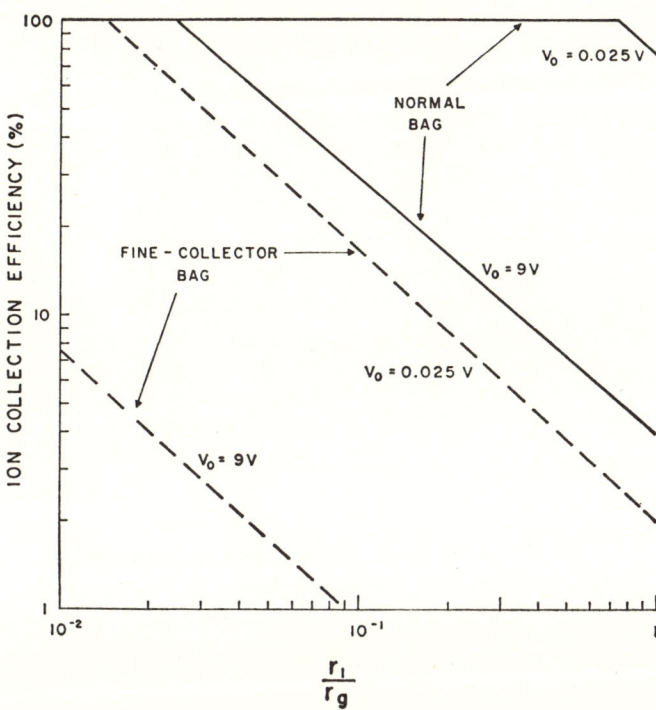

Fig. 8.4 Ion collection efficiency (f) as function of radius at which ion formed (r_1), r_g is grid radius. Solid curves: gauge with 200-micron diam collector; dashed curves: gauge with 4-micron diam collector.

collector BAG can be quite low, especially if the ions are formed with significant initial kinetic energy. By closing the ends of the grid to prevent axial ion loss, and by increasing the grid-collector voltage, it is possible to recover most of this loss of sensitivity. For example, Van Oostrom's (1961) gauge is operated with a grid-collector voltage of 270 V to obtain a sensitivity factor of 12 torr^{-1} for nitrogen. The ions formed inside the closed grid of a fine-collector BAG which do not strike the collector are trapped into Kingdon orbits until they either (a) are scattered into a new untrapped orbit and are collected, or (b) build up sufficiently positive space-charge to modify the electric field so that they are collected.

The very complex situation outlined above has not been analysed in detail but it is possible that nonlinear collector currents vs pressure characteristic may result. In a BAG with a normal size collector (150 to 200 microns diam) only a relatively few ions are trapped and these are all formed near the grid; thus they can escape the Kingdon orbits via the irregularities in the field near the grid wires. With a 25-micron diam collector, considerable nonlinearity has been observed, as indicated in Fig. 8.5, where the relative sensitivity factors for neon of a BAG with a normal (175-micron diam collector) and for argon with

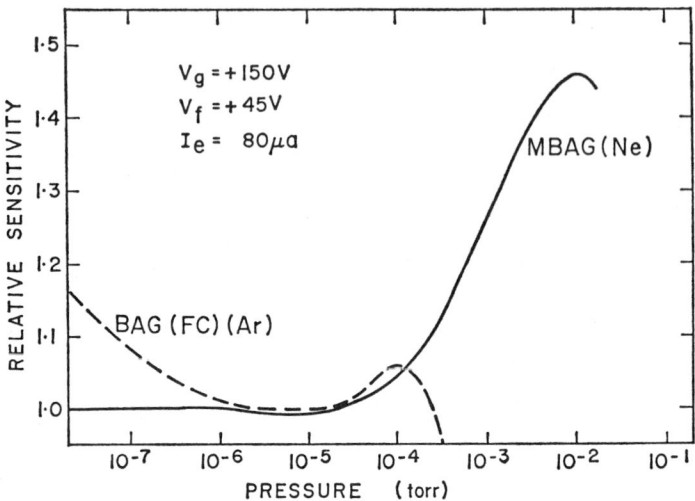

Fig. 8.5 Relative gauge sensitivity as a function of pressure for modulated Bayard-Alpert gauge with 175-micron diam collector (MBAG), and Bayard-Alpert gauge with 25-micron diam collector (F.C.)

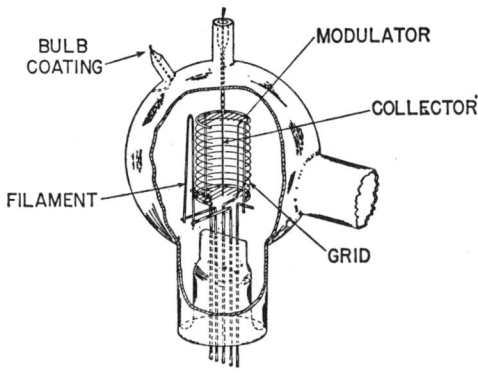

Fig. 8.6 Schematic diagram of modulated Bayard-Alpert gauge.

a fine-collector (25-micron diam collector) are compared as a function of pressure.

Another useful modification to the BAG is the addition of an electrode to modulate the ion current (Redhead 1960b). Figure 8.6 shows a schematic diagram of a modulated Bayard-Alpert gauge. The modulator consists of a wire placed parallel to the ion-collector inside the grid of a BAG. By switching the potential of the modulator wire from grid to ground potential the ion current to the collector can be modulated by about 30—40% while the residual current is almost unaltered. Measurements with the modulator allow the residual current and the true ion current to be calculated as described in Section 8.1.7. The modulation method allows the determination of residual current more accurately and rapidly than by any other method. The matter of rapidity is important, since the residual current can change considerably in a short time if chemically active gases are present.

8.1.2 Suppressor gauge

The x-ray induced photo-emission of electrons from the ion-collector of a gauge can be suppressed by a sufficiently large negative field at the collector surface. The first design of such a suppressor gauge (Metson 1951) was unsuccessful because the electrode producing the negative field (the suppressor) could intercept x-rays; the photo-electrons liberated from the suppressor were then attracted to the ion-collector and constituted a negative collector current. More recently a design of suppressor gauge, in which the suppressor electrode is shadowed from the x-ray source, has been described (Schuemann 1963) which has reduced the x-ray limit to below 10^{-12} torr*. Figure 8.7 shows a

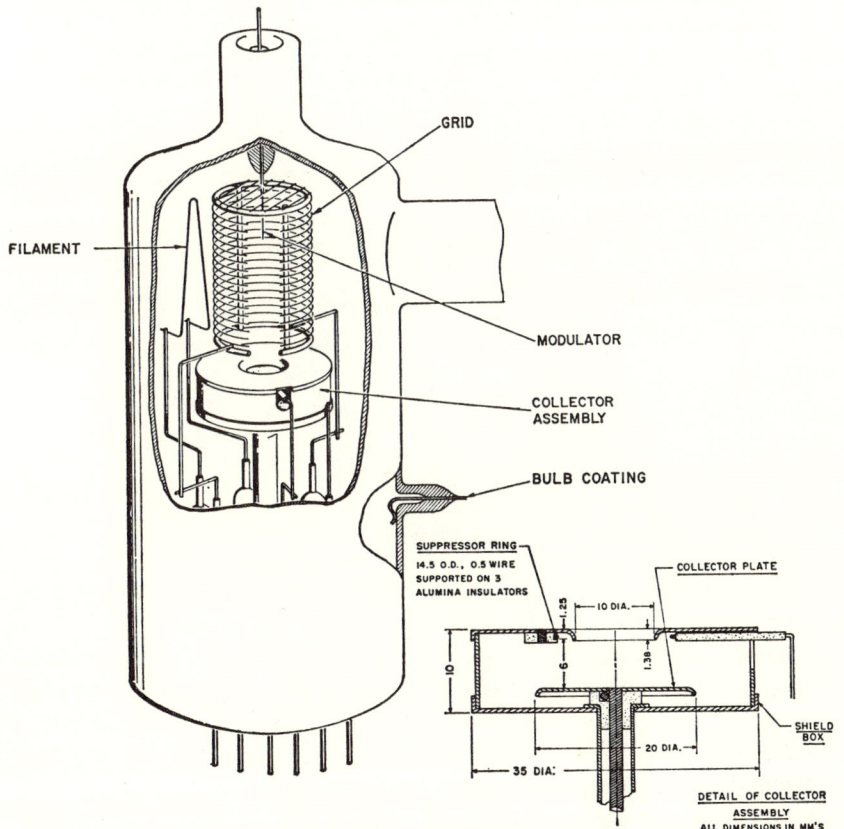

Fig. 8.7 Schematic diagram of modulated suppressor gauge (Redhead & Hobson 1965).

* A gauge similar to Schuemann's design is commercially available from Radio Corporation of America.

modified form of the suppressor gauge, with a modulating electrode, developed in the authors' laboratory.

The important feature of Schuemann's design that results in a lowered x-ray limit is that the suppressor electrode is shielded by another electrode from the x-rays produced at the grid. Thus photocurrents from the negative suppressor to the grounded ion-collector are minimized. Although the suppressor electrode is completely shielded from direct x-rays, it is still possible for x-rays to reach the suppressor electrode by reflection. The reflection of soft x-rays is discussed in more detail in Section 7.4, where it is shown that the reflection coefficient of soft x-rays from 100 eV electrons is a few tenths for most metals. To minimize the photocurrent of electrons from the suppressor to the collector, caused by x-rays reflected from the collector, it is necessary to minimize the solid angle subtended by the suppressor ring at the collector surface (Redhead & Hobson 1965).

The ion-collector and shield box (see Fig. 8.7) are operated at ground and the suppressor at about -600 V. The dimensions of the collector-suppressor-shield region must be carefully chosen to ensure that (a) the suppressor electrode is completely shadowed from the grid, and (b) the potential minimum on the axis is sufficiently low to suppress photo-electrons from the central region of the collector where the suppression field is weakest.

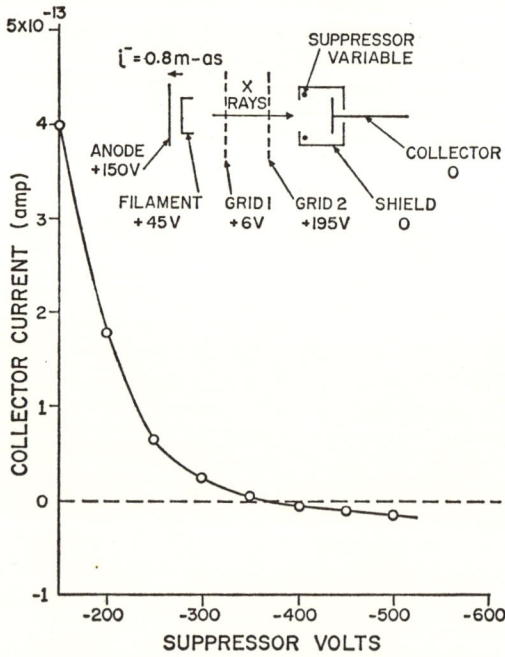

Fig. 8.8 Collector current as function of suppressor voltage for a special tube in which all ion-current could be cut off by grids 1 and 2. Electron current 0·8 mA, electron energy 105 eV.

At zero suppressor voltage (V_S) the collector current is constant for all pressures below about 10^{-10} torr; in the gauge shown in Fig. 8.7 this constant current is about 10^{-11} A for $I^- = 8$ mA and $V_{gf} = 105$ V. This current is the unsuppressed photocurrent of electrons from the collector, i_{x_1}. As V_S is increased negatively at relatively high pressures (say 10^{-11} torr), i_{x_1} is suppressed and the collector current reaches a constant value which is the true ion current, i^+. Adequate suppression of the forward photocurrent, i_{x_1}, can be obtained at $V_S = -400$ V at these pressures.

The suppression characteristic at extremely low pressure is complicated by the presence of a current of positive ions desorbed from the grid by electron bombardment. To avoid this problem a special tube was constructed, shown schematically in Fig. 8.8, which used a collector-suppressor structure identical to that in the gauge and had two grids interposed between the electron collector (anode) and the collector structure. The potential of the grids was set up to prevent any positive ions reaching the collector structure. With this arrangement the curve of Fig. 8.8 was obtained showing that the collector current went *negative* at high values of V_S. This negative photocurrent (i_{x_2}) results from photo-electrons, ejected from the suppressor by reflected x-rays, reaching the collector. This reverse x-ray photocurrent can be minimized by reducing the area of the suppressor electrode.

The sensitivity of the suppressor gauge is very similar to that of a BAG. The x-ray limit of the design shown in Fig. 8.7 is about $1 \cdot 5 \times 10^{-13}$ torr. When modulation is used to measure residual currents the lowest measurable pressure is established by the noise level in the amplifiers and is typically about 10^{-14} torr (Redhead & Hobson 1965).

Separation of currents, resulting from electronic desorption of ions from the grid, by means of the modulator is not as clear cut in the suppressor gauge as the BAG (see Section 7.4.2).

8.1.3 Extractor gauge

The extractor gauge design (Redhead 1966) resulted from attempts to overcome some of the difficulties inherent in the suppressor gauge. The basic drawback to the suppressor gauge is that the ion-collector is of large area and that the photocurrent from the collector is greater than the ion current to be measured at pressures below 5×10^{-11} torr. This implies that for low pressure measurements the suppression of this photocurrent must approach 100% and that minor variations in suppressor electrode position become crucially important. The extractor gauge uses very similar ion optics to the suppressor gauge except that the ions are collected on a short, fine wire rather than a flat plate.

Figure 8.9 shows the electrode arrangement in the extractor gauge. Ions formed within the grid are attracted towards the shield, which is at filament potential. Most of the ions pass through the hole in the shield and are focused onto the grounded ion-collector by the action of the hemispherical ion-reflector (at grid potential). The x-ray limit of this gauge is reduced because the collector wire subtends a very small solid angle at the grid. The extractor gauge is normally operated with filament, bulb and shield at $+200$ V, grid to filament voltage of 105 V, ion-collector at ground, and an electron current of 2 mA or less. The sensitivity factor under these conditions is similar to that of a BAG. For the gauge of Fig. 8.9 the sensitivity factor is 13 torr^{-1} for nitrogen.

The x-ray limit of the extractor gauge has not yet been measured but it is estimated to be about 3×10^{-13} torr. The lowest pressure measured with this gauge is 7×10^{-13} torr.

The extractor gauge is much less affected by electronically desorbable gases on the grid (see Section 7.4.2) than either the BAG or the suppressor gauge. Since this is frequently a more serious limitation to the measurement of very low pressures than the x-ray limit, this represents a very distinct advantage of the extractor gauge.

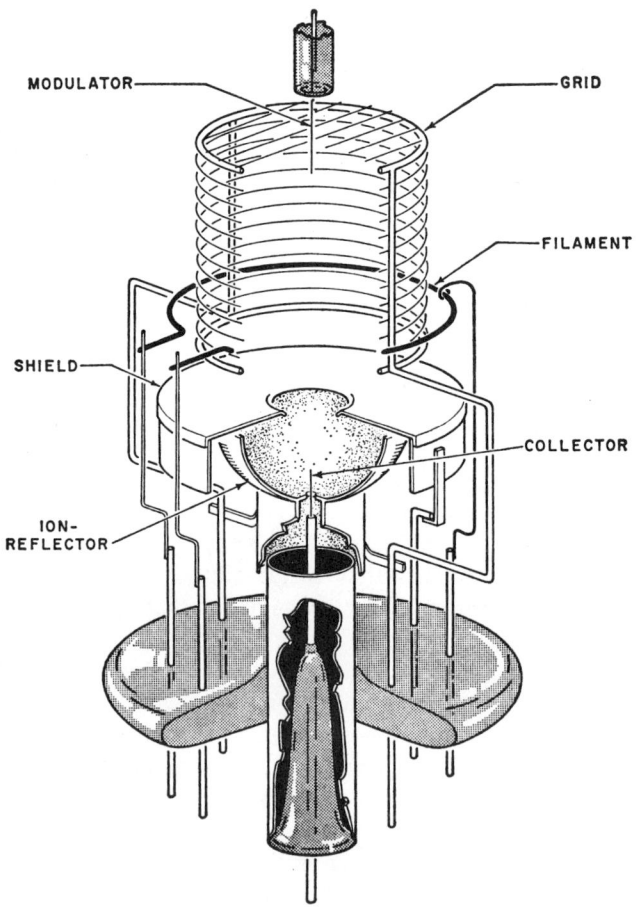

Fig. 8.9 Schematic diagram of extractor gauge (Redhead 1966).

Total Pressure Gauges

Gauges of similar design to the extractor gauge have been described by Clay and Melfi (1966) and Groszkowski (1966c).

8.1.4 Orbitron gauge

The orbitron gauge depends on the production of very long electron path lengths by launching electrons into Kingdon orbits (Kingdon 1923) between two coaxial cylinders. When electrons are launched, with appropriate angular

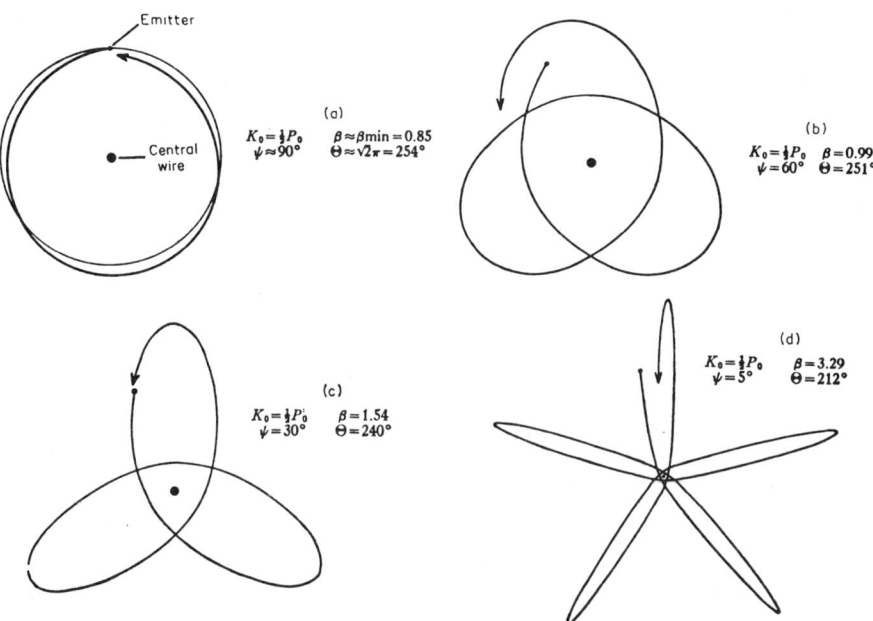

Fig. 8.10 Examples of Kingdon orbits that may occur in the orbitron. All electrons emitted with kinetic energy $K_0 = \frac{1}{2}P_0 = eV/ln\,(r_1/r_2)$, where V is voltage between two concentric cylinders and r_1 and r_2 their radii. ψ angle of emission with radius vector. θ is angular separation between successive apogees (or perigees). $\beta < 0.8466$ and is a dimensionless parameter governing the orbit shape (Hooverman 1963).

momentum, into the space between two coaxial cylinders, they perform complex orbits around the central, positive cylinder without striking it. The nature of these orbits has been analysed by Hooverman (1963) and some typical orbits are shown in Fig. 8.10.

The possibility of using electrons trapped into Kingdon orbits as the basis of a vacuum gauge or pump was suggested by Gabor (1962). However, it was not until Herb and his colleagues showed that the electrons could be launched from a small filament, rather than an elaborate electron gun, that the use of Kingdon orbits in pumps and gauges became practical (Herb et al. 1963).

A typical orbitron gauge is shown in Fig. 8.11 (Meyer & Herb 1967). The filament is made from very fine wire to reduce the number of orbiting electrons that are intercepted. The filament support wire is placed between filament and anode to reduce the number of electrons reaching the anode directly. The anode is a fine wire of 25- to 250-micron diam and operated at about +500 V. The reflecting tubes and end-plates are at ground potential and serve to reflect the orbiting electrons. A description of several orbitron gauges with different dimensions has been given (Mourad *et al.* 1964).

The gauge of Fig. 8.11 with an anode diameter of 230 microns operated at an anode voltage of 540 V, an anode current of $1 \cdot 2 \times 10^{-6}$ A, and a reflector bias of -52 V had a sensitivity factor K of 9×10^4 torr^{-1} (Meyer & Herb 1967). The mean electron path length is given approximately by $\bar{l} = K/10$ cm; thus

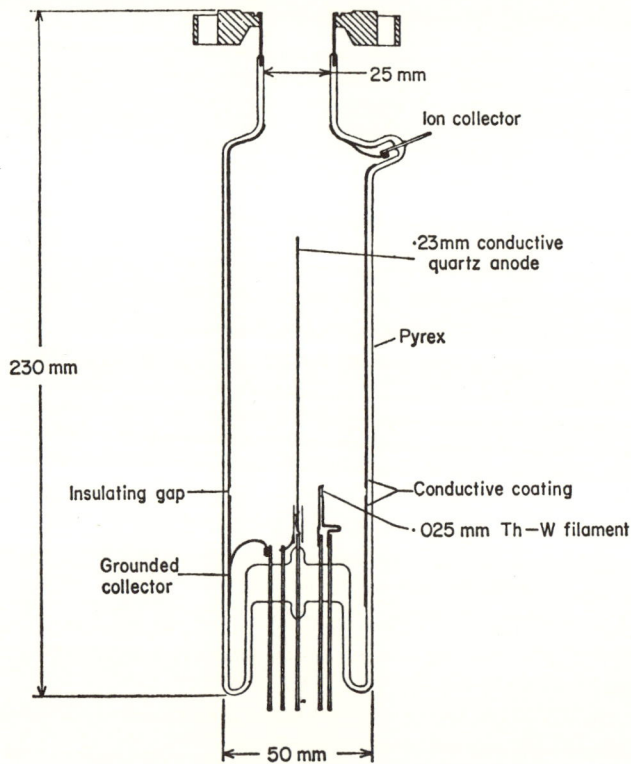

Fig. 8.11 Orbitron gauge (Meyer & Herb 1967).

the above gauge has an $\bar{l}$ of 9×10^3 cm. These extremely high sensitivity factors allow the orbitron to be operated at microampere electron currents, and still produce ion currents comparable to those produced by a BAG with milliampere electron currents. Operation at low electron currents is advantageous on two counts, (*a*) the filament temperature is much reduced and thus chemical effects due to the hot filament are minimized and (*b*) effects caused by electronic desorption from the anode are minimized. The above gauge requires a total power of only 88 mW compared with 21·5 W required for a BAG of identical sensitivity.

In principle the x-ray limit of the orbiton gauge can be made quite low if the anode diameter is very small. The probability of a photo-electron, liberated from the outer cylinder by x-rays from the anode, reaching the anode is given approximately by eqn. (8.3) with a value of V_0 appropriate to the mean energy of the photo-electrons. This probability can be of the order of 10^{-4} for an anode to collector diameter ratio of 2×10^{-3}. It can be estimated that with a 100-micron diameter anode at an anode voltage of 400 V and $K = 10^4$ torr^{-1} the x-ray limit would be about 10^{-12} torr, *provided that photo-electrons from the ion-collector go only to the anode*. In practice there will be other electrodes (grounded shields, etc.) with which photo-electrons could be readily exchanged in an axial direction where the considerations of eqn. (8.3) do not apply. It is not possible to analyse this situation in detail. Meyer and Herb (1967) have found the x-ray limit of the orbitron gauge described above to be about 9×10^{-12} torr with an anode voltage of 500 V.

8.1.5 Hot-cathode gauges with magnetic field

A magnetic field may be applied to a total pressure gauge to achieve two characteristics:

(i) increased electron path length and hence increased sensitivity factor, or

(ii) suppression of photo-electrons from the ion-collector surface.

The hot-cathode magnetron gauge was first described by Conn and Daglish (1954). Another design was developed by Houston (1956) which has been described in more detail in Dushman and Lafferty (1962, p. 337). These two designs of gauge do not appear to have been widely used and have been superseded by a design due to Lafferty (1960, 1961a).

Figure 8.12 shows schematically the hot-cathode magnetron gauge developed by Lafferty. It consists of a cylindrical magnetron operated with a magnetic field beyond cut-off (about 250 gauss, 2·5 times the cut-off field). The electrons are prevented from escaping axially by end-plates which are negative with respect to the filament.

Electrons emitted from the central filament spiral around the magnetic field lines within the anode cylinder and the effective electron path length can be made extremely long. The positive ions produced by electron collision are collected on one of the end-plates. Typically the anode is 24 mm diam and 32 mm long. The 200-micron diam hairpin filament is 19 mm long, separated 1 mm at the base. Very small electron emission currents are used, in the range 10^{-7} to 10^{-9} A, to ensure stable operation. At higher electron emission the temperature of the trapped electrons increases drastically leading to instabilities and nonlinearities. This behaviour is common to all discharges where the electrons are trapped and will be referred to as the 'hot-electron' effect.

Below about 10^{-8} torr the electron current to the anode remains constant, independent of pressure, and the ion-collector current is a linear function of

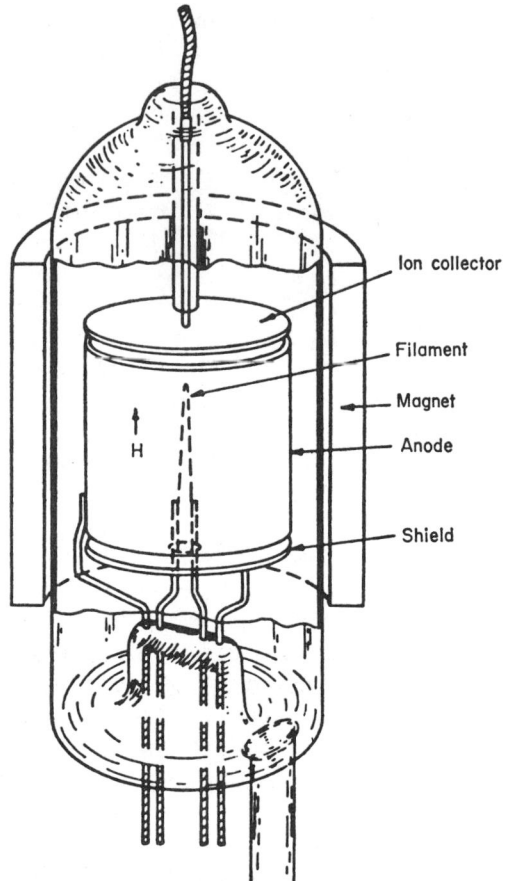

Fig. 8.12 Schematic diagram of hot-cathode magnetron gauge (Lafferty 1960).

pressure. With an anode potential of 345 V, shield potential of 35 V, filament potential of 45 V, and an anode current of 3.5×10^{-9} A (10^{-7} at zero magnetic field), the gauge sensitivity is 9×10^{-2} A/torr. The x-ray induced photocurrent is given by

$$i_x \simeq 10^{-6} I_A,$$

where I_A is the anode current. Thus, with the above condition, the x-ray limit is about 3×10^{-14} torr.

The sensitivity has been increased and the x-ray limit reduced by the addition of an electron multiplier (Lafferty 1961b, 1962, 1963). The electrode arrangement in this gauge is shown schematically in Fig. 8.13. Ions are extracted from the magnetron through an aperture and focused, by an electrostatic lens, onto the first dynode of an electron multiplier. The aperture plate also serves to limit the x-rays incident on the first dynode. With this

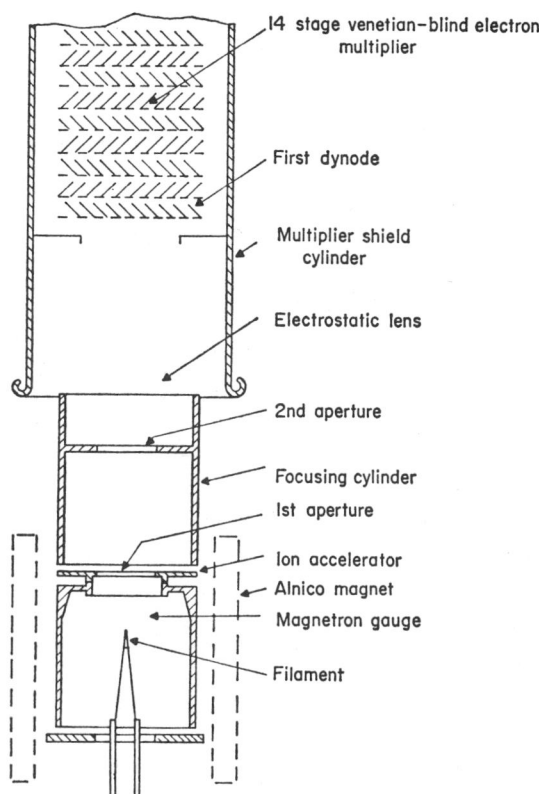

Fig. 8.13 Schematic diagram of hot-cathode magnetron gauge with electron multiplier. Ions are focused through apertures 1 and 2 onto first dynode (Lafferty 1963).

arrangement the x-ray photo-emission is reduced to such a low level that pressures as low as 3×10^{-18} torr should be measurable by counting the pulses from the multiplier produced by individual ions at the first dynode.

The second use of a magnetic field, mentioned at the start of this section, was the suppression of photo-electrons. The second hot-cathode, magnetic gauge we wish to consider is of this type (Klopfer 1961b). This gauge is shown schematically in Fig. 8.14, where it can be seen that it is essentially an omegatron without any rf field applied. The ribbon-shaped electron beam is focused by a strong magnetic field (1000 gauss). Ions formed within the ionization box ($G_3 - G_4$) are attracted to the ion-collector (J), which is negative with respect to the box. The x-ray limit of this type of gauge is reduced by two effects:

(i) the ion-collector is shielded to some extent from the area of impact of the electrons; thus not many x-rays can reach the collector directly, and

(ii) most photo-electrons released from the ion-collector are returned to it

Pressure Measurement

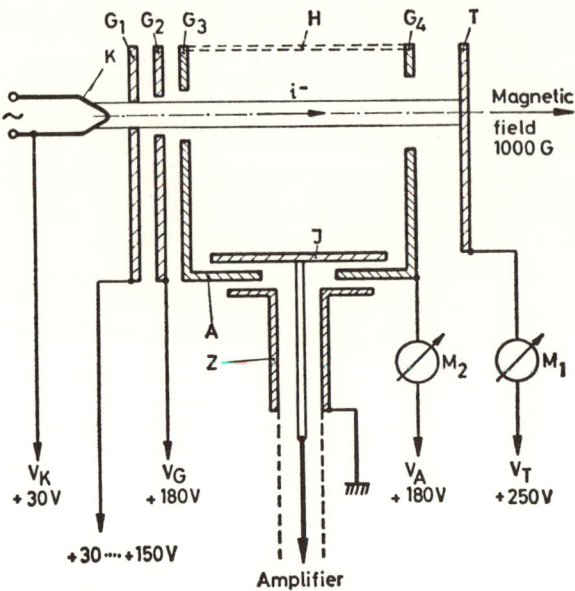

Fig. 8.14 Schematic diagram of Klopfer gauge (Klopfer 1961b).

by the magnetic field. Some photo-electrons can escape along the magnetic field.

This gauge is operated with the potentials as indicated in Fig. 8.14 and has a sensitivity factor of 15 torr^{-1} for nitrogen. In the pressure range below 10^{-4} torr the ion current is linearly related to the electron current up to electron currents of 1 mA. If currents of 10^{-15} A could be measured then pressures of about 6×10^{-14} torr could be detected if the x-ray limit were sufficiently low. Measured x-ray limits on this gauge have not been published.

8.1.6 Other hot-cathode gauges

The hot-cathode gauges described in previous sections are the only types proved useful for uhv applications. Some other hot-cathode gauges which have been described in the literature have potentialities for uhv measurements, but sufficient measurements have not yet been made to establish their usefulness.

Alexeff (1961) has described a gauge in which the ultraviolet radiation from the electron bombarded gas is passed through a thin window (lithium fluoride) into a G-M counter. The x-ray limit of this arrangement can be made very low in principle. With an electron current of 10^{-5} A at 150 V a counting rate of about 1.5×10^7 counts/sec/torr was obtained for air.

Gauges in which the pressure is measured by effects resulting from the neutralization of electron space-charge by positive ions have some potential for ultrahigh vacuum use. These space-charge gauges may be divided into two types:

(i) those without magnetic field, where the effect of positive ions is to decrease the transverse spread of the electron beam; called an 'ion-focusing gauge' (Atkinson 1963; Atkinson & Banks 1966), and

(ii) those with axial magnetic field where transverse motion of the electrons is prevented. In this case a virtual cathode is formed, positive ions are then trapped at the virtual cathode neutralizing the electron space-charge until finally the virtual cathode is destroyed. This type is called a 'virtual cathode gauge'. A gauge of this type has been developed by Lloyd (1966) based on the results of Volosok and Chirikov (1957).

Pressure measurement with both types of gauge can be made by measuring the time rate of change of the electron current or by observing the displacement current when the virtual cathode suddenly dissipates. Since the ion current is not directly measured there is no 'x-ray limit' in space-charge gauges. The fact that an amplifier for measuring small ion currents is not necessary with these gauges is claimed to be an advantage (Lloyd 1966; Atkinson & Banks 1966) in itself. This claim is of doubtful validity since the electronic circuits necessary to measure pressure from the time-variation of electron current are more complex and expensive than a low-current d.c. amplifier.

The advantages of space-charge gauges are:

(i) Absence of an x-ray limit.

(ii) The virtual cathode gauge can be operated in high magnetic fields which may be an unavoidable part of certain experiments.

For uhv use there are two corresponding disadvantages of these gauges:

(i) The ion focusing gauge requires a very high electron current ($\sim$30 mA, Atkinson & Banks 1966).

(ii) The (pressure) × (relaxation time) product for the virtual cathode gauge is about 3×10^{-10} torr. sec. (Lloyd 1966). Thus at a pressure of 10^{-11} torr the period of relaxation (and hence the interval between pressure measurements) would be 30 seconds. The inter-measurement period becomes impractically long for pressures below 10^{-11} torr. Similar limitations apply to the ion focusing gauge. Lloyd (1966) has operated a virtual cathode gauge down to 5×10^{-10} torr. Linear behaviour was observed down to 4×10^{-9} torr; at lower pressures the measurements were not reproducible. Atkinson and Banks (1966) claim that their ion focusing gauge is satisfactory to 'below 10^{-8} torr'.

8.1.7 Measurement of residual currents

All hot-cathode gauges, other than space-charge gauges, have a lower limit of measurable pressure which is reached when the true positive ion current (from ionization of molecules in the gas phase) equals the residual current. The processes giving rise to the residual current (soft x-rays and electron-impact desorption) have been discussed in Section 7.4. Whenever the residual current is an appreciable fraction of the true ion current then accurate measurements of the residual current must be made to permit calculations of the pressure.

As was pointed out in Section 7.4 the residual current is the sum of two principal components: (a) the x-ray photocurrent, which is approximately constant for a given gauge with constant electrical condition, and (b) the electronic desorption ion current, which may vary by several orders of magnitude for any given gauge depending on the amount and nature of gases adsorbed on the electron collector. Thus (a) above is a characteristic of the type of gauge and the electrical operating conditions only, whereas (b) depends also on the state of 'cleanliness' of the electron-collector and depends on the prior history of the gauge. In certain gauges (the BAG for example) the electronic desorption current may greatly exceed the x-ray photocurrent under appropriate conditions of contamination of the grid surface. Reliance on a single measurement of the residual current or on the manufacturers' published value of the 'x-ray limit' may be very misleading particularly if the electron-collector is not scrupulously outgassed or if chemically active gases are present in the system.

The original method of measuring the x-ray photocurrent of a hot-cathode gauge was described by Bayard and Alpert (1950) and depends on the difference in shape of the curves of (a) x-ray photocurrent and (b) positive ion current, as a function of electron energy. The x-ray photocurrent varies as the 1·1 to 1·5 power of the electron energy in the range 10 to 500 eV. The positive ion current increases with electron energy to a maximum at 100 to 200 eV and then decreases slowly.

The ion-collector current is measured as a function of the grid-to-filament voltage (V_{gf}) and the results plotted on a log-log scale. Typical results are shown in Fig. 8.2. At high pressures, where the positive ion current greatly exceeds the x-ray photocurrent, the curve is similar in shape to an ionization probability curve. At low pressures, where the positive ion current is comparable to the x-ray photocurrent, the curve approximates a straight line at higher values of V_{gf}. The x-ray photocurrent is estimated by extrapolating the straight line portion of the curve back to the operating value of V_{gf}. Subtraction of the extrapolated photocurrent curve from the curve of ion-collector current should yield a curve whose shape is identical to the curve of positive ion current versus V_{gf}.

Various values of the exponent of the x-ray photocurrent curves have been reported by different experimenters varying from 1·1 to 2·0. It is probably unwise to attach any physical significance to the value of the exponent but rather treat this measurement as an empirical method of determining the x-ray photocurrent.

The following characteristics of this method of measurement should be noted:

(i) Only the x-ray photocurrent is measured. Any ion current resulting from electronic desorption at the grid surfaces varies with V_{gf} in the same general way as ion current from the gas phase and is difficult to distinguish therefrom.

(ii) The measurement takes considerable time to complete and conditions may change during the course of the measurement; in particular, electronically desorbed ion currents may change considerably.

(iii) Variation in V_{gf} changes the gauge pumping speed and grid temperature which may result in pressure changes. Accurate determination of true pressures may be difficult under these circumstances.

Another method of measuring residual currents was proposed by Redhead (1960b). This method depends on being able to modulate the positive ion current from the gas phase while causing little or no change in the residual current. A diagram of a BAG to which a modulating electrode has been added is shown in Fig. 8.6. In general, the modulation method can be applied to any type of hot-cathode gauge. Modulation electrodes have been added to Bayard-Alpert gauges (Redhead 1960b; Appelt 1962; Reich 1966), suppressor gauges (Redhead & Hobson 1965), and extractor gauges (Redhead 1966).

The modulating electrode is switched between two potentials, usually the upper is grid potential (V_g) and the lower, ground potential. Other modes of modulation are discussed later. With the modulator potential (V_m) equal to grid potential (V_g), the ion-collector current is given by

$$I_1 = i^+ + i_R \tag{8.4}$$

where i^+ is the true ion current from the gas phase and i_R is the residual current defined in Section 7.4. For the modulation method to be applicable, it is necessary, that either (a) the ratios of the various component currents making up I_R remain constant during the measurements, or (b) one of the components of i_R greatly exceed the sum of all other components.

When the potential of the modulator electrode is switched to the modulated condition (usually ground) the ion-collector current is given by

$$I_2 = \alpha i^+ + \gamma i_R \tag{8.5}$$

where α and γ are the modulation factors for the true ion current and residual current respectively. It is convenient to set

$$\alpha = 1-\beta \text{ and } \gamma = 1-\varepsilon. \tag{8.6}$$

The gas-phase modulation factor β for any particular type of gauge is a function of the electrical operating condition and may be determined by measurement at high pressures where $i^+ \gg i_R$; then

$$\beta = \frac{I_1 - I_2}{I_1} \equiv \frac{\Delta I}{I_1}, \tag{8.7}$$

or by making modulation measurements at two different pressures, indicated by subscripts a and b, then

$$\beta = \frac{\Delta I_a - \Delta I_b}{I_{1a} - I_{1b}}. \tag{8.8}$$

When $\varepsilon = 0$, equations (8.4), (8.5) and (8.6) can be readily solved for the ion current and the residual current,

$$i^+_{\varepsilon=0} = \frac{\Delta I}{\beta}, \tag{8.9}$$

and

$$i_{R\varepsilon=0} = I_1 - i^+. \tag{8.10}$$

With the modulated BAG the assumption that $\varepsilon = 0$ does not cause appreciable errors in the measurement of pressures higher than 3×10^{-11} torr (Redhead & Hobson 1965; Groszkowski 1966b).

If $\varepsilon \neq 0$, it is not possible to solve equations (8.4), (8.5) and (8.6) for the three unknowns i^+, i_R and ε by modulation methods alone and it is necessary to introduce a third piece of experimental evidence. Appelt (1962) has suggested that ε can be measured at high values of the grid-filament voltage so that she residual current (presumed to be predominantly an x-ray photocurrent) it emphasized in relation to the ion current. Hobson (1964) has shown that this method is not suitable for the measurement of ε in the modulated BAG and leads to errors of the same order as the assumption that $\varepsilon = 0$. Another method for measuring ε can be used when it is possible to estimate the true ion current in the gauge by some other method of pressure measurement. In the system used by Hobson it was possible to estimate the pressure by a dynamic method when a portion of the system was immersed in liquid helium. Equations (8.4), (8.5) and (8.6) can be manipulated to yield

$$\varepsilon i_R = \Delta I - \beta i^+, \tag{8.11}$$

and

$$i^+ = \frac{\Delta I - \varepsilon i_R}{\beta}. \tag{8.12}$$

If ΔI and β are measured and i^+ found from an auxiliary measurement (such as the dynamic method of Hobson), then εi_R can be found from eqn. (8.11). The true ion current can then be found at any later time by the use of eqn. (8.12). This method is only valid if εi_R is constant during the course of the measurements. In general, this method is only workable in extremely clean systems.

Measurements show β to be constant and typically about 36%. ε varied between $-2\cdot 5$ and $+5\cdot 9\%$ depending on processing and stray magnetic fields. ε did not change if conditions were held constant.

Modulation as described above in which the modulator potential is switched from grid potential to ground is called mode I modulation and has two disadvantages:

(i) the change in electron current to the modulator electrode can cause bursts of gas to be desorbed, and

(ii) modulation of the x-ray photocurrent occurs (i.e. $\varepsilon \neq 0$).
Lange and Singleton (1966) have shown that switching the modulator over another range of voltage (mode II) greatly reduces the two above effects. For Bayard-Alpert gauges with open-end grids the ion current to the collector *increases* as the modulator potential is reduced below grid potential, when $V_m \simeq V_g - 30$ the collector current reaches a maximum and then decreases. Modulation factors as large as 40% could be obtained by switching V_m from V_g to $(V_g - 30)$. Operation in mode II eliminated any gas bursts on modulation. Since changes in electrical conditions in the gauge are greatly reduced in mode II operation, Lange and Singleton predict that modulation of the x-ray photocurrent will be reduced.

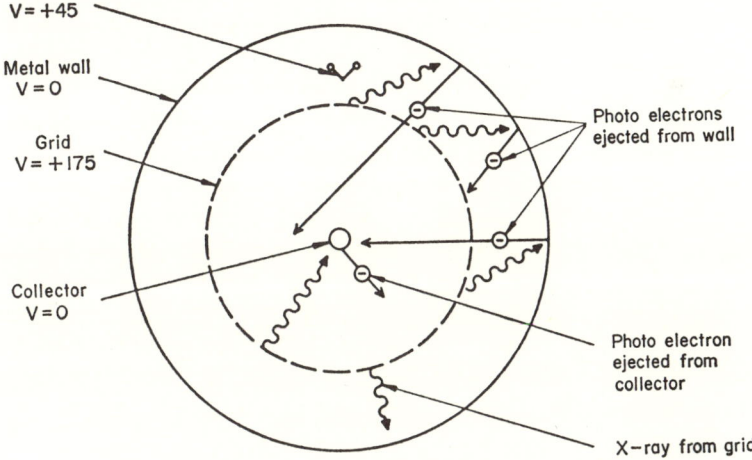

Fig. 8.15 Schematic of reverse x-ray effect (Hayward *et al.* 1963).

Various modes of modulation of BAGs have been compared (Redhead (1967a) and yet another mode (mode III) proposed to minimize the two undesirable effects of mode I operation noted above. Mode III operation can be used with BAGs having closed-end grids where the maximum in the ion collector current, used in mode II operation, does not occur. Mode III modulation is achieved by switching V_m from $(V_g - 20)$ to $(V_g - 60)$. Reasonably large modulation factors can be obtained in this mode (about 20%) with only small disturbance of conditions at the modulator.

The familiar 'x-ray limit' of a hot-cathode gauge results from the ejection of photo-electrons from the ion-collector by soft x-rays. We will call this the 'forward' x-ray effect. Another less familiar process can occur in any hot-cathode gauge which has electrodes at the same potential as the ion-collector. The process in a BAG is illustrated schematically in Fig. 8.15. Soft x-rays produced at the electron-collector strike the gauge envelope and eject photoelectrons therefrom. If the envelope and the ion-collector are both at ground potential, it is energetically possible for some of these photo-electrons to pass through the grid and reach the ion-collector. Although the fraction of the photo-electrons that reach the ion-collector is extremely small, this is more than compensated by the fact that the flux of x-rays striking the envelope of a BAG is much larger than that striking the ion-collector. This effect was first reported by Hayward et al. (1963) and has been called the reverse x-ray effect since the resulting photocurrent is in the opposite direction to the forward x-ray effect. Although the reverse x-ray effect has only been studied in any detail in a BAG, the effect occurs to some extent in all hot-cathode gauges having an electrode or element at the same potential as the ion-collector.

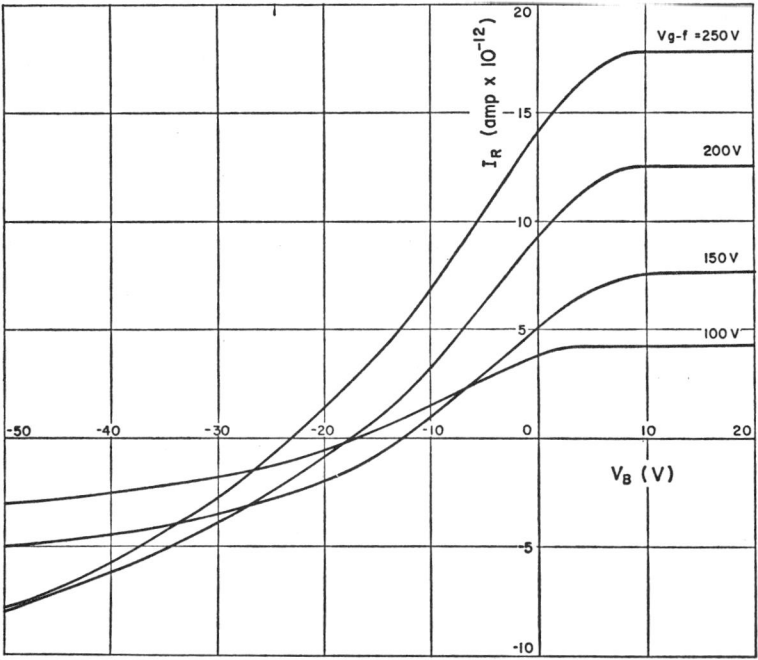

Fig. 8.16 Residual current (i_R) as a function of voltage between collector and envelope (V_B) of modulated Bayard-Alpert gauge for various grid-filament voltages, V_{gf}. Electron current was 8 mA.

The reverse x-ray effect in a BAG is strongly dependent on the voltage between the ion-collector and the envelope since this voltage controls the fraction of the photocurrent from the bulb which reaches the ion-collector. Figure 8.16 shows the residual current measured by the modulation method as a function of the envelope-to-collector voltage (V_B). When V_B is positive no photo-electrons can reach the collector and the residual current is constant at a value determined by the forward x-ray effect. As V_B is made negative the residual current decreases and finally goes negative; this is the condition where the reverse x-ray photocurrent exceeds the forward x-ray photocurrent.

From the above it can be seen that the x-ray photocurrent measured in a gauge may vary considerably depending on the collector-envelope voltage and the geometry of the envelope. The x-ray photocurrent measured in a BAG with a conductive envelope may be quite different from that observed when the same gauge structure is used 'nude' inside a large chamber. To prevent this variation in total x-ray photocurrent, it is best to operate the gauge envelope at a few volts positive with respect to the ion-collector to eliminate any reverse x-ray photocurrent.

Typical residual currents for various types of gauge are listed in Section 8.3 and the limitations on the lowest measurable pressures compared.

8.2 CROSSED-FIELD, COLD-CATHODE GAUGES

The electron current to the anode of a crossed-field, cold-cathode discharge is equal to the positive ion current to the cathode, and is thus a function of pressure. The absence of a constant electron current, independent of pressure (as exists in a hot-cathode gauge), leads to two significant advantages for crossed-field, cold-cathode gauges:

(i) there is no x-ray limit, and

(ii) electronic desorption effects are small and cause little error.

Crossed-field, cold-cathode gauges suffer from three generic disadvantages not shared by hot-cathode gauges:

(i) in general, the output varies non-linearly with pressure,

(ii) the very dense electronic space-charge trapped in these gauges leads to instabilities associated with mode-jumping of the high frequency oscillations that occur,

(iii) their pumping speed is usually high and uncontrollable.

Because of the non-linear response and unstable behaviour, cold-cathode gauges are seldom used for very accurate pressure measurement. However, their simplicity and absence of x-ray limit have led to their widespread use in uhv systems where highly accurate pressure measurements are not required. Calibration measurements on crossed-field, cold-cathode gauges show wide variations between various experimenters; this presumably reflects the difficulty of obtaining repeatable measurements on an intrinsically unstable device.

8.2.1 Penning gauge

The basic geometry of the Penning gauge with a tubular anode (Penning & Nienhuis 1949) has been widely applied to the design of ionic pumps (see Section 11.4), but has not been much used for uhv pressure measurements. It might appear, at first glance, that the electrical and magnetic fields in a Penning gauge are parallel and not crossed. However, the potential on the axis of an operating Penning discharge is depressed to near cathode potential by electronic space-charge resulting in a potential distribution similar to that in a magnetron.

A modified form of the original Penning geometry, with the addition of a small filament that can be flashed to start the discharge, has been developed by Young and Hession (1963) for uhv use.* A schematic diagram of this gauge, called a trigger discharge gauge, is shown in Fig. 8.17. This gauge consists of a cylindrical anode 1·25 cm diam and 1·25 cm long. The ion-current is measured at the two interconnected cathode discs. A small filament is placed behind an off-axis hole in one of the cathode plates. The filament is heated to provide electrons to start the discharge and turned off once the discharge is running.

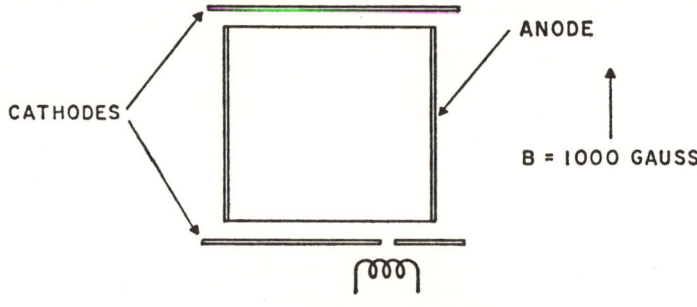

Fig. 8.17 Schematic diagram of trigger discharge gauge. (Young & Hession 1963).

This gauge is normally operated with 2 kV on the anode and a magnetic field of $1,000 \pm 50$ gauss.

The trigger discharge gauge has been calibrated against a mass-spectrometer and a Bayard-Alpert gauge over the pressure range from 10^{-7} to 5×10^{-12} torr (Young 1966a). The ion-current pressure relation is non-linear and is given by

$$I^+ = kp^n, \qquad (8.13)$$

with $n = 1\cdot 2$. The sensitivity is approximately 1 A/torr at a pressure of 10^{-11} torr rising to 6·3 A/torr at 10^{-7} torr.

Bryant and Gosselin (1966) have calibrated the trigger-discharge gauge in the range 10^{-3} to 10^{-11} torr. In the range 10^{-5} to 5×10^{-8} torr the response followed a power law with $n = 1\cdot 15$. The response from 10^{-8} to 10^{-11} torr was non-linear and can be approximated by a power law with $n = 1\cdot 4$. In the pressure range 10^{-10} to 10^{-11} torr the discharge extinguished and a pressure-independent residual current was observed in the gauge.

Lange et al. (1966) have calibrated the trigger-discharge gauge over the range 10^{-4} to 10^{-12} torr for Ar and H_2, and to 10^{-10} torr for He, N_2 and CO. The sensitivity was found to decrease with pressure and the calibration could

* Commercially available from General Electric (U.S.A.).

be represented by a series of linear segments separated by sudden steps. The linear segments probably correspond to various stable discharge modes. Hysteresis effects cause overlapping of the modes and hence the accuracy of pressure measurement is difficult to predict even for a particular gauge. Lange's calibration method permitted a continuous record of ion current vs pressure rather than the more usual point-by-point calibration. It is possible that the apparent power law behaviour observed by other experimenters was actually a number of short linear segments.

Hayashi (1966) has reported the use of a beta-emitting radioisotope (Ni^{63}) to trigger a Penning discharge at low pressures. The 50 keV betas from the Ni^{63} release slow secondary electrons from the envelope of the device. The slow electrons can then enter the discharge region and initiate the discharge.

8.2.2 Magnetron gauge

A cold-cathode magnetron gauge suitable for uhv use was developed by Redhead (1959a) and is shown schematically in Fig. 8.18. The anode consists of a cylinder (20 mm long and 30 mm diam) which may be perforated to improve

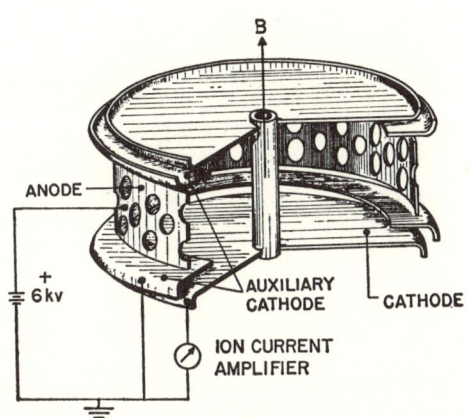

Fig. 8.18 Schematic diagram of cold-cathode magnetron gauge (Redhead 1959a).

gas flow through the gauge. The cathode is shaped like a spool, consisting of an axial cylinder (3 mm diam and 20 mm long) joined to two circular end-discs. The end-discs are shielded from high electric fields by two annular shield electrodes operated at cathode potential. Any field emission that may occur from the shield electrodes is not measured by the ion current amplifier. This gauge normally operates with an anode voltage of 4·5 to 6 kV and a magnetic field of 1000 gauss.* The ion current vs pressure relationship for this gauge in the range 10^{-4} to about 10^{-10} torr is linear (Redhead 1959a; Hobson 1959; Feakes & Torney 1963; Torney & Feakes 1963; Bryant et al. 1966). In the

*Commercially available from Norton Company (U.S.A.).

linear range of this gauge the sensitivity to nitrogen is about 10 A/torr for an anode voltage of 6 kV and about 5 A/torr for $V_A = 4.8$ kV.

Hobson (1959) first demonstrated that the reponse of this design of magnetron gauge departs from linearity below about 10^{-10} torr and follows a power law where the value of the exponent is dependent on the operating conditions. With $V_A = 5$ kV and $B = 1060$ gauss, $n = 1.7$ has been observed for He (Hobson 1959). With $V_A = 4.8$ kV, $n = 1.35$ for $B = 1050$ gauss and $n = 1.90$ for $B = 900$ gauss (Feakes & Torney 1963). With $V_A = 4.8$ kV and $B = 1000$ gauss, $n = 1.62$ for He and N_2 (Torney & Feakes 1963).

Bryant and his colleagues (1966) have studied the very low pressure behaviour of several types of magnetron gauge. Their results for the magnetron gauge described above are shown in Fig. 8.19. The response deviates from linearity at a pressure of about 2×10^{-10} torr and then follows a power law with $n \simeq 1.4$. At a pressure of 3×10^{-12} torr the discharge extinguished and a pressure independent background current of about 5×10^{-14} A remains at

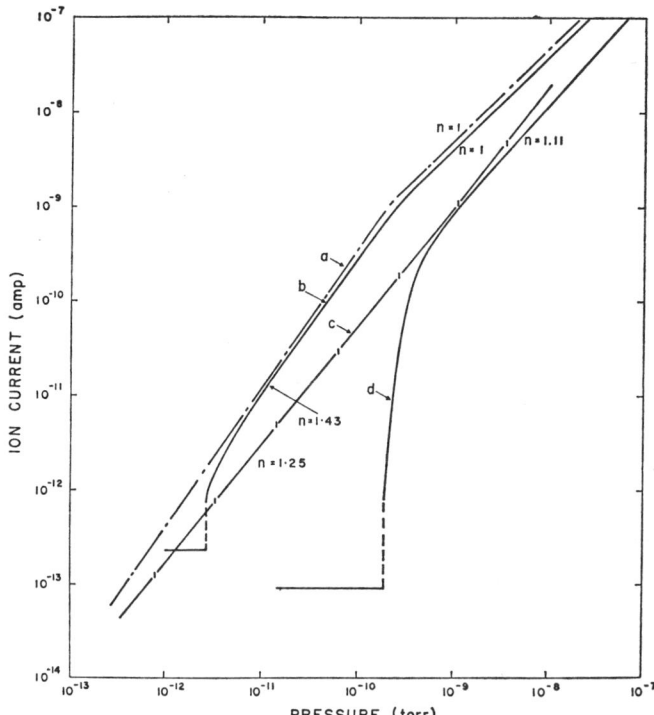

Fig. 8.19 Ion current vs pressure for various cold-cathode, crossed-field gauges. (a) Magnetron gauge, $B = 1$ kG, $V_a = 4.5$ kV, helium (Feakes et al. 1965); (b) Magnetron gauge, $B = 1$ kG, $V_a = 4.8$ kV, helium (Bryant et al. 1966); (c) Inverted-magnetron gauge, $B = 2.1$ kG, $V_a = 6$ kV, helium (Bryant et al. 1966); (d) Magnetron gauge (Kreisman type without shield electrodes), $B = 1$ kG, $V_a = 4$ kV, helium (Bryant et al. 1966). Exponent of power law relation designated by n.

lower pressures (presumably field emission). Results obtained by Hobson (1959) during the measurement of the helium adsorption isotherm with a magnetron gauge can also be interpreted as if the discharge extinguished at about $1-2 \times 10^{-12}$ torr and that the gauge current thereafter was independent of pressure. The above measurements are in conflict with those of Feakes, Torney and Brock (1965) who have reported calibration of the same type of magnetron gauge. They find that below 2×10^{-10} torr the magnetron gauge is non-linear but that the discharge did not extinguish down to the lowest pressures achieved (3×10^{-13} torr). The results of this experiment are summarized in Fig. 8.19.

Bryant *et al.* (1966) has also reported measurements on another commercial version of the magnetron gauge which has no shield electrodes.* The ion current pressure characteristics of this gauge are also shown in Fig. 8.19. These data indicate that this gauge is non-linear at all pressures and that the discharge goes out at the relatively high pressure of 2×10^{-10} torr.

8.2.3 Inverted magnetron gauge

A crossed-field, coaxial cold-cathode gauge in which the anode is on the axis is called an inverted-magnetron gauge. This type of discharge geometry was first studied by Beck and Brisbane (1952) and Haefer (1954). A design suitable for uhv work was reported by Hobson and Redhead (1958) and Redhead (1958) and is shown in Fig. 8.20. The cathode is a cylinder (30 mm diam by 20 mm long) partially closed at both ends, with its axis parallel to the magnetic

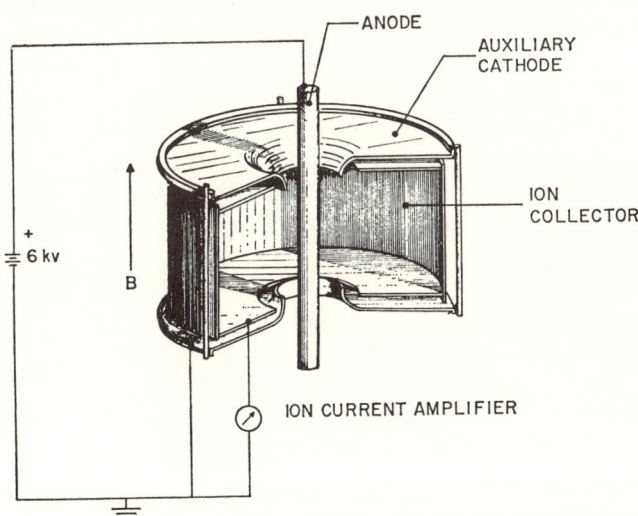

Fig. 8.20 Schematic diagram of cold-cathode inverted magnetron gauge. (Hobson & Redhead 1958).

*Geophysical Corporation of America, Kreisman gauge.

field. The anode is a rod (1 mm diam) passing through holes in the end-plates of the cathode. The shield electrodes are circular discs with short spouts placed between the anode and the end-plate which prevent field emission from the edges of the holes in the cathode end-plates. This gauge is normally operated with an anode voltage of 5–6 kV and a magnetic field of 2000 gauss.

The ion current was found to follow a power law relation (as eqn. (8.13)), where $n \simeq 1 \cdot 10$ to $1 \cdot 15$ (Hobson & Redhead 1958) in the pressure range 10^{-4} to 10^{-10} torr. Feakes, Torney and Brock (1965) found that $n \simeq 1 \cdot 25$ in the pressure range 10^{-8} to 3×10^{-13} torr for He and N_2 (see Fig. 8.19). No sign of any change in slope of the ion-current vs pressure characteristics was observed, in contrast to the behaviour of the magnetron gauge. Grigor'ev (1959) has reported on a gauge of very similar dimensions to that of Fig. 8.20; with $V_A = 6$ kV and $B = 3 \cdot 53$ kG, a power law dependence of ion current on pressure was observed with an exponent of $1 \cdot 1$.

The inverted-magnetron gauge appears to be less subject to mode-jumping and noisy behaviour than the magnetron gauge.

8.2.4 Current-pressure characteristics and oscillatory behaviour

Theoretical descriptions of the striking characteristics of crossed-field Townsend discharges have been published by several authors (Haefer 1953a and b; Redhead 1958 and 1959a; Reikhrudel & Sheretov 1966a), which are in reasonable agreement with measurements. These theories ignore the effects of space-charge which is only of secondary importance in determining striking characteristics. However, when we examine the state of theories of the *operating* characteristics of cross-field discharges, where space-charge effects dominate, we find the situation to be far less satisfactory. No theory has yet been advanced which even approximately explains the observed non-linear current-pressure relationship, nor is it possible to predict with any certainty the effects of dimensional or electrical changes on the current-pressure or current-voltage relationships. This lack of adequate theoretical understanding of the factors controlling crossed-field Townsend discharges has resulted in a cut-and-try method of designing crossed-field, cold-cathode gauges and pumps.

Several recent papers have elucidated various aspects of the behaviour of these discharges. The dominance of electron space-charge in controlling the discharge behaviour at higher pressures has been demonstrated by Helmer and Jepsen (1961), Knauer (1962) and Knauer and Lutz (1963). More recently it has been shown that electron space-charge continues to dominate the discharge behaviour down to at least 10^{-10} torr (Redhead 1965b). A theory of discharge characteristics has been presented by Jepsen (1961) which makes the simplifying assumption that the electron space-charge is uniform; though admittedly an over-simplification, this theory does lead to some useful scaling laws which have been shown to be approximately correct.

It was suggested by Knauer (Knauer 1962; Knauer & Lutz 1963) that electrons reach the anode by classical cross-field mobility. Calculations using this approach have been made by several Russian authors (e.g. Kervalishvili & Zharinov 1966) and reasonable agreement obtained between measured and calculated current-voltage and current-magnetic field characteristics. A more detailed calculation has been made by Reikhrudel and Sheretov (1966b) without using the concept of cross-field mobility.

All the theoretical treatments of current-pressure and current-voltage characteristics produced so far have ignored the dynamic characteristics of the dense electron space-charge. The electrons in a crossed-field discharge at low pressures are trapped for extremely long times, allowing large amplitude rf oscillations of the electron cloud to build up. These oscillations modify the static characteristics of the discharge, because of the production of excess-energy electrons, and also cause low-frequency instabilities which are associated with mode-jumping of the rf oscillations. The rf oscillations may also cause serious measurement errors if unintentionally rectified in the ion-current amplifier. In many cases the rf oscillations dominate the behaviour of the discharge and any theory that ignores their presence can only be marginally useful in predicting the behaviour of a practical gauge.

Coherent rf oscillation in crossed-field discharges have been observed by several experimenters. Most observations have been made on Penning gauges at pressures higher than 10^{-6} torr (Dumas 1955; Honig & Parzen 1955; Helmer & Jepsen 1961; Wasa & Hayakawa 1965). More detailed measurements have been made by Reikhrudel and his co-workers (Reikhrudel *et al.* 1952, 1962; Reikhrudel & Smirnitskaya 1958; Smirnitskaya & Reikhrudel 1957) and a theory of rf oscillations developed which assumes that electrons oscillate isochronously in the parabolic potential distribution that is shown to exist in the absence of space-charge in Reikhrudel's experimental tubes. Minor modifications have been made to include the effects of space-charge. (Vasil'eva & Reikhrudel 1962). This assumption has been questioned by Redhead (1965b) and it is shown that an approximately parabolic potential distribution will exist, in the presence of electronic space-charge, in virtually any geometry, leading to isochronous oscillations in the space-charge cloud. The frequency of these rf oscillations is typically in the range 15 to 100 Mc/s and is given approximately by

$$f = \frac{nV_A}{\pi r_A^2 B}, \tag{8.14}$$

where V_A is the anode voltage, r_A the anode radius and B the magnetic field. Pressure has little effect on f other than a sudden change in mode number (n) at higher pressures. These oscillations have been observed in a magnetron gauge to pressures as low as 10^{-10} torr (Redhead 1965b).

Low frequency oscillations (500 c/s to 100 kc/s) have been observed in the inverted magnetron gauge (Hobson & Redhead 1958) in the pressure range 10^{-4} to 5×10^{-8} torr. The frequency of these oscillations was proportional to

V_A/B and to pressure (in contrast to the rf oscillations which were pressure independent). Similar low-frequency oscillations have been observed by Kaganskii *et al.* (1964) in a Penning gauge at high pressures (5×10^{-4} to 4×10^{-3} torr). Low-frequency oscillations of this type, with frequency proportional to pressure, may result from the relaxation oscillation of a potential minimum which is periodically neutralized by positive ions (Volosok & Chirikov 1957).

The oscillatory behaviour of crossed-field, cold-cathode discharges is important to the measurement of pressure with such gauges because:

(i) mode-jumping of the rf oscillations and the pressure dependent low-frequency oscillations may result in large current fluctuation within the passband of the ion-current amplifier. Unstable regions are frequently observed in all types of crossed-field gauges as pressure, voltage, or magnetic field is varied;

(ii) the rf signal on the cathode lead may be large enough to interfere with the proper operation of the ion-current amplifier. Accidental resonance in the external circuit of the gauge can increase this effect.

8.3 COMPARISON OF TOTAL PRESSURE GAUGES

The choice of the most suitable type of gauge to be used for a particular experiment or system depends mainly on the following gauge characteristics:

(i) Lowest measurable pressure. Can the gauge measure to a sufficiently low pressure for the particular experiment?

(ii) Gauge pumping speed. Is the gauge pumping speed sufficiently low to prevent serious errors in pressure measurement? This is particularly important in experiments where the system pumping speed is very low (see Table 7.3 for listing of gauge pumping speeds).

(iii) Effect of chemically active gases. When measuring pressure in systems with chemically active gases present it is important to reduce chemical effects at hot filaments and to minimize the effects of electronic desorption.

(iv) Sensitivity.

(v) Mechanical ruggedness.

(vi) Interference by stray magnetic fields. In some experiments the stray fields from magnetic gauges cannot be tolerated.

(vii) Re-emission of previously pumped gas.

Tables 8.1 and 8.2 list some of the characteristics and limits of various uhv gauges in an attempt to assist in the problems of choosing the most suitable gauge type. It should be noted that there may be considerable variation between different manufacturers' versions of particular types of gauges. The gauges whose characteristics are listed in the above tables are those used in the noted references; there is no guarantee that commercial versions will have the same characteristics.

Some of the gauges listed in the above tables are fairly complex devices (the hot-cathode magnetron gauge with electron multiplier is one such). At some

TABLE 8.1
Comparison of UHV Gauges; Hot-cathode, non-magnetic

Gauge		(a) P_x (torr) Measured	(a) P_x (torr) Estimated	(b) P_{MIN} measured	(c) Electronic desorption	(d) I^- (A)	(e) V_{gf} (V)	(f) K_{eff} (torr^{-1})	(g) Sensitivity A/torr	Ref. No.
Bayard-Alpert		3×10^{-11}			YES	8×10^{-3}	105	25	0.2	1
Modulated BAG	Mode I			3×10^{-12}	NO	8×10^{-3}	105	8	6.4×10^{-2}	2
	Mode II			$\sim 8 \times 10^{-12}$	NO	10×10^{-3}	105	~ 4	4×10^{-2}	3
	Mode III				NO	8×10^{-3}	105	5	4×10^{-2}	4
Suppressor		1.5×10^{-13}		$\leq 3 \times 10^{-13}$	YES	8×10^{-3}	105	25	0.2	1
	175 μ diam Collector		3×10^{-13}	7×10^{-13}*	NO	8×10^{-3}	105	6	4.8×10^{-2}	5
Extractor	500 μ diam Collector		$\sim 10^{-12}$		NO	8×10^{-3}	105	12	0.1	
Buried collector			7×10^{-12}	$\sim 1 \times 10^{-11}$	?	10×10^{-3}	105	7	7×10^{-2}	6
Orbitron		$\sim 9 \times 10^{-12}$			NO	1.2×10^{-6}	540	9×10^4	0.1	7

TABLE 8.2
Comparison of UHV gauges; magnetic types

	(a) P_X (torr)	(b) P_{MIN} (torr)	(c) Electronic Desorption	(d) I^- (A)	(h) V_a (V)	(i) B (kgauss)	(f) K (torr^{-1})	(g) Sensitivity A/torr	(j) Linearity	Ref. No.
HOT-CATHODE										
Hot-cathode magnetron	4×10^{-14}	4×10^{-13}*		3.5×10^{-9}	300	0.25	3×10^7	0.09	1	8
Hot-cathode magnetron with multiplier	3×10^{-17}			3.5×10^{-9}	300	0.25		10^4	1	9
Klopfer	10^{-11}	10^{-11}*		10^{-3}	150	1	15	0.015	1	10
COLD-CATHODE										
Trigger discharge	0	5×10^{-12}			2,000	1	—	1.0, $P=10^{-8}$ 0.4, $P=10^{-10}$	1.2	11
Magnetron	0	3×10^{-13}			4,500	1	—	5.0, $P > 2 \times 10^{-10}$	1, $P > 2 \times 10^{-10}$ 1.44, $P < 2 \times 10^{-10}$	12
Inverted magnetron	0	3×10^{-13}			6,000	2	—	1.0, $P=10^{-9}$ 0.3, $P=10^{-12}$	1.2–1.4	13

Notes and References for Tables 8.1 and 8.2

Columns

(a) P_X Pressure at which ion current equals x-ray photocurrent.
(b) P_{MIN} Minimum pressure measured. Values marked (*) are lowest indicated pressures, not checked by other means.
(c) Electronic Desorption   Indicates whether gauge is subject to serious errors caused by electronic desorption of ions from the electron collector. Electronic desorption is not a problem in any magnetic type of gauge.
(d) I^- Typical operating electron current to anode.
(e) V_{ef} Grid-filament voltage.
(f) K_{eff} Effective sensitive factor for nitrogen. For unmodulated gauges, $K_{eff} = K = \dfrac{I^+}{I^-} \dfrac{1}{P}$. For modulated gauges $K_{eff} = K\beta$ where β is modulation factor (see eqn. (8.6)).
(g) Sensitivity   Measured at noted electron current.
(h) V_a Anode voltage.
(i) B Magnetic field.
(j) Linearity, n Exponent in power law, $I^+ = kp^n$.
All pressures are equivalent nitrogen.

References

1. Redhead and Hobson (1965)
2. Hobson (1964)
3. Lange and Singleton (1966)
4. Redhead (1967)
5. Redhead (1966a)
6. Clay and Melfi (1966)
7. Meyer and Herb (1967)
8. Lafferty (1960, 1961a)
9. Lafferty (1962)
10. Klopfer (1961b)
11. Young (1966b)
12. Feakes and Torney (1963)
13. Feakes et al. (1965)

level of cost and complexity it may be more appropriate to invest in a mass spectrometer rather than a total pressure gauge, and gain a great increase in information. Two factors make total pressure gauges useful even when they approach a mass spectrometer in cost and complexity:

(i) Total pressure gauges are nearly always easier to outgas than a mass spectrometer, and

(ii) the higher sensitivity of most total pressure gauges is frequently an important factor.

CHAPTER NINE

PARTIAL PRESSURE GAUGES

The basic objective in partial pressure measurement is to provide a signal (usually an ion current or a change in an ion current) which can be accurately related to the number density of a *particular species* of molecule in that region of a vacuum system where an experiment is being performed. In this chapter, we will discuss three techniques which have been used to give such partial pressure information in uhv systems: Mass-spectrometers, which identify molecules by separating ions according to their charge-to-mass ratio, are discussed in Section 9.1; spectrometers which detect the release of chemisorbed and physisorbed gases upon heating of the adsorbent are the subject of Section 9.2; and in Section 9.3 we mention devices which estimate the adsorbed gas coverage on surfaces by the changes produced in field emission and work function characteristics, and which relate these coverages to the pressure of the adsorbing gas.

Problems arising from local pressure variations due to high pumping speeds are comparable to those encountered with total pressure gauges, and have been discussed earlier (Section 7.3). Interpretation of the results obtained with desorption spectrometers (Section 9.2) in terms of partial pressures is particularly difficult under these conditions since both the adsorption rate and the desorption signal are dependent on the magnitude and location of the pumps.

Mass spectrometric studies have shown that the residual gases in most uhv systems are extremely simple molecules of low molecular weight, except in oil diffusion pumped systems with inadequate trapping (see Table 9.1). The origin and behaviour of these gases will be discussed in detail in Chapter 12. The relatively small number of species found and their low molecular weight greatly simplifies the task of analyzing the residual gas composition and has a strong influence on the design of uhv partial pressure measuring instruments.

9.1 MASS-SPECTROMETERS

A review of mass-spectrometer partial pressure measurement in uhv systems has been given by Huber (1963a, 1963b). The operation of mass-spectrometers can be divided into four functional steps:

TABLE 9.1
Fragment ions of common residual gases

Typical percentage of parent ion current appearing at the indicated mass number

Gas	1	2	4	12	13	14	15	16	17	18	19	20	28	29	32	36	40	44	45
H_2	—	100																	
	2·1																		
He			100																
CH_4	—	0		1·8	5·7	12·5	81	—	2·7										
	—	3		2·4	7·7	15·6	86	100	1·2										
	3·4	1·1		3·1	8·4	17	84		0										
	—	0·2		2·8	8	16	86		1·1										
H_2O	—	0·6–1·5						1·8	21		11–17								
	—	0·7						2·8	25	100	—								
		x						x	x		x								
N_2						7·4							100	0·75					
						7·2								0·77					
						7·9								0·77					
						5·2								0·73					
CO				3·3		0·55		1·3						0·88					
				4·5		0·6		1·0					100	1·1					
				5		0·7		1·1						0					
				4·7		0·75		1·7						1·2					
Ar												14·2				0·38	100		
												10·7				0·31			
												19				0·43			
												12·9				0·32			
CO_2				3·5				7·8					11·5		0·4			0	
				6				8·5					11·4		0				1·3
				7·1				9·2					9·2		0			100	1·2
				4·6				7·1					7·1		0				1·1

Notes on Table 9.1
0 column exists but no number given.
— column does not exist.
x gas not reported.

Sources of data in order listed:
1. Klopfer and Schmidt (1960), $V_e = 90V$ (omegatron).
2. MS 10 catalogue, $V_e = 70 V$ (180° sector).
3. Baily (1963), $V_e = 70 V$ (180° sector).
4. Mass spectral data (Petroleum Institute), $V_e = 70–75 V$ (mostly 60° sector).

(i) Ions are created from the gas, usually by electron impact.

(ii) The ions are accelerated to known kinetic energies in a chosen direction, and focused onto the entrance aperture of an analyser.

(iii) The ions entering the analyser are subjected to an arrangement of electric and/or magnetic fields which separate them on the basis of their charge-to-mass ratio.

(iv) The separated ions are detected upon arrival at a collector. Ions of a selected charge-to-mass ratio are brought to the collector by adjustment of the ion accelerating potential and/or the analyser fields.

The thermal kinetic energy which most ions have when produced is small compared to that imparted by the acceleration (typically a few hundred electron volts). If, however, the ions are produced with much higher kinetic energies, either by dissociative ionization (see Section 3.2.1) or from molecular beams of unusually high velocity, the resulting uncertainties in energy and flight direction of the ions after acceleration can lead to errors in the analysis. Mass-spectrometers of most types give some peaks which are not directly related to partial pressures but rather to electron impact desorption of ions from surfaces in the ion source (see, for example, Davis 1962). Under certain circumstances peaks of this type dominate the residual spectrum.

For a general discussion of 'static' mass-spectrometers which employ magnetic field deflection see, for example, Duckworth (1958), McDowell (1963a). 'Dynamic' mass-spectrometers (i.e. with time-varying fields) are reviewed by Blauth (1966).

9.1.1 General characteristics

There are three characteristics of primary importance in mass-spectrometers for uhv partial pressure measurement:

Sensitivity. The sensitivity in amperes of *ion* current per torr must be fairly high. It is difficult to obtain sensitivities $> 10^{-4}$ A torr using aperture or slit systems in the acceleration and analyser regions which are of the dimensions necessary to obtain the desired resolving power (usually < 1 mm diam or width). Since partial pressures of $\sim 10^{-13}$ torr must be detected for most uhv analysis, even the above 'high' sensitivity requires the measurement of currents $\sim 10^{-17}$ A or ~ 60 ions/sec. Multiplier-type detectors are necessary for the measurement of such currents. Even with high-gain multipliers, however, the statistical 'shot noise' (i.e. the counting statistics) of the ion current sets a fundamental limit to the accuracy with which a given current can be determined in a given measuring time (see Section 7.5.3). Any large reduction in sensitivity below 10^{-4} A/torr makes it impossible to detect the required partial pressure (10^{-13} torr) in a normally acceptable measuring time ($\sim$ a few seconds). This limitation is independent of multiplier gain above $\sim 10^3$.

Outgassing rate. The outgassing rate of the mass-spectrometer when operating must be low. A gas evolution rate of about 10^{-11} torr litre/sec, which would lead to a pressure increment of $\sim 10^{-11}$ torr in the presence of a pumping speed of a few litres/sec, should be considered a maximum acceptable value. Outgassing effects have been discussed in Section 2.5. The requirement for low gas evolution implies that all component parts of the spectrometer should be outgassed in vacuum to as high a temperature as possible (800°C–1000°C) without structural damage; and that the completed instrument should be bakeable to at least 400°C. With the proper choice of materials and such heat treatment, total operating pressures below 10^{-12} torr have been attained (Davis 1962). It is, nevertheless, probable that mass-spectrometers themselves contribute a large fraction of the residual outgassing rate in many systems in which they are used.

Resolving power. The resolving power must be adequate to separate clearly the gases which are likely to occur in the system either as residual gases or as those deliberately introduced. In Table 9.1 the first column lists the residual gases which have been reported to appear in well processed uhv systems. Many systems have been reported with peaks appearing at nearly every mass number. This is indicative of contamination of the system with oil diffusion pump fluids, forepump fluids and/or gases evolved from elastomer seals. In properly processed and cold-trapped systems, however, gases from among the eight of Table 9.1 *alone* appear in various proportions regardless of the types of pumps, envelopes or internal components used in their construction. Since the highest mass number encountered is 44 (CO_2), a resolving power of 50 at this mass number is adequate for most partial pressure measurements. Somewhat higher resolving power is an advantage when very large amplitude ratios occur between neighbouring peaks in the spectrum, and the smaller one is of interest. When heavier gases such as Kr and Xe are introduced into the system, the same resolving power still allows their identification even though their isotopes are not clearly resolved, since no residual gas peaks appear in the pertinent portions of the mass spectrum. If the instrument cannot be swept above mass 50, the heavier masses can be identified from their multiply charged components.

There are a number of other mass-spectrometer characteristics which are significant but usually less important than the above three. Some of these are discussed briefly below.

Peak shape. If quantitative monitoring of an individual mass peak is required, the peak shape (i.e. the profile of collected ion current vs the independent variable used to scan the spectrum) should be as nearly as possible flat-topped. Small inadvertent variations in the scanning variable or other electrical or mechanical parameters will not then cause the collected ion current to depart from its peak value. For the identification of extremely small peaks in the neighbourhood of much larger ones, the 'wings' on the peaks are also important. Such wings can result from the scattering of ions by neutral atoms or surfaces in the analyser region, or from ions with anomalous kinetic energy.

Magnet requirements. The size, weight and stray magnetic field of the mass-spectrometer magnet can in some cases be a decisive factor in the choice of an instrument for a specific experiment. Studies involving low-energy electrons or ions require that stray magnetic fields be particularly small. The choice of a non-magnetic mass-spectrometer may in some such cases be preferable to providing adequate magnetic shielding.

Complexity of construction and circuitry. In general, increased complexity results in higher probability of failure, and requires more technically skilled maintenance and testing. Unless some compensating advantage can be gained, the simpler of two alternatives is to be preferred. It has, however, been found that the 'total system' complexity (i.e. spectrometer structure plus electronic circuits) does not vary greatly from one mass-spectrometer type to another. A choice of instrument type is thus often based on the kind of experience and technical support available. Radio frequency mass-spectrometers usually require somewhat more complex electronic circuits than do magnetic deflection instruments. The pulse circuit requirements for time-of-flight mass-spectrometers are fairly stringent, whereas their mechanical tolerances are quite modest.

Background currents. Because collected ion currents in mass-spectrometers are small, it is important that spurious currents appearing at the collector be minimal. X-rays from the electron collector, light from the electron emitting filament, and unfocused ions scattered onto the collector can cause significant spurious currents. In most magnetic spectrometers, adequate shielding against varying electric fields, stray charged particles, and light is not hard to achieve since the ions approach the collector through relatively small apertures or slits. Shielding in radio-frequency and time-of-flight mass-spectrometers is often more difficult, particularly in those types where the ion collector is in direct line-of-sight with the ion source. Leakage currents and spurious electron emission from electron multiplier dynodes can cause significant spurious multiplier output currents (see Section 7.5).

Accessibility of the ionization volume. Some experiments involving highly reactive molecules or fragments may require that the ionization volume of the mass-spectrometer be directly (i.e. in the line-of-sight) accessible to the experimental volume. The molecules of interest may then be ionized and detected before they have had a chance to encounter a surface. Ion sources satisfying this requirement can be fitted to any mass-spectrometer in which the ion production region and the analyser region are completely separate.

Scanning speed. If the experiment requires the rapid comparison of the signals at different mass numbers, the maximum scanning speed may be important. Magnetic deflection instruments in which the magnetic field is varied can be swept only slowly ($\sim$ a few scans/sec). Variations of the accelerating voltage can be much more rapid since the magnet inductance is eliminated. Davis and Vanderslice (1960), for example, report the use of a 30 kHz voltage scanning rate. Time-of-flight instruments are particularly suitable for very fast scan rates, since the flight time of even the heaviest ions from the source to the collector seldom exceeds a few tens of microseconds. Kendall (1962) has used scan rates as high at 100 kHz in a time-of-flight mass-spectrometer with a 60 cm long drift tube.

9.1.2 Examples of ultrahigh vacuum mass-spectrometers

The mass-spectrometers discussed in the subsections below are representatives of various types which have been used for uhv partial pressure analysis. A great many instruments of these and related types have been reported in the literature over the past decade, and no attempt will be made to include them all. In no cases are the ones described here meant to indicate the limits of what can be achieved with that particular type of instrument, but only to give some of the better performance figures actually reported and to direct the reader to papers detailing their construction and characteristics. Where possible, advantages or disadvantages of the various types for specific purposes have been mentioned.

A selection of published papers describing the use of the various types of mass-spectrometers in uhv systems appears in Table 9.2. These examples were chosen primarily because they mention the examination of residual gases in reasonably good uhv systems. In many cases the cited papers give no technical details of the mass-spectrometers used, and in some the performance figures (resolving power and sensitivity) are not the best reported for that design. The relatively small variations of sensitivity and the similarity of the residual spectra are, nevertheless, noteworthy.

90° Magnetic sector. One design of this type, reported originally by Davis and Vanderslice (1960), has become one of the most widely used uhv partial pressure analysers. The sector radius is 5 cm and the maximum envelope dimensions about 30 cm. The instrument can be used either with a permanent magnet and voltage scanning ($B = 3000$ gauss, $MV \approx 10,000$ amu volt where M is the ion mass number and V the ion accelerating voltage) or with a constant ion energy and a scanning electro-magnet.

Both the ion source and the electron multiplier collector are mechanically separate from the analyser region. A 'nude' ion source suitable for experiments with highly reactive fragments (see Section 9.1.1) can be incorporated without difficulty.

The fabrication of the envelopes of sector mass-spectrometers in glass is

TABLE 9.2
Some typical UHV mass spectrometers

Type	Basic equation (a) $M/n =$	Salient dimension (b) (cm)	Electron current (A)	Resolving power (c)	Sensitivity (d) (A/torr)	Minimum detectable partial press (torr)	System residual pressure (e) (torr)	Reference
90° magnetic sector	$\dfrac{\eta_H B^2 R^2}{2V}$	$R = 5$	1×10^{-3}	65	5×10^{-4}	$\sim 5 \times 10^{-18}(f)$	1×10^{-12}	Davis (1962)
60° magnetic sector	$\dfrac{\eta_H B^2 R^2}{2V}$	$R = 11\cdot5$	5×10^{-3}	150	$\sim 10^{-4}$	1×10^{-14}	5×10^{-10}	Reynolds (1956)
180° magnetic deflection	$\dfrac{\eta_H B^2 R^2}{2V}$	$R = 5$	5×10^{-5}	75	5×10^{-5}	2×10^{-11}	5×10^{-10}	Farrar et al. (1964)
trochoidal or 'cycloidal'	$\dfrac{\eta_H B^2 d}{2\pi E}$	$L = 8\,(g)$	1×10^{-4}	150	2×10^{-5}	2×10^{-14}	1×10^{-10}	Lange (1965)
quadrupole	$\begin{cases} 0\cdot144\,\dfrac{\eta_H V_{rf}}{r_0^2 f^2};\ \dfrac{U}{V_{rf}} = \dfrac{1}{6} \\ \text{up to } M = 50 \end{cases}$	$L = 20\,(j)$	$2\cdot5 \times 10^{-4}$	$\Delta M = 0\cdot75(h)$ up to $M = 50$	2×10^{-5}	2×10^{-13}	1×10^{-10}	Bültemann & Delgmann (1965)
monopole	$\begin{cases} \dfrac{\eta_H V_{rf}}{2\pi^2 r_0^2 f^2};\ V_{rf} = \dfrac{a}{2q}; \\ a<0\cdot116 \\ q<0\cdot7 \end{cases}$	$L = 18$	1×10^{-3}	$\Delta M \sim 1$ up to $M = 600$	—	—	$2 \times 10^{-7}(k)$	Hudson & Watters (1966)
time-of-flight	$\dfrac{Lt^2}{2\eta_H V}$	$L = 60$	1×10^{-3}	50	4×10^{-3}	5×10^{-13}	2×10^{-10}	Kendall (1962)
omegatron	$\dfrac{\eta_H B}{2\pi f}$	$L = 2$	3×10^{-5}	30 at $M = 30$ (i)	3×10^{-4}	1×10^{-12}	6×10^{-12}	Klopfer (1961a)

(a) Only the basic first-order expressions for the resolved mass number are given. For restricting conditions see papers and general references cited in Section 9.1.

The following symbols appear:
$\eta_H = e/m$ for the hydrogen atom ($9\cdot58 \times 10^7$ coul/kg)
R = magnetic deflection radius
B = magnetic field
V = ion accelerating voltage
E = electric field
V_{rf} = peak high frequency voltage
U = dc quadrupole or monopole voltage
r_0 = axis to rod distance
f = rf frequency
L = flight tube length.

(b) R = magnetic deflecting radius, L = analyser length.
(c) Definitions vary somewhat from one author to another. Usually $M/\Delta M$ for 10% of peak height.
(d) Collected ion current *without* multiplication.
(e) System at room temperature.
(f) For the total pressures $< 10^{-12}$ torr.
(g) Overall length of E-field forming plates.
(h) Peak width in mass units is nearly constant.
(i) Resolving power $\propto (1/M)$.
(j) L is the length of the analyser structure.
(k) Bakeable instrument but this system not baked.

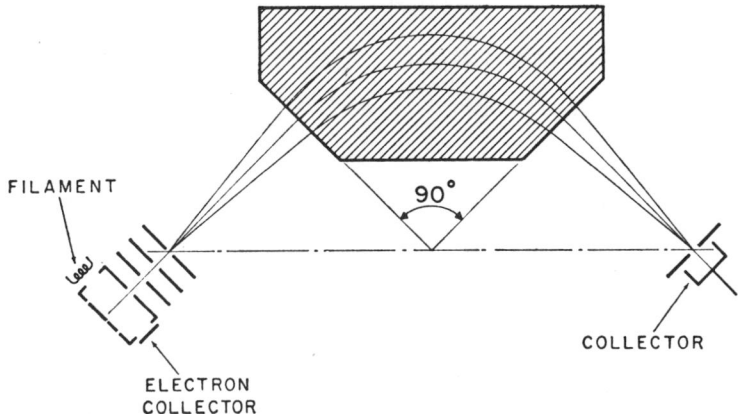

Fig. 9.1 Schematic diagram of a 90° magnetic sector mass-spectrometer.

difficult because of the required mechanical tolerances. The analyser region must have small external dimensions in the direction of the magnetic field to make efficient use of the magnet. With modern welding techniques, fabrication in metal presents no serious problems.

60° Magnetic sector. This geometry has been exploited less for uhv pressure analysis, probably because of its relatively large overall size. The four regions (ion source, acceleration region, analyser and collector) are similar in construction to those of the 90° deflection instrument mentioned above except for the deflection angle. The entrance and exit slits are, for the 60° case, separated by $4R$ as compared to $2\sqrt{2}R$ for the 90° case (R is the mean deflection radius). The inter-slit distance for Lichtman's (1963) instrument is ~ 60 cm whereas that of Davis (1962) is ~ 15 cm.

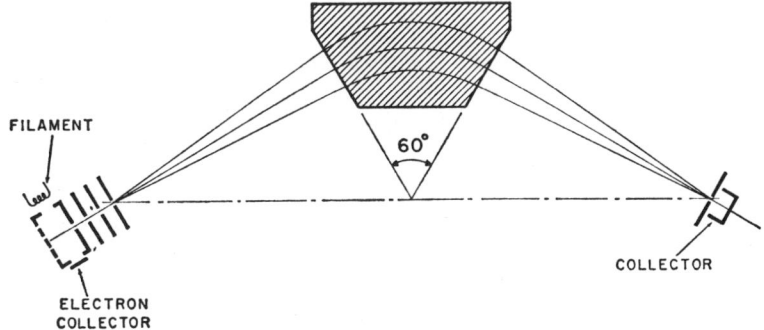

Fig. 9.2 Schematic diagram of a 60° magnetic sector mass-spectrometer.

Reynolds (1956) reported the use of a 60° sector instrument of high sensitivity in work related to the inert gas dating of minerals. For a survey of this important application see, for example, Herzog *et al.* (1961).

180° Magnetic deflection. A widely used instrument of this type with $R = 5$ cm has been described by Craig and Harden (1966). With a simple current collector, the minimum detectable partial pressure was reported to be $\sim 1 \times 10^{-11}$ torr. The addition of an electron multiplier to a mass-spectrometer of this type is difficult but has been reported on one commercial instrument.*

The low ion energies for heavy gases (~ 100 eV for Ar^+) give very low ionic pumping speeds and small disturbance of previously pumped gas. These can be important advantages in the analysis of small static gas samples. The same arguments apply to voltage-scanned sector instruments, usually.

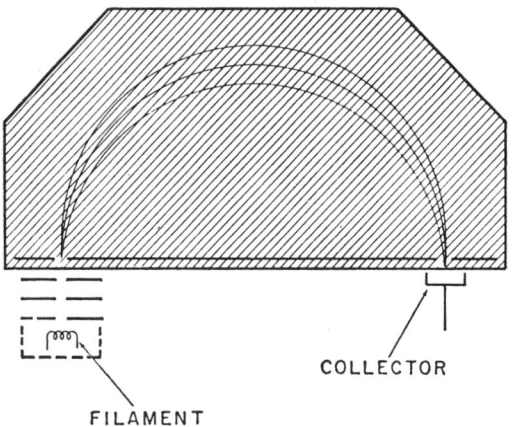

Fig. 9.3 Schematic diagram of a 180° magnetic deflection mass-spectrometer.

Although the transmission probability implied from the sensitivity is fairly high ($\sim 10\%$) only low electron currents are usually used. At the same transmission probability, 1 mA of electron current should yield a sensitivity of $\sim 10^{-3}$ A torr, but this has not been experimentally verified.

Using ion trapping, Redhead (1967c) has obtained a sensitivity of $\sim 10^{-3}$ A torr for neon in a small ($R = 1$ cm) 180° deflection spectrometer using ~ 1 mA of electron current.

Trochoidal path or 'cycloidal'. An important feature of this type of spectrometer is the combined velocity and direction focusing which results from the crossed electric and magnetic fields. This results in excellent peak shape and relatively high transmission probability. The low electron currents generally usable, however, ($\sim 10^{-4}$ A) give only moderately high sensitivity. The instrument used by Lange (1965) was described by Robinson and Hall (1956) except for the addition of an electron multiplier which required the use of a Wien filter to extract the ions from the magnetic field region.

*Sold by Compagnie Francais Thomson-Houston, Type TH-N-205.

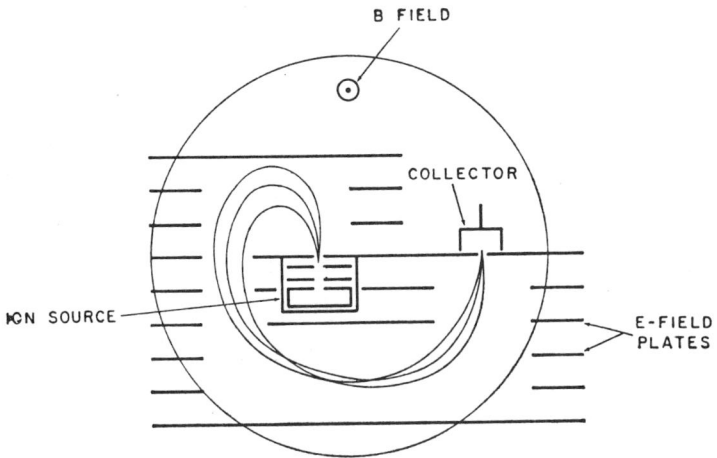

Fig. 9.4 Schematic diagram of a trochoidal focusing or 'cycloidal' type mass-spectrometer.

If voltage scanning is used, both the ion source voltages and electric field-plate voltages must be swept in synchronism to prevent the ions being intercepted on the electric field-plates. Dimensions of the ion source are small and construction correspondingly difficult. Again, ion energies are generally low and pumping speeds small except for the light gases.

Quadrupole and monopole. The absence of an analyser magnet is one of the most important characteristics of spectrometers of these types. Ion source and collector are easily separated from the analyser and thus the addition of an electron multiplier is not difficult. Transmission probability can be varied at the expense of resolving power. Bültemann and Delgmann (1965) give a transmission of 16% at peak width $\Delta M \sim 0.75$ amu for a quadrupole spectrometer scanning the mass range 1 to 50.

The ion source and collector are connected in a straight line path. The transmission of x-rays and light from source to collector can thus cause spurious (pressure independent) currents.

The monopole (von Zahn 1963; Hudson & Watters 1966) depends on the same general mass separation principle as the quadrupole but use as single cylindrical rod and a reflecting 90° angle (the electric fields have the same form). In theory, the resolving power is twice as high as that of a quadrupole analyser of the same length.

Anomalous line shapes of as yet unexplained origin have been observed by several workers in monopole instruments. These result in serious reduction of the resolving power as well as amplitude variations not simply related to the partial pressures.

Potentially, both monopole and quadrupole mass spectrometers have

M*

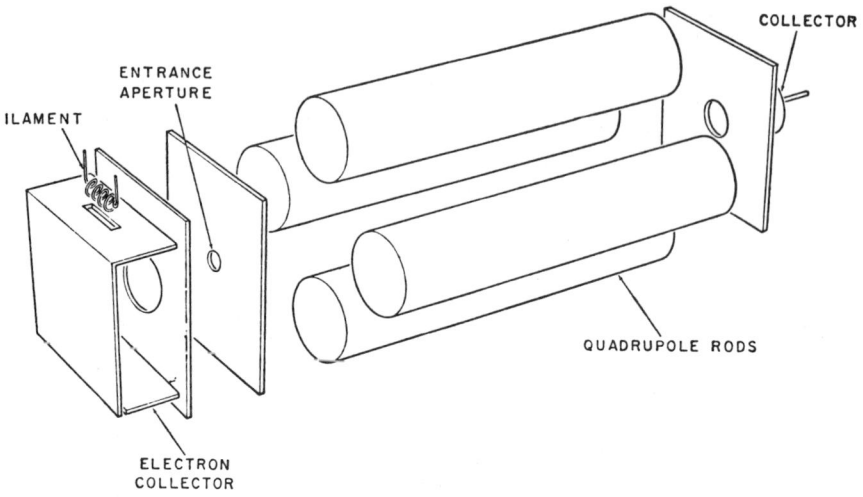

Fig. 9.5 Schematic view of a quadrupole mass-spectrometer. The focused ions travel along the symmetry axis of the four-rod structure.

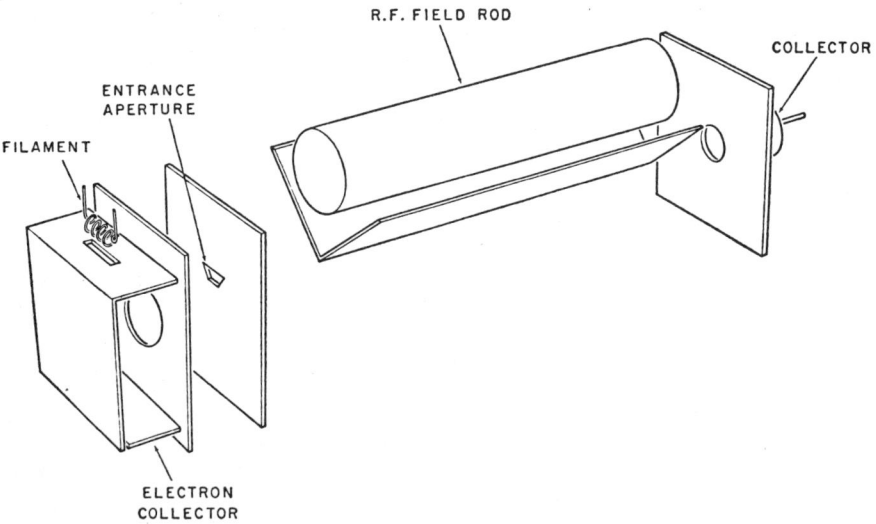

Fig. 9.6 Schematic view of a monopole mass-spectrometer. The focused ions travel in the space between the field rod and the right angle plate.

advantages for the high mass region of the spectrum. To a first approximation the peak width in mass units, rather than the resolving power, is independent of mass number. Resolving powers of several hundred appear to be possible at high mass numbers (Grande *et al.* 1966).

Time-of-flight. For applications requiring very rapid response, the time-of-flight instrument has an inherent advantage. The flight time of an ion from source to collector in the instrument described by Kendall (1962) is, for mass 28, about 17 μsec. Thus an entire spectrum can be displayed in this time. By using modulation of the pulse repetition frequency, scanning rates as high as 100 kHz were successfully used.

Mechanical tolerances in the ion source and analyser are not severe, and the ion beam cross-section is large. Length and total volume of the instrument are fairly large ($\sim$1 metre and 2–3 litres respectively).

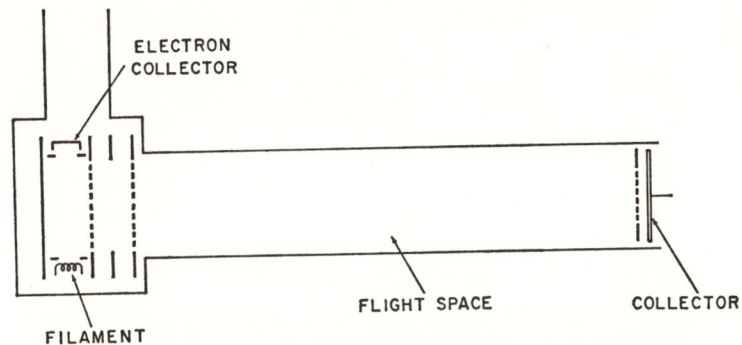

Fig. 9.7 Schematic diagram of a time-of-flight mass-spectrometer.

Even though the pulse duty cycle is $\sim 5 \times 10^{-3}$, proper potential configurations in the ion source make it possible to accumulate ions in the ionization region between pulses and yield a high overall sensitivity (Kendall 1962; Studier 1963). To measure low partial pressures quantitatively, some form of integration technique is necessary (Kendall 1962). This is because the number of ions arriving per sweep at a given peak is small.

Omegatron. Mainly because of its very small size, the omegatron has been one of the most widely used uhv mass-spectrometers. Typical designs are reported by Klopfer and Schmidt (1960) and Lawson (1962). The size advantage is partially off-set, however, by the rather large magnet required to give magnetic field of adequate magnitude and uniformity.

It has been found that calibrations are difficult to reproduce over extended periods, probably because of the interfering effect of varying surface potentials on electrodes. Space-charge effects by trapped non-resonant ions are also capable of affecting both the resolving power and sensitivity.

The collection efficiency of the resonant ions is close to 100%. Electron currents, however, are limited to a few microamperes at high pressure, and $< 10^{-4}$ A even at low pressure. The resulting sensitivity is therefore not high. It is impractical to add multiplier type ion detectors to an omegatron because the collector is immersed in the high magnetic field. The minimum detectable partial pressure reported by Klopfer (1961) (1×10^{-12} torr) is thus probably close to a fundamental limit for the omegatron.

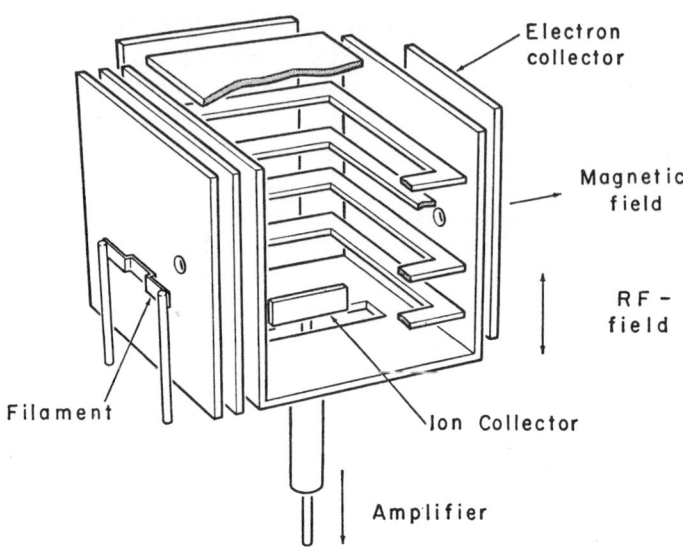

Fig. 9.8 Schematic diagram of an omegatron mass-spectrometer.

9.1.3 Measurement problems in partial pressure analysis

There are several processes which make it difficult to relate mass analysed ion currents to partial pressures. Some of the more common ones will be discussed briefly below.

Fragmentation and multiple ionization. From a particular molecule, ions of various charge-to-mass ratios can be created by multiple ionization and, for polyatomic molecules, by dissociative ionization (see Table 9.1). While the fragmentation patterns or 'cracking patterns' are reasonably well known for certain types of mass-spectrometers operating under specified conditions (see, for example, Bailey 1963) the relative ion currents vary with the ion source conditions (draw-out voltage, electron space charge, and electron energy). The four sets of numbers given in Table 9.1 indicate the variations to be expected in the fragment ion current amplitudes in different instruments. The variations are enhanced in some cases by excess kinetic energy of the fragment ions. The most detailed data can be found in the volume *Mass Spectral Data* of the American Petroleum Institute. Quantitative interpretation of the cracking patterns in terms of the partial pressures of the parent molecules thus becomes difficult. Such processes occur also in total pressure gauges but are not observable in detail.

Ion trapping. The potential distributions typically used in mass-spectrometer ion sources often favour the trapping of ions in the space charge potential well of the ionizing electron beam. This effect is most noticeable in magnetically confined ion sources. Multiply charged ions can then be created by *successive* electron impacts because the ions spend relatively long times in the ionizing region (Redhead 1967c; Baker & Hasted 1966). The anomalous cracking patterns which result are dependent both on the ion source conditions (potentials and electron current) and on the pressure and may make quantitative analysis difficult.

Variations of transmission probability. The fraction of the produced ions which are transmitted through the analyser region is not usually large. For most types of spectrometers, values of a few per cent or less are typical. In trochoidal path and time-of-flight instruments, transmission probabilities of 20 – 30 % have been achieved, and in the omegatron the fraction can approach unity.

When the transmission probability is small, its dependence on the focusing conditions in the ion accelerating and the analysing regions is usually strong. Small variations in supply voltages or in the positions of electrodes or deflecting magnets can cause serious variations in the transmission probability and therefore in the spectrometer sensitivity. For accurate quantitative measurements, frequent calibration checks against a reliable total pressure gauge are desirable.

Non-linearity at high pressures. The effects of ionic space charge and ion-atom scattering in the accelerating and analysing regions begin to cause reductions in the transmission probability of most mass-spectrometers in the pressure range 10^{-6} to 10^{-4} torr. Although the ion currents are not large ($< 10^{-8}$ A, usually) the low ion velocities make their volumetric charge density high and coulomb repulsion significant. At a given current, the space charge spreading in a beam of ions of mass m_i and energy E_i can be calculated from the space charge equation (Spangenberg 1948) to be the same as that of an electron beam of the same current and of energy E_e where

$$E_e = \left(\frac{m_e}{m_i}\right)^{1/3} E_i. \tag{9.1}$$

Thus an argon ion beam of 100 eV will spread like an electron beam of 2·4 eV of the same current. Especially in that part of the acceleration region where the ion energy is not yet large, this can lead to serious space charge spreading effects and consequent reductions in transmission.

The decrease in beam intensity caused by ion-atom scattering can be expressed by the usual attenuation formula

$$I = I_0 \exp\left[\frac{-d}{\lambda(p)}\right] = I_0 \exp\left[\frac{-pd}{\lambda_0}\right] \tag{9.2}$$

where d is the beam flight distance and $\lambda(p)$ the collision mean free path (see von Ardenne (1956), p. 620) $\lambda(p) = \lambda_0/p$. For example, in argon $\lambda_0 = 6\cdot7 \times 10^{-3}$ cm torr, so for a 30 cm flight and a pressure 10^{-5} torr $I = 0\cdot973\, I_0$ and for 10^{-4} torr $I = 0\cdot76\, I_0$.

Interference effects from ion scattering. The scattering of ions by gas molecules (Section 2.3) can also lead to the arrival at the collector of ions whose charge-to-mass ratio is completely unrelated to that of the ions being focused. The scattered ions from a very large peak can cause significant spurious currents over a large portion of the spectrum when peaks of smaller amplitude are being detected. The spurious currents become proportionally larger with increasing pressure since the probability of scattering increases linearly with pressure. Even at very low pressures, however, the scattering of unfocussed ions from surfaces in the analyser can produce similar spurious currents. Davis (1962), for example, reports a background current at low pressure more or less evenly distributed over the whole mass range with an amplitude about 10^{-6} to 10^{-5} of that of the largest peak in the spectrum.

Mass discrimination. Variation of transmission probability with the charge-to-mass ratio leads to a mass dependent sensitivity or 'mass discrimination'. Spectrometers in which the ion accelerating voltage is varied to scan the spectrum are particularly vulnerable to this effect for the following reason: When the accelerating voltage is changed, the electric fields which extract the ions from the ionization region and/or focus them onto the entrance aperture of the analyser are changed. The potentials within the ionizing region, however, are usually dominated by the electron space charge and remain almost constant. The trajectories of the ions (i.e. the focal lengths of the ion lenses) can thus be considerably altered and the transmission probability consequently changed. Calibration with gases of widely varying mass can be used to obtain correction factors for this discrimination, but the corrections will be valid only under the operating conditions for which the calibrations were done.

Discrimination effects can occur in electron multiplier detectors

(i) because the ion-electron conversion ratio is not independent of mass number at constant energy,

(ii) because the multiplier gain may vary with varying fringe magnetic fields in magnetically swept spectrometers.

9.1.4 Mass numbers typically observed

The most common residual gases found in uhv systems are listed in the first column of Table 9.1. The remainder of the table gives typical values of the relative ion currents appearing at various mass number positions for each gas. Only amplitudes $>0.4\%$ of the parent ion current have been included and the values should be taken as representative only. The relative partial pressure sensitivities for various gases are, in the absence of mass discrimination and ion trapping effects (Section 9.1.3), proportional to the gas ionization cross-sections at the electron energy used. (See Table 7.2 for *total* ionization cross-sections.)

In addition to the mass peaks mentioned above, others not directly related to gas partial pressures appear at or near mass numbers 1, 16, 19, 35 and 37.

These seem to be the result of electron impact desorption (see Section 4.2) of H^+ (Lichtman 1965a), O^+, F^+ and Cl^+ (Davis 1962) from the electron collector in the ion source. Peaks have also been observed at mass numbers 6 and 7, 23, 39 and 41 corresponding to the alkali metal ions Li^+, Na^+ and K^+. The alkalis are believed to originate in the heated filaments (see Section 4.4).

9.2 DESORPTION SPECTROMETERS

A total pressure gauge may be used in conjunction with a surface of variable temperature to yield an approximate analysis of the partial pressures of the gases in an uhv system. Such a combination is termed a desorption spectrometer and two types have been used, the chemical desorption spectrometer and the physical desorption spectrometer. Both types are characterized by simplicity and under some circumstances they can be used in place of mass spectrometers. Both methods employ a surface whose temperature increases in a controlled way releasing, at certain temperatures, adsorbed gases whose presence in the gas phase is detected by the gauge. The relations between the activation energies of desorption and the temperature of desorption are presented in Section 2.4.4. For a surface heated at a programmed rate in a particular system each gas will produce a certain characteristic desorption spectrum which can be recognized and serves as the means of identification of that gas. If the desorption spectrum were independent of the specific details of each system then no calibration *in situ* would be required. While this desirable situation is possible in principle, it has not yet been achieved in practice and it is necessary at present to calibrate desorption spectrometers by operating them in known pressures of known gases in order to establish their characteristics.

9.2.1 Chemical desorption spectrometers

The usual surface is a wire or ribbon of a refractory metal (usually tungsten) which is first cleaned by flashing to about 2400°K. The wire is then cooled to 300°K and some of the gases in the system chemisorb on the wire surface. After an interval (t_c) the wire is 'flashed', i.e. the temperature of the wire is increased to about 2400°K in a controlled fashion, and the total pressure recorded during the flash. A feedback circuit to heat a wire at a controllable rate is described by Redhead *et al.* (1962a). Gases that have chemisorbed on the wire during the cold interval are desorbed and produce a series of peaks on the total pressure record (the desorption spectrum). The temperature at which a peak occurs in the desorption spectrum increases with increasing activation energy of desorption (binding energy) of the desorbed species.

The desorption spectrometer method, which is based on the old flash filament technique (Apker 1948), was first described by Hagstrum (1953). An essentially similar technique has been used by many experimenters for the study of adsorption processes of gases on metals; the primary objective of these experiments has been the study of chemisorption kinetics rather than the analysis of gas mixtures which is our interest here. The theory, techniques and results of flash desorption methods for the study of surface kinetics have been reviewed by Ehrlich (1963a).

We first consider the spectra observed when essentially only one chemically active gas is present. Figure 9.9 shows the desorption spectrum observed for nitrogen on tungsten (Kornelsen & Domeij 1966), with a linear temperature-time increase of 40°K/sec during the flash. The nitrogen pressure was $1 \cdot 25 \times 10^{-11}$ torr and the adsorption time at 300°K was 4 minutes. The total pressure in the system was 4×10^{-11} torr but the only gases present, other than N_2, were inert and did not adsorb at 300°K and do not appear in the desorption spectrum. Figure 9.9 shows several features which are representative of most desorption spectra:

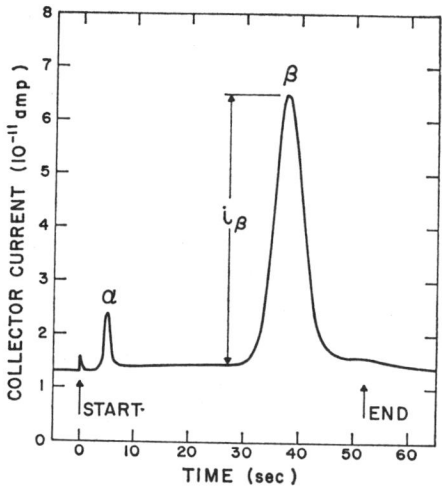

Fig. 9.9 Nitrogen-tungsten desorption spectrum for linear temperature rise rate of 40°K/sec. The nitrogen pressure was $1 \cdot 25 \times 10^{-11}$ torr and the adsorption time at 300°K was 4 min. Pumping speed was 3·8 litres/sec. (Kornelsen & Domeij 1966).

(i) Single gases frequently adsorb in more than one binding state with different binding energies. In the case of N_2, the α and β states can be clearly distinguished (see Fig. 9.9).

(ii) The peak amplitude of one state (usually the peak at highest temperature because it is most strongly bonded) is linearly related to the partial pressure of the chemisorbing gas for small coverages of the surface. For the situation of Fig. 9.9 a linear relation between i_β (the gauge current corresponding to the β peak) and $p(N_2)$ was observed up to 10^{-9} torr indicating that the sticking probability of N_2 on W remained constant to exposures of about 10^{14} molecules incident/cm². In this linear range $i_\beta/p(N_2) = 4 \cdot 0$ A/torr for a 4-minute adsorption time; a BAG was used to measure pressure with

$K = 25 \text{ torr}^{-1}$ and $I^- = 8$ mA, and the system pumping speed was about 3·8 litres/sec.

The amplitude of the peaks and their shape is dependent on the pumping speed in the system. This problem has been analysed in some detail (Ehrlich 1961c; Redhead 1962b). For very small pumping speed the desorption spectrum would appear like an integral of Fig. 9.9 and the amplitude of the 'peaks' (which would now be only inflexion points) would be a maximum. The pressure at any instant is proportional to the total amount of gas desorbed. As pumping speed is increased, the peak amplitude decreases and

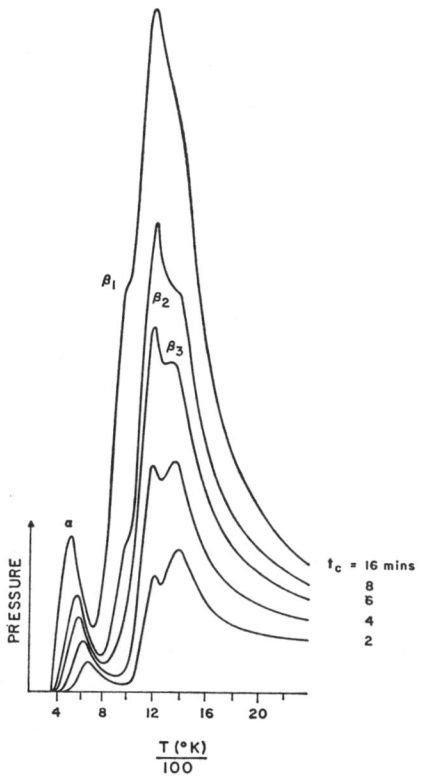

Fig. 9.10 Carbon monoxide-tungsten desorption spectra for linear temperature rise rate of 35°K/sec. Pressure about 5×10^{-9} torr. (Redhead 1962c)

the pressure-time curve approaches the desorption rate-time curve. For optimum gas analysis the pumping speed should be adjusted, if possible, to give good peak separation and shape with reasonable amplitude.

Figure 9.10 shows the desorption spectra of CO on W (Redhead 1962c) obtained by adsorption at 300°K and at a pressure of about 5×10^{-9} torr. The various spectra are for increasing adsorption time t_c. This figure illustrates the increase of peak amplitude with adsorption time. This relation is usually linear for small coverages and becomes non-linear for relative coverages of about 1/3 in most cases and approaches saturation near monolayer coverage. Again we see the presence of several adsorbed phases, the low temperature α-phase is thermally reversible at room temperature and *decreases* with time if the CO pressure is reduced (Rigby 1964).

Figure 9.11 shows the desorption spectra of H_2 on W for various ad-

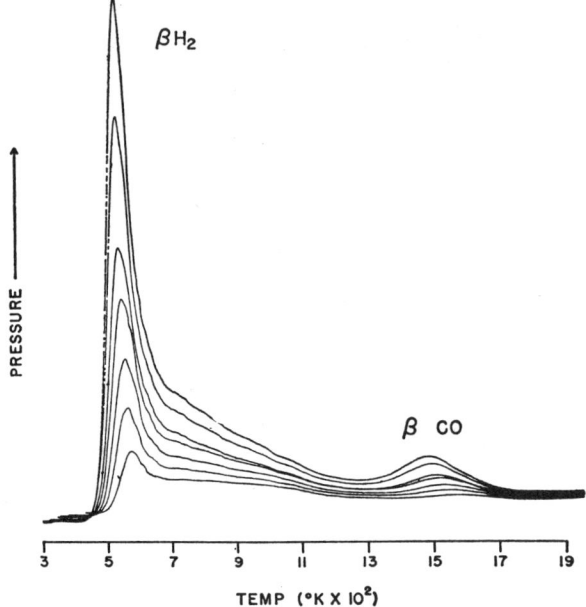

Fig. 9.11 Hydrogen-tungsten desorption spectra for linear temperature rise rate of $35°K/sec$. Pressure about 5×10^{-10} torr. Adsorption time of 10, 20, 30, 40, 60, 90 and 120 seconds. (Rigby 1965).

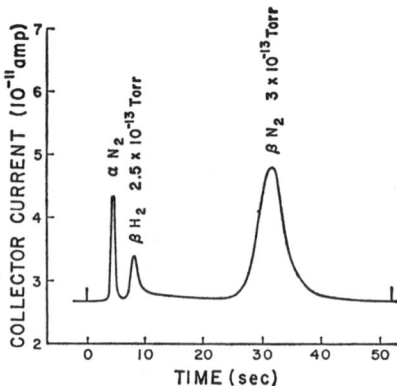

Fig. 9.12 Desorption spectrum of residual gas at total pressure of 8×10^{-11} torr. Linear temperature rise rate of $40°K/sec$. Adsorption time 218 minutes.

sorption times at a pressure of 5×10^{-10} torr. The presence of a small amount of CO is clearly evident.

When more than one chemically active gas is present the spectra are more complex and it is not always possible to obtain complete separation of the various peaks. Figure 9.12 shows the desorption spectrum obtained from the residual gas in a system operating at a total pressure of about 8×10^{-11} torr; the only chemically active gases present were H_2 and N_2. The adsorption time was 218 minutes and the estimated pressures are $p(H_2) = 2 \cdot 6 \times 10^{-13}$ torr and $p(N_2) = 3 \cdot 1 \times 10^{-13}$ torr; these estimates are no better than a factor two. Desorption spectra of residual gases consisting of H_2 and CO have been published (Redhead 1959b; Reich 1961; Hoch 1966) and complete separation of these two gases is relatively simple. Figure 9.13 shows a typical example (Redhead et al. 1962a). Separation of CO and N_2 is much more difficult because their high-temperature desorption peaks (the β-phases overlap). Van Oostrom (1966) has solved this difficulty by using wires of rhodium and tungsten and comparing the desorption spectra obtained.

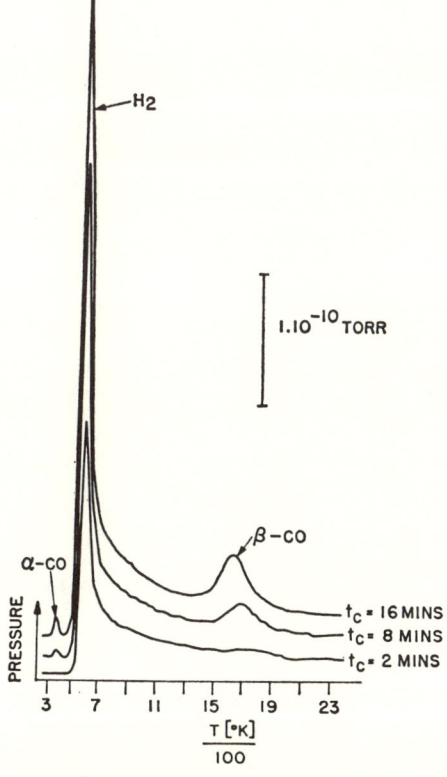

Fig. 9.13 Desorption spectrum of residual gas at total pressure of about 1×10^{-10} torr. Linear temperature increase of 40°K/sec. (Redhead et al. 1962a).

Nitrogen does not absorb on rhodium at 300°K and hence the desorption spectrum from the rhodium wire shows the CO only. The desorption spectrum for the W wire shows both CO and N_2.

Analysis of gas mixtures by desorption spectrometry is further complicated by the phenomenon of replacement, wherein gas B replaces gas A previously adsorbed. This subject is discussed in Section 2.3.2. This effect results in the relative abundances of the gases desorbed differing from their relative abundances in the gas phase.

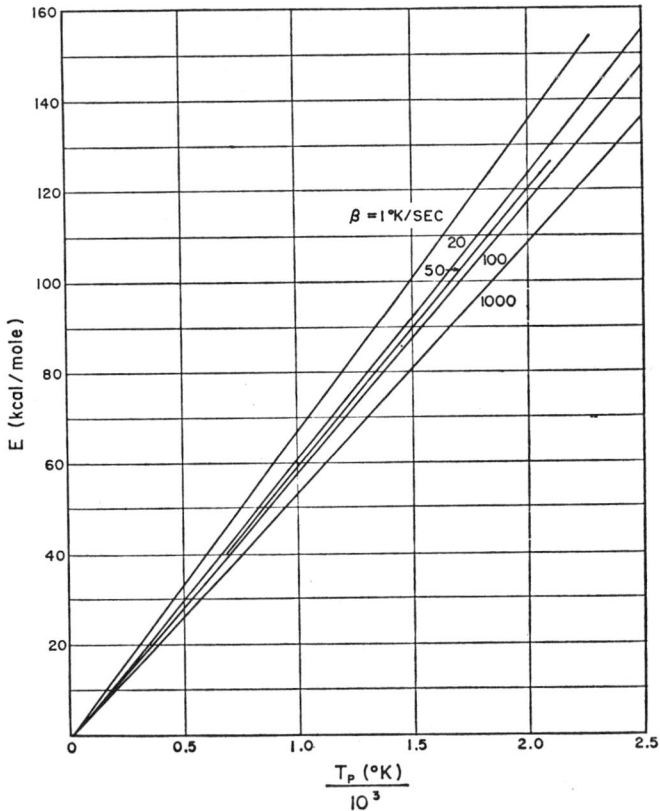

Fig. 9.14 Activation energy of desorption (E) as a function of temperature at peak (T_p) for a first-order reaction and linear temperature sweep ($T = T_0 + \beta t$) taking $v_1 = 10^{13}$ sec^{-1}. (Redhead 1962b).

The position of the peaks along the temperature scale is a function principally of the activation energy of desorption (E_d). A detailed discussion of methods for calculating E_d and other parameters from desorption spectra may be found in Ehrlich (1961c) and Redhead (1962b). For a first-order desorption reaction the temperature at which the peak occurs (T_p) is given approximately by

$$\frac{E_d}{RT_p} = \ln\left[\frac{v_1 T_p}{\beta}\right] - 3\cdot 64 \qquad (9.3)$$

for a linear temperature sweep ($T = T_0 + \beta t$), v_1 is the first-order rate constant (a reasonable value is $v_1 = 10^{13}$ sec^{-1}). Figure 9.14 shows this relation for β from 1 to 10^{3} °K/sec. For a second-order relation T_p decreases with increasing coverage; at low coverages the data of Fig. 9.14 still give a reasonable approximation to E_d.

Identification of the gases present from the peak positions must be done by prior calibration of the particular desorption spectrometer with known gases or with a mass spectrometer. Desorption spectra vary considerably between different samples of wire or ribbon.

The sensitivity of the desorption spectrometer for partial pressure analysis cannot be simply stated because it depends, *inter alia*, on sample area, heating rate, adsorption time, pumping speed, gauge sensitivity and the concentration of other desorbable gases. Figure 9.13 is representative of the best sensitivity presently attainable. With an adsorption time of about three hours in a system of 4 litres volume it was possible to distinguish clearly partial pressures of $2\cdot 6 \times 10^{-13}$ torr (H_2) and $3\cdot 1 \times 10^{-13}$ torr (N_2) in a total pressure of 1×10^{-10} torr (mainly He). It is clear from this figure that partial pressures about an order of magnitude lower could be determined in the same adsorption time. The accuracy of partial pressure measurements by means of a desorption spectrometer is poor, being no better than a factor 2. However, the extreme simplicity of the apparatus and its high sensitivity makes the desorption spectrometer a very valuable tool in some experiments where the complexity of a mass-spectrometer is not desirable.

Desorption spectrometry methods have been used for outgassing studies of materials (Oldal & Tahy 1965) and for the study of semiconductor surfaces (Kleint & Moldenhauer 1965).

9.2.2 Physical desorption spectrometers

All gases will adsorb on all surfaces if the surface temperature is sufficiently low. For a given energy (a given gas and a given homogeneous surface) this physical adsorption will take place over a small temperature range (Section 2.4) and in principle a measurement of pressure as the surface temperature is changed permits an identification and approximate partial pressure measurement of each gaseous component. Thus the physical adsorption spectrometer provides, in principle, a partial pressure analyser for all gases. There are several problems which have prevented the full utilization of this promise to date:

(*a*) During desorption the temperature of the surface must rise from liquid helium to room temperature if all gases are to be included. The upper temperature limit is that of the apparatus in general and we have chosen room temperature as the most usual case. The general apparatus temperature is the upper bound because gases which adsorb at temperatures above this have, generally speaking, already done so and do not enter into the residual gas components.

(b) When the pressure is the observed variable there must not be any surface as cold as the spectrometer between the spectrometer and the gauge. This is because heats of physical adsorption are similar for many surfaces and there will be a strong tendency for adsorption to take place on the coldest surface available. The flash filament desorption technique above liquid nitrogen temperature has been used (Ehrlich 1961b) but the interpretation of the results is complicated by the presence of the cold glass walls which remain at liquid nitrogen temperature and thus partially pump the desorbed gases. The field emission microscope permits measurements on the adsorbed phase and in principle removes the restriction that the desorbing surface should be the coldest in the system. The time taken to collect an observable amount ($\theta \sim 0.01$) at very low pressure may be exceedingly long and a partial pressure analysis involves a measurement of differences in this minute amount.

(c) The restriction that the energy of adsorption be single-valued is difficult to meet experimentally. It can probably be done with a single crystal surface but restrictions (a) and (b) have so far limited the operative surfaces in physical desorption spectrimeters to heterogeneous surfaces. One can obtain some quantitative idea of the effect of heterogeneity from the isosteric heats (q_{st}) measured by Hobson and Armstrong (1963) from the physical adsorption isotherms of nitrogen and argon on Pyrex. For nitrogen at $\theta = 10^{-3}$, $q_{st} = 6000$ cal/mole; at $\theta = 0.15$, $q_{st} = 4000$ cal/mole; at $T = 77.4°K$ these correspond to equilibrium pressures of 5×10^{-10} torr and 5×10^{-4} torr respectively. For argon at $\theta = 10^{-5}$, $q_{st} = 5300$ cal/mole; at $\theta = 10^{-2}$, $q_{st} = 3600$ cal/mole; at $T = 77.4°K$ these correspond to pressures of 5×10^{-9} torr and 10^{-5} torr. These values of q_{st} are converted to temperatures of desorption in Table 9.3 by using the simple relation

$$\frac{T}{E_d(\text{kcal/mole})} \sim 20 \qquad (9.4)$$

and identifying q_{st} with E_d. Equation (9.4) corresponds to the desorption of an adsorbed molecule in 10^{-2} sec.

These figures are representative only; but they serve to illustrate that while it might be possible to construct a satisfactory physical desorption spectrometer in one pressure range it would be difficult to construct one which would unambiguously distinguish between gases like N_2 and Ar with substantially different pressures. Four of the gases commonly found in uhv systems: N_2, Ar,

TABLE 9.3
Temperatures of desorption from $T°K = 20\ E_d$ (kcal mole^{-1})

E_d (cal mole^{-1})	3600	4000	5300	6000
$T°K$	72	80	106	120

CO, CH$_4$ (as well as O$_2$) have desorption temperatures around 100°K and are difficult to distinguish. However, gases such as He, H$_2$, Ne, CO$_2$ have desorption temperatures which are well-spaced and their identification is less ambiguous.

(d) The important temperature is of course the surface temperature, from which desorption is taking place, but it is difficult to measure this temperature and also to maintain it constant over the whole surface.

(e) While a physical desorption spectrometer can control the rate of desorption of gases, the resulting pressure in a system will depend upon the pumps present. Since the pressure is the simplest variable to measure, it is difficult to design a physical desorption spectrometer which can be used with a variety of pumping systems.

These difficulties, however, do not prevent physical desorption spectrometers from being calibrated in a system similar to the one in which they are to be used, although the calibration procedure may involve the use of

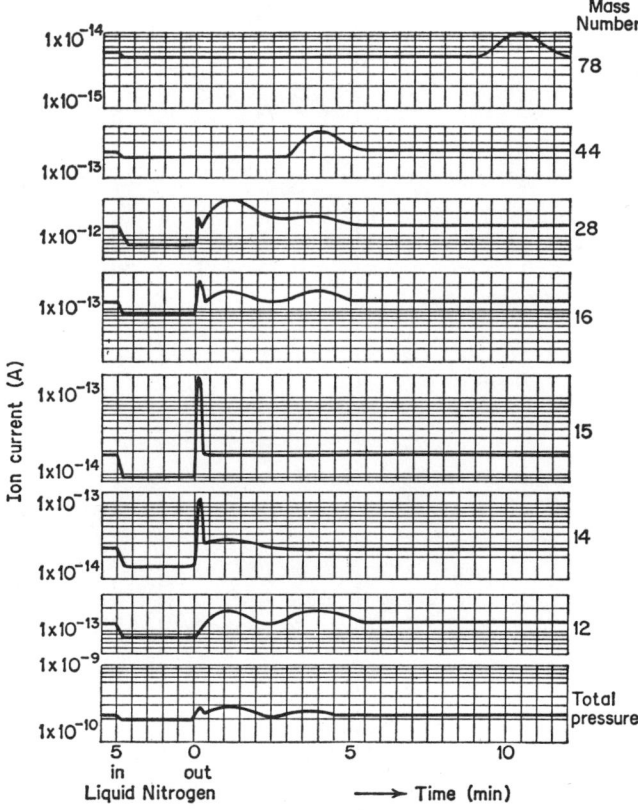

Fig. 9.15 Partial pressures as a function of warming time of liquid nitrogen trap. (Hengevoss & Huber 1961).

TABLE 9.4
Temperatures of desorption maxima °K for physical desorption spectrometers

Adsorbent	Approximate pressure during adsorption (torr)	Ar	O_2	CH_4	N_2	CO	CO_2	H_2O	Reference
Glass	2×10^{-8}			108		133	168		Hengevoss & Huber (1961)
Glass	1×10^{-9}	97		137	103	136	187		Kornelsen (1965)
Molecular sieve	$\sim 10^{-3}$	130	141	173	186	206	298	493	Murakami & Okamoto (1963)
Molybdenum film	3×10^{-10}			100					Richardson & Strehlow (1963)
Glass	3×10^{-8}			109		135	177	153	Bas (1965)*

*Bas (1965) identifies his gases CH_4, CO, CO_2 by comparing his desorption spectra with those of Hengevoss and Huber (1961).

a mass-spectrometer or the admission of known gases. Several examples of this procedure have been published. Figure 9.15 shows the results of Hengevoss and Huber (1961) demonstrating the behaviour of various mass peaks as observed with a mass-spectrometer as a function of time after the liquid nitrogen trap on their uhv (oil diffusion pump) system was warmed. It is quite clear that different components desorb at different times. Table 9.4 shows the temperature of desorption maxima obtained by several workers using physical desorption spectrometers of various designs. While the general ordering of the gases is similar for all workers, the absolute values of the temperature maxima differ, reflecting the influence of the factors discussed above. The most extensive instrumental work was that of Murakami and Okamoto (1963) but despite the fact that their pressures were much higher than those of the others (which should give a lower heat of desorption and a lower desorption temperature) their actual desorption temperatures are well above those of the others. The most likely reason for this is that their sample temperature lagged their measured temperature.

9.3 GAUGES USING FIELD EMISSION OR WORK-FUNCTION CHANGES

The adsorption of gases on the surface of a metal results in changes of work-function, which cause changes in the field emission from the surface. Measurement of these changes in work-function or of the resulting changes in field emission can be used as an indication of the partial pressure of the gases which adsorb on the metal surface. Using a metal surface at room temperature restricts this method to the chemically active gases.

The field emission methods make use of the field-emission microscope (see Good and Müller (1956) for a general review of field emission). Kleint (1960, 1962, 1963) has shown that the low-frequency noise ($10-10^3$ Hz) in the emission current of a field-emission microscope increases with the amount of gas adsorbed until approximately a monolayer is adsorbed when the noise current saturates. It was found (Kleint 1960) that the time to reach saturation of the noise after cleaning the field-emitter was given by $t = 10^{-6}/p$ sec.

The time taken for the pattern in a field-emission microscope to change to some recognizable, stable form after cleaning the point by flashing is also a function of the partial pressure of adsorbable gas. This technique appears to have been first used by Drechsler and Hess (referred to in Koenig 1953) and has since been employed by many experimenters (e.g. Kirchner & Kirchner 1956; Schlier 1958). A linear relation has been observed between the time for a selected pattern change and pressure for O_2, H_2 and N_2 in the pressure range 5×10^{-10} to 10^{-7} torr (Little & Whitney 1962).

The change in work-function ($\Delta\phi$) caused by the adsorption of gas on a clean metal surface can be measured by several methods and is an indication of the partial pressure of adsorbable gases. A simple method of measuring $\Delta\phi$ is the retarding-field diode method (Crowell & Armstrong 1959) where the adsorption sample, usually a tungsten wire or ribbon, is the anode of a thermionic diode. By use of a simple constant-current circuit the change in work-function of the anode may be directly measured with a reproducibility of about $\pm 1\cdot 5$ mV (Redhead 1962c). The maximum change in work-function ($\Delta\phi_{max}$), which occurs near monolayer coverage, depends on the gas and the metal. A compilation of experimental values of $\Delta\phi_{max}$ is given in Eberhagen (1960).

The work-function change is extremely sensitive to small amounts of chemically active gases such as O_2, CO, N_2 or H_2. This method is very useful for determining whether chemically active impurities are present in rare gases at low pressures. It is also an extremely sensitive method for observing the presence of CO impurities in N_2 or H_2. In pure N_2 or H_2 $\Delta\phi_{max}$ remains constant once monolayer coverage is achieved. If minute quantities of CO are present they displace the H_2 or N_2 adsorbed on the sample and cause a slow drift of the work-function.

Part C
PRODUCTION OF ULTRAHIGH VACUUM

CHAPTER TEN

PROCESSING TECHNIQUES FOR ULTRAHIGH VACUUM

The choice of materials for use in uhv systems is governed largely by the requirement that it be possible to reduce the outgassing rate to a value sufficiently low that the desired background pressure conditions can be obtained. For detailed considerations concerning the choice of materials for uhv, reference should be made to Roberts and Vanderslice (1963); Lewin (1965) and Trendelenburg (1963). An excellent compendium of the properties of materials used in vacuum technology can be found in Espe (1966). Data on outgassing rates of materials used in vacuum technology can be found in Basalaeva (1958); Schittko (1963) and Diels and Jaeckel (1966). Some plastics and elastomers have sufficiently low outgassing rates to be used in uhv systems; for data on outgassing rates of these materials, see Barton and Govier (1965). This chapter is concerned with the various processing techniques used to reduce the outgassing rate from the materials of uhv systems.

Two sources of gas from constructional materials can be distinguished; gas adsorbed on surfaces, and gas dissolved in the bulk. The outgassing rate can be reduced by either (*a*) reducing the concentration of gas on the surface or in the bulk (by thermal or chemical methods), (*b*) treating the surface chemically or electrochemically to build up a layer which impedes the evolution of gas from the bulk, or (*c*) cooling.

The reduction of the amount of gas in an adsorbed layer by heat treatment is a simple process and has been treated in detail elsewhere. The rate of desorption is given by eqns. (2.16) or (2.71) and representative values of heats of adsorption, which govern the desorption rates, can be found in Tables 2.13 and 2.14. In general, there is only a limited range of the activation energy of desorption (E_d) which leads to troublesome outgassing rates. For very high values of E_d the rates of desorption are too small to be of any consequence; for small values of E_d the desorption rates are so high that there is no significant surface concentration. The troublesome range of E_d is approximately 15–45 kcal/mole for a first-order desorption process at 300°K. Hobson (1961a) has evaluated a simple example assuming a first-order desorption rate in a system of 1 litre volume, with 100 cm^2 surface area and a pumping speed of 1 litre/sec. Figure 10.1*a* shows the change in pressure

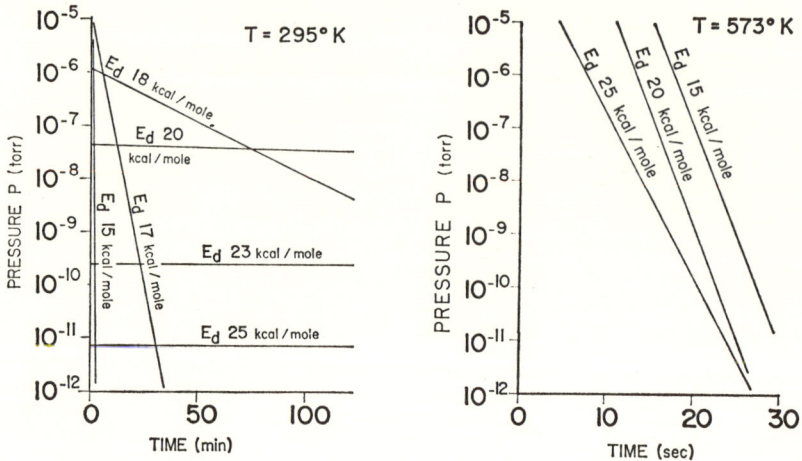

Fig. 10.1 Pressure vs time with $V = 1$ litre, $S = 1$ litre sec^{-1}, $A = 100$ cm^2 covered with a monolayer at $t = 0$.
(a) $T = 295°K$
(b) $T = 573°K$

(Hobson 1961a).

with time of this hypothetical system for various values of E_d, assuming that the surfaces of the system are covered with a monolayer of gas at time zero and $T = 295°K$. We observe that for $E_d > 25$ kcal/mole the pressure remains well in the uhv range at all times, for $E_d < 17$ kcal/mole the pressure drops into the uhv region in times less than one hour. In the intervening range ($25 > E_d > 17$) the system cannot be pumped into the uhv range in reasonable time. Figure 10.1b shows the same system at a temperature of 573°K, here the pressure drops rapidly into the uhv range for E_d in the range of values that were troublesome at 295°K. The results of this simple calculation, which should not be taken as directly applicable to a practical system because of replenishment of the surface layers from gas in the bulk, illustrate the profound effect that heat treatment can have on reducing outgassing from surface layers alone.

In practical systems, being used for the first time, the largest source of outgassing is from the gas dissolved in the bulk of internal parts and the vacuum envelope. The outgassing rates are now controlled by diffusion of the gas from the bulk to the surface; these problems have been considered in some detail in Section 2.5. After the gas from the bulk has been removed, by suitable processing, subsequent pump-downs will be dominated by outgassing from adsorbed layers which have been built up by exposure of the system to atmosphere.

In the following, we consider processing procedures in three categories, viz: pre-treatment of component parts before system assembly; baking of complete systems; the subsequent outgassing of portions of the system.

10.1 PRE-TREATMENT OF MATERIALS

Vacuum firing of the component parts (metal and ceramic) of an uhv system is a simple and effective method of reducing the amount of gas in the bulk of the material and hence decreases the subsequent outgassing rate. The dominant gas evolved during vacuum firing is hydrogen. Vacuum firing, at a temperature as high as the material will permit (e.g. 1300°K for Nichrome or steels), for about one hour, is used in the authors' laboratory as a pretreatment for all metal and ceramic components before mounting in an uhv system. After subsequent baking of the assembled system at 720°K the outgassing rates routinely observed for hydrogen are about 10^{-14} torr litre $\sec^{-1}$ cm^{-2} at temperatures near 300°K. Outgassing rates of this general order are obtained with metals such as W, Mo, Ti, Pt, Ni, Cu, Nichrome V and stainless steel (type 304) and high-alumina ceramics. If a metal part operates above 300°K (such as filament supports or other electrodes near hot filaments) and outgassing rates approaching 10^{-12} torr litre sec^{-1} cm^{-2} are desired, then it is essential to use a refractory metal which can be rigorously degassed at temperatures above 1300°K. It has been found in our laboratories that Nichrome and stainless steel components that are subsequently operated at temperatures above 900°K cannot be adequately degassed at any temperature low enough to avoid significant evaporation.

Calder and Lewin (1966) have reported measurements of the outgassing rates of stainless steel samples after various heat treatments in vacuum. Table 10.1 summarizes some of the data obtained from a 300 series, low carbon stainless steel of 2 mm thickness; 99% of the gas released was

TABLE 10.1
Average hydrogen outgassing rates of stainless steel specimens after various heat treatments

Treatment	Measurement temperature (°K)	Hydrogen outgassing rate (Torr litre sec^{-1} cm^{-2})
Vacuum baked for total of 75 hours at 570°K	293	$1 \cdot 5 \times 10^{-12}$
	320	$5 \cdot 3 \times 10^{-12}$
	345	$1 \cdot 5 \times 10^{-11}$
	378	$3 \cdot 9 \times 10^{-11}$
	418	$9 \cdot 2 \times 10^{-11}$
	573	$2 \cdot 5 \times 10^{-11}$
Degas at 1270°K for 3 hours, followed by in situ bake at 630°K for 25 hours	293	$1 \cdot 3 \times 10^{-14}$
Exposed to 1 torr of H$_2$ for 12 days, then pumped for 3 days at 300°K	293	5×10^{-12}

hydrogen. Calculations based on a diffusion model suggest that it should be possible to achieve outgassing rates for hydrogen as low as 10^{-16} torr litre $\sec^{-1}$ cm^{-2}. Experimentally it was found that after degassing 2-mm thick stainless steel for 2 hours at 1300°K the hydrogen outgassing rate was at least as low as 1×10^{-14} torr litre $\sec^{-1}$ cm^{-2} at 300°K and, because of limitation of the apparatus, was estimated to be much lower.

Varadi (1960) has determined the gases released from nickel samples during vacuum firing at 1130°K; the largest constituent was H_2, followed in order by CO, H_2O, and CO_2.

Mechanical, chemical and electrochemical methods have been widely used for the pre-treatment of materials to reduce subsequent outgassing. Quantitative data on the effects of these pre-treatments on outgassing rates are very scarce. Doré (1963) has discussed pre-treatment in general terms. Varadi (1961) has measured the outgassing rate of nickel samples after various chemical and thermal pre-treatments, including chemical degreasing, hydrogen firing, and acid cleaning. It was found that if the nickel were properly stored between pre-treatment and use, no increase in gas content occurred. Handling did cause an increase in measured gas content. Basalaeva (1958) has shown that the outgassing rate of Duraluminum could be reduced by a factor of ten by mechanical scouring and washing in benzol and acetone. Blears et al. (1960) show that washing aluminum in detergent reduced the outgassing rate by a factor of about ten, they also show that sandblasting of rusty steel decreases the rate by a factor of about four.

Milleron (1967) has reported on the outgassing rates observed after chemical treatment of several metals in DS-9 (Diversey Co., Chicago, Illinois). The outgassing rates for stainless steel, copper, invar, and mild steel were measured by a rate of rise method after 12 hours of pumping and no bakeout; the gas observed was principally water and the rates were less than 10^{-12} torr litre $\sec^{-1}$ cm^{-2}.

Electrochemical treatments of stainless steel and other metals are reported to produce a 'passive' surface (presumably an oxide layer) with reduced outgassing rates. Solutions for these electrochemical passivating processes are now commercially available (e.g. Suma process, available from Molectrics Inc., Inglewood, California).

Pétermann (1965) has developed a pre-treatment process for stainless steel which consists of heating the material in air at atmospheric pressure at temperatures between 700 and 820°K for a few hours. The resultant surface layer of oxide passivates the surface and reduces the outgassing rate of hydrogen, which is the principal gas evolved.

The estimates of outgassing rates quoted above were all made without any allowance for the effects of adsorption-desorption of chemically active gases in the system. Hobson and Earnshaw (1967) have analysed this problem and show that the outgassing rates measured without accounting for these effects may be as much as a thousand times lower than the true outgassing rates. This effect is particularly troublesome with H_2 and CO and is negligible for the rare gases which are not significantly adsorbed at 300°K.

10.2 BAKING PROCEDURES

After assembly and evacuation of the complete uhv system it is usual, where possible, to bake the system at a temperature in the range 400 to 900°K. The primary purpose of this bakeout is to remove adsorbed water. Water is particularly troublesome in uhv systems both because of its ubiquity and because the activation energy of desorption on most solids is about 20 kcal/mole (see Fig. 10.1).

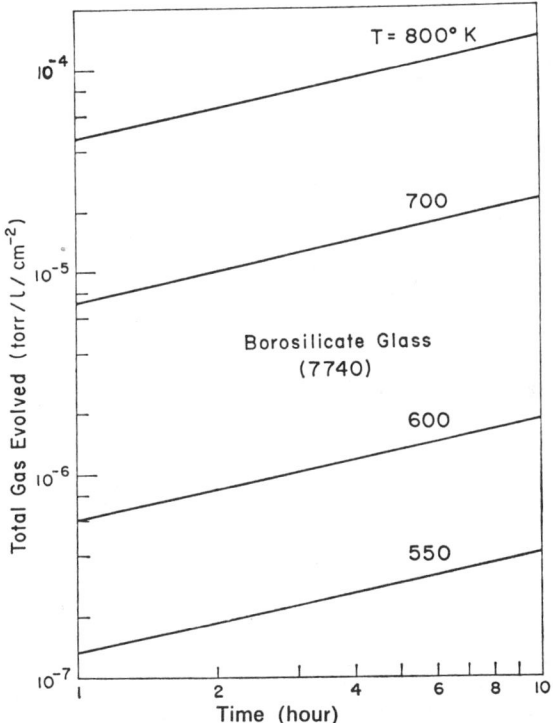

Fig. 10.2 Amount of water evolved from borosilicate glass as a function of time for various temperatures. (From the data of Todd 1955).

Rigorous baking is particularly important for glass systems because glass contains large quantities of dissolved water. Todd (1955) has studied the outgassing of glasses and has shown that the gas evolved below the softening point is principally water. Above 600°K the total amount of water evolved is proportional to the square root of the baking time. Figure 10.2 shows the amount of water evolved from the bulk of borosilicate glass as a function of time for various temperatures. The variation of amount of water evolved as a function of temperature can be expressed as

$$\log_{10} m = B - A/T, \qquad (10.1)$$

where A and B are constants and m is in torr litre (at 298°K) cm^{-2} hour$^{-1/2}$.

Table 10.2 lists the values of A and B for various glasses and the corresponding activation energy (ΔH_a). Garbe et al. (1960) give qualitatively similar results for the desorption of water from lime glass and show that the adsorption is reversible, i.e. a glass surface once cleaned acts as a rapid adsorption pump for water. Todd (1956) also demonstrates that water adsorption on glass is reversible. Garbe and Christians (1962) have made a detailed study of the gases evolved from glass on heating and find, in order of abundance, H_2O, CO, CO_2, N_2, Ar and CH_4.

TABLE 10.2
Constants in equation for gas evolution from glass

Glass Code No.	Glass type	A (°K)	B	ΔH_a (kcal/mole)
7911	Vycor brand 96% silica	6230	0·397	57
7910	Vycor brand 96% silica	8240	4·772	75
1720	Aluminosilicate	7000	2·952	64
7740	Borosilicate	4510	1·310	41
7720	Lead-borosilicate	4150	0·983	38
0080	Soda-lime	5420	3·153	50
0120	Potash-soda-lead	3910	1·208	36
9014	Potash-soda-barium	4840	2·799	44

$\text{Log}_{10} m = B - A/T$ (torr litre (at 298°K) cm^{-2} hr$^{-1/2}$)
From Todd (1955)

Hickmott (1960b) has measured the composition of the residual gases in a borosilicate glass uhv system pumped with a mercury diffusion pump after various baking treatments; the results are shown in Table 10.3. It can be seen that a bake at 400°K reduces the amount of water tenfold; after baking at 700°K only CO and a trace of water could be detected. It can also be seen that the difference, both in total pressure and active gas content, in the results of a 500 and 700°K bake is very marked. In contradiction to Hickmott's results, Young and Hession (1964) have reported that they can readily obtain pressures below 10^{-11} torr in a predominantly glass system pumped with a mercury diffusion pump with a bake at only 525°K. The partial pressure of H_2O after a 15-hour bake at 525°K was less than 10^{-13} torr. No improvement was noticed in the low pressure limit, or the time necessary to reach the lowest pressure, by baking at 725°K. The reasons for the difference between the results of Hickmott and Young are not known but experience in the authors' laboratories with borosilicate glass systems indicates that systems not previously baked require a temperature of about 675°K for the first bake. Subsequent bakes, after brief exposure to atmosphere, can be at lower temperatures.

It has been pointed out (Singleton & Lange 1965) that when carbon dioxide is a significant constituent of the gases evolved during bake, it is necessary to prevent its accumulation in a liquid nitrogen cooled trap. The vapour pressure of CO_2 at 77°K is 10^{-8} torr, which will be approached if multilayers of CO_2 build up in the trap. Singleton and Lange (1965) recommend warming the trap to room temperature while pumping, with the previously baked system isolated by a valve, to remove CO_2 from the trap. With this procedure ultimate pressures of about 5×10^{-11} torr were obtained.

TABLE 10.3
Change in residual gases in a Pyrex vacuum system with bakeout

Gas	Mass peak	Unbaked system	Baked at 400°K	Baked at 500°K	Baked at 700°K, metal outgassed
Hg	100	1280	530	—	—
CO_2	44	790	69	24	—
CO or N_2	28	2580	570	152	2
H_2O	18	6100	430	36	1
CH_4 or O	16	300	33	7	—
H_2	2	860	240	64	—
Total pressure (torr eq. N_2)		8×10^{-7}	4×10^{-7}	5×10^{-9}	2×10^{-10}

Mass spectrometer ion currents in units of 3×10^{-16} A.
(from Hickmott 1960b).

Steinrisser (1967) has obtained pressures of about 1×10^{-11} torr with a glass system pumped by an oil diffusion pump and a zeolite trap, with a valve between the system and the trap. A detailed sequence of procedures is described in the paper which is designed to prevent accumulation of CO_2 in the liquid nitrogen cooled zeolite trap. The lowest partial pressures observed were He, 5×10^{-12}; H_2, 4×10^{-12}; CO, 6×10^{-12} torr. It was also noted that the conversion of O_2 to CO in the presence of a hot filament (see Section 7.2.2) was greatly reduced by the processing technique described in the paper.

A typical pressure-vs-time curve during processing for a small borosilicate glass system pumped by ion pumps is shown in Fig. 10.3. Very similar behaviour is observed when the baking temperature is lowered to 575°K. The system is pumped during bakeout by a sputter-ion pump (nominally 8 litres sec^{-1}) outside the baked portion of the system. At the end of the bake the pressure is typically about 10^{-6} torr. The gauges, etc.

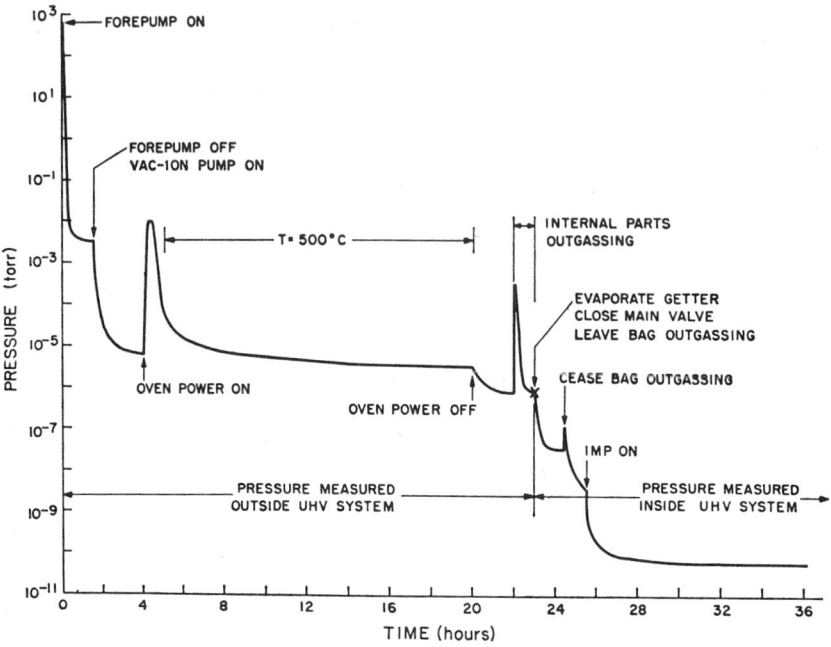

Fig. 10.3 Typical curve of pressure vs time during processing of a new glass system. (Redhead *et al.* 1962a).

are then degassed and the bakeable valve closed, at this point titanium is evaporated in the baked getter-ion pump (magnetron or inverted magnetron pump, see Section 11.5.2) and the getter-ion pump turned on. Pressures in the range $3–7 \times 10^{-11}$ torr are achieved after about 24 hours.

The total outgassing rate in a carefully processed glass system can be made very low. Crawley (1965) has estimated the outgassing rate for a borosilicate glass system to be about 5×10^{-15} torr litre sec^{-1} cm^{-2}; this figure includes the outgassing from all surfaces, glass and metal, in the system. After extensive processing of a system consisting of aluminosilicate glass and stainless steel, the following outgassing rates have been measured in the authors' laboratory (all in torr litre cm^{-2} sec^{-1}): H_2, 4×10^{-15}; He, $<2 \times 10^{-18}$; CH_4, 5×10^{-16}; $CO < 10^{-15}$; Ar, 6×10^{-18}. These figures represent actual desorption rates rather than net desorption rates as are often reported (see Hobson & Earnshaw (1967) for a discussion of the difference between these two rates).

The gases evolved when various metals are baked to 725°K have been examined with a mass-spectrometer by Flecken and Nöller (1961). Samples of Cu, Al, stainless steel and iron were tested and the total amount of gas evolved during bake at 725°K lay between 3×10^{-3} and 1×10^{-2} torr litre cm^{-2} sec^{-1} for all these metals in the untreated state. These quantities of gas correspond to a few hundred atomic layers. Analysis of the gases released from the iron sample showed that hydrocarbons greatly exceed the other gases (CO_2, H_2O, CO, H_2) when the sample was untreated. For a degreased sample the gases evolved were in the following order of abundance; H_2O, CO_2, CO, hydrocarbons, H_2.

Baking of metal uhv systems not only removes the adsorbed water, as with glass systems, but also assists in the removal of hydrogen that is the predominant gas in solution in most metals. Crawley (1965) has reported outgassing rates observed after baking stainless steel (18/8) systems at 675°K. These data are shown in Table 10.4. Once again it must be pointed out that these estimates of outgassing rates may be low because of neglect of adsorption-desorption effects. Even if the absolute values of outgassing rates are too low it is still possible to see the effects of varying the pre-treatment or the baking time. It is clear that prolonged baking (up to 450 hours in this case) significantly reduced the net outgassing rate and that vacuum firing (during brazing) at 1300°K has a very dramatic effect in reducing the net outgassing rate.

TABLE 10.4
Degassing rates from stainless steel vacuum envelope given various pre-treatments

Pre-treatment	Baking time at 700°K (hours)	Outgassing rate (torr litre sec^{-1} cm^{-2})
Welded, mechanically polished and degreased	4	10^{-11}
	16	5×10^{-12}
	450	5×10^{-14}
Machined components vacuum brazed at 1300°K	1	2.5×10^{-12}
	4	1.5×10^{-13}

Envelope material mainly 18/8 stainless steel.
(From Crawley 1965).

Davis (1962) has reported measurements on a system of a few litres volume constructed almost entirely of stainless steel and pumped by a sputter-ion pump of 5 litres sec^{-1} nominal speed. The bakeout schedule was 700°K for at least 24 hours. After degassing the gauge and mass-spectrometer the lowest partial pressures observed at room temperature were: H_2, 6×10^{-13}; N_2+CO, 1×10^{-13}; Ar, 1×10^{-13}; CH_4, 1×10^{-13}; He, 2×10^{-14} torr, total pressure 1×10^{-12} torr (see also Table 12.1). From leak-up measurements reported in Davis' paper, a rough estimate of the outgassing rate of the system for helium is 1×10^{-18} torr litre sec^{-1} cm^{-2}. The estimated outgassing rate for hydrogen is about 3×10^{-19} torr litre sec^{-1} cm^{-2}; this latter figure is almost certainly too low because of readsorption of the hydrogen which is evidenced by the non-linear increase of hydrogen pressure with time.

Although the absolute values of outgassing rates of chemically active gases may be in doubt, the investigation described above indicates that properly baked and degassed metal systems can achieve ultimate pressures and outgassing rates that are similar to those obtained with glass systems.

Blahnik and Shoulders (1966) have described a double-wall vacuum system in which the inner uhv chamber can be heated to 1175°K. The thin-walled inner bell-jar is sealed to its base-plate by a boron oxide seal which melts at 875°K. The inner chamber is raised and both chambers pumped to 10^{-5} torr. The oven is turned on and the uhv chamber heated to 1175°K for 15 minutes. The inner chamber is then lowered into the channel containing molten boron oxide and allowed to cool; making the seal as the boron oxide freezes. An ion pump is then turned on in the uhv chamber and pressure in the 10^{-10} torr range rapidly achieved. The fastest cycle time achieved with minimum thermal mass in the uhv chamber was 1 hour.

10.3 DEGASSING PROCEDURES AFTER BAKING

After completion of system bakeout it is necessary to degas all portions of the system which normally operate at elevated temperatures. This includes all hot filaments (in gauges, mass-spectrometers and other instruments), getters and all electrodes which are heated by conduction or radiation from hot filaments or by electron or ion bombardment. It is also necessary to rigorously degas all electrodes which are struck by electrons, even if the resulting temperature rise is insignificant; this is to minimize gas evolution by electron impact desorption (see Section 4.2).

Heating for degassing purposes can be done by resistive heating, rf heating, radiant heating or electron bombardment. Of these four methods the first and last methods are the most convenient and widely used. Degassing is accomplished by heating the electrode or component above its normal operating temperature until the outgassing rate, under normal operating conditions, is sufficiently low. It is advisable to degas all surfaces simultaneously so that gas does not transfer from one surface to another.

The degassing of electron-bombarded electrodes in gauges and mass-spectrometers is particularly important. Electron bombardment heating is particularly suitable for this purpose since the bombarding electrons can desorb gases by electron impact (such as oxygen) which cannot be easily removed by heating alone. As an example of the rigorous degassing of gauge electrodes necessary for consistent uhv measurements, in the authors' laboratory it is standard practice to degas the grid of Bayard-Alpert gauges by electron bombardment (1 kV, 15 watt cm^{-2} of grid area) for about 30 minutes.

In some experiments it is necessary to remove the gas from the surface of materials which would melt before properly degassing. Low melting point materials can frequently be satisfactorily degassed by ion bombardment; this is discussed in Section 4.3.2.

CHAPTER ELEVEN

PUMPS FOR ULTRAHIGH VACUUM

The design and performance of vacuum pumps has been treated in detail by many authors, e.g. Power (1966); Dushman and Lafferty (1962), and pumps for uhv use have been discussed in several books, e.g. Power (1966), Roberts and Vanderslice (1963), Trendelenburg (1963). This chapter is confined to a brief review of the literature on uhv pumps and a discussion of those aspects of uhv pump design and performance which have not been treated in detail in other books. The chapter concludes with a comparison of the properties of the various types of uhv pumps.

11.1 MOLECULAR DRAG AND TURBO-MOLECULAR PUMPS

Designs of both molecular drag and turbo-molecular pumps have been developed which are capable of reaching pressures in the uhv range. The principle of molecular drag and turbo-molecular pumps is that a molecule can be given momentum in a desired direction by repeated collisions with a rapidly moving solid surface. The rapidly moving surface is a flat disc, in the case of molecular drag pumps, and is a series of rotating blades, in the case of a turbo-molecular pump. The design and performance of these pumps have been well reviewed by Power (1966).

Several designs of molecular drag pumps with magnetically suspended rotors have been developed by Beams and his colleagues (Beams 1959, 1960; Williams & Beams 1961). The rotor in this type of pump is magnetically suspended without bearings and is rotated at high speed by a rotating magnetic field applied from outside the vacuum system. With a backing pressure of 6×10^{-8} torr the lowest pressure observed was 4×10^{-10} torr. The delicate nature of these pumps and their sensitivity to vibration and shock has restricted their use.

At least two types of axial-flow turbo-molecular pumps are commercially available which are capable of achieving pressures in the uhv range. The first type has been described in several papers by Becker (1960, 1966). Models with pumping speeds from 140 to 4,000 litres/sec are available and ultimate pressures of below 10^{-10} torr have been obtained. The second type of turbo-pump is of different mechanical construction and is quite large

(70 cm diam intake, 8,000 litres/sec speed); ultimate pressures below 5×10^{-11} torr are claimed (Rubet 1966).

The theory of the turbo-molecular pump has been developed in several papers (e.g. Becker 1962; Maurice & Sagot 1964; Becker 1966) and reasonable agreement with measurements obtained. The pumping speed for different gases varies slowly with atomic weight while the maximum compression ratio increases very rapidly with atomic weight. This behaviour is indicated by the data of Table 11.1 measured on a Pfeiffer model TVP 500 turbo-pump.

TABLE 11.1
Pumping speed and maximum compression ratio for various gases in a turbomolecular pump

Gas	Speed (litres/sec)	Maximum compression ratio
H_2	160	750
He	170	10^4
Ne	145	10^8
N_2	140	10^9
Ar	130	6×10^{10}
Kr	120	10^{14}
Xe	110	$\sim 10^{17}$

(from Becker 1966).

The major advantage of turbo-molecular pumps for the production of uhv is the freedom from pump fluids and the ability to produce very low pressures without traps or baffles. In spite of the absence of pump fluids the turbo-pump may cause system contamination with hydrocarbons originating in the oil used to lubricate the bearings. Care must be taken to cool the bearings and to use oils of very low vapour pressure. The ultimate pressure produced by a turbo-pump is particularly dependent on outgassing from the very large surface area of the rotor and stator blades. In most cases outgassing temperatures can only be raised to 150°C. Extreme care must be taken to throughly outgas the pump if low pressures are desired. The major disadvantage is the low compression ratio for hydrogen which requires that the partial pressure of hydrogen be kept low in the backing line; for example, if it is desired to keep the partial pressure of hydrogen below 10^{-11} torr in the system it is necessary to reduce the hydrogen pressure in the backing line to 10^{-8} torr (see compression ratios listed in Table 11.1). Because of the necessary mechanical precision of the pump the cost per unit pumping speed is high.

11.2 DIFFUSION PUMPS

Both oil and mercury pumps can be used satisfactorily for the production of uhv provided care is taken to ensure that the pumps are properly trapped.

Diffusion pumps make use of several jets of oil or mercury vapour. The pumping action of each jet results from the diffusion of gas molecules through the low-density periphery of the jet into the dense core of the freely expanding jet. By collision with molecules of the vapour jet, most gas molecules acquire momentum directed towards the high-pressure end of the pump. Descriptions of pump designs and performance have been given by several authors (e.g. Power 1966; Dushman & Lafferty 1962) and the theory of diffusion pumps is well developed (see Nöller 1966 for review).

We first examine mercury diffusion pumps which have three main advantages for uhv use, (a) the pump fluid is easily trapped at liquid nitrogen temperatures and if the trap accidentally warms up, the mercury can be removed by baking, (b) a mercury pump will operate against high backing pressures permitting the backing pump to be turned off for long periods, and (c) the pump fluid is thermally stable. The attendant disadvantages of mercury pumps are (a) the need for a constant liquid nitrogen supply and (b) the toxicity of the pump fluid.

Venema (1959) showed that it was possible to achieve pressures below 10^{-12} torr with a mercury diffusion pump and properly designed cold-traps. This system used three glass traps, both the traps and the upper portion of the mercury pump were baked. Several authors (see Power 1966 for references) have described glass and metal systems basically similar to Venema's design. Multiple cold-traps are necessary because of the high density of mercury vapour at the first trap. The first trap or baffle can conveniently be cooled to some temperature above the freezing point of mercury by solid CO_2, or thermo-electrically, to condense most of the back streaming mercury. This mercury can be returned to the pump either by periodically warming this trap or by designing the trap so that the lowest portion is warmer than the rest, allowing mercury to drip back into the pump.

Mercury pumps are frequently chosen because of the absence of any organic fluids which might cause contamination of the system with hydrocarbons. However, this advantage is lost if oil contamination from the backing pump is not prevented by suitable trapping (see Section 11.3). Replacing the oil in the backing pump with a low vapour pressure diffusion pump oil has also been suggested (Haas and Jackson 1966). Alternatively, the oil-filled mechanical pump can be replaced by a sorption or aspirator pump.

Oil diffusion pumps have been widely used for the production of uhv with properly designed traps to prevent contamination of the system with the back-streaming oil vapour. Simple one-hit traps are adequate because of the low vapour pressure of pump oils. It is essential, for lengthy operation, that the cold-trap be provided with a barrier at low temperatures to prevent the surface migration of oil molecules along the uncooled walls of the trap. The same problem of migration of oil from backing pump to diffusion pump applies to oil pumps as was discussed above for mercury pumps.

The problems of back-streaming of pump oil into the system can be minimized by choosing a pump design with minimum back-streaming (see Power 1966) and by using very low vapour pressure oils such as the polyphenyl ethers (Hickman 1961).

Eruptive boiling of the fluid in oil diffusion pumps can cause pressure fluctuations in the vacuum system and increased oil contamination (Smith 1959). Eruptive boiling can be prevented by using (a) a flash boiler (Stevenson 1959), (b) a stirring device (which may be driven by the vapour stream, Bachler 1962), or (c) an external ring heater placed higher than the fluid surface (Okamoto and Murakami 1967). Of these three solutions the third one appears the simplest and most effective.

The decision to use a mercury diffusion pump rather than an oil diffusion pump is frequently based on the assumption that hydrocarbon contamination will be absent with the mercury pump. As shown above, this is not always the case unless care is taken to prevent contamination of the mercury by oil from the mechanical backing pump. However, if the diffusion pump is adequately trapped and, in the case of oil pumps, the proper operating procedure is followed (see Power 1966) then it is possible to obtain very similar residual gas conditions with either mercury or oil pumps. Table 11.2 shows

TABLE 11.2
Residual gases in a system pumped with different pump fluids

Pump fluid	Calculated effective speed for N_2 (litres/sec)	Ultimate pressure (torr)	% composition					
			H_2	CO	CO_2	H_2O	O_2	CH_4
Octoil–S	0·4	$1·6 \times 10^{-10}$	58	17	22	3	0·4	0·7
DC 704	1·1	$5·4 \times 10^{-11}$	15	27	45	9	2	1·0
Mercury	1·0	$3·7 \times 10^{-11}$	57	23	17	2	0·6	0·6

C_2 hydrocarbons (estimated from 27 peak) and Hg were not detectable.
(from Singleton 1966).

the results of Singleton (1966b) on the observed partial pressures obtained in the same vacuum system pumped either by an oil diffusion pump (with two different types of oil) or by a mercury pump. It can be seen that, with proper trapping, the hydrocarbon content of the system is much the same in all three cases. The mercury pump was trapped with one re-entrant trap cooled with solid CO_2 directly above the top jet and a single re-entrant, glass trap cooled to liquid nitrogen temperatures. The oil pump was trapped with the single, liquid nitrogen cooled, glass trap.

Failure of the trap in a mercury pumped system is not catastrophic; a rebake of the system will completely remove mercury (except when amalgamation occurs, in which case a rigorous heat treatment is required). Failure of the trap in an oil-pump system causes contamination of the system with oil which is very difficult or impossible to remove completely. Baking the system only results in oil decomposition leaving solid residues.

11.3 CRYOPUMPS

Cooled surfaces have been used for many years in vacuum systems (Dushman 1949) but the combination of cryogenic and vacuum techniques has advanced greatly in the past decade (Holland 1965b).

In Chapter 2 it was shown that, over a wide range of practical conditions, the condensation coefficient for a gas striking a surface was greater than 0·1 and frequently close to unity (Section 2.4.3). This result is the basis of all cryopumps and also of traps, which may be regarded as special forms of these pumps. The physical basis of traps is the same as that of cryopumps and the distinction between the two is mainly one of function. In the following they will be treated together. The net rate of pumping of molecules by 1 cm² of flat surface is given by:

$$\frac{d\sigma}{dt} = cvp - \frac{\sigma}{\tau} \text{ molecules cm}^{-2} \text{ sec}^{-1} \tag{11.1}$$

where σ is the number of molecules adsorbed per cm², ν is the specific arrival rate, c the condensation coefficient, and τ the mean time of sojourn of molecules on the surface. The central problem of cryopump design is to maintain the difference between the condensation rate (first term on right) and the desorption rate (second term on right) sufficiently large to satisfy the pumping requirements of the problem, within the constraints imposed by other special features of the problem. Since c, as noted above, is not a quantity that varies greatly the condensation term is not subject to wide changes by design, and it is the desorption term that is primarily affected by design. The desorption term can be expressed as the equilibrium pressure p_{eq} which would exist over the same condensate in an isothermal system at the temperature T_s of the surface, in the absence of any net flow of gas to the surface. Assuming c is not thereby greatly changed, we can write

$$cv\sqrt{\left(\frac{T}{T_s}\right)}p_{eq} = \frac{\sigma}{\tau} \text{ molecules cm}^{-2} \text{ sec}^{-1} \tag{11.2}$$

and

$$\frac{d\sigma}{dt} = cv\left[p - \sqrt{\left(\frac{T}{T_s}\right)}p_{eq}\right] \text{ molecules cm}^{-2} \text{ sec}^{-1} \tag{11.3}$$

where T is the effective temperature of the gas striking the cryopump. Eqn. 11.3 can be expressed as the ratio of the actual pumping speed S to the maximum possible pumping speed S_{max} (at temperature T)

$$S/S_{max} = c\left[1 - \sqrt{\left(\frac{T}{T_s}\right)\frac{p_{eq}}{p}}\right]. \tag{11.4}$$

Cryopumps divide themselves into two classes depending on the appropriate value of p_{eq}: (i) if $p_{eq} < p_0$ (the vapour pressure of the bulk condensed gas at the surface temperature) then the quantity of gas pumped is insufficient to form a bulk layer of condensed gas. Such pumps are termed cryosorption pumps and often, but not always, use high area adsorbents such as molecular sieves, silica gel, activated alumina, or activated charcoal, and relatively low gas loads (Kindall & Wang 1962; Stern et al. 1965; Bauer & Jeffers 1965). For a discussion of the properties of molecular sieves see Hersh (1961) and Espe and Hybl (1965). In these pumps the adsorbent is normally covered with less than a monolayer of gas. The low pressure limit of these pumps (p_{eq}) is the pressure determined by the physical adsorption isotherm at the surface temperature. Cryosorption pumps are thus characterized by a saturation quantity of pumped gas, which increases as the surface area increases (Table 2.19). (ii) If $p_{eq} = p_0$ the quantity of gas pumped is more than sufficient to form a bulk layer of condensed gas. Such pumps are termed cryogenic pumps, and are usually characterized by non-porous pumping surfaces (porous surfaces having no special advantage when covered by bulk condensate) and large gas loads (Borovik et al. 1962; Dawson & Haygood 1965; Roussel et al. 1965). In these pumps the adsorbent is covered with a bulk layer of condensate. The low pressure limit is the vapour pressure (p_0) of the condensate at the surface temperature. Cryogenic pumps are characterized in principle by the ability to pump an indefinitely large quantity of gas. In practice, effects arising from the thickness of the condensed layer limit the quantity of gas that can be pumped. The process of pumping is always accompanied by the transfer to the surface of the heat of adsorption and the heat must be conducted away if the surface temperature is to be maintained.

Almost all cryopumps have surface temperatures below room temperature, and usually this reduces outgassing and makes these pumps particularly suitable in uhv systems. Cryopumps also have the property of being storage pumps, whose total pumped load can be released for examination by raising the surface temperature. On the other hand, the need for cooling pump parts to low temperatures (often liquid helium) in the interior regions of large vacuum systems frequently poses complex technical problems. However, Santeler (1959) estimates that the installation cost of a large cryogenic pumping system ($\sim 10^8$ litre sec^{-1}) for a space simulator is some two orders of magnitude less than for a comparable diffusion pump system. Borovik et al. (1962) and Balwanz et al. (1960) agree that operating costs of large cryogenic pumps ($\gtrsim 10^4$ litres sec^{-1}) at low pressures ($p \lesssim 10^4$ torr) are less than those of equivalent diffusion pumps.

In the discussion below we draw extensively from the results of Chapter 2, and disregard, in the main, detailed considerations of the non-uniform temperature environment normally present in cryogenic pumps (Section 7.1). Power (1966) has described many practical details of cryopumps. Cryotrapping (Section 2.4.5) is not discussed below because it has not yet been developed into a practical pumping method.

11.3.1 Cryosorption pumps

Three classes of cryosorption pump which have been used in uhv systems and which cover the whole sub-atmospheric range will be described.

Loose-adsorbent pumps. 100 gm of molecular sieve 13X(Linde) have a surface area of $5 \cdot 14 \times 10^8$ cm^2 (Table 2.19). If this sieve is cooled to 77°K, then at $p = 10^{-3}$ torr the relative coverage of N_2 is $\theta = 10^{-1}$ (Fig. 2.11), assuming that the isotherm expressed as $\theta(p, T)$ is essentially the same for any material as suggested by Hobson (1965b). A coverage of $\theta = 10^{-1}$ corresponds to $3 \cdot 2 \times 10^{22}$ N_2 molecules adsorbed, which is approximately the number of molecules in a volume of 1 litre at S.T.P. Thus modest quantities of porous materials can act as forepumps to reduce the pressure from atmosphere to a range where other pumps can be operated. Manes and Grant (1963) have given design calculations for cryosorption pumping systems. Jepsen et al. (1959) have described an arrangement in which the system was sealed from the atmosphere and a finger containing activated charcoal was immersed in liquid nitrogen. The pressure fell to 10^{-2} torr in a few minutes, at which time a valve to the system was closed and the pump-down continued to the uhv range with a Vac-Ion pump. Similar systems using molecular sieves have been described by Bannock (1962) who used sieves type 4A (sodium alumino-silicate), type 5A (calcium alumino-silicate) and type 13X (sodium alumino-silicate), and by Knor (1963) who used activated charcoal and sieve type 13X. Sieve pumps of this type have been used in cascade (Windsor 1963); the first pump is cooled and adsorbs the majority of the gas present before it is sealed off from a second sieve pump which is then cooled. Windsor (1963) achieved a pressure of 10^{-9} torr in a small glass system in this way after flushing his sieve with dry nitrogen gas. Sorption pumps cooled to liquid N_2 temperature do not adsorb H_2 or He (Stern et al. 1965) or Ne (Hunt et al. 1963b) and thus other pumps suitable for these gases must supplement the sorption pump. Read (1963) has achieved a pressure of 5×10^{-11} torr in a small metal system, using only a small diffusion pump which pumped the system through a 13X zeolite sorption pump.

Sorption pumps operating from atmospheric pressure require safety valves because dangerously high pressures are generated when they are warmed. Pumps and traps with high area sorbents generally require baking at 300–400°C under vacuum or in the presence of a dry gas to remove the adsorbed water (Windsor 1963; Read 1963).

Simple loose-sieve systems may also be used at room temperature as a substitute for conventional liquid nitrogen traps above oil diffusion pumps. Biondi (1959) has operated a zeolite trap at room temperature between a system at $p = 2 \times 10^{-10}$ torr and an oil diffusion pump for 60 days without reconditioning the trap by baking. Goerz (1960) has described a similar trap with built-in heater. These loose-sieve systems have the disadvantage that the sieve material can be accidentally transmitted through the vacuum system causing possible damage to valve seats. If warmed too quickly the sieve particles tend to disintegrate explosively. The earlier copper foil traps described by Alpert (1953b) and Carmichael and Lange (1958) did not have this disadvantage and could also trap oil diffusion pumps satisfactorily for many days. Haller (1964) and Milleron (1965) have combined mechanical rigidity with high sorption area by the use of self supporting discs of porous glass and porous stainless steel respectively in the pumping line. Performance was comparable to Biondi's (1959) trap.

Two physical processes in loose-adsorbent pumps make them inefficient at low pressures: (i) Below $p = 10^{-4}$ torr thermal conduction through the gas phase tends to zero and it becomes difficult to maintain the regions of the adsorbent remote from the bath at the bath temperature. A discussion of the various heat sources and sinks for dispersed adsorbents has been given by Hobson (1967). (ii) At low pressures and coverages sojourn times become longer (Table 2.28) and the rate at which gas penetrates the adsorbent through the gas phase and by surface diffusion may become long. Some combination of these effects is probably responsible for the general nature

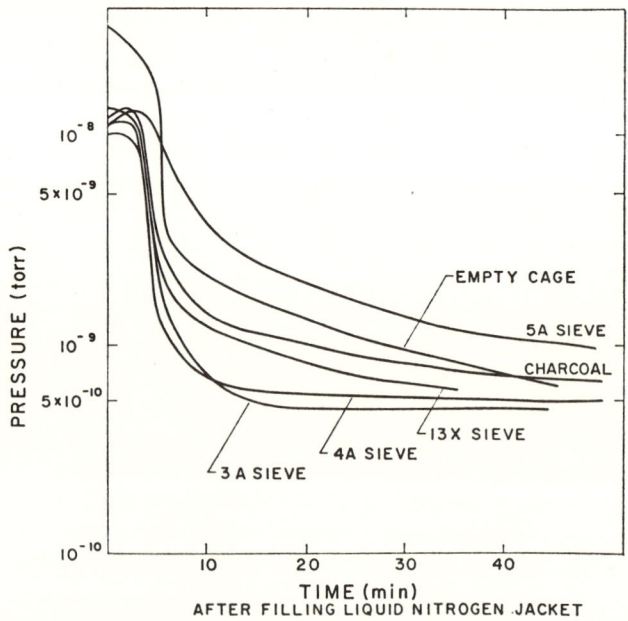

Fig. 11.1 Pressure vs time after cooling (to 77°K) finger containing loose adsorbent held mechanically against wall. Small system had been pumped to ~10^{-8} torr by diffusion pump which was valved off at zero time. (Inkley & Coleman 1965).

of the results of Inkley and Coleman (1965) shown in Fig. 11.1. A small system of both metal and glass was pumped down to pressure of about 10^{-8} torr by a diffusion pump which was then valved-off and a cold finger containing various sieve materials was immersed in liquid nitrogen at zero time as shown in Fig. 11.1. Note that the pressure drops relatively slowly and the empty finger gave a result commensurate with that of the sieves.

Both these problems of loose-adsorbents can be reduced by appropriate mechanical design (Grant & Davey 1963) or more effectively by bonding the adsorbent to the pump wall. Some reduction in total pump capacity will, however, result.

Bonded-adsorbent pumps. If molecular sieve 5A (Linde) is bonded to 1000 cm^2 of surface to produce a roughness factor of $1 \cdot 3 \times 10^5$ (Table 2.19) then the total available adsorption area is $1 \cdot 3 \times 10^8$ cm^2. Since the adsorbent is bonded, the surface will be useful as a pump at very low pressures because cooling does not depend upon thermal conduction in the gas phase. From Fig. 2.10b, for an equilibrium H_2 pressure $p_{eq} = 10^{-7}$ torr at $20 \cdot 4°K$, 150 cm^3 (NTP) gm^{-1} will be adsorbed on coconut charcoal. This corresponds to 4×10^{21} molecules gm^{-1}. Since coconut charcoal has an area of 889 $metre^2$ gm^{-1} (Table 2.19), this corresponds to $4 \cdot 5 \times 10^{14}$ molecules cm^{-2} which approaches the monolayer coverage of $6 \cdot 95 \times 10^{14}$ molecules cm^{-2} (Table 2.17), giving $\theta = 0 \cdot 65$. Once again we assume (Hobson 1965b) that the isotherm expressed as $\theta(p, T)$ will be the same for molecular sieve 5A as for coconut charcoal and hence that at $p = 10^{-7}$ torr, $5 \cdot 9 \times 10^{22}$ molecules of H_2 will be adsorbed. A typical H_2 outgassing rate for a baked uhv system might be 10^{-13} torr litre cm^{-2} sec^{-1} or $3 \cdot 5 \times 10^6$ molecules cm^{-2} sec^{-1}. Even for a system with internal surface area of 10^7 cm^2 (a sphere of 60 ft diam) the bonded-adsorbent pump will hold the pressure below 10^{-7} torr for $1 \cdot 7 \times 10^9$ sec (54 years).

Table 11.3 lists various techniques that have been used for bonding sieves to surfaces. Bauer and Jeffers (1965) bonded 40 gm of activated charcoal on to 140 cm^2 of geometric area and estimated their total adsorption area as 4×10^8 cm^2 which has not checked experimentally. From observations of the rate of pumping of gas pulses they estimated that the time for redistribution of the pumped gas over the whole adsorbing area at $p \sim 10^{-5}$ torr was less than 1 minute, which is much less than the times shown in Fig 11.1 for loose adsorbents. Figure 11.2 shows the behaviour of H_2 pressure observed by Stern et al. (1966) when a hydrogen leak was turned on and off in a system containing a cryosorption pump of molecular sieve 5A bonded on an aluminum plate. Equilibration times are shorter than those of Fig. 11.1. Stern et al. (1966) report $0 \cdot 16$ for the condensation coefficient of H_2 on their panel cooled with liquid hydrogen, which is surprisingly low compared to the condensation coefficient for He on a similar panel at $4 \cdot 2°K$ (Table 2.23). Stern et al. (1966) argue that their low value was probably caused by their

TABLE 11.3
Bonded cryosorption surfaces

Substrate	Bonded material	Method of bonding	Roughness factor	Temperature (°K)	Test gas	Reference
Aluminum	Molecular Sieve 5A	Slurry	$1\cdot3\times10^5$	20	H_2	Stern et al. (1966)
Aluminum	Molecular sieve 5A	Slurry	$1\cdot3\times10^5$	4·2	He	Grenier & Stern (1966)
Copper	Activated charcoal	Water base emulsion Polymer Cl 303	3×10^6	77	Air, N_2 H_2	Bauer & Jeffers (1965)
Stainless steel	Molecular sieve 13X	(a) Viton coating (b) Silica sol (c) Plain plasma spray (d) Modified plasma spray	Not given	77	N_2	Sands & Dick (1966)
Copper	Copper felt	Mechanical	~32	<6	H_2	Milleron (1965)
Copper	Porous copper	Mechanical	~16	<6	H_2	Milleron (1965)
Copper	Copper	Electroplating at high current density	Not given	<6	H_2	Milleron (1965)
Copper	Copper	Sputtering in Argon	~20	<6	H_2	Milleron (1965)
Brass	Copper	Distillation of zinc from brass	>11	<6	H_2	Milleron (1965)
Stainless Steel	Zeolite 5A	Epoxy resin	Not given	77	Air	Narcisi et al. (1962)

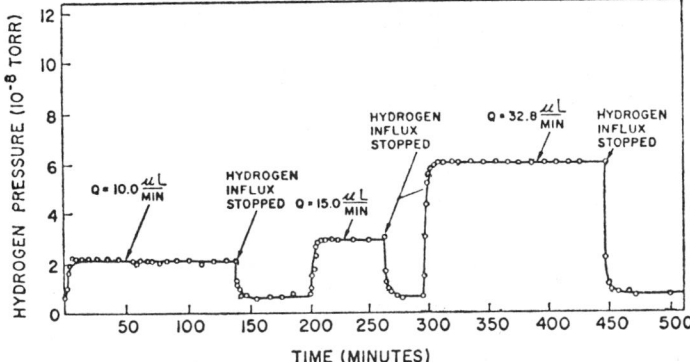

Fig. 11.2 Time dependence of pressure during opening and closing of a H₂ leak in a system containing cryosorption panel of molecular sieve 13X bonded to aluminum. (Stern et al. 1966).

panel temperature being 24°K rather than 20°K, in turn caused by the temperature of a chevron shield protecting the cryosorption surface being 170°K rather than 80°K as anticipated. Sands and Dick (1966) tested several bonding methods at pressures of 10^{-2} torr and found substantial differences in pumping performance for the different methods (Table 11.3). The modified plasma spray was found to give the most satisfactory bonding.

Cryosorption pumps with nearly flat surfaces. If a surface of area 100 cm² at $T = 4\cdot2°K$, restricted by a conductance, presents a pumping speed of 5 litres sec⁻¹ to the small Pyrex system discussed in Section 2.5.4(b), how long will it be before the helium pressure caused by helium permeation through the walls causes the pressure to rise above 10^{-11} torr.

The system has walls of total area 500 cm² and thickness 1·5 mm and an external helium pressure of 4×10^{-3} torr. Assuming the walls to be at room temperature the permeation constant from Fig. 2.36a is about 6×10^{-11} cm² sec⁻¹, which yields a total permeation leak into the system of $2\cdot8 \times 10^{7}$ molecules sec⁻¹ or 8×10^{-13} torr litres sec⁻¹. A speed of 5 litres sec⁻¹ will hold the pressure initially at $1\cdot6 \times 10^{-13}$ torr. Thus, when the pressure rises to 10^{-11} torr, the leak will be overwhelmed by the re-emission from the surface and the pressure will be determined by the helium adsorption isotherm, and we require the coverage of this isotherm at 4·2°K. Figure 2.13 shows this to be $\theta = 7 \times 10^{-2}$. 100 cm² of surface can accommodate $7\cdot22 \times 10^{16}$ atoms of He in the monolayer (Table 2.17). Thus coverage at $\theta = 7 \times 10^{-2}$ is 5×10^{15} atoms. At an input leak rate of $2\cdot8 \times 10^{7}$ molecules sec⁻¹ the pump will hold the pressure below 10^{-11} torr for $1\cdot8 \times 10^{8}$ sec or 57 years. Thus flat surface cryosorption pumps have special uses in uhv systems.

Hobson and Redhead (1958) and Hobson (1964) have used a simple finger immersible in liquid helium as a reversible cryosorption pump for the residual gases in a small uhv system. Pressures in the range 10^{-12} to 10^{-14} torr were obtained. Caswell (1959) has used a liquid helium trap for fast pumping during thin film evaporation. Figure 11.3 shows the behaviour of the pressure when the walls surrounding a small space chamber were cooled with liquid helium. The absolute values of the pressures in Fig. 11.3 are subject to some uncertainty because of non-linearities in cold-cathode magnetic gauges at very low pressures (Chapter 8.2) but there is no doubt that a very large pressure drop occurred upon cooling the walls. We have classed this system as pumped by cryosorption rather than cryogenically (Section 11.3.2) because gas loads were small. Feakes and Torney (1963) have calibrated gauges in a chamber whose walls were cooled to approximately 10°K. The background

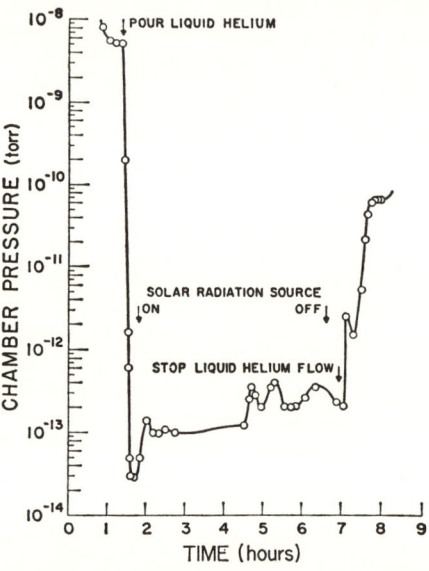

Fig. 11.3 Pressure response of small space chamber when walls of 'space volume' were cooled with liquid helium (Mark & Sommers 1962).

pressure in this chamber was estimated to be below 10^{-14} torr. The temperature was chosen so that all gases except helium would be pumped by cryosorption on the chamber walls. This permitted quantitative gauge calibration in helium to be carried out at pressures as low as 10^{-13} torr (see Fig. 8.19). Hunt et al. (1963a) have deposited frozen films of various gases (10–100 monolayers thick) upon a copper substrate of area 5×10^3 cm^2 cooled to 11°K and have measured the pumping speed of the films for H_2. At H_2 coverages well below a monolayer they obtained the following H_2 speeds on the indicated substrates (all expressed in litre sec^{-1}): Ar, $1 \cdot 4 \times 10^4$; N_2, 2×10^4; O_2, 2×10^4; H_2O, $1 \cdot 3 \times 10^5$; N_2O, 8×10^4; CO_2, 6×10^4. These speeds, which are net pumping speeds, correspond to condensation coefficients between 0·1 and 1 and declined as a coverage of 10^{15} molecules cm^{-2} of geometric area was approached.

Busol and Yuferov (1966) have carried out similar measurements for the pumping of H_2 on solid layers of CO_2 ($2 \times 10^3 - 4 \times 10^4$ monolayers thick) in the temperature range $14-20\cdot4°K$ at pressures of 10^{-7} to 10^{-3} torr. Their pumping speeds were comparable to those of Hunt et al. (1963a) but they reported that the quantities of hydrogen pumped were proportional to the quantity of CO_2 deposited and was as high as 1000 monolayers of H_2 on a CO_2 layer 4×10^4 monolayers thick, although the time required for adsorption equilibrium increased with the thickness of the deposit. These results suggest either a porous CO_2 deposit or solution of H_2 in the solid CO_2. Qualitatively similar results were obtained for deposits of alcohol, gasoline, acetone and water.

Figure 2.27 demonstrates that a reduction in surface temperature can increase the sticking probability and capacity of a getter for an active gas. This result is the basis for a series of studies in which evaporated metal films were cooled to liquid N_2 temperature and the pumping properties of the resulting 'cryogetter' pumps examined. Examples of these studies are: Mo film, H_2 (Hunt et al. 1963b); Ti film, H_2 (McCracken 1965); Ti film, N_2, H_2, D_2 (Elsworth et al. 1965); Ti film, mainly H_2 (Rivière et al. 1965). In most cases cooling raised the initial sticking probability and the capacity of the films for the gas under study, the result being controlled in part by the evaporation conditions for the film. A variety of pumping effects took place for other gases in the system. Rivière et al. (1965) found that the partial pressures of all gases were reduced by the Ti film at liquid N_2 temperature, but the major effect of cooling was for the gases H_2, Ar, CO_2, C_2H_4 and H_2O. Rivière et al. (1965) measured a background pressure of $1\cdot6 \times 10^{-12}$ torr in this system with a triggered magnetron gauge (see Chapter 8).

For forepumping the lack of a large adsorption area may be compensated by a lower temperature and Ames et al. (1958) have used a liquid helium trap, similar to that of Caswell (1959), as a forepump to reduce atmospheric pressure to 10^{-3} torr, whereupon an ion pump subsequently pumped a small glass system to 10^{-10} torr.

11.3.2 Cryogenic pumps

All cryogenic pumps have factors in common, but in practice two main groupings are found. These are (a) cryogenic pumps which have been designed as a unit for attachment to, or incorporation into, a vacuum system and which we call below localized cryogenic pumps, (b) cryogenic pumps which have been designed as an inseparable part of the vacuum system. Such pumps have had wide application in space simulation chambers.

Factors common to all cryogenic pumps. Cryogenic pumps differ from cryosorption pumps in that p_{eq} in eqn. 11.3 is the vapour pressure of the condensate at the pump temperature. Since the measured pumping speed will follow the difference $p - \sqrt{(T/T_s)}p_{eq}$ in eqn. 11.3 the speed will rise to a constant value, determined mainly by c (see Table 2.25) at $p \gg p_{eq}$ and will fall to zero at $p = \sqrt{(T/T_s)}p_{eq}$. At liquid helium temperature the only gaseous vapour pressure above the uhv range is that of helium itself (see Table 2.16). Helium is not usually a large gas load and may be removed by other methods such as diffusion pumping or cryosorption pumping. While cryogenic pumps at 4·2°K have had application and will be discussed below, economic considerations have kept the temperature of many cryogenic pumps at 20°K thus necessitating auxiliary pumps for He, H_2, and Ne.

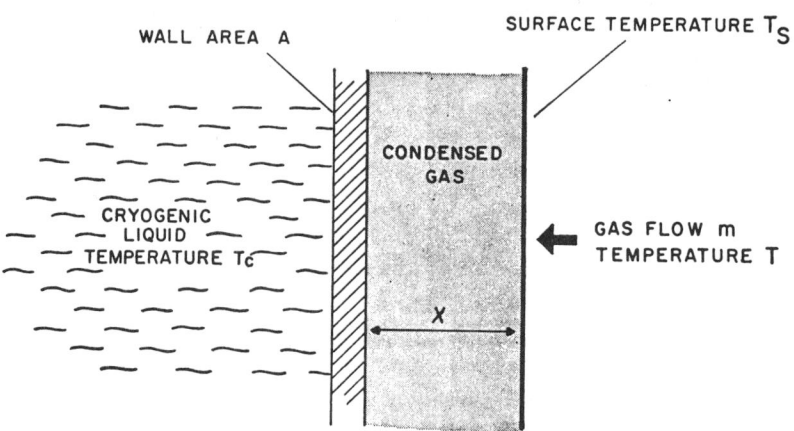

Fig. 11.4 One dimensional model of cryogenic pump. (Coy & Ricketson 1963).

Figure 11.4 shows a simple model for the action of a cryogenic pump presented by Coy and Ricketson (1963). The gas at a flow rate $\dot{m}$ (moles cm^{-2} sec^{-1}, assumed constant) is allowed to condense into the solid phase and a layer of thickness x with a free surface temperature T_s is built up on the wall. The steady-state heat flow to the liquid bath is given by

$$H = \dot{m}\,[C_p(T-T_s)+q_s]\,\text{cals cm}^{-2}\,\text{sec}^{-1} \tag{11.5}$$

where C_p is the specific heat per gm mole of the gas at fixed p and q_s is the latent heat of solidification per gm mole. This heat flow develops a temperature drop across the condensed solid given by

$$T_s - T_c = \frac{xH}{k} \tag{11.6}$$

where k is the thermal conductivity of the solidified gas. There will be some limiting surface temperature T_l at which re-evaporation and possibly changes in the condensation coefficient will render the pump unsatisfactory. The factors determining T_l will be specific to each situation. This value of T_l will define a running-time t given by

$$t = \frac{k\rho}{\dot{m}^2} \frac{T_l - T_c}{C_p(T - T_s) + q_s} \text{ sec} \quad (11.7)$$

where ρ is the density of the solidified gas in moles cm^{-3}. Coy and Ricketson (1963) found reasonable agreement with this model for the condensation of CO_2 on a brass surface at 77°K. A detailed analysis of mass and heat transfer at a frost surface has been given by Smith et al. (1964).

The gas striking the surface of a cryogenic pump will have an equivalent temperature higher than the surface temperature. For surface temperatures near 20°K and for gas temperatures T above 300°K Dawson (1966), using a method first suggested by Buffham et al. (1962), has shown that the condensation coefficient follows a relation of the form

$$1 - c = \exp\left(-\frac{T^\dagger}{T}\right). \quad (11.8)$$

Values of $T^\dagger$ for various gases measured empirically are given in Table 11.4 where it may be seen that for the pure gases $T^\dagger$ is either about 290 or 532°K. Equation 11.8 is a useful relation in the form

$$(1 - c_1)^{T_1} = (1 - c_2)^{T_2} \quad (11.9)$$

since only one measured pair of values c_1, T_1, is required to predict c_2 at any T_2.

The model of Fig. 11.4 has disregarded the radiant flux load on the cryogenic pump. For many cryogenic pumps, in particular space simulation chambers, the radiant heat load may far exceed the condensation heat load and the design problem becomes one of reducing the radiant heat load without serious reduction of the pumping efficiency (see below). This is normally done by interposing between the pump surface and the main source of the radiant heat flux a shield or shroud at an intermediate tempera-

TABLE 11.4

Temperature dependence of condensation coefficient for gases at $T \sim 300°K$ impinging on Cryogenic pump surface

$T_s \sim 20°K. \ 1 - c = \exp\left(-\frac{T^\dagger}{T}\right)$

Gas	$T^\dagger$
N_2	290
Ar	293
CO_2	289
CO	532
N_2O	288
O_2	532
90% N_2/10% O_2	361
80% N_2/20% O_2	416

(from Dawson 1966).

ture. Typically a pumping surface at 4–20°K uses a shroud at 77°K. It is thus desirable that the low temperature surface have a low absorptance and the shroud have a high absorptance (or emissivity) for radiation from the main radiation source. It is relatively easy to achieve the desired conditions for clean prepared surfaces, but as condensate builds up on these surfaces the values of the absorptance change. Figure 11.5 shows the results of Moore (1962) for changes in emissivity for condensates of H_2O, CO_2 and N_2 on a polished copper surface at 20°K. Note that all condensates cause the emissivity to rise. The large changes caused by H_2O suggest that condensates of H_2O on a shroud on the side facing the vehicle should not cause serious degradation of performance. This has been confirmed by Caren et al. (1964) for a condensate of H_2O on a blackened surface (cat-o-lac) on the substrate aluminum. The rate of deposition of the condensate appears to influence the results obtained in these measurements, which is probably the reason that the results of Cunningham and Young (1963) are in only qualitative agreement with those of Moore (1962) and Caren et al. (1964).

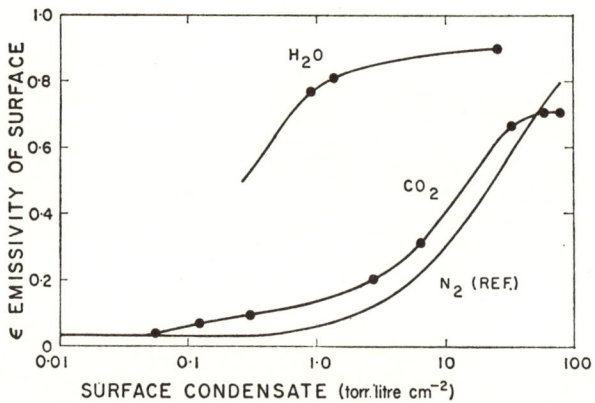

Fig. 11.5 Variation of total emissivity of condensates on a polished copper sphere. 1 torr litre at 273°K = 3.5×10^{19} molecules (Moore 1962).

Localized cryogenic pumps. These pumps are characterized by a cold surface of limited extent somewhere in the vacuum system. They have been built both with and without radiation shielding between the cold surface and room temperature. Figure 11.6 shows an example of an unshielded cryogenic pump which was designed to provide an assessment of cryogenic pumping of hydrogen and deuterium for controlled fusion research. The system volume was 60 litres, background pressures with the cryogenic pump cold were in the low 10^{-11} torr range. Condensation parameters were varied over the ranges: gas incidence rate 10^{12} to 10^{14} molecules cm^{-2} sec^{-1}, surface coverage 10^{13} to 10^{17} molecules cm^{-2}, surface temperature 2·1 to 3·7°K, gas temperature 290 and 370°K, and radiation load 9×10^{-3} watt cm^{-2}

and 30×10^{-3} watt cm^{-2}. (It may be noted here that Borovik et al. (1963) found a radiation load of $3-5 \times 10^{-4}$ watt cm^{-2} sufficient to cause both boiling and poor conduction of heat to the bath.) Central to the measurements of Chubb et al. (1967) was a Monte Carlo calculation of the pressure reading at various points in the system as a function of the condensation coefficient at the cold surface. When this calculation was used together with a pulsed gas input both the condensation coefficient and desorption rate were found, under the dynamic conditions of pump operation. It was found that the condensation coefficient increased from about 0·5 at 10^{13} molecules cm^{-2} to between 0·7 and 0·95 at 10^{16} molecules cm^{-2}. The condensation coefficient was found to increase with increasing gas incidence rate and was

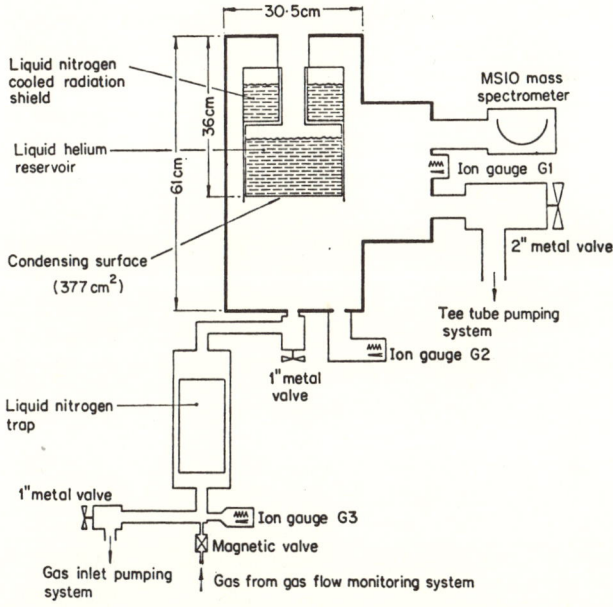

Fig. 11.6 Example of unshielded localized cryogenic pump. (Chubb et al. 1967)

relatively insensitive to surface temperature. The desorption rates indicated excessive radiation incident on the condensed layer and it was concluded that radiation shielding was necessary for pumping of H_2 at liquid helium temperatures. Systems similar to that of Chubb et al. (1967) have been described by Bachler et al. (1962), $2·5 < T < 30°K$; Forth (1965), $2·5 < T < 4·2$; Mullen and Jacobs (1962), T = 20, 77°K.

Pumps have been described which are similar to the pump shown in Fig. 11.6 except for radiation shielding of the pumping surface. Borovik et al. (1965) have described an arrangement in which a sphere was cooled to 20°K and was shielded by louvered baffles at 77°K. The assembly was inserted into the vacuum line between a system of volume 60 litres and a diffusion pump. Surface area of the sphere was 200 cm². The cryogenic pump was included in the baked region of the system, but was valved off from the diffusion pump before operation. A pressure of 3×10^{-10} torr was reached about 1·5 hours after start of the cryogenic pump. The pumping speed for N_2 was 600 litres $\sec^{-1}$ and the hydrogen consumption was 0·05 litre $hour^{-1}$. Other shielded cryogenic pumps comparable to that of Borovik (1965) have been described by Lazarev et al. (1957), 20°K; Moore (1962), 20°K; Turner and Hogan (1966), <20°K; Hoenig (1964), 4·2°K; Kienel and Wutz (1966), 1·5–4·2°K. The latter pump had a speed of 2900 litres $\sec^{-1}$ for air and 11000 litres $\sec^{-1}$ for H_2. Borovik et al. (1963) have described a liquid helium pump for H_2 of concentric annular design to enclose the beam in a thermonuclear machine.

Localized cryogenic pumps with integral refrigerators have been described by Borovik et al. (1961), $T = 20°K$; and by Turner and Hogan (1966), $T < 20°K$.

Space simulation chambers. Two of many requirements of space simulation for a space vehicle (Santeler 1959; Garwin 1963; Goethert 1964; Choumoff 1965) are met by the cryogenic pump. These are (i) the flux of molecules leaving a space vehicle at 300°K will be relatively high while the flux returning to the vehicle will correspond to the pressure of space which Garwin (1963) estimates to be 10^{-13} to 10^{-14} torr within the solar system and 10^{-16} torr outside the solar system. Santeler (1959) showed that if this situation is to be simulated a high condensation coefficient for molecules striking the walls of the simulator is required (see eqn. 7.21). (ii) the radiant flux leaving a space vehicle at 300°K will be relatively high while the return flux not too near a star is estimated to correspond to enclosing the vehicle in a container at 3°K. However, surrounding a vehicle at 300°K by an absorbing surface at 100°K reduces the return flux by a factor $3^4 = 81$, which is considered adequate for most purposes. The surrounding surfaces are generally treated with epoxy or oxide coatings, special black paint, or anodizing, to maintain their absorptance above 0·9 (Garwin 1963). Since low temperatures are a necessary property of cryogenic pumps, the radiation requirement of a space simulator is automatically fulfilled by these pumps. The surfaces at 100°K serve to shield the surfaces at lower temperature (usually 20°K) from an excessive heat load.

The design considerations of a space simulator have been computed by Pinson and Peck (1962) using the model shown in Fig. 11.7. The cylinders are infinite in extent in a direction perpendicular to the plane of the diagram. Monte Carlo computations are made for molecules and radiation leaving the vehicle envelope. P and η are the probabilities that molecules and radiation respectively are absorbed at the cryopumping panel without returning to the vehicle. The shielding panel is assumed to have no condensation coefficient for molecules and an absorptance of 0·9 for radiation,

Pumps for Ultrahigh Vacuum

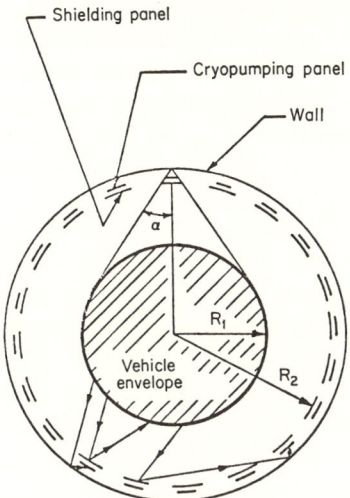

Fig. 11.7 Model for calculation of pumping and radiation shielding in a space simulation chamber. (Pinson & Peck 1962).

while the pumping surface is assumed to have a condensation coefficient of unity for molecules and an absorptance of unity for radiation. Design objective is to maximize P with a tolerable value of η, the latter value depending in the main upon economic considerations. Results of calculations comparable to those of Pinson and Peck (1962) have been published by Goethert (1964) and by Hurlbut and Mansfield (1963). Balwanz (1964) concludes from experimental studies that non-condensable gases striking condensed water frost at 77°K are accommodated to a temperature of 77°K after two bounces. Theoretical studies on the shapes of the radiation shields, in order to focus the molecular flux on the surfaces at 20°K, have shown that only a small gain in pumping efficiency is achieved in this way (Ballance et al. 1963).

The fourth array in Fig. 11.8 (Santeler 1959) has been used by Richman and Hood (1962) in each of three large space simulation chambers of 32,000 ft³ each. These stainless steel systems were spherical, of diameter 38·5 ft. A cross-section is shown in Fig. 11.9. The radiant infrared source was between the test object and the cryopanels. This system could subject the vehicle to a radiant flux of 390 watts ft^{-2}. The room temperature walls of the chamber were shielded from the vehicle by panels cooled by liquid nitrogen (4300 ft² in all) and estimated to operate at 100°K. These panels had a water pumping speed of $4 \cdot 2 \times 10^7$ litre sec^{-1}. The panels were polished on the outside and coated with epoxy paint on the inside to give an absorptance of more than 0·90. The 100°K panels provided radiation shielding for 1000 ft² of panels operated at 20°K cooled with helium gas. The estimated speed of the 20°K panels for N_2 was $2 \cdot 8 \times 10^6$ litre sec^{-1}. Pumping of He,

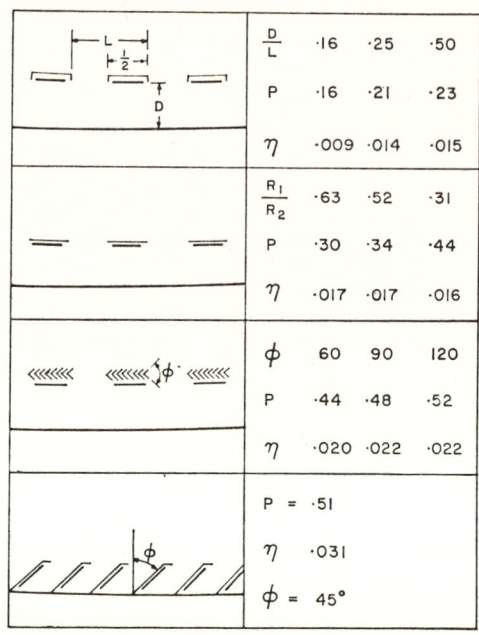

Fig. 11.8 Results of calculation of pumping and radiation shielding for cryogenic array in a space simulation chamber. R_1 and R_2 as shown in Fig. 11.7 (Pinson & Peck 1962).

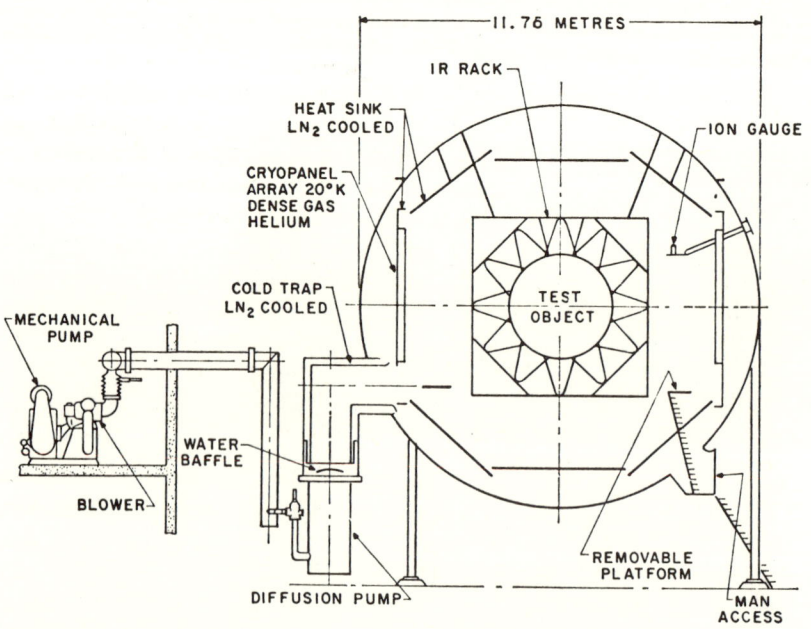

Fig. 11.9 Example of large space simulation chamber (Richman & Hood 1962).

H_2 and Ne was done with four diffusion pumps rated at a total speed of 2×10^5 litre sec^{-1}.

With only the 100°K panels operating in this system a pressure of 5×10^{-9} torr was achieved and with the 20°K panels operating a pressure of 5×10^{-10} torr was achieved.

11.4 GETTER PUMPS

Chemically active gases are pumped in a getter pump at an evaporated getter film or a bulk getter by a combination of (a) chemisorption, (b) formation of chemical compounds, and (c) solution.

Since the rare gases are not pumped by a getter pump it is usually necessary to combine a getter pump with some other type of pump capable of removing rare gases. Getter pumps have been combined with cryopumps (Section 11.3), ion pumps (11.5), turbo-pumps (11.1) and diffusion pumps (11.2) to give increased pumping speeds to the chemically inactive gases. In experiments involving rare gases, getter pumps are particularly useful since they will not pump the rare gases but have a high pumping speed for any chemically active contaminant gases.

The chemisorption of gases on solids has been discussed in Chapter 2; in particular, data on the binding energy of chemisorbed gases on solids can be found in Section 2.3.2 and data on sticking probabilities in Section 2.4.3.

11.4.1 Getter materials

Getters for uhv use normally consist of a high melting-point metal, such as titanium or molybdenum, which can be evaporated or sublimed by resistive heating, or by electron bombardment heating, onto a surface at room temperature or below. The pumping speed of a freshly prepared getter surface to any gas that can be chemisorbed thereon is given by

$$S = 3 \cdot 64 \, s \left(\frac{T}{M} \right)^{\frac{1}{2}} A \text{ litres/sec} \qquad (11.10)$$

where s is the sticking probability (a function of surface coverage),
A the surface area in cm^2,
M the molecular weight of the gas;
for nitrogen at 300°K

$$S(N_2) \approx 12 \, s \, A \text{ litres/sec}. \qquad (11.11)$$

Since maximum sticking probabilities for most simple gases lie between 0·1 and 0·8 it can be seen that very substantial pumping speeds to the chemically active gases can be obtained with getter surfaces of a few square centimetres.

The material most widely used for getters in uhv is titanium, though a few other metals (Zr, Mo and Nb) are occasionally used. The properties of bulk titanium metal as a getter have been reported by Stout and Gibbons (1955) who show that above 970°K, O_2, N_2 and CO_2 were rapidly sorbed. Hydrogen was sorbed in the range 300 to 670°K and released at higher temperatures; it was found to be the only gas released on heating. Gibbons and Stout (1959) have measured the gettering properties of alloys of titanium and zirconium in bulk form. These alloys are capable of dissolving any oxide film and of sorbing hydrogen at temperatures below 670°K. Kindl (1963) has developed a highly active Zr—Al alloy which can be bonded onto a thin metal ribbon; a non-evaporable getter cartridge using this material has been developed (Pisani & della Porta 1967) for use in uhv systems. Pumping speeds of about 60 litres/sec for CO were observed at 670°K which remained constant up to sorbed quantities as high as 2 torr litre. At room temperature the getter cartridge adsorbed about 0·25 torr litre of CO before pumping ceased; this was similar to the behaviour of a 5-mg evaporated titanium film.

Evaporated films of metals such as Ti are more widely used as getters for uhv than getters in the form of bulk metal. The parameters of most importance in definining the pumping characteristics of getter films are the initial sticking probability (i.e. the sticking probability for a freshly prepared film) and the maximum capacity of the film. Both these parameters are strongly dependent on the way in which the films are prepared and thus there is considerable disagreement between measurements of different experimenters. The results we consider below are representative of the best measurements and indicate the spread of values that may be expected in practice.

Clausing (1961) has measured the sticking probability as a function of amount of gas adsorbed for various gases on evaporated titanium films. Evaporation of titanium was done in two different ways: (i) flash evaporation or (ii) continuous evaporation. The values of initial sticking probability for various gases on titanium films prepared in different ways are shown in Table 11.5. It can be seen that films prepared by evaporation through a high pressure (2×10^{-3} torr) of He showed an increase in initial sticking probability for most gases. Table 11.6 lists a compilation of data, measured under uhv conditions, on initial sticking probability and maximum capacity of flash-evaporated titanium films for various gases. These films were produced under a variety of conditions and reference should be made to the original papers for details. It has been found that many evaporations are necessary before the evaporated films show repeatable properties. Kornelsen (1967) shows that with titanium films evaporated on glass the initial sticking probability and maximum capacity increase with each film evaporated until 30–40 evaporations have been made. Table 11.6 also illustrates the large increase in initial sticking probability and capacity that occurs for most gases when the temperature of the titanium film is reduced to liquid nitrogen temperature (see Section 11.3.1).

When the getter material is evaporated continuously, rather than in

TABLE 11.5
Initial sticking probability for various gases on titanium films deposited under several conditions

Gas	High vacuum, 280°K		2×10^{-3} torr He, 280°K	High vacuum, 78°K		2×10^{-3} torr He, 78°K
	Continuous evaporation	Flash evaporation	Flash evaporation	Continuous evaporation	Flash evaporation	Flash evaporation
H_2	0·07	0·05	0·19	0·14	0·24	0·85
D_2	—	—	0·14	—	—	0·78
N_2	>0·20	0·08	0·17	<0·5	0·85	0·93
CO	0·86	0·38	0·66	—	0·95	—
O_2	0·63	0·85	0·82	—	0·86	—
CO_2	>0·5	<0·4	0·92	—	0·98	—
He	<0·0005	—	—	—	—	—
Ar	<0·0005	—	—	—	—	—
CH_4	<0·0005	—	—	—	—	—

(from Clausing 1961).

TABLE 11.6
Initial sticking probability and maximum capacity of various gases on titanium films prepared by flash evaporation

Gas	Substrate temperature during deposition (°K)	Film temperature during adsorption (°K)	Initial sticking probability	Maximum capacity*		Reference
				Torr litre mg^{-1}	Molecules cm^{-2}†	
N_2	300	300	—	$5·1 \times 10^{-3}$		Klopfer & Ermrich (1959a)
N_2	300	300	0·1		$\sim 10^{15}$	Elsworth et al. (1965)
N_2	300	83	0·22		$\sim 1·7 \times 10^{15}$	
N_2	83	83	0·22		‡ $\begin{cases} 2·7 \times 10^{15} \\ 6·8 \times 10^{15} \end{cases}$	
N_2	300	300	0·3		10^{14}	Harra & Hayward (1967)
N_2	350	300	0·5		10^{15}	Kornelsen (1967)
H_2	300	300	—	0·6		Klopfer & Ermrich (1959a)
H_2	300	300	$\sim 0·01$		$7·8 \times 10^{15}$	Elsworth et al. (1965)
H_2	83	83	$\sim 0·4$		$5·75 \times 10^{16}$	
H_2	300	83	—		$8·4 \times 10^{14}$	
O_2	300	300		$4·5 \times 10^{-2}$		Klopfer & Ermrich (1959a)
CO_2	300	300		$7·5 \times 10^{-5}$		Klopfer & Ermrich (1959a)
CO	300	300		10^{-1}		Klopfer & Ermrich (1959a)
H_2O	300	300		5×10^{-3}		Klopfer & Ermrich (1959a)

* When sticking probability has decreased to about 10^{-2}.
† Assuming film area equals the geometric area.
‡ Film thicknesses of 60Å and 170Å respectively.

intermittent flashes, the sticking probability remains at its initial value provided that the rate of evaporation is sufficiently high. If the pressure is increased, so that the rate of arrival of gas molecules at the getter surface becomes comparable to the rate of arrival of evaporated metal atoms at the surface, then the sticking probability drops. Table 11.7 lists measurements with continuously evaporated titanium of (*a*) the sticking probability at low gas impingement rates (s_0) and (*b*) the ratio of the arrival rates of gas molecules to titanium molecules when the sticking probability has decreased to 10^{-2}.

Only a few measurements have been made of the characteristics of metals other than titanium for use as getter materials in uhv. A comparison of the residual gas spectrum in an uhv system using several different getter materials (Ta, Ti, Mo, V, Nb, Zr) has been made (Jackson & Haas 1967). The authors conclude that:

(i) Ta is best overall but requires high temperature for evaporation, precluding use of a support wire and thus filament burn-out is a problem.

(ii) Ti is cheap, effective, easily evaporated. CH_4 and H_2 given off during evaporation.

(iii) Mo is similar to Ti but higher evaporation temperature.

(iv) Zr is difficult to evaporate and not as effective in pumping H_2 as Ti, Ta or Mo.

(v) V easily evaporated but not very effective in pumping H_2 and CH_4 and gives off these two gases in large amounts when heated.

(vi) Nb is very poor in pumping CH_4 and intermediate in H_2 pumping, also has high evaporation temperature.

TABLE 11.7
Sticking probability at low pressures (s_0) and gas/metal ratios for various gases on continuously evaporated titanium films

Gas	Temp. (°K)	s_0	Gas molecules sec^{-1}/ Ti molecules sec^{-1}	Reference
N_2	300	0.47	1.0	Harra & Hayward (1967)
N_2	283	>0.2	0.6	Clausing (1961)
N_2	78	0.55	1.0	Clausing (1961)
H_2	283	0.07	0.9	Clausing (1961)
H_2	78	0.14	1.0	Clausing (1961)
CO	283	0.86	0.9	Clausing (1961)
CO_2	283	>0.5	1.0	Clausing (1961)

It has been shown (della Porta & Giorgi 1966) that niobium will reach an atomic ratio of hydrogen/metal of about 0.45 at both 77 and 300°K. In contrast titanium reaches an atomic ratio of 1.0 at 300°K but only 0.04 at 77°K. The quantity of H_2 sorbed by a niobium film is linearly proportional to the mass of the film and independent of temperature (in the range 77 to 300°K) indicating that the gas has complete accessibility to the metal. Thus for getter pumping of H_2 at 77°K niobium would have a very significantly larger capacity than titanium.

Evaporation of titanium by resistive heating of titanium wire presents some difficulties. To overcome the burn-out problem the titanium wire may be overwound on a tungsten support wire. An 85% Ti, 15% Mo alloy wire has several advantages and the properties of this alloy as an evaporative getter have been reported in detail (Kuzmin 1963; McCracken & Pashley 1966; Lawson & Woodward 1967).

The residual gas composition over an evaporated titanium getter was found to contain considerable quantities of methane (Klopfer & Ermrich 1960). It has been shown (Holland *et al.* 1961; Mikhailov & Kulakov 1963) that methane and other hydrocarbons are produced by interaction of hydrogen with carbon impurities in the titanium. This process can be a serious source of hydrocarbon contamination in an uhv system using no oils, and which is thus free of hydrocarbon contamination from other sources. The methane production rate can be minimized by using carbon-free materials for the getter and its supports or by cooling the evaporated titanium film to liquid nitrogen temperature (Simonov & Mileshkin 1962).

11.4.2 Types of getter pumps

Getter pumps are of two general types, those that sublime the getter material (usually titanium) by resistance heating of a wire, and those which sublime or evaporate titanium by electron bombardment. We shall consider briefly examples of both types.

The simplest form of getter pump consists of a loop of titanium evaporator wire from which titanium can be sublimed onto a glass or metal envelope. A simple and convenient titanium evaporator can be made by simultaneously winding a 0·25 mm diam titanium wire and a 0·1 mm diam tungsten wire onto a 0·25 mm tungsten support. Approximately half the titanium can be evaporated from a getter of this type.

Many getter pumps have been described with speeds (for air) up to several thousand litres/sec. Typical of these is a series of getter pumps (described by Nazarov *et al.* 1965), having pumping speeds for air of 500, 2000 and 5000 litres/sec and using resistively-heated Ti-Mo alloy wire. This series of pumps also contains a hot-filament source for ionizing the gas by electron bombardment to provide pumping of the rare gases; the pumping speed for argon is about 1% of the pumping speed for nitrogen. The ultimate pressure of these pumps is about 10^{-9} torr consisting of H_2, CO, Ar and smaller amounts of H_2O and CH_4. Several references to papers describing similar getter pumps can be found in Nazarov *et al.* (1965).

Pumps in which the getter material is evaporated onto a cooled surface (cryogetter pump) are particularly effective for pumping hydrogen which is usually the predominant gas in an uhv system (see also Section 11.3.1). Rivière *et al.* (1965) have described a pump in which titanium could be evaporated from a resistively-heated wire source over a stainless steel surface cooled to liquid nitrogen temperature. The sticking probability of H_2 on this surface was about unity. After outgassing at only 250°C the base pressure was reduced from 10^{-8} to $<10^{-11}$ torr by the cryogetter pump. McCracken (1965) has reported a similar pump with Ti, from Ti—Mo alloy evaporator wires, evaporated onto a surface at 78°K. Pumping speeds in excess of 10^4 litres/sec for hydrogen were obtained and pressures of the order of 10^{-10} torr obtained.

The second type of getter pump, in which titanium is evaporated from a molten ball, usually by electron bombardment, requires a means for mechanically feeding in the titanium wire. Several designs of such pumps have been reported (e.g. Herb 1960) but mechanical difficulties and maintenance problems have prevented their widespread use; Gould and Dryden (1961) have summarized the practical difficulties.

The advantages and disadvantages of getter pumps will now be summarized. Getter pumps are a very simple and cheap way of obtaining high pumping speeds to chemically active gases. When air or rare gases have to be pumped the getter pump must be combined with some other type of pump (e.g. cold-cathode discharge or hot-cathode ionization pump, diffusion pump, turbo-pump, cryopump). Unless mechanical means for feeding titanium wire is used, which is generally unreliable, the life of a getter pump is limited by the amount of titanium contained in the pump. This is not a serious problem on small systems with limited gas loads. Unless care is taken to minimize impurities and/or cool the titanium film a considerable amount of hydrocarbons (predominantly CH_4) may be generated at the getter surface.

11.5 ION PUMPS

An electrical discharge (hot- or cold-cathode) is established in an ion pump to ionize and excite the gas; the interaction of the ions with a solid surface results in binding of the gas to the solid and hence a pumping action. The pumping process can be divided into two principal mechanisms:

(i) The entrapment in solids of molecules accelerated as positive ions; this is called ionic pumping (Section 4.3.4) and is the only mechanism for pumping rare gases.

(ii) The sputtering of the cathode surfaces by positive ion bombardment creates 'clean' surfaces on which chemically active gases can chemisorb or form chemical compounds. (See Section 4.3.2 for discussion of sputtering.)

In some ion pumps a getter is included which can be evaporated over the cathode surfaces of the pump; this type of pump is called a getter-ion pump and was mentioned in Section 11.4. The evaporated getter film serves two purposes; firstly, it chemisorbs active gases and secondly, it covers the surfaces at which ion pumping has occurred and prevents the subsequent re-emission of the ion-pumped gas. Some of the problems of pumping at, and re-emission from solid surfaces have been reviewed by Hobson (1963).

The pumping speed resulting from ion-entrapment alone is given by

$$S_i \approx 0\cdot 19 \frac{I^+}{p} s \quad \text{(litres/sec)}, \qquad (11.12)$$

where I^+/p, the ratio of ion current to pressure, is called the discharge intensity (A/torr),
and s the sticking probability of the ions is a function of ion impact energy, ion species, nature and condition of the cathode surface (see Section 4.3.4).

Eqn. 11.12 assumes that all the ions are singly charged. For ions with energies greater than about 1 keV the sticking probabilities approach unity on most surfaces (see Section 4.3.4).

The steady-state pumping speed for chemically active gases resulting from sputtering is directly proportional to the discharge intensity (when pumping a single gas); in most cases the proportionality constant cannot be readily calculated as in the case of ionic pumping. Bachler (1965) has shown experimentally that the pumping speed of a Penning discharge pump for nitrogen was given by

$$S(N_2) = 7 \times 10^{-2} \frac{I^+}{p} \quad \text{(litres/sec)}. \qquad (11.13)$$

The speed was varied from 10 to 3×10^3 litres/sec by varying the anode voltage (4 to 7 kV) and the magnetic field. The proportionality constant for chemically active gases will depend on many factors including the previous history of the pump.

The effective speed of an ion pump is reduced if re-emission of previously pumped gas is significant (see Section 4.3.4 for discussion of the re-emission process). This process is sometimes called 'memory' or 'historical' effect since it is the regurgitation of previously pumped gas. We can distinguish two types of re-emission:

(i) spontaneous re-emission; the thermal emission of previously pumped gas without the action of the discharge (see Section 4.3.4);

(ii) induced re-emission; the emission of previously pumped gas resulting from an ion bombardment of the cathode. The re-emitted gas may be the same species as the bombarding ion (homogeneous re-emission) or the emitted gas may differ from the bombarding ion species (heterogeneous re-emission) (see Section 4.3.2).

Spontaneous re-emission can be sufficiently rapid at room temperature for a rare gas that it may constitute the dominant gas source in certain uhv systems. It has been shown experimentally (Varnerin & Carmichael 1957;

Carmichael & Knoll 1958) that the spontaneous re-emission rate of ionically pumped gas is given by

$$r = kn/t, \qquad (11.14)$$

where n is the number of molecules pumped at time $t = 0$. Values of the re-emission factor (k) can be found in Section 4.3.4. As an example of the effects of re-emission on the ultimate pressure of a system we consider a 5-litre volume pumped by a 15-litres/sec (N_2) ion pump which has been evacuated from atmospheric pressure by a liquid-nitrogen cooled sorption pump. At time zero there will be about 4×10^{-3} torr of helium, left from the air, which cannot be removed by the sorption pump. After the ion-pump has been on for one hour, pumping He at 1·5 litres/sec (10% of N_2 speed), we estimate the equilibrium pressure of He from $p_{eq} = r/S$ with $k = 4 \times 10^{-3}$ ($He^+ \rightarrow Mo$ at ion energy of 2·1 keV, Varnerin & Carmichael (1957)) and obtain $p_{eq} = 1·5 \times 10^{-8}$ torr. In this example we see that re-emission from the ion-pump may control the equilibrium pressure for several hours; indeed, we note that it will take several hundred hours before the helium partial pressure reaches an equilibrium level. Spontaneous re-emission can be very substantially reduced and in most cases eliminated, if a metal such as titanium is evaporated over the cathode surfaces to trap the pumped gas. Re-emission rates decrease rapidly as the temperature of the surface is lowered (Section 4.3.4).

Heterogeneous induced re-emission is particularly troublesome in experiments where the nature of the gas in the system is frequently changed. After every change of gas the ion pump tends to re-emit the gas that was previously in the system. This effect has been studied in a triode sputter-ion pump by Bance and Craig (1966) for the gases H_2, N_2, Ar and C_2H_2. Heterogeneous re-emission was greatest after pumping H_2 and was less after pumping the other gases, decreasing in the order indicated above.

In the following subsections we examine the uhv performance of particular types of ion pumps.

11.5.1 Sputter-ion pumps

Most sputter-ion pumps make use of a cold-cathode crossed-field discharge (such as a Penning, magnetron or inverted-magnetron discharge) to ionize the gas as efficiently as possible. These types of discharge were discussed in Section 8.2 in their application to gauges. The positive ions produced in the discharge are accelerated into the cathode surfaces (usually titanium) where pumping occurs by ion entrapment and by chemical binding at the fresh titanium surfaces generated by sputtering. Detailed descriptions of these pumps may be found in Power (1966). The majority of sputter-ion pumps are based on Penning cells, each cell consisting of a tubular anode between two flat cathodes with the magnetic field directed along the axis of the anode. The general arrangement is similar to the Penning gauge structure shown in Fig. 8.17.

Diode sputter-ion pumps consist of a multiple-anode structure (usually in the form of an 'egg-crate') between two cathode plates, as indicated in Fig. 11.10a. The structure of a triode pump is indicated in Fig. 11.10b; two honeycombed electrodes, the sputter-cathodes, are placed between the anodes and the collector plates. Ions strike the surface of the sputter-cathode at oblique incidence producing a much higher yield of sputtered titanium than is obtained at normal incidence (see Section 4.3.2). The sputtered titanium lands mainly on the collector where chemical binding with gas molecules can occur. The collector is at the same potential as the anode so that ions cannot strike the collector electrode unless the state of charge of the ions is reduced upon interaction with the sputter-cathode.

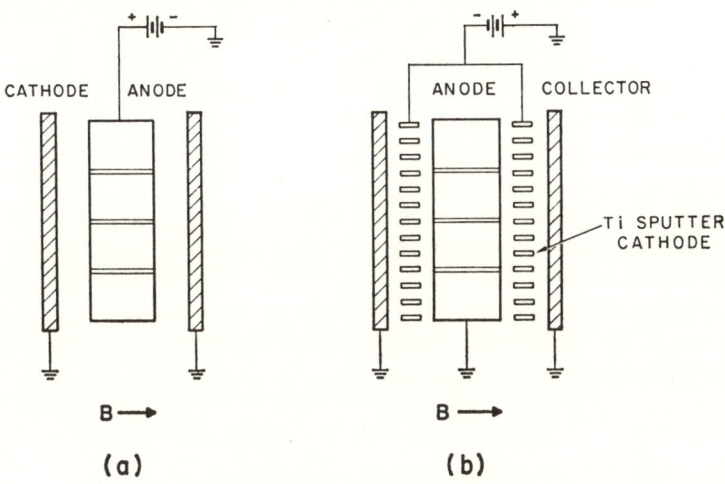

Fig. 11.10 (a) Schematic of diode ion pump.
(b) Schematic of triode ion pump.

The most important characteristics of a sputter-ion pump for uhv use are:
(i) the shape of the pumping speed vs pressure curve at low pressures,
(ii) the degree to which the pump-re-emits previously pumped gas, and
(iii) the ability to pump rare gases without pressure instabilities (if the system has to be re-cycled to high pressures frequently).
We now examine the performance of diode and triode pumps with respect to the above characteristics.

Unlike diffusion, turbo and getter-ion pumps, the pumping speed of a crossed-field discharge pump is not independent of pressure. This results from the fact that the ion-current vs pressure characteristic of crossed-field discharges are non-linear and, in general, I^+/p, and hence S, decreases with p. This problem was discussed in Section 8.2. Rutherford (1963) has measured the discharge intensity as a function of pressure (10^{-3} to 10^{-11} torr) for diode pumps with various size and anode cells. The anode diameters were 1·25, 2·5 and 5 cm and the aspect ratio of the anodes (L/D) was kept constant at 1·5; the anode voltage was 3 kV and the magnetic field was varied from 1 to 2 kilogauss. In all cases it was found that the discharge intensity decreased with pressure. With the larger diameter cells and higher magnetic fields, the reduction of I^+/p at low pressures was much less. The same general behaviour has been observed by Bachler (1965). So-called 'large-cell' pumps are now commercially available which are reported to maintain their rated pumping speeds down to lower pressures than the small-cell pumps.

The pumping speeds of a diode and a triode sputter-ion pump have been measured as a function of pressure for several gases (Dallos & Steinrisser 1967). The diode pump was rated at 15 litres/sec (N_2) and was operated at 7·2 kV in a magnetic field of 1·4 kilogauss. The triode was rated at 8 litres/sec and was operated at 5 kV in 1·35 kilogauss. Figure 11.11a shows the discharge intensity and pumping speed for H_2, He and N_2 for the diode pump. The pumping speeds are shown for pumping times of zero and one day. Figure 11.11b shows the pumping speed of the triode pump. With both the diode and triode pumps it can be seen that the pumping speeds fall dramatically with pressure until the speeds are near zero at 10^{-11} torr. At 10^{-10} torr the pumping speeds, for both types of pump, were about 20% of that at 10^{-6} torr. It should be noted that these data are for small-cell pumps.

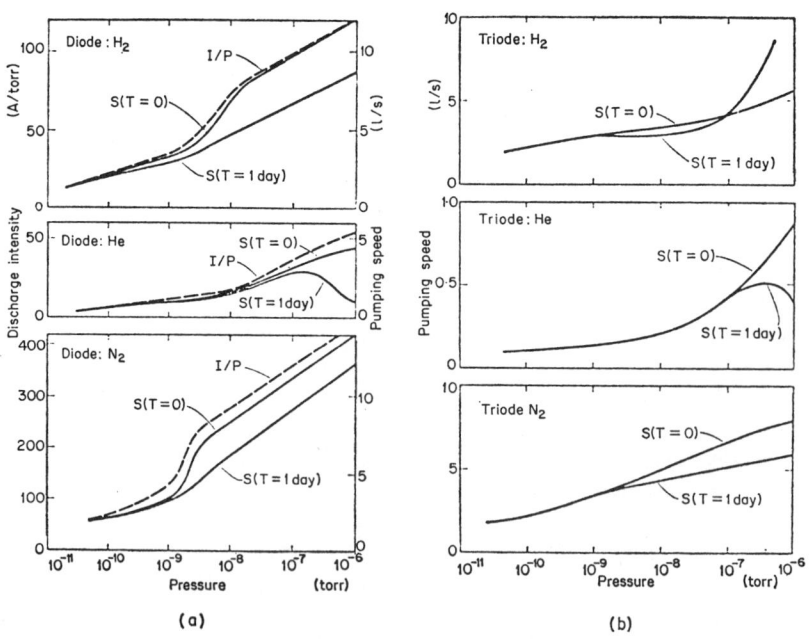

Fig. 11.11 (a) Pumping speed S and discharge intensity I^+/P vs pressure P in a diode pump. (Dallos & Steinrisser 1967).
(b) Pumping speed S vs pressure in a triode pump (Dallos & Steinrisser 1967).

A design of diode sputter-ion pump has been reported (Komiya et al. 1966) which uses a radioactive source (Ni^{63}, a beta emitter) to release slow electrons from another electrode, the slow electrons are able consistently to trigger the discharge even at pressures down to 10^{-11} torr (Hayashi 1966). The measured pumping speed for nitrogen is reportedly constant from 10^{-8} to 10^{-10} torr. No reasons have been presented, at the time of writing, to explain why the pumping speed of this particular design of pump does not decrease with pressure in contrast to the behaviour of all other crossed-field pumps. The reported pumping speed characteristics of this type of pump would make it eminently suitable for uhv applications.

The decrease of pumping speed with decreasing pressure for crossed-field pumps is a serious limitation to their ability to achieve very low pressures. Perhaps more important is the decrease in effective pumping speed at low pressures resulting from the re-emission of previously pumped gas acting as a source within the pump. In the majority of cases it is this effect which prevents the production of very low pressures by sputter-ion pumps in systems where sizeable quantities of gas are pumped. Under these circumstances an equilibrium pressure is reached where the pumping speed becomes equal to the rate of gas re-emission and the *effective* speed of the pump goes to zero. When very low pressures are desired with a sputter-ion pump, one simple expedient is to reduce the amount of gas pumped by the sputter-ion pump to a minimum by suitable design of the system and the experiment. It is particularly important to minimize the amount of rare gases pumped. When the quantities of gas pumped cannot be reduced, other methods are required. Some sputter-ion pump designs permit cooling the cathodes and/or collectors, which is a very effective method of reducing re-emission. Vinogradov and Rudnitskii (1966) have described a triode sputter-ion pump in which the anode and collector electrodes can be cooled by liquid nitrogen. Pressures as low as 5×10^{-12} torr have been obtained with liquid nitrogen cooling of the anode and collector electrodes.

The simplest and most effective method of preventing re-emission of previously pumped gas is by the deposition of a metal film over the pumping surfaces to bury the pumped gas. Kornelsen (1960) has shown that previously pumped rare gases can be buried, and prevented from re-emitting, by evaporating titanium over the cathode surfaces of a small inverted-magnetron pump. Pumps of similar design based on the magnetron geometry have now been in use in the authors' laboratory for some years. In small systems (a few litres) it is possible to repeatedly pump rare gases from pressure as high as 10^{-4} torr and consistently achieve ultimate pressures of about 7×10^{-11} torr and 4×10^{-11} torr with the inverted-magnetron and magnetron pumps respectively.

Chemically active gases are, in general, less troublesome than rare gases as far as re-emission is concerned; the exception to this rule is hydrogen. To obtain very low pressures with large sputter-ion pumps it is usually necessary to provide additional pumping speed for hydrogen, usually by means of a titanium getter external to the sputter-ion pump.

The pumping speed of most sputter-ion pumps to the rare gases tends to be only a small fraction of the speed to nitrogen. Moreover, many sputter-ion pumps will not pump rare gases stably and develop the so-called 'argon instability'. Argon instabilities are the cyclical pressure fluctuations that may occur when pumping considerable quantities of rare gases. The periodicity of the bursts may be several minutes with an amplitude of about 10^{-4} torr. These matters, discussed in detail in Power (1966) are not specifically uhv problems unless it is necessary to pump rare gases continuously in an uhv system or if the system has to be frequently cycled up to atmospheric pressure.

Table 11.8 shows the pumping speed of a diode pump for various gases relative to the nitrogen pumping speed. It can be seen that the rare gases speeds are all less than 10% of nitrogen. The behaviour of the diode pump to argon can be improved by grooving the cathode to give a densely ribbed surface (Jepsen et al. 1960). This change reduces the argon instability and increases the argon speed by a factor of about 5 to about 10% of the nitrogen speed. The useful life of the grooved cathodes is less than that of smooth cathodes.

TABLE 11.8
Pumping speeds of sputter-ion pumps relative to nitrogen

Gas	Diode pump[a]	Triode pump[b]
H_2	2·7	2·1
D_2	1·9	
CO_2	1·0	1·0
H_2O	1·0	1·0
O_2	0·57	0·59
Light hydrocarbons	0·9 to 1·6	0·9 to 1·0
He	0·1	0·3
Ne	0·04	
Ar	0·01	0·3

(a) From Barrington (1963).
(b) From Bance and Craig (1966).

Another solution to the problem of low argon speeds in diode pumps has been reported by Tom and James (1966). Differential sputtering is achieved by the use of one titanium cathode and one tantalum cathode. Stable argon pumping speeds of about 0·2 that of air can be obtained and argon instabilities do not occur below 10^{-5} torr. Instabilities do occur at higher pressures.

As can be seen from Table 11.8 the relative pumping speeds of triode pumps for rare gases is considerably higher than with diode pumps. Triode pumps are thus preferable for uhv when considerable quantities of rare gases must be pumped, or when frequent cycling of the system to atmospheric pressure occurs. Triode pumps do not usually show argon instabilities.

11.5.2 Getter-ion pumps

Getter-ion pumps rely on chemisorption (or chemical binding) at evaporated metal surfaces to pump chemically active gases; rare gases are pumped by the action of an electrical discharge (ionic pumping). The prime advantage of getter-ion pumps is that they can provide very high pumping speeds to chemically active gases very simply and cheaply. Their main disadvantage is their rather low rare gas speeds, relative to the speeds for chemically active gases. Getter pumps were discussed in Section 11.4 and some simple getter-ion pumps were also briefly discussed. In this section we briefly examine some other types of getter-ion pumps suitable for uhv use.

Small cold-cathode getter-ion pumps (Penning, magnetron or inverted-magnetron geometry) have been in use in several laboratories for many years and have proved themselves capable of producing ultimate pressures of about 5×10^{-11} torr in systems of a few litres volume. The main advantage of these pumps over sputter-ion pumps is that the evaporation of getter material over the cathode surfaces completely eliminates the historical or memory effect discussed in Section 11.5.1. In spite of the advantages of these pumps for small laboratory systems, there are none commercially available, at the time of writing.

Klopfer and Ermrich (1959b) have described a small getter-ion pump based on a Penning geometry consisting of a ring anode and two cathodes made of several hairpin shape titanium wire evaporators. The pump operates at a magnetic field of 400 gauss and 2 kV anode voltage. Titanium can be evaporated when required by passing current through the evaporators. With a typical load of titanium of 50 mg the pump has a maximum capacity of about 5×10^{-1} torr litre. Approximate pumping speeds are: CO, 50 litres/sec; CH_4, 0·1 litre/sec; inert gases 10^{-3} to 10^{-2} litre/sec.

A small getter-ion pump based on the geometry of the inverted-magnetron gauge (see Section 8.2.3) was developed by Kornelsen (1960). This pump is very similar in design to the inverted-magnetron gauge (see Fig. 8.20) except that the auxiliary cathodes are omitted and the anode is replaced by a composite titanium evaporator. The evaporator consists of a 0·25 mm diam titanium wire and a 0·1 mm diam tungsten wire wound together on a 0·25 mm diam tungsten support wire. This composite getter wire is then wound on a 1 mm diam sapphire rod. The anode is grounded and the cathode is operated at -6 kV with a magnetic field of 2 kilogauss. Pumping speeds to the chemically active gases (CO, N_2, H_2, etc.) are high and are essentially limited by the conductance of the connection to the pump; the speed for methane is low (see Section 11.4). The discharge intensity (I^+/p) is about 1 A/torr giving a pumping speed to argon of 0·25 litre/sec and to helium of 0·03 litre/sec. Ultimate pressures of 7×10^{-11} torr can be obtained with this pump, even after repeatedly pumping rare gases at pressures as high as 10^{-3} torr, if the ionically pumped rare gases are covered with evaporated titanium. With a typical titanium charge of about 30 mg the maximum capacity of the pump is about 0·5 torr litre of argon.

A pump of similar design based on the design of the magnetron gauge (see Section 8.2.2) has been used in the authors' laboratory for some time. The pump is similar to the magnetron gauge shown in Fig. 8.18 with the omission of the auxiliary cathodes and the addition of a titanium getter placed in a slit around the centre of the anode. The pump is operated with the anode at ground, a cathode voltage of -6 kV and a magnetic field of 2 kilogauss. The discharge intensity of this pump is about 5 A/torr, giving a pumping speed to argon of about 1 litre/sec and to helium of about 0·1 litre/sec. Ultimate pressures of about 4×10^{-11} torr can be obtained with this pump in systems of a few litres volume.

The orbitron pump (Maliakal *et al.* 1964; Douglas *et al.* 1965) combines (*a*) a titanium evaporator to provide getter-pumping of chemically active gases with (*b*) an efficient ionizer, depending on electrostatic containment of electrons in the electrostatic field between two coaxial cylinders (see Section 8.1.4 for a description of the orbitron principle in connection with the orbitron gauge), to provide ionic pumping of the rare gases. Two general types of orbitron pump have emerged; those in which the titanium is evaporated by bombardment heating caused by electrons from the discharge and those where the titanium is evaporated from a resistively heated source.

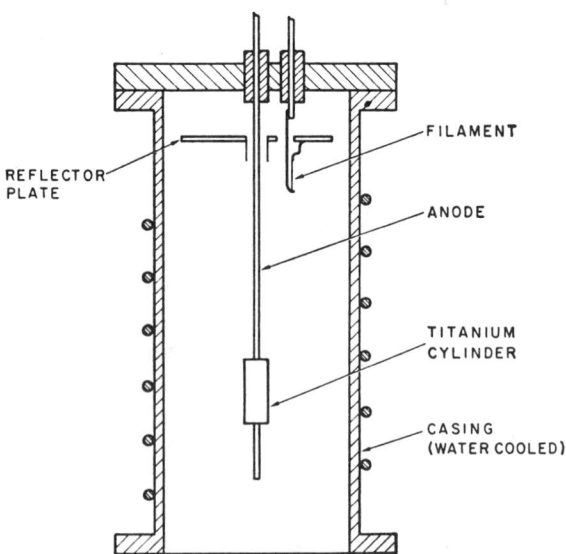

Fig. 11.12 Schematic diagram of orbitron pump employing electron bombardment heating of the titanium.

Orbitron pumps have the following advantages: (a) no magnetic field required, (b) historical effects and argon instability are minimized, and (c) discharge intensity (and hence pumping speed) does not decrease at low pressure. The attendant disadvantages must be recorded as: (a) the presence of a heated titanium evaporator causes the formation of CH_4 from carbon and hydrogen in the titanium, (b) water cooling is necessary in the larger sizes, and (c) the control of the rate of titanium evaporation as a function of pressure is not possible in some designs and, when it is possible, is not simple. The problem of CH_4 formation can be reduced by premelting the titanium (Holland et al. 1966).

Figure 11.12 shows a schematic diagram of an orbitron pump in which titanium evaporation is achieved by electron-bombardment heating of a titanium cylinder fixed to the anode. A detailed description of this type of pump can be found in Power (1966). A pump of 10 cm diam (Douglas et al. 1965) operated with an anode voltage of 5 kV and current of 25 mA gave a nitrogen pumping speed of 500 litres/sec and relative speeds for other gases of: H_2, 1·8; Ar, 0·014. A 30 cm diam pump (Maliakal et al. 1964) was operated at 20 kV and 50 mA giving a nitrogen speed of 6,000 litres/sec and relative speeds of: H_2, 1·5; Ar, 0·017. Evaporation of the titanium by electron bombardment of the anode has the virtue of simplicity but has the drawback that the rate of evaporation cannot be readily controlled. In particular, it would be desirable to reduce the rate at lower pressures to conserve the titanium supply. This can only be easily done by reducing the power in the discharge which results in an undesirable reduction in rare gas pumping speed. The other disadvantage of this type of orbitron pump is that the presence of the titanium slug on the anode tends to reduce the electron path lengths and hence the ionization efficiency and rare gas pumping speed.

Figure 11.13 shows a cross-section of the second type of orbitron pump in which titanium is evaporated from a resistively heated evaporator. In this pump design (Bills 1967; Denison 1967) there are four separate orbitron 'cells' with a central titanium evaporator outside the cells. Electrons are

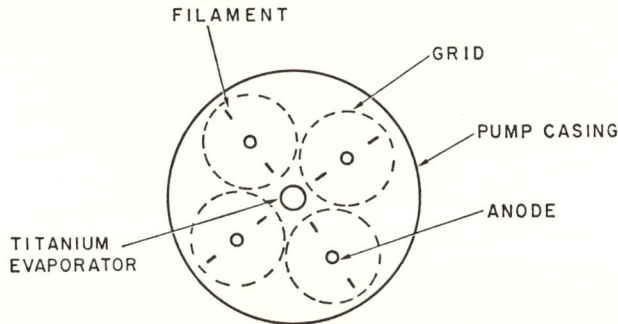

Fig. 11.13 Schematic cross-section of orbitron pump employing resistive-heating of the titanium

TABLE 11.9
Comparison of properties of uhv pumps

Type of pump	Traps required	Maximum backing or starting pressure (torr)	Typical pumping efficiency (a) (%)			Pump cooling required (b)	Hydrocarbon contamination problem (c)	Advantages	Disadvantages
			Chemically active gases (N_2)	Rare gases (Ar)					
Turbo-pump	None	$<10^{-2}$ (d)	12–18	12–18		W	D	No trap. No working fluids	Low compression ratio for H_2. Difficult to outgas. Some noise and vibration
Diffusion pumps:									
(a) Mercury	L N_2	up to 100	5–10	5–10		W	A	Fluid thermally stable	Toxic fluid, amalgamates with some metals. Requires L N_2 trap.
(b) Oil	L N_2 or uncooled sorbent	~0.5	10–20	10–20		W, F	B	Fluid non-toxic. Trapping at room temperatures possible	Failure of trap may cause permanent contamination of system with oil
Cryopumps	None	Atmospheric	50–90	50–90		L	A	Almost 100% of system wall area can pump	H_2 and He difficult to pump. Requires constant supply of cryogenic fluid
Getter pumps	None	~10^{-2}	~45	0		None	C	Simple. No trap.	No pump speed for rare gases. Limited getter life
Ion pumps:									
(a) Getter-ion	None	5×10^{-3}	~45	0.6		W, F	C	No trap	Low rare gas speeds. Limited getter life
(b) Sputter-ion	None	2×10^{-2}	~15	0.15–5		None (e)	C	No trap	Pumping speed decreases at very low pressures. Memory effect. Magnetic interference. Difficult to start at high pressure. Low rare gas speeds.
(c) Orbitron	None	5×10^{-2}	~45	2		W	C	No magnet. No trap. No memory effect	Limited getter life. Low rare gas speeds

(a) Ratio of observed pumping speed to maximum theoretical speed of pump orifice with necessary traps in position.
(b) W, water; F, forced air; L, cryogenic fluid.
(c) A. No oils present in pump itself, possibility of oil contamination from auxiliary apparatus.
   B. System may be severely contaminated if traps fail.
   C. Some formation of light hydrocarbons (CH_4) etc. possible from H_2—C reaction at titanium.
   D. Oil contamination from bearings unless properly cooled.
(d) Compression ratio for H_2 less than 10^3, thus H_2 partial pressure in backing line must be very low.
(e) Forced air or water cooling may be required with some pumps at high pressures.

provided for each cell from specially shaped tungsten cathodes. The electrons are then trapped in the fields between a central rod-shaped anode and a co-axial grid near cathode potential. Ions formed in the orbitron cell are accelerated through the grid and then receive further acceleration by the voltage between the grid and the case of the pump (post-acceleration). Ions are pumped at the water-cooled case of the pump where they are covered by titanium evaporated from the central evaporator. The post-acceleration ensures that the ions strike the case with sufficient energy so that their sticking probability is close to unity (see Fig. 4.41). A pump with an orifice of 15 cm diam gave a nitrogen speed of 1700 litres/sec and a relative argon speed of 0·015. The evaporation rate of the titanium is decreased in three steps, as pressure decreases, to conserve titanium. The changes in power to the titanium evaporator to change evaporation rate are automatically controlled from a pressure signal obtained from the orbitron pump. This arrangement for controlling titanium evaporation rate conserves titanium and reduces the power dissipated in the pump at low pressures. However, these advantages are achieved with some loss of simplicity. Pressures below 3×10^{-11} torr have been obtained with this type of pump.

The two types of oribtron pump have very similar pumping speed characteristics. The first type (electron bombardment) is simple and reasonably cheap with disadvantages, mentioned above, resulting from the lack of control of evaporation rate. The second type (resistively-heated) is more complex and costly but has the advantages noted above.

11.6 COMPARISON OF PROPERTIES OF UHV PUMPS

Some important properties of uhv pumps are tabulated in Table 11.9. The parameters chosen for inclusion in this table are those which principally determine the choice of pump for any particular system or experiment. The values given are intended to be representative and there may be considerable variation between different manufacturers' versions of the same type of pump. Some information about the approximate price of various types of pump may be found in Santeler *et al.* (1966).

CHAPTER TWELVE

EXAMPLES OF ULTRAHIGH VACUUM SYSTEMS

The processes which determine the overall performance of practical uhv systems are a complex combination of the processes treated in Part A of this book. In view of the great number of possible combinations, it is perhaps surprising that practical uhv systems have many similarities in performance.

12.1 SMALL GLASS AND METAL SYSTEMS

12.1.1 Apparatus

Systems of volume < 10 litres are used for many experimental purposes including surface studies, very pure gas studies, gauge calibration, and thin film evaporation. The envelopes of the systems are either all-glass, all-metal, or some combination of the two. The most common materials are borosilicate glass and stainless steel. Some of the system characteristics depend upon the wall material (e.g. residual helium arising from permeation of atmospheric helium is present in borosilicate glass systems pumped with ion pumps, and residual hydrogen is prominent in stainless steel systems), but the general level of pressure achieved in small systems appears to be independent of envelope material, and is usually in the range 10^{-12} to 2×10^{-10} torr. Special small systems have been described, in which lower total pressures have been obtained ($< 10^{-12}$ torr in glass using a Hg diffusion pump (Venema & Bandringa 1958); $\sim 10^{-14}$ torr in aluminosilicate glass using a cryosorption pump (Hobson 1964)) but these systems did not contain mass spectrometers. The discussion below is restricted to systems with mass spectrometers (Chapter 9), since a more detailed description of vacuum conditions is then possible. Figures 12.1 and 12.2 show representative small vacuum systems of glass and metal respectively. The processing of such systems has been discussed in Chapter 10.

12.1.2 Residual conditions achieved

Residual partial pressures obtained in small low pressure systems have been given in Table 9.2 and Table 11.2 to illustrate the performance of mass spectrometers and pumps respectively. Similar data have been collected in Table 12.1 to provide a comparison between the background conditions achieved in

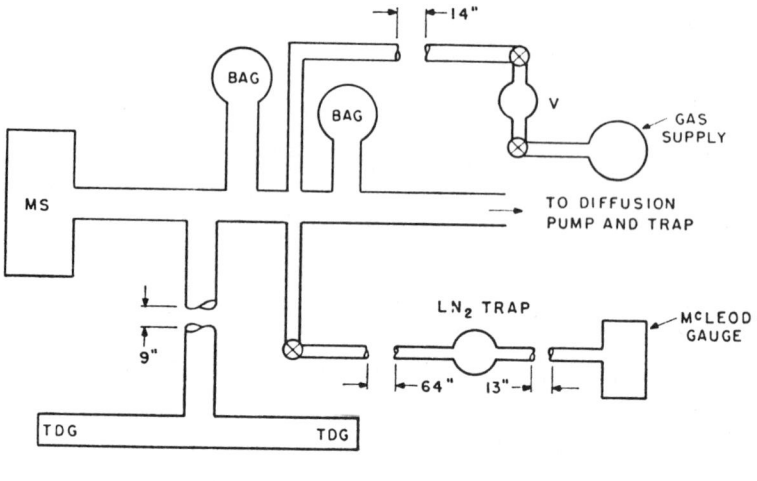

Fig. 12.1 Small uhv system, primarily glass, used for calibration of gauges from 10^{-10} to 10^{-3} torr. Connections and tubulations are shown to scale. BAG denotes Bayard-Alpert ion gauges; MS, the mass spectrometer (cycloidal); TDG, the trigger discharge gauges under test. Whole system, except McLeod gauge and gas supply, baked 14 hours at 400°C. McLeod gauge baked 1 hour. $P_{ult} \sim 2 \times 10^{-10}$ torr. (Lange et al. 1966).

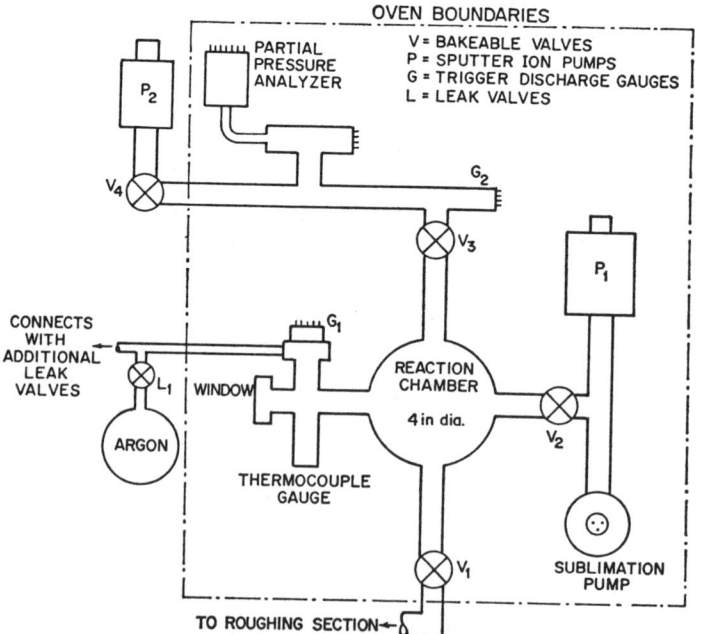

Fig. 12.2 Small metal uhv system, stainless steel, used for thin film deposition. $P_{ult} \sim 5 \times 10^{-11}$ torr (Rozgonyi et al. (1966).

TABLE 12.1

Comparison of background conditions in small uhv systems at pressures $<10^{-10}$ torr

Total background pressure (Units of 10^{-12} torr)	Partial pressures (units of 10^{-12} torr, equivalent N_2)								Type of mass spectrometer	Main pumps	System envelope	Reference
	H_2	He	CH_4	H_2O	CO	N_2	O_2	Ar	CO_2				
0·3	0·2	0·02	0·01		0·05(a)			0·01		90° Magnetic sector	Sputter ion & cryopump	Stainless steel	Davis (1962)
1	0·3	0·02	0·1		0·1(a)			0·1		90° Magnetic sector	Sputter ion & getter	Stainless steel	Davis (1962)
5	3·3	0·06			1				0·6	90° Magnetic sector	Hg Diffusion pump	Glass and metal	Young & Hession (1963)
6	5	~0·1		<1	<1	1				Omegatron	Sputter ion pump	Glass and metal	Klopfer (1961a)
30	10			12	6(a)					Omegatron	Oil diffusion pump	Stainless steel	Hoch (1966)
37	21		0·3	0·9	9		3	6		Cycloidal	Hg diffusion pump	Glass and metal	Singleton (1966b)
54	8		0·6	5	15		1	24		Cycloidal	Oil diffusion pump	Glass and metal	Singleton (1966b)
50	7	0·02	2	0·07	20(a)			20(b)	0·06	90° Magnetic sector	Sputter ion & cryogetter	Stainless steel	Rozgonyi et al. (1966)
~10	1·5				6					Quadrupole	Oil diffusion pump & cryogetter	Stainless steel	Huber (1966)
90	72		3	3	14					Quadrupole	Oil diffusion pump	Stainless steel	Huber (1966)

(a) Represents partial pressure of $CO+N_2$.
(b) Argon artificially high because apparatus used for argon sputtering. Partial pressure estimated from spectrum. Other minor partial pressures present.
Incomplete data for total $p<10^{-10}$ torr may also be found in Rivière et al. (1965); Tseitlin (1965); Young (1966a).
Data similar to the above for $10^{-10}<p<10^{-9}$ torr may be found in Reynolds (1956); Kornelsen (1959); Kendall (1962); Farrar et al. (1964); Bültemann and Delgmann (1965); Gosselin and Bryant (1965); Lichtman (1965a); McCracken (1965); Lange et al. (1966); Theodorou et al. (1966).

small uhv systems at total pressures below 10^{-10} torr. General conclusions based upon Table 12.1 are:

(i) The main residual components are relatively few in number. H_2 and CO are always present, with H_2 usually dominant. All other gases are below the limit of detection in some systems. O_2 is reported only in the two systems of Singleton (1966b), suggesting some special process in these systems in which H_2O, CO, CO_2 are also relatively high.

(ii) He and Ar are absent or very low in all systems pumped with diffusion pumps. This reflects increased speed of diffusion pumps over ion pumps for the rare gases at very low pressures. With less certainty, the same conclusion may be drawn for CH_4. The He partial pressure may be expected to increase with the fraction of the envelope made of glass. Davis (1962) notes that for a system made entirely of borosilicate glass, He is the major gas for pressures below 10^{-10} torr.

(iii) The pressures of H_2 and CO do not appear to have any correlation with the type of pump, the type of mass spectrometer, or the type of system envelope, although it may be noted that all the systems of Table 12.1 had some metal in their envelopes. All the mass spectrometers had conventional ion sources with hot filaments.

(iv) Before baking a system, H_2O is a prominent peak in the mass spectrum (Hickmott 1960b; Tuzi 1962), but Table 12.1 demonstrates that H_2O is not normally a major component of the residual uhv spectrum.

(v) Since real leaks from the atmosphere are characterized by large N_2 and Ar peaks, it is clear from Table 12.1 that the present limitations of small uhv systems are not the result of real leaks.

Typical residual spectra observed in small uhv systems are shown in Figs. 12.3 and 12.4. In Fig. 12.3 the peaks at mass 16 and 19 are surface ionization peaks created at the grid cage in the ionization source (see also Figs. 4.20 and 4.21) of the 90° magnetic sector mass spectrometer. These surface peaks are much less prominent in Fig. 12.4, showing the residual spectrum taken with a quadrupole mass spectrometer in which less of the surface ions reached the collector.

The relationship between H_2 and CO, and the surface peaks associated with these gases, is further illustrated in Fig. 12.5 which shows the behaviour of the gas phase H_2 and CO peaks in a mass spectrum as a molybdenum ribbon is heated near the ion source of a mass spectrometer. The ribbon is simultaneously bombarded with electrons (100 eV) causing the H^+ and O^+ peaks. Note that as H_2 is desorbed the H^+ peak declines, and as CO is desorbed the O^+ peak declines. The second H_2 peak which appears simultaneously with the CO desorption peak is assumed to be the result of replacement of H_2 by CO at the walls of the system (Section 2.4.5).

12.1.3 Residual processes

A detailed description of the physical processes contributing to the residual

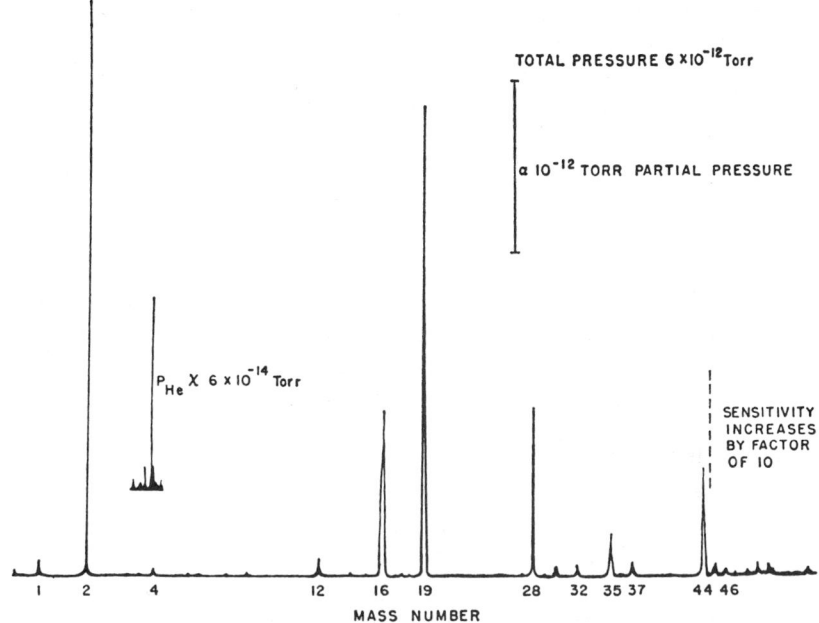

Fig. 12.3 Mass spectrum at a pressure near 5×10^{-12} torr. 90° magnetic sector mass spectrometer (Young & Hession 1963).

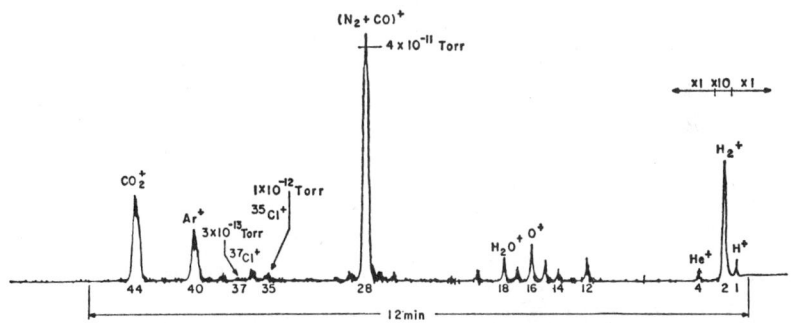

Fig. 12.4 Residual gas spectrum at a pressure of $1 \cdot 2 \times 10^{-10}$ torr. Quadrupole mass spectrometer (Bültemann & Delgmann 1965).

gas composition in an uhv system is usually not possible because of the complexity of conditions. The magnitude of the equilibrium partial pressure of a gas is only the resultant of the sources and sinks of unknown magnitude. The study of residual processes has usually involved the changing of some system parameter and the observation of the subsequent transient behaviour of the parial pressure of interest. Reynolds (1956) reported that simply turning on the ion beam (electron beam on continuously) in his mass spectrometer raised the pressure from 5×10^{-11} torr to 5×10^{-10} torr in 3 hours. He assigned this

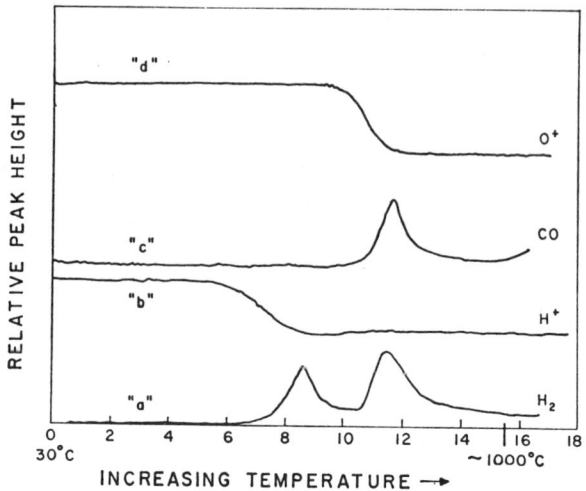

Fig. 12.5 Behaviour of H_2 and CO peaks in a mass spectrum, together with surface peaks from H^+ and O^+ due to electron bombardment of a Mo ribbon, as ribbon is heated. Amplitudes of each curve have been adjusted to make a convenient figure and relative amplitudes between curves are not related. The CO peak is considerably larger in absolute magnitude than the H_2 peak. Numbers on abscissa are indicative only (Lichtman 1965a).

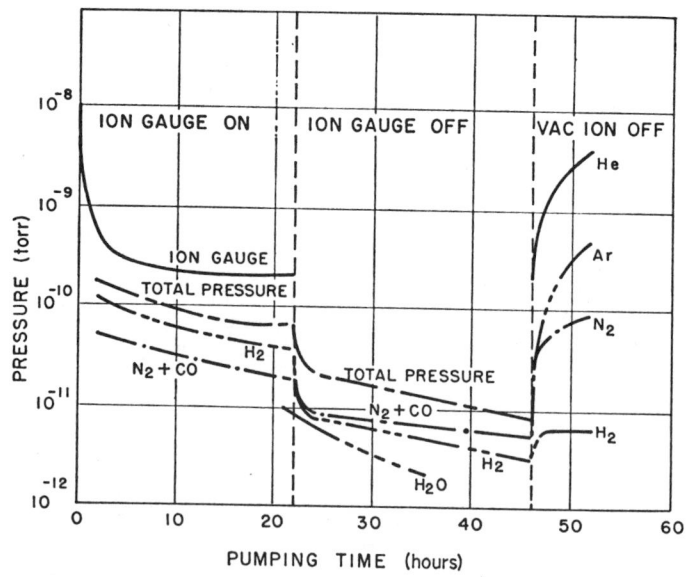

Fig. 12.6 Behaviour of main residual gases when ionization gauge and ion pump turned off in small glass and metal uhv system (Klopfer 1961a).

result to heterogeneous re-emission (Section 11.5) resulting from ion bombardment. It is generally true that the sources for the residual gases are temperature dependent and that the cooler an uhv system is operated the lower is the ultimate pressure. It is frequently observed (Klopfer 1961a; Young 1966a) that the operation of an ionization gauge causes a general rise in system pressure. Figure 12.6 shows a result of Klopfer (1961a) which is typical of the transient behaviour of gases in an uhv system. When the ionization gauge was switched off, the partial pressures of H_2, $N_2 + CO$, and H_2O fell slowly (note time scale); when the ion pump was switched off, the partial pressure of the rare gases He and Ar dominated the pressure rise (see also Fig. 7.4), while the partial pressure of H_2 remained low. Klopfer (1961a) measured the leak rates in this system to be 2×10^{-13} torr litre sec^{-1} for He, 2×10^{-14} torr litre sec^{-1} for Ar, 3×10^{-12} torr litre sec^{-1} for N_2, and 6×10^{-12} torr litre sec^{-1} for H_2. These are representative figures for a small glass and metal system.

For each of the gases H_2, He, CH_4, N_2, CO and Ar in turn, Hobson and Earnshaw (1967) have tested whether the partial pressures followed the differential equation

$$\frac{V dp}{dt} = -pS + L \text{ torr litre sec}^{-1}, \qquad (12.1)$$

with S (pumping speed) and L (leak rate) considered as constants. It was found that this was a satisfactory procedure for He, CH_4, N_2 and Ar, but not for H_2 and CO. For the latter two gases the presence of a surface phase was postulated, in equilibrium with the gas phase. At present the physical identity of this surface phase is not established: it could be an adsorbed phase (Section 2.4.2) or it could be a dissolved phase (Section 2.5). The postulate of the surface phase requires the replacement of eqn. (12.1) by two simultaneous differential equations

$$\frac{V dp}{dt} = -pS - pS_s + L + \frac{n_s}{\tau_s} \qquad (12.2)$$

$$\frac{dn_s}{dt} = pS_s - \frac{n_s}{\tau_s}, \qquad (12.3)$$

where n_s is the quantity of gas (torr litre) in the surface phase whose pumping speed was S_s, and in which a molecule had a mean sojourn time τ_s. The equilibrium solutions to eqn. (12.1) and to the combination of eqns. (12.2) and (12.3) are the same, namely

$$p_\infty = L/S \qquad (12.4)$$

where p_∞ is the equilibrium pressure. However, the transient solutions of eqn. (12.1) and the combination of eqns. (12.2) and (12.3) are not the same. Soo and Huang (1962) have reported transient solutions for a surface phase governed by surface diffusion. After application of eqns. (12.2) and (12.3) to transient data on H_2 and CO in an uhv system, Hobson and Earnshaw

(1967) concluded that the quantity of these gases in their respective surface phases could be much larger than the quantities in the gas phase. The consequence of this conclusion was that experiments designed to establish the origin of residual H_2 or CO in an uhv system generally measure the transient effects of the surface phase. For example, it should not be unambiguously concluded from Fig. 12.6 that the source of H_2 was in the ionization gauge; an alternative explanation is that the cooling of the gauge caused an additional pump in the ionization gauge which in time would have saturated and permitted the H_2 partial pressure to rise to its original value of about 5×10^{-11} torr. This experiment could, in principle, have been done, but equilibration times at these low pressures could be even longer than those shown in Fig. 12.6. Davis (1963) reported that the main difficulty in calibrating gauges in N_2 at pressures below 10^{-10} torr was the length of time required for equilibration. In an effort to reduce the interaction of gases with the walls of a glass uhv system, Singleton (1963) deposited Teflon bakeable to 373°C on the walls. The ultimate pressure of 2×10^{-10} torr was not reduced by this treatment, but less wall adsorption for CO was measured. Hobson and Earnshaw (1967) also concluded that localized surface phases could cause gauges located in different regions of an uhv system to read differently (see Section 7.2.3). Lange et al. (1966) concluded that locally high pressures of CO, CO_2, and H_2O were present in their mass spectrometer at a total pressure near 1×10^{-11} torr. Similarly, Rivière et al. (1965) concluded that locally high pressures of one or more of the gases H_2O, CO, CH_4, and H_2 were present in their mass spectrometer at a total pressure $\sim 1 \cdot 2 \times 10^{-12}$ torr.

The origins of the various residual gases will be briefly reviewed. General agreement is not found among various authors.

Rare gases. Small partial pressures of the rare gases are to be expected, particularly in ion pumped system, as a result of thermal re-emission and of both homogeneous and heterogeneous re-emission (Section 11.5 and 4.3.4). The same processes occur in gauges (Section 7.2.1). Davis (1962) found that, following baking of an evacuated pump, the re-emission rates of He and Ar were reduced. Reynolds (1956) reported that much, but not all, of the memory of his mass spectrometer for argon was eliminated by a complete bake. Rivière and Allinson (1964) concur that heating of ion pumps enhances the re-emission of gas previously pumped and that gas can be transferred more or less reversibly from one pumping location to another. Davis (1962) found that a general increase of system pressure from 3×10^{-11} torr to 10^{-9} torr increased the Ar partial pressure from 2×10^{-13} to 10^{-11} torr, presumably as a result of heterogeneous re-emission in the ion pump.

Leak rates of the rare gases arising either from permeation (Alpert 1953a), or re-emission, can provide a pressure rising linearly with time over a short range which can be used for checking the linearity of gauges (Alpert 1953a; Davis 1962). This method has been discussed in Section 7.1

Hydrogen. The mechanisms suggested by various authors as the cause of residual hydrogen in their uhv systems are varied. Gosselin and Bryant (1965) found the H_2 peak to increase when the valve to the oil diffusion pump was opened, and suggest that H_2 originates in the unbaked portions of the pump. Huber (1966) also regards the oil diffusion pump as the source of residual hydrogen in his system. Tseitlin (1965) found that the limiting pressure of 10^{-10} torr in a stainless steel system arose from hydrogen entering the system by permeation through some corroded stainless steel bellows as a result of the action of atmospheric water vapour. He suggests the reaction

$$Fe + H_2O \rightarrow FeO + H_2. \tag{12.5}$$

The hydrogen so formed diffused through the walls of the bellows. Painting the bellows with anti-corrosion silicone layers lowered the residual pressure from 10^{-10} torr to 4×10^{-12} torr, as measured on an inverted-magnetron gauge. This pressure was maintained for a year. In contrast to the mechanism described by Tseitlin (1965), Young (1963a) has described a process in which H_2 diffuses from low to high pressure through palladium, and to a lesser degree through titanium and nickel. The requirements for this type of diffusion pump are: (a) any H_2 permeable membrane, (b) the high pressure side must be exposed to an oxidizing atmosphere, and (c) the membrane must catalyse the reaction between H_2 and O_2 to yield water. Young (1963a) found that a pressure ratio of 4×10^9 could be maintained across hot palladium by this mechanism, the low pressure being the interior of the system. Young's (1963a) mechanism might be considered as a reversal of the arrow in eqn. (12.5) and might tend to reduce the permeation of atmospheric hydrogen through the metal walls of a uhv system.

Davis (1962) located two sources of H_2, one in the mass spectrometer itself and another in the body of the uhv valve, and noted that permeation of atmospheric hydrogen through the bellows in the valve might account for the source of residual hydrogen. This conclusion is similar to our conclusion in Section 2.5.4 that outgassing of H_2 from stainless steel can account for residual H_2 pressures of 3×10^{-13} torr. These considerations are invalid if the mechanism of Young (1963a) noted above plays a role.

Buser and Sullivan (1966) have also reported that opening and closing all-metal uhv valves containing copper as the sealing material caused pressure bursts in the range 10^{-10} to 10^{-7} torr in a small system consisting mostly of H_2 with trace amounts of masses 16, 18, 28 and 40. The effect was larger for the valves employing a cutting action rather than a deformation of the copper. This suggested the presence of H_2 dissolved in the bulk copper. However, motion of valve bellows also gave pressure bursts, suggesting desorption of surface H_2 as well.

Methane. Davis (1962) found little CH_4 in glass systems, but located its source in a metal system at the metal walls of the ionization gauge. He suggested that it might arise from the interaction of carbon in the stainless steel with hydrogen atoms or ions in the gauge, or alternatively its source might be the Ti or Ta used in the gauge. Lichtman (1964) concludes that CH_4 arises from the combination of hydrogen and carbon in his ion pump. At H_2 pressures in the range 10^{-7} to 10^{-6} torr, Holland et al. (1961) have measured the rate of generation of methane from the interaction of H_2 with carbon impurities in Ti (see Section 11.4.1). Gosselin and Bryant (1965) found no CH_4 in a glass-metal system pumped with a mercury diffusion pump with a liquid nitrogen trap and using a cold-cathode gauge at a pressure of 2×10^{-10} torr, but CH_4 was a major residual component in a similar system pumped with an oil diffusion pump and employing a hot-cathode gauge at a pressure of 5×10^{-10} torr. It was established that CH_4 was not being formed at the tungsten filament of the ionization gauge, but was apparently enhanced by increasing the grid bombardment potential from 130 to 350 volts. Gosselin and Bryant (1965) concluded that CH_4 was an adsorbed species originating from the oil diffusion pump during bakeout. However, Davis (1962) and also Hobson and Earnshaw (1967) show that CH_4 is not readily chemisorbed on system parts, and behaves much like a rare gas.

Carbon monoxide and nitrogen. These two common gases have molecular mass 28 and are both readily chemisorbed on system parts. Their physisorption properties are also similar (see Table 2.18), but they can be readily distinguished in chemical desorption spectrometers (see Section 9.2.1). In conventional mass spectrometers they are relatively difficult to distinguish in an uhv system, and Table 12.1 shows that only in about half the studies below 10^{-10} torr has the distinction been made. The simplest method of distinction is on the basis of their fragmentation patterns: Table 9.1 shows that 2% of the CO peak at mass 28 will be found at mass 12, while 5% of the N_2 peak at mass 28 will be found at mass 14. Since these fractions are relatively small, sensitivity difficulties may prevent a clear-cut distinction at very low pressures (Klopfer 1961a; Davis 1962). Significant amounts of CH_4 also confuse the distinction. Following the bake, Davis (1962) found CO dominated N_2 but after continued outgassing, baking and prolonged operation of his mass spectrometer filament the two gases were reduced to approximately the same level, the total pressures being given in Table 12.1. Kornelsen and Domeij (1966) report that the chemisorbable gas in a small glass system was reduced to a pressure of $\sim 1 \times 10^{-12}$ torr and was predominantly N_2, based on identification with a desorption spectrometer (Section 9.2). The detailed origins of these two residual gases has not been described. It seems probable that N_2 represents some residual adsorption or solution of atmospheric N_2 in system parts. There is no doubt from the results of Schuemann et al. (1963), shown in Fig. 7.7, that the introduction of oxygen to a small glass uhv system causes the production of CO, presumably at a hot filament, and it may be inferred that small amounts of

O_2 remaining from the original atmosphere could create residual CO in a similar process. Gosselin and Bryant (1965) regard hot tungsten filaments as the source of their CO. Reynolds (1956) and also Singleton (1966a) report that CO can arise from the interaction of CO_2 at hot filaments, but this cannot be the generally dominant process because CO is normally in excess of CO_2 (Table 12.1).

Carbon dioxide. CO_2 is not normally found as a major residual gas in small uhv systems, but Singleton and Lange (1965) report it as such, under particular conditions of processing of a Pyrex system. CO_2 was released from glass during bake and accumulated on the cold trap of the diffusion pump as an adsorbed layer or as bulk condensate. Unless the system was sealed off, and this condensate removed from the trap by warming the trap to room temperature, a major residual gas in the system was found to be CO_2, whose pressure was sensitive to the level of liquid in the trap (see Section 10.2).

12.1.4 Admission of gas

Frequently the purpose of achieving uhv is to provide background conditions of adequate simplicity to permit operations to be performed at higher pressures with controlled atmosphere. Thus gas is often admitted to uhv systems and it is necessary to inquire to what extent the background composition is changed, while gas is being admitted, and whether the original background conditions can be recovered after the admission has ceased.

The admitted gas must, of course, be of adequate purity. A study of various methods of admitting the gases O_2 (Whetten & Young 1959), He (Young & Whetten 1961), H_2 (Young 1963b) has been made. Primarily these studies were concerned with the purities which could be obtained by permeating O_2 through Ag, He through quartz, and H_2 through palladium respectively, but comparisons with other methods of gas introduction were made. Table 12.2 compares three common methods of introducing helium into an uhv system. It is evident that the permeation method provides the purest sample. It is not clear from Young and Whetten's (1961) paper whether method II, as used by them, could have introduced sufficient impurities to account for the discrepancy between the supplier's specification and the observed impurities. In their studies of O_2 permeating silver, Whetten and Young (1959) found no difference between methods I and II, although impurities of CO and CO_2 were present at about 1% of the O_2 peak. These were thought to arise inside the uhv system by oxygen interacting with carbon present in the incandescent Ta filament in the mass spectrometer (Sections 7.2 and 7.6). Young (1963b) found that method II for H_2 gave impurities of 8 ppm, while H_2 permeating Pd and Pd 5% Ag always gave impurity levels <1 ppm and with very careful processing the impurity level was 10^{-4} ppm. With H_2 permeating nickel the impurity level was always >1 ppm.

TABLE 12.2
Impurities observed in helium introduced into vacuum system by three different methods

Method I He *introduced by permeation through quartz using tank helium in outer jacket*		Method II Pure He *from 1-litre Pyrex flask introduced through bakeable metal valve*			Method III Tank He *introduced through a valve*	
		Observed		(Suppliers' specifications)		
	ppm		ppm	ppm		ppm
Ar	<1	Ar	2		Ar	280
O_2	<1	O_2	<1	<1	O_2	65
N_2	<1	N_2	3		N_2	950
Ne	<1	Ne	25		Ne	160
H_2	10	H_2	10	<1	H_2	120
CO	<1	CO	30	<1	CO	1
CO_2	<1	CO_2	2	<1	CO_2	1

(from Young & Whetten 1961).

Impurities, however, may be introduced by the uhv system itself even when a rare gas is admitted. Figure 12.7 shows the background spectrum measured by Rozgonyi (1966), and also the spectrum after the introduction of argon to a pressure of 4×10^{-8} torr. Only the mass spectrometer and two trigger discharge gauges were operating in the system. The peak at mass 20, which is 10% of the main peak at mass 40, is in reasonable agreement with the fragmentation pattern given in Table 9.1, but there are several peaks at 1% of the main peak and these are much larger than they were in the background spectrum. Rozgonyi (1966) estimated his original argon impurity level as <50 ppm (in reasonable agreement with method II of Table 12.2), and hence could account for the high level of impurities only by processes taking place within the uhv system itself. It appears likely that the main desorption process is that of heterogeneous re-emission (Section 11.5) but secondary processes also take place because the surface peaks marked S in Fig. 12.7 have also been increased by the introduction of Ar. It has been found in the authors' laboratory that the admission of chemically active gases, in particular oxygen, can lead to the appearance of the rare gases.

During the deposition of thin films, the background pressure of an uhv system is raised, and the critical properties of the film are frequently determined by the ratio of the arrival rate at the substrate of a particular impurity gas to that of the film material (Caswell 1963). Specific gases have more severe effects than others, the order of difficulty frequently being O_2, H_2O and CO_2, as shown in Table 12.3, which gives the upper limits of the partial pressures of these gases which may exist during film deposition of tin and indium, to give sharp changes with magnetic field between a normal and a superconducting state. At a deposition rate of 50 Å sec^{-1} and an oxygen

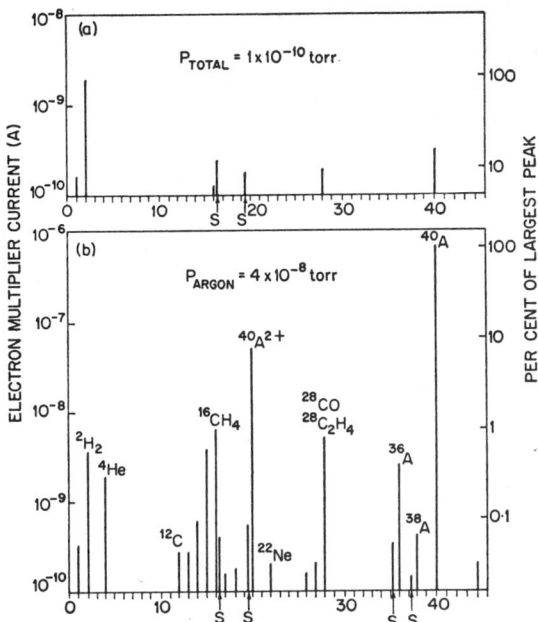

Fig. 12.7(a) Mass spectrum at a total pressure of 1×10^{-10} torr in a small metal system with mass spectrometer and two trigger discharge gauges only operating.
Fig. 12.7(b) Same conditions as for Fig. 12.7(a) after introduction of Ar to raise pressure to 4×10^{-8} torr (Rozgonyi 1966).

pressure of 10^{-7} torr, approximately 200 atoms of metal strike the substrate for every molecule of oxygen. The requirements set forth in Table 12.3 can be met, with care, in small uhv systems (Caswell 1963).

The process of heterogeneous re-emission can take the form of raising the minimum detectable leak of a helium leak detector (Barrington *et al.* 1965). Normal procedure in setting up and testing a leak detector might be to evacuate it to the lowest possible pressure and test it with a source of He.

TABLE 12.3

Pressure limitations (torr) on residual gases required for sharp, reproducible magnetic field transitions from normal to superconducting state

Evaporation rate of film 50 Å sec^{-1}

Gas	Tin	Indium
O_2	5×10^{-8}	5×10^{-8}
H_2O	4×10^{-7}	4×10^{-6}
CO_2	8×10^{-7}	10^{-5}
Other	10^{-5}	10^{-5}

(from Caswell 1963).

Some ionic pumping of He will occur both in the mass spectrometer and in ionization gauges. In subsequent operation, however, the leak detector will operate probably several orders in pressure above the test conditions. The process of heterogeneous re-emission may well raise the partial pressure of helium so that the minimum detectable leak is raised above that measured under test conditions.

While it is clear that very low pressures can be achieved in small glass and metal systems, mass spectrometers are essential for many types of experiments.

12.2 LARGE UHV SYSTEMS

The application of uhv techniques in large systems has lagged the development of small systems by several years. However, since late in the 1950's, expanding work in the fields of space flight, thin-film electronics, plasma physics research and, more recently, storage rings for high energy particle accelerators has led to the development of a variety of large systems (10 to 10^6 litre) operating in the uhv range. In most of the above applications, large gas loads during operating have made it necessary to provide very large pumping speeds to maintain pressures in the uhv range. The reduction of outgassing rates to an absolute minimum is thus usually of less importance than in the small systems described in Section 12.1.

In addition to the high speed-high outgassing rate condition described above, two other factors make detailed assessment of the conditions in large systems difficult:

(i) Surfaces cooled to cryogenic temperatures are very commonly used in the pumping systems, making thermomolecular corrections to pressure indications necessary between different parts of the system (see Section 7.3).

(ii) Practically no partial pressure information, of the type used extensively in Section 12.1, has yet been reported for large systems.

In the few cases where partial pressures were measured, H_2, H_2O, $(CO+N_2)$ and CO_2 were the only gases found. Since the inert gases He, CH_4 and Ar would be removed relatively efficiently in systems with high speed diffusion pumps, the spectra suggest that the outgassing properties of large systems do not differ greatly from those of smaller systems (Section 12.1) except for the greater prevalence of water.

The use of stainless steel as the envelope material in large uhv systems has become almost universal. It combines reasonable outgassing properties and machinability with resistance to corrosion, strength at bakeout temperature, and welding properties. Demountable sealing gaskets are generally of soft metal (Cu, Al or Au) for sizes < 50 cm diam, and of elastomer material (neoprene, butyl rubber, or fluorinated hydrocarbons) for larger sizes. In the latter case, the gaskets are frequently the dominant source of background gas evolution. When not too many cycles of opening and reclosing are required, welding along feathered edges integral with the flanges can provide an alternative to elastomer seals of large diameter (Milleron 1967). Such welds must be ground off to open the seal and re-welded upon closure. In cases where the systems are exposed to the radiation from nuclear reactors (e.g. space simulators testing components for vehicles which contain reactors), stainless steel envelopes are not acceptable because long-lived radioactivities may be induced in components or impurities in the envelope. Hard aluminum alloys have been used in place of stainless steel in such applications.

We will describe briefly below a selection of systems which use a wide variety of techniques to achieve pressures below 10^{-9} torr.

12.2.1 Conventional systems of moderate size ($< 10^3$ litre)

Chambers with volumes up to several hundred litres pumped with either sputter-ion pumps or cold-trapped diffusion pumps have been reported by several experimenters. In all these systems, the working volume is at or near room temperature. Miller and Geiger (1962) describe two systems, 100 and 800 litre, which were pumped by liquid N_2 trapped oil diffusion pumps, baked at 250–300°C and achieved base pressures of $\sim 3 \times 10^{-10}$ torr. They found it necessary to cool their elastomer gaskets (neoprene) to $-20°C$ to reduce their outgassing rate.

Farkass et al. (1962) have reported that, in systems of similar design, (500 to 2000 litre), they were able to obtain pressures near 1×10^{-11} torr without *any* bakeout when their elastomer seals were cooled to $-25°C$. Since corrections for cold-cathode gauge non-linearity (see Section 8.2.2) were not made, the pressure estimate is undoubtedly too low, although probably not by more than a factor ten.

Additional pumping speed at an evaporated metal film has been used in conjunction with trapped oil diffusion pumps to achieve uhv. Holland (1960) used Ti evaporation to achieve $\sim 2 \times 10^{-10}$ torr in a film evaporation unit of 30 litre volume sealed with Al gaskets and baked at 370°C. Hunt et al. (1961) obtained $\sim 4 \times 10^{-10}$ torr in an 85-litre unbaked steel chamber after repeated evaporation of Mo onto the chamber walls.

A sputter-ion pump and additional Ti evaporation has given 4×10^{-11} torr in a baked 100 litre system pre-evacuated with sorption pumps (Zaphiropoulos & de Taddeo 1966). Outlaw (1966) reports a pressure of 6×10^{-12} torr in a ~ 10 litre system evacuated first with a turbomolecular pump, then sealed and pumped by Ti deposited on a 77°K surface. (Ionic pumping was provided by a magnetron gauge.) In this case, the entire wall of the working volume was at 77°K (liquid N_2).

Attempts have been made to reduce both the background pressure and the processing time required to reach it by using double-walled vacuum systems. An uhv vessel is surrounded by a guard vacuum to reduce leak rate and facilitate rapid bakeout. Metcalfe and Trabert (1961) surrounded a 450 litre Inconel chamber with a guard vacuum system capable of producing

~1×10^{-6} torr. Flat-faced flanges, 3 cm wide, polished to an '8μ inch' finish provided the seal between the two chambers. After 8 hours bakeout at 400°C, a pressure of less than 1×10^{-10} torr was attained in about 24 hours. The uhv region was pumped by an oil diffusion pump with a chevron-type liquid N_2 trap. Blahnik and Shoulders (1966) describe an 80-litre stainless steel system which was baked at 900°C for 15 minutes in a guard vacuum of 10^{-5} torr. The system is described in more detail in Section 10.2.

12.2.2 Cryogenic systems

A vacuum enclosure such that every molecule released within it was pumped on its first impact with the wall would constitute a theoretically perfect pumping system. Systems have been reported which, for many practical purposes, approach this theoretical limit by lowering the entire wall of a chamber to such low temperatures that practically all gases physisorb on it with high efficiency (see Section 11.3). An unavoidable consequence of this approach is that any apparatus within such a system will tend to cool, by radiation and conduction, to a very low temperature. Conversely, excessive radiation from the apparatus falling on the pumping system may induce desorption and seriously degrade the pumping performance (Chubb et al. 1968).

Mark and Sommers (1962) describe a large liquid helium cryostat containing a 150-litre volume surrounded by walls at 4·2°K, which has been mentioned earlier (Section 11.3.1). The system was pumped to ~10^{-6} torr by an oil diffusion pump, and cooled to 20°K by gaseous He before liquid He was transferred. With the walls at 4·2°K, the pressure recorded by a magnetron gauge was below 10^{-13} torr (see Fig. 11.3). Even correcting for the gauge non-linearity (see Section 8.2.2) pressures were probably well below 10^{-11} torr, and may have been much lower. Introduction of radiant flux equivalent to that from the sun to a model suspended in the chamber had only a minor effect on the indicated pressure.

A somewhat similar system, with a 100-litre copper chamber cooled to 10°K by a gaseous helium refrigerator, is described by Feakes et al. (1965). The copper chamber can be heated to 'several hundred °C' while the system is evacuated by a large trapped oil diffusion pump. The pressure before cooling to 10°K is nominally 1×10^{-6} torr, and is deduced to be ~10^{-14} torr at 10°K although direct indications on the magnetron gauge did not fall below 2×10^{-12} torr. The system was used to calibrate cold-cathode gauges in He at very low pressures, using a single-stage pressure division technique (see Section 7.1.3).

A larger system (~2000 litre) of similar design is reported nearing completion (Stephens (1966). In this case, the wall is composed entirely of wedge-shaped aluminium fins cooled to 20°K by helium gas. Titanium will be evaporated onto the fins to increase the hydrogen pumping speed, and the entire assembly will be mounted in a 3-metre diam steel chamber evacuated by a trapped diffusion pump. It is estimated by the author that a molecule (of O_2, N_2, CO or CO_2) striking the wall will have a 99·97% probability of capture.

12.2.3 Accelerator vacuum systems

Storage rings used in conjunction with high energy particle accelerators offer two significant advantages:

(i) They allow beam intensities to be greatly increased over those available directly from accelerators.

(ii) With counter-rotating beams in intersecting rings, they make the entire kinetic energy of two colliding particles available for reactions (i.e. the centre-of-mass is stationary).

To obtain sufficiently long storage times in such rings, pressures must be reduced to the uhv range to avoid collision of the circulating particles with residual gas molecules.

A proposal for a pair of intersecting 28 GeV proton storage rings is reported by Fischer (1965). The rings would each accept 600 pulses from the CERN proton synchrotron and would have circulating proton currents of 30 A. A pressure of 1×10^{-9} torr would, by single and multiple scattering, result in a reduction of 4% in the circulating current and a 44% increase in its cross-sectional area in a 12 hour period. It is proposed to pump the rings, 5×16 cm in cross-section and a total of 2 km long, with 240 sputter-ion pumps of 400 litre/sec each. In the intersection regions, additional cryogenic pumping would be used to lower the pressure to $\sim 10^{-11}$ torr to avoid excessive background counting rates.

Designs have also been proposed for electron-positron storage rings of 1·5 GeV energy, 94 mA circulating current (Bernardini & Malter 1965) and 3 GeV with 1 A circulating current (Fischer & Mack 1965), both requiring uhv. For the case of electrons and positrons, in contrast to heavy particles, a very severe problem is caused by the synchrotron radiation emitted tangentially by the circulating particles. The radiation covers a very broad spectrum, down to wavelengths of less than 10 Å, and in the largest rings considered (Fischer & Mack 1965) amounts to a continuous power of 500 kW. The energetic photo-electrons ejected from the solids struck by this radiation will in turn release adsorbed gases by electron induced desorption (see Section 4.2) to give the predominant gas load in the system. Fischer and Mack (1965) propose to pump the 172 metre long, 2500 litre ring, with fifty 500 litre/sec sputter-ion pumps supplemented by titanium evaporation on a large fraction of the internal surface area (the beam volume would be shielded from the evaporation).

12.2.4 Systems for plasma physics research

High density, fully ionized hydrogenic plasmas are at present the main approach toward producing a controlled thermonuclear fusion reaction.

Ultrahigh vacuum is required to reduce energy losses due to atomic excitation in such plasmas to acceptable values. Since the loss rates are approximately proportional to Z^2, the heavier residual gases are the most serious contaminants.

The vacuum system of the C-Stellarator was one of the earliest large ($\sim$500 litre) uhv systems (Henderson *et al.* 1959; Mark & Dreyer 1959). It consisted of a stainless steel tube 20 cm in diameter, about 12 metres long, with many gold wire seals, viewing ports and electrical feedthroughs. The system was connected to a mercury diffusion pump through a large bakeable valve and two liquid N_2 traps in series. Following 450°C bakeout, the system reached a pressure of 2×10^{-10} torr. The cleanest conditions (as far as plasma losses are concerned) have been attained by closing the bakeable valve and repeatedly running a high density He plasma in the tube (currents up to 900 A, Martin & Lewin 1966). The sputtering action of the plasma ions striking the tube walls reduced the outgassing rate by a large factor.

Wells and Postma (1962) have described the vacuum system of the DCX-1 thermonuclear plasma experiment. A 120-litre copper liner is mounted within a larger (750 litre) system pumped to 3×10^{-8} torr by oil diffusion pumps and Ti evaporation. The liner is heated to 400°C by superheated steam, then cooled to 77°K and coated internally with titanium by evaporation. The best background pressure observed was 7×10^{-10} torr, and 2×10^{-9} torr was maintained in the presence of a 10 mA beam of H_2^+ ions.

12.2.5 Space simulators

Although many shall uhv systems have been used for tests on space vehicle components, the testing of entire vehicles at uhv is considered justifiable in spite of the considerable cost of construction of vacuum systems of adequate size. For two reasons, cryogenic pumping systems are the only type feasible for this application:

(i) Cold walls surrounding the vehicle are necessary to simulate the thermal environment of space.

(ii) A large fraction of the wall area must provide pumping action in order to maintain uhv in the presence of the unavoidable large outgassing load of the unbaked vehicle. Adequate diffusion pumping speed would be prohibitively expensive (Santeler 1959).

Systems of this type have been discussed in Section 11.3.2.

REFERENCES

A

ABRAHAMSON, A. A. (1963), *Phys. Rev.* **130,** 693.
ABRAHAMSON, A. A. (1964), *Phys. Rev.* **133,** A990.
ABRIKOSOVA, I. I. and DERYAGIN, B. V. (1957), *Sov. Phys. JETP* **4,** 2.
ABROYAN, I. A. (1961), *Sov. Phys. Solid State* **3,** 431.
ABROYAN, I. A. and LAVROV, V. P. (1963), *Sov. Phys. Solid State* **4,** 2382.
ABROYAN, I. A. and MOVNIN, S. M. (1961), *Sov. Phys. Solid State* **3,** 416.
ACKLEY, J. W., LOTHROP, C. F. and WHEELER, W. R. (1962), *Trans. AVS Vac. Symp.* **9,** 452.
ADAMS, R. O. (1964), Ph.D. Thesis, Washington State University.
ADAMS, R. O. and DONALDSON, E. E. (1965), *J. Chem. Phys.* **42,** 770.
ADAMSON, A. W. and LING, I. (1961), *Adv. Chem. Series* **33,** 51.
AFROSIMOV, V. V. and FEDORENKO, N. V. (1957), *Sov. Phys. Tech. Phys.* **2,** 2391.
AFROSIMOV, V. V., GORDEEV, YU. S., PANOV, M. N. and FEDORENKO, N. V. (1965a), *Sov. Phys. Tech. Phys.* **9,** 1248.
AFROSIMOV, V. V., GORDEEV, YU. S., PANOV, M. N. and FEDORENKO, N. V. (1965b), *Sov. Phys. Tech. Phys.* **9,** 1256.
AFROSIMOV, V. V., GORDEEV, YU. S., PANOV, M. N. and FEDORENKO, N. V. (1965c), *Sov. Phys. Tech. Phys.* **9,** 1265.
AGEEV, V. N. and IONOV, N. I. (1966), *Sov. Phys. Tech. Phys.* **10,** 1614.
AGEEV, V. N., IONOV, N. I. and USTINOV, YU. K. (1965), *Sov. Phys. Tech. Phys.* **9,** 1581.
AGISHEV, E. I. and IONOV, N. I. (1956), *Sov. Phys. Tech. Phys.* **1,** 201.
AGISHEV, E. I. and IONOV, N. I. (1958), *Sov. Phys. Tech. Phys.* **3,** 1638.
ALEXEFF, I. (1961), *Trans. AVS Vac. Symp.* **8,** 472.
ALLEN, F. G., BUCK, T. M. and LAW, J. T. (1960), *J. Appl. Phys.* **31,** 979.
ALLEN, J. S. (1939), *Phys. Rev.* **55,** 966.
ALLEN, J. S. (1950), *Proc. I.R.E.* **38,** 346.
ALMÉN, O. E. and BRUCE, G. (1961a), *Nuclear Inst. and Methods* **11,** 257.
ALMÉN, O. E. and BRUCE, G. (1961b), *Nuclear Inst. and Methods* **11,** 279.
ALPERT, D. (1953a), *J. Appl. Phys.* **24,** 860.
ALPERT, D. (1953b), *Rev. Sci. Instr.* **24,** 1004.
ALPERT, D. (1958a), *Handbuch der Phys.* **12,** 609.
ALPERT, D. (1958b), *Proc. 1st Int. Cong. on Vac. Tech.* (Pergamon Press), p. 31.
ALPERT, D. (1962), *Le Vide* **17,** 19.
ALPERT, D. and BURITZ, R.S. (1954), *J. Appl. Phys.* **25,** 202.
ALTEMOSE, V.O. (1961), *J. Appl. Phys.* **32,** 1309.
AMDUR, I. and GUILDNER, L. A. (1957), *J. Amer. Chem. Soc.* **79,** 311.
AMDUR, I., JORDAN, J. E. and COLGATE, S. O. (1961), *J. Chem. Phys.* **34,** 1525.
AMES, I., CHRISTENSEN, R. L. and TEALE, J. (1958), *Rev. Sci. Instr.* **29,** 736.

ANDERSON, G. S. (1962), *J. Appl. Phys.* **33,** 2017.
ANDERSON, G. S. (1963), *J. Appl. Phys.* **34,** 659.
ANDERSON, G. S. (1966), *J. Appl. Phys.* **37,** 2838.
ANDERSON, G. S. and WEHNER, G. K. (1960), *J. Appl. Phys.* **31,** 2305.
ANDERSON, H. U. (1963), *Rev. Sci. Instr.* **34,** 703.
ANDERSON, J. and ESTRUP, P. J. (1967), *J. Chem. Phys.* **46,** 563.
ANDERSON, P. A. (1935), *Phys. Rev.* **47,** 958.
ANDREEN, C. J. and HINES, R. L. (1967), *Phys. Lett.* **24A,** 118.
APGAR, E. (1963), *Proc. 2nd European Vac. Symp.* p. 223.
APKER, L. (1948), *Ind. Eng. Chem.* **40,** 846.
APPELT, G. (1962), *Vakuum-Technik* **11,** 174.
ARIFOV, U.A., RAKHIMOV, R. R. and KHOZINSKII, O. V. (1962), *Bull Acad. Sci. USSR Phys. Ser.* **26,** 1422.
ARMSTRONG, R. A. (1965), *Proc. 25th Phys. Electronics Conf.*, M.I.T., Cambridge, Mass., p. 209.
ARMSTRONG, R.A. (1966), *Can. J. Phys.* **44,** 1753.
ASH, R., BARRER, R. M. and NICHOLSON, D. (1963), *Zeit. fur Phys. Chem.* **37,** 257.
ASUNDI, R. K., CRAGGS, J. D. and KUREPA, M. V. (1963), *Proc. Phys. Soc. (London)* **82,** 967.
ATKINSON, H. H. (1963), *Trans. AVS Vac. Symp.* **10,** 213.
ATKINSON, H. H. and BANKS, P. H. T. (1966), *J. Sci. Instr.* **43,** 511.
AUBRY, B. and CHOUMOFF, S. (1965), *Comptes Rendues Acad. Sci. Paris* **261,** 1803.
AUBRY, R. and DELBART, R. (1965), *Le Vide* **20,** 194.
AXILROD, B. M. (1966), *Nat. Bur. Stand. Tech. Note* 246.

B

BÄCHLER, W. (1962), *Trans. AVS Vac. Symp.* **9,** 395.
BÄCHLER, W. (1965), *Trans. 3rd Int. Vac. Cong.*, Vol. 2, p. 609.
BÄCHLER, W., KLIPPING, G. and MASCHER, W. (1962), *Trans. AVS Vac. Symp.* **9,** 216.
BAILEY, J. R. (1963), *Nuovo Cimento*, Suppl. 1, 494.
BAKER, F. A. (1962), *Le Vide* **17,** 256.
BAKER, F. A. (1965), *Vacuum* **15,** 578.
BAKER, F. A. and GIORGI, T. A. (1960), *Brit. J. Appl. Phys.* **11,** 433.
BAKER, F. A. and HASTED, J. B. (1966), *Phil. Trans. Roy. Soc.* **A261,** 33.
BAKER, F. A. and PÉTERMANN, L. A. (1966), *J. Vac. Sci. and Technol.* **3,** 285.
BALLANCE, J. O., ROBERTS, W. K. and TARBELL, D. W. (1963), *Adv. in Cryogenic Eng* **8,** 57.
BALWANZ, W. W. (1964), *Adv. in Cryogenic Eng.* **9,** 451.
BALWANZ, W. W., SINGER, J. R. and FRANDSEN, N. P. (1960), *Trans. AVS Vac. Symp.* **7,** 182.
BANCE, U. R. and CRAIG, R. D. (1966), *Vacuum* **16,** 647.
BANNOCK, R. R. (1962), *Vacuum* **12,** 101.
BARNES, G., GAINES, J. and KEES, J. (1962), *Vacuum* **12,** 141.
BARONETZKY, E. and KLOPFER, A. (1960), *Advances in Vac. Sci. and Technol.* **1,** 401.
BARRER, R. M. (1951), *Diffusion in and through Solids* (Cambridge University Press, New York).
BARRER, R. M. (1954a), *Brit. J. Appl. Phys. Suppl.* **3,** 41

BARRER, R. M. (1954b), *Brit. J. Appl. Phys. Suppl.* **3**, 49.
BARRER, R. M. and GROVE, D. M. (1951), *Trans. Far. Soc.* **47**, 826.
BARRINGTON, A. E. (1963), *High Vacuum Engineering* (Prentice-Hall, Englewood Cliffs, N.J.).
BARRINGTON, A. E., HERZOG, R. F. K. and SAUERMANN, G. O. (1965), *J. Vac. Sci & Technol.* **2**, 89.
BARTON, R. S. and GOVIER, R. P. (1965), *J. Vac. Sci. and Technol.* **2**, 113.
BAS, E. B. (1965), *Vakuum-Technik* **14**, 65.
BASALAEVA, N. (1958), *Sov. Phys. Tech. Phys.* **3**, 1027.
BATANOV, G. M. (1961), *Sov. Phys. Solid State* **3**, 409.
BATES, D. R. (1962), *Atomic and Molecular Processes* (ed. D. R. Bates; Academic Press, New York), p. 550.
BATES, D. R. and GRIFFING, G. (1953), *Proc. Phys. Soc. (London)* **A66**, 961.
BATES, D. R. and MCCARROL, R. (1958), *Proc. Roy. Soc. (London)* **A245**, 175.
BATZER, T. H. and BUNSHAH, R. F. (1967), *J. Vac. Sci. and Technol.* **4**, 19.
BAUER, E. (1965), *Editions du Centre National de la Recherche Scientifique*, p. 19.
BAUER, E. (1966), *Surface Science* **5**, 152.
BAUER, S. H. and JEFFERS, P. (1965), *J. Phys. Chem.* **69**, 3317.
BAYARD, R. T. and ALPERT, D. (1950), *Rev. Sci. Instr.* **21**, 571.
BEAMS, J. W. (1959), *Science* **130**, 1406.
BEAMS, J. W. (1960), *Trans. AVS Vac. Symp.* **7**, 1.
BEAMS, J. W., SPITZER, D. M. and WADE, J. P. (1962), *Rev. Sci. Instr.* **33**, 151.
BEATTIE, J. A. and STOCKMAYER, W. H. (1951), in *A Treatise on Physical Chemistry*, Chapter II, (Van Nostrand Co., New York; ed. H. S. Taylor and S. Glasstone).
BECK, A. H. and BRISBANE, A. D. (1952), *Vacuum* **2**, 137.
BECKER, J. A. (1955), *Advances in Catalysis* **7**, 135.
BECKER, J. A. (1958), *Solid State Physics* **7**, 379.
BECKER, J. A., BECKER, E. J. and BRANDES, R. G. (1961), *J. Appl. Phys.* **32**, 411.
BECKER, W. (1960), *Adv. in Vac. Sci. and Technol.* **1**, 173.
BECKER, W. (1962), *Le Vide* **17**, 350.
BECKER, W. (1966), *Vacuum* **16**, 625.
BEECK, O. (1950), *Disc. Far. Soc.* **8**, 118.
BEELER, J. R. and BESCO, D. G. (1963), *J. Appl. Phys.* **34**, 2873.
BEHRISCH, R. (1964), *Ergeb. der exact. Naturwiss.* **35**, 295.
BELYAKOV, YU I. and IONOV, N. I. (1961), *Sov. Phys. Tech. Phys.* **6**, 146.
BENNETTE, C. J., STRAYER, R. W. and SWANSON, L. W. (1965), *Quarterly Progress Report No. 3*, Field Emission Corp. Contract NAS3-5902. STAR Index No. N65-24124.
BENNEWITZ, H. G. and DOHMANN, H. D. (1965), *Vakuum-Technik* **14**, 8.
BERNARDINI, M. (1965), *Trans. 3rd Int. Vac. Cong.*, Vol. 2, p. 481.
BERNARDINI, M. and MALTER, L. (1965), *J. Vac. Sci. and Technol.* **2**, 130.
BERNHARD, F., KREBS, K. H. and ROTTER, I. (1961), *Zeit. fur. Physik* **161**, 103.
BERNSTEIN, R. B. (1963), *Proc. 3rd Int. Conf. on the Physics of Electronic and Atomic Collisions*, (London), p. 895.
BILLINGTON, D. S. and CRAWFORD, J. H. (1962), *Radiation Damage in Solids* (Princeton University Press).
BILLS, D. G. (1967), *J. Vac. Sci. and Technol.* **4**, 149.
BILLS, D. G. and CARLETON, N. P. (1958), *J. Appl. Phys.* **29**, 692.

BILLS, D. G. and EVETT, A. A. (1959), *J. Appl. Phys.* **30**, 564.
BIONDI, M. A. (1959), *Rev. Sci. Instr.* **30**, 831.
BIONDI, M. A. (1963), *Advances in Electronics and Electron Physics* **18**, 67.
BLAHNIK, C. E. and SHOULDERS, K. R. (1966), *J. Vac. Sci. and Technol.* **3**, 301.
BLANKENFELD, G. (1951), *Ann. Physik* **9**, 48.
BLAUTH, E. W. (1966), *Dynamic Mass Spectrometers*, (Elsevier, Amsterdam).
BLAUTH, E. W. and VENUS, G. (1965), *Trans. 3rd Int. Vac. Cong.*, Vol. 2, p. 523 (Stuttgart).
BLEAKNEY, W. (1930), *Phys. Rev.* **36**, 1303.
BLEARS, J. (1947), *Proc. Roy. Soc. London* **A188**, 62.
BLEARS, J., GREER, E. J. and NIGHTINGALE, J. (1960), *Advances in Vac. Sci. and Technol.* **2**, 473.
BLODGETT, A. J. and SPICER, W. E. (1966), *Phys. Rev.* **146**, 390.
BOERSCH, H., GEIGER, J., HELLWIG, H. and MICHEL, H. (1962), *Zeit. fur Physik* **169**, 252.
BOERSCH, H., GEIGER, J., IMBUSCH, A. and NIEDRIG, N. (1966), *Phys. Lett.* **22**, 146.
BØGH, E. and UGGERHØJ, E. (1965), *Phys. Lett.* **17**, 116.
BØGH, E., DAVIES, J. A. and NIELSEN, K. O. (1964), *Phys. Lett.* **12**, 129.
BOHM, D. and PINES, D. (1953), *Phys. Rev.* **92**, 609.
BOHR, N. (1948), *Mat. Fys. Medd. Dan. Vid. Selk.* **18**, #8.
BOND, G. C. (1962), *Catalysis by Metals* (Academic Press, London).
BOROVIK, E. S., GRISHIN, S. F. and GRISHINA, E. Ya. (1960), *Sov. Phys. Tech. Phys.* **5**, 506.
BOROVIK, E. S., LAZAREV, B. G. and MIKHAILOV, I. F. (1961), *Journal of Nuclear Energy: Part A* **13**, 194.
BOROVIK, E. S., LAZAREV, B. G. and MIKHAILOV, I. F. (1962), *Le Vide* **17**, 231.
BOROVIK, E. S., BUSOL, F. I. and KOVALENKO, V. A. (1963), *Sov. Phys. Tech. Phys.* **8**, 68.
BOROVIK, E. S., NIKOLAEV, G. T. and SHAREVSKII, B. A. (1965), *Sov. Phys. Tech. Phys.* **9**, 957.
BOWDEN, F. P. and HANWELL, A. E. (1964), *Nature* **201**, 1279.
BOWDEN, F. P. and TABOR, D. (1964), *The Friction and Lubrication of Solids II* (Clarendon Press, Oxford).
BRACKMANN, R. T. and FITE, W. L. (1961), *J. Chem. Phys.* **34**, 1572.
BRANDT, W. and LAUBERT, R. (1967), *Nuclear Inst. and Methods* **47**, 201.
BRENNAN, D. and GRAHAM, M. J. (1965), *Phil. Trans. Roy. Soc.* **A258**, 325.
BRENNAN, D. and HAYES, F. H. (1965), *Phil. Trans. Roy. Soc.* **A258**, 347.
BRENNAN, D. and HAYWARD, D. O. (1965), *Phil. Trans. Roy. Soc.* **A258**, 375.
BRENNAN, D., HAYWARD, D. O. and TRAPNELL, B. M. W. (1960), *Proc. Roy. Soc. (London)* **A256**, 81.
BRINKMAN, J. A. (1954), *J. Appl. Phys.* **25**, 961.
BROMBACHER, W. G. (1961), *Bibliography and Index on Vacuum and Low Pressure Measurement*, NBS Monograph 35 (U.S. Dept. of Commerce).
BROMBACHER, W. G. (1967), *NBS Tech Note No.* 298.
BRONSHTEIN, I. M. and DENISOV, S. S. (1965), *Sov. Phys. Solid State* **6**, 1515.
BRONSHTEIN, I. M. and FRAIMAN, B. S. (1962), *Radio Engineering & Electronic Phys.* **7**, 1530.

References

BRONSHTEIN, I. M. and SCHUCHINSKIY, Ya. M. (1964), *Radio Engineering & Electronic Phys.* **9**, 738.
BRONSHTEIN, I. M. and SEGAL, R. B. (1960), *Sov. Phys. Solid State* **1**, 1365.
BROWN, R. D. and BURTON, R. A. (1966), *Rev. Sci. Instr.* **37**, 1699.
BROWN, E. and LECK, J. H. (1955), *Brit. J. Appl. Phys.* **6**, 161.
BRUINING, H. (1938), *Physica* **5**, 913.
BRUINING, H. and deBOER, J. H. (1939), *Physica* **6**, 834.
BRUNAUER, S. (1945), *The Adsorption of Gases and Vapors* (Princeton University Press).
BRUNAUER, S., EMMETT, P. H. and TELLER, E. (1938), *J. Amer. Chem. Soc.* **60**, 309.
BRUNÉE, C. (1953), *Zeit. fur Physik* **147**, 161.
BRYANT, P. J. and GOSSELIN, C. M. (1966), *J. Vac. Sci. and Technol.* **3**, 350.
BRYANT, P. J., LONGLEY, W. W. and GOSSELIN, C. M. (1966), *J. Vac. Sci. and Technol.* **3**, 62.
BRYANT, P. J., TAYLOR, L. H. and GUTSHALL, P. L. (1963), *Trans. AVS Vac. Symp.* **10**, 21.
BUCKINGHAM, J. D. (1965), *Brit. J. Appl. Phys.* **16**, 1821.
BUCKMAN, R. W. and HETHERINGTON, J. S. (1966), *Rev. Sci. Instr.* **37**, 999
BUFFHAM, B. A., HENAULT, P. B. and FLINN, R. A. (1962), *Trans. AVS Vac. Symp.* **9**, 205.
BÜLTEMANN, H. J. and DELGMANN, L. (1965), *Vacuum* **15**, 301.
BUREAU, A. J., LASLETT, L. S. and KELLER, J. M. (1952), *Rev. Sci. Instr.* **23**, 683.
BURKE, V. M. and SEATON, M. J. (1961), *Proc. Phys. Soc. (London)* **77**, 199.
BURNS, J. (1960), *Phys. Rev.* **119**, 102.
BURTT, R. B., COLLIGON, J. S. and LECK, J. H. (1961), *Brit. J. Appl. Phys.* **12**, 396.
BUSER, R. G. and SULLIVAN, J. J. (1966), *Vacuum* **16**, 421.
BUSOL, F. I. and YUFEROV, V. B. (1966), *Sov. Phys. Tech Phys.* **11**, 125.

C

CAIRNS, R. B. and SAMSON, J. A. R. (1966), *J. Opt. Soc. Amer.* **56**, 1568.
CALDER, R. and LEWIN, G. (1966), Private Communication.
CALDER, R. and LEWIN, G. (1967), *Brit. J. Appl. Phys.* **18**, 1459.
CAMPBELL, K. C. and THOMSON, S. J. (1959), *Trans. Far. Soc.* **55**, 306.
CARBONE, R., FULS, E. N. and EVERHART, E. (1956), *Phys. Rev.* **102**, 1524.
CAREN, R. P., GILCREST, A. S. and ZIERMAN, C. A. (1964), *Adv. in Cryogenic Eng.* **9**, 457.
CARLETON, N. P. and LAWRENCE, T. R. (1958), *Phys. Rev.* **109**, 1159.
CARLSTON, C. E., MAGNUSON, G. D., COMEAUX, A. and MAHADEVAN, P. (1965), *Phys. Rev.* **138**, A759.
CARMICHAEL, J. H. and KNOLL, J. S. (1958), *Trans. AVS Vac. Symp.* **5**, 18.
CARMICHAEL, J. H. and LANGE, W. J. (1958), *Trans. AVS Vac. Symp.* **5**, 137.
CARMICHAEL, J. H. and TRENDELENBURG, E. A. (1958), *J. Appl. Phys.* **29**, 1570.
CARR, P. H. (1964), *Vacuum* **14**, 37.
CARSLAW, H. S. and JAEGER, J. C. (1959), *Conduction of Heat in Solids* (Clarendon Press, Oxford).
CARTER, G. (1959a), *Vacuum* **9**, 190.
CARTER, G. (1959b), *Nature* **183**, 1619.
CARTER, G. and LECK, J. H. (1959), *Brit. J. Appl. Phys.* **10**, 364.

CARTER, G. and LECK, J. H. (1961), *Proc. Roy. Soc.* **A261**, 303.
CASIMIR, H. B. G. and POLDER, D. (1948), *Phys. Rev.* **73**, 360.
CASWELL, H. L. (1959), *Rev. Sci. Instr.* **30**, 1054.
CASWELL, H. L. (1963), *Physics of Thin Films* (ed. G. Hass), Vol. I, p. 1 (Academic Press, New York).
CAVALERU, A., COMSA, G. and IOSIFESCU, B. (1964), *Brit. J. Appl. Phys.* **15**, 161.
CHOUMOFF, P. S. (1965), *Trans. 3rd Internat. Vacuum Congress*, Vol. 1, p. 111 (Stuttgart).
CHRISTIAN, R. G. and LECK, J. H. (1966), *J. Sci. Instr.* **43**, 229.
CHUBB, J. N. (1966), *Vacuum* **16**, 681.
CHUBB, J. N., GOWLAND, L. and POLLARD, I. E. (1968), *J. Phys. D.* **1**, 361.
CLARKE, K. C. (1952), *Phys. Rev.* **87**, 271.
CLAUSING, P. (1930a), *Ann. der Physik* **7**, 489.
CLAUSING, P. (1930b), *Ann. der Physik* **7**, 521.
CLAUSING, P. (1932), *Ann. der Physik* **12**, 961.
CLAUSING, P. (1962), *Physica* **28**, 298.
CLAUSING, R. E. (1961), *Trans. AVS Vac. Symp.* **8**, 345.
CLAY, F. P. and MELFI, L. T. (1966), *J. Vac. Sci. and Technol.* **3**, 167.
CLOSE, K. J. and YARWOOD, J. (1966), *Brit. J. Appl. Phys.* **17**, 1165.
COBIC, B., CARTER, G. and LECK, J. H. (1961a), *Brit. J. Appl. Phys.* **12**, 288.
COBIC, B., CARTER, G. and LECK, J. H. (1961b), *Brit. J. Appl. Phys.* **12**, 384.
COBIC, B., CARTER, G. and LECK, J. H. (1961c), *Vacuum* **11**, 247.
COLLIGON, J. S. and LECK, J. H. (1961), *Trans. AVS Vac. Symp.* **8**, 275.
COMES, F. J. and WENNING, U. (1966), *Phys. Lett.* **23**, 537.
COMPTON, K. T. and LANGMUIR, I. (1930), *Rev. Mod. Phys.* **2**, 123.
COMPTON, K. T. and VAN VOORHIS, C. C. (1925), *Phys. Rev.* **26**, 436.
COMSA, G. (1966), *J. Appl. Phys.* **37**, 554.
COMSA, G. and IOSIFESCU, B. (1961), *Trans. AVS Vac. Symp.* **8**, 413.
CONN, G. K. T. and DAGLISH, H. N. (1954), *J. Sci. Instr.* **31**, 412.
CONSTABARIS, G. and HALSEY, G. D. (1957), *J. Chem. Phys.* **27**, 1433.
CONSTABARIS, G., SINGLETON, G. and HALSEY, G. D. (1959), *J. Phys. Chem.* **63**, 1350.
CONSTABARIS, G., SAMS, J. R. and HALSEY, G. D. (1961), *J. Phys. Chem.* **65**, 367.
COY, J. D. C. and RICKETSON, B. W. A. (1963), *Adv. in Cryogenic Eng.* **8**, 65.
CRAIG, R. D. and HARDEN, E. H. (1966), *Vacuum* **16**, 67.
CRAMER, W. H. and SIMONS, J. H. (1957), *J. Chem. Phys.* **26**, 1272.
CRANK, J. (1956), *The Mathematics of Diffusion* (Clarendon Press, Oxford).
CRAWLEY, D. J. (1965), *Vacnique* **5**, No. 16, p. 2.
CROWELL, A. D. and YOUNG, D. M. (1953), *Trans. Far. Soc.* **49**, 1080.
CROWELL, C. R. and ARMSTRONG, R. A. (1959), *Phys. Rev.* **114**, 1500.
CUNNINGHAM, T. M. and YOUNG, R. L. (1963), *Adv. in Cryogenic Eng.* **8**, 85.
CUTLER, P. H. and DAVIS, J. C. (1964), *Surface Science* **1**, 194.
CZAVINSKY, P. (1959), *J. Chem. Phys.* **31**, 178.

D

DACEY, J. R. (1961), *Advances in Chemistry Series* **33**, 172.
DAHL, P. and MAGYAR, J. (1965), *Phys. Rev.* **140**, A1420.

References

DALGARNO, A. and YADAV, H. N. (1953), *Proc. Phys. Soc. (London)* **A66**, 173
DALLOS, A. and STEINRISSER, F. (1967), *J. Vac. Sci. and Technol.* **4**, 6.
DANFORTH, W. E. (1961), *Vacuum* **11**, 80.
DAS, D. K. (1962), Technical Doc. Report No. AEDC (U.S.) TDR 62–19 Doc. No. 66/5498. January.
DATZ, S. and MINTURN, R. E. (1964), *J. Chem. Phys* **41**, 1153.
DATZ, S. and SNOEK, C. (1964), *Phys. Rev.* **134**, A347.
DATZ, S. and TAYLOR, E. H. (1956), *J. Chem. Phys.* **25**, 389.
DATZ, S., MINTURN, R. E. and TAYLOR, E. H. (1960), *J. Appl. Phys.* **31**, 880.
DATZ, S., MOORE, G. E. and TAYLOR, E. H. (1962), *Proc. 3rd Conf. on Rarefied Gas Dynamics*, Vol. I, p. 347 (Academic Press, New York).
DAVIES, J. A., BROWN, F. and MCCARGO, M. (1963), *Can. J. Phys.* **41**, 829.
DAVIES, J. A. and JESPERSGAARD, P. (1966), *Can. J. Phys.* **44**, 1631.
DAVIES, J. A., BALL, G. C., BROWN, F. and DOMEIJ, B. (1964), *Can. J. Phys.* **42**, 1070
DAVIS, W. D. (1962), *Trans. AVS Vac. Symp.* **9**, 363.
DAVIS, W. D. (1963), *Trans. AVS Vac. Symp.* **10**, 253.
DAVIS, W. D. and VANDERSLICE, T. A. (1960), *Trans. AVS Vac. Symp.* **7**, 417.
DAWSON, J. P. (1966), *Jour. of Spacecraft and Rockets* **3**, 218.
DAWSON, J. P. and HAYGOOD, J. D. (1965), *Cryogenics* **5**, 57.
DAYTON, B. B. (1955), *Trans. AVS Vac. Symp.* **2**, 91.
DAYTON, B. B. (1956), *Trans. AVS Vac. Symp.* **3**, 5.
DAYTON, B. B. (1959), *Trans. AVS Vac. Symp.* **6**, 101
DAYTON, B. B. (1961), *Trans. AVS Vac. Symp.* **8**, 42.
DAYTON, B. B. (1962), *Trans. AVS Vac. Symp.* **9**, 293.
DAYTON, B. B. (1965), *J. Vac. Sci. and Technol.* **2**, 290.
de BOER, J. H. (1953), *The Dynamical Character of Adsorption* (Clarendon Press, Oxford).
de BOER, J. H. and KRUYER, S. (1955), *Proc. K. Ned. Akad. Wetensch. B*, **58**, 61.
DEGRAS, D. A. (1963), *Il Nuovo Cimento*, Suppl. 1, 663.
DEGRAS, D. A. (1965), *Trans. 3rd Int. Vac. Cong.* Vol 2, p. 673.
DEGRAS, D. A. and LECANTE, J. (1965), *Proc. 7th Int. Conf. on Ionized Gases, Belgrade*, Vol. I, p. 124.
DEKKER, A. J. (1958), *Solid State Physics* **6**, 251.
DELCHAR, T. A. and EHRLICH, G. (1965), *J. Chem. Phys.* **42**, 2686.
de LEEUW, J. H. (Ed.) (1964), *4th Int. Symp. on Rarefied Gas Dynamics* (Toronto).
DELLA PORTA, P. and GIORGI, T. (1966), *Vacuum* **16**, 379.
DELLA PORTA, P., GIORGI, T. A. and SORMANI, G. (1963), Suppl. *Nuovo Cimento*, *1*, 557.
DENISON, D. R. (1967), *J. Vac. Sci. and Technol.* **4**, 156.
DENISON, D. R., WINTERS, H. F. and DONALDSON, E. E. (1963), *Trans. AVS Vac. Symp.* **10**, 218.
DE POORTER, G. L. and SEARCY, A. W. (1963), *J. Chem. Phys.* **39**, 925.
DEVONSHIRE, A. F. (1937), *Proc. Roy. Soc. (London)* **A158**, 269.
devries, A. E. and ROL, P. K. (1965), *Vacuum* **15**, 135.
DIELS, K. and JAECKEL, R. (1966), *Leybold Vacuum Handbook* (Pergamon Press, Oxford).
DIENES, G. J. and VINEYARD, G. H. (1957), *Radiation Effects in Solids* (Interscience, New York).

DILKE, M. H., ELEY, D. D. and MAXTED, E. D. (1948), *Nature* **161**, 804.
DILLON, J. A. (1961), *Trans. AVS Vac. Symp.* **8**, 113.
DITCHBURN, R. W. and ÖPIK, U. (1962), in *Atomic and Molecular Processes*, (ed. D. R. Bates; Academic Press, New York) p. 79.
DOBRETSOV, L. N. (1952), *Electronnaya i Ionnaya Emissiya (Gostekhizdat)* Eng. Trans. NASA report TTF-73 (1963).
DOBRETSOV, L. N. and MATSKEVICH, T. L. (1957), *Sov. Phys. Tech. Phys.* **2**, 663.
DOMEIJ, B. (1966), *Arkiv. fur Fysik.* **32**, 179.
DOMEIJ, B. and BJÖRKQVIST, K. (1965), *Phys. Lett.* **14**, 127.
DOMEIJ, B., BROWN, F., DAVIES, J., PIERCY, G. R. and KORNELSEN, E. V. (1964a), *Phys. Rev. Lett.* **12**, 363.
DOMEIJ, B., BROWN, F., DAVIES, J. A. and MCCARGO, M. (1964b), *Can. J. Phys.* **42**, 1624.
DONALDSON, E. E. (1962), *Vacuum* **12**, 11.
DORE, R. (1963), *Proc. 2nd Eur. Vac. Symp.* p. 179.
DORMAN, F. H. and MORRISON, J. D. (1961), *Can. J. Phys.* **35**, 575.
DOUGLAS, R. A., ZABRITSKI, J. and HERB, R. G. (1965), *Rev. Sci. Instr.* **36**, 1.
DOWDEN, D. A. (1950), *J. Chem. Soc.* p. 242.
DRAIN, L. E. and MORRISON, J. A. (1953), Private Communication.
DRAWIN, H. W. (1961), *Zeit. fur Physik* **164**, 513.
DUBININ, M. M. and RADUSHKEVICH, L. V. (1947), *Dokl. Akad. Nauk. SSSR.* **55**, 331.
DUCKWORTH, H. E. (1958), *Mass Spectroscopy* (Cambridge University Press).
DUFFY, W. E. and RUIZ, C. P. (1966), *Proc. ASTM Annual Conference on Mass Spectrometry* **14**, 302.
DUMAS, G. (1955), *Rev. Gen. d'Elec.* **64**, 331.
DUNKELMAN, L., FOWLER, W. B. and HENNES, J. (1962), *Appl. Optics* **1**, 695.
DUNN, G. H. and KIEFFER, L. J. (1963), *Phys. Rev.* **132**, 2109.
DUNNING, W. J. (1967), *The Solid Gas Interface*, Vol. I, p. 271 (Marcel Dekker, New York).
DUPP, G. and SCHARMANN, A. (1966), *Zeit. fur Phys.* **192**, 284.
DUSHMAN, S. (1949), *The Scientific Foundations of Vacuum Technique* (Wiley, New York).
DUSHMAN, S. and LAFFERTY, J. M. (1962), *Scientific Foundations of Vacuum Technique* (Wiley, New York).

E

EBERHAGEN, A. (1960), *Fortschritte der Physik* **8**, 245.
EDMONDS, T. and HOBSON, J. P. (1965), *J. Vac. Sci. and Technol.* **2**, 182.
EGGLETON, A. E. J. and TOMPKINS, F. C. (1952), *Trans. Far. Soc.* **48**, 738.
EHRENREICH, H. and PHILIPP, H. R. (1962), *Phys. Rev.* **128**, 1622.
EHRLICH, G. (1956), *J. Phys. Chem.* **60**, 1388.
EHRLICH, G. (1959), *Int. Conf. on Structure and Properties of Thin Films* (New York), p. 423.
EHRLICH, G. (1961a), *J. Chem. Phys.* **34**, 29.
EHRLICH, G. (1961b), *J. Chem. Phys.* **34**, 39.
EHRLICH, G. (1961c), *J. Appl. Phys.* **32**, 4.

EHRLICH, G. (1963a), *Adv. in Catalysis* **14**, 255.
EHRLICH, G. (1963b), *Ann. New York Acad. Sci.* **101**, 722.
EHRLICH, G. (1964a), *Proc. 3rd Int. Cong. on Catalysis*, Vol. I, p. 113.
EHRLICH, G. (1964b), *Brit. J. Appl. Phys.* **15**, 349.
EHRLICH, G. and HUDDA, F. G. (1959), *J. Chem. Phys.* **30**, 493.
EISINGER, J. (1957), *J. Chem. Phys.* **27**, 1206.
EISINGER, J. (1958a) *J. Chem. Phys.* **28**, 165.
EISINGER, J. (1958b), *J. Chem. Phys.* **29**, 1154.
EISINGER, J. (1959), *J. Chem. Phys.* **30**, 412.
ELSWORTH, L., HOLLAND, L. and LAURENSON, L. (1965), *Vacuum* **15**, 337.
ENDOW, N. and PASTERNAK, R. A. (1966), *J. Vac. Sci. and Technol.* **3**, 196.
ERENTS, K. and CARTER, G. (1966), *Vacuum* **16**, 523.
ERENTS, K. and CARTER, G. (1967), *Vacuum* **17**, 97.
ERENTS, K., GRANT, W. A. and CARTER, G. (1965), *Surface Science* **3**, 480.
ERGINSOY, C., VINEYARD, G. H. and ENGLERT, A. (1964), *Phys. Rev.* **133**, A595.
ERGINSOY, C., VINEYARD, G. H. and SHIMIZU, A. (1965), *Phys. Rev.* **139**, A118.
ERIKSSON, L. (1967), *Phys. Rev.* **161**, 235.
ERIKSSON, L., DAVIES, J. A. and JESPERSGAARD, P. (1967), *Phys. Rev.* **161**, 219.
ERMRICH, W. (1965), *Phil. Res. Rep.* **20**, 94.
ESCHBACH, H. L. (1960), *Advances in Vac. Sci. and Technol.* **1**, 373.
ESCHBACH, H. L., JAECKEL, R. and MÜLLER, D. (1961), *Trans. AVS Vac. Symp.* **8**, 1110.
ESPE, W. (1959a), *Werkstoffkunde der Hochvakuumtechnik* (p. 659). Veb. Deutscher Verlag der Wissenschaften (Berlin).
ESPE, W. (1959b), *Werkstoffkunde der Hochvakuumtechnik* (p. 136). Veb Deutscher Verlag der Wissenschaften (Berlin).
ESPE, W. (1966), *Materials of High Vacuum Technology* (Pergamon Press, Oxford).
ESPE, W. and HYBL, C. (1965), *Vakum Technik* **14**, 108.
ESTERMANN, I. and STERN, O. (1930a), *Zeit. fur Physik* **61**, 95.
ESTERMANN, I. and STERN, O. (1930b), *Zeit. fur Physik* **61**, 114.
ESTRUP, P. J. and ANDERSON, J. (1967), *J. Chem. Phys.* **46**, 567.
EVANS, D. S. (1965), *Rev. Sci. Instr.* **36**, 375.
EVDOKIMOV, I. N., MASHKOVA, E. S., MOLCHANOV, V. A. and ODINSTOV, D. D. (1967), *Phys. Stat. Sol.* **19**, 407.
EVERHART, E. (1963), *Phys. Rev.* **132**, 2083.
EVERHART, E., CARBONE, R. J. and STONE, G. (1955), *Phys. Rev.* **98**, 1045.
EVERHART, E. and KESSEL, Q. C. (1965), *Phys. Rev. Lett.* **14**, 247.

F

FANO, U. and LICHTEN, W. (1965), *Phys. Rev. Lett.* **14**, 627.
FARKASS, I., GOULD, P. R. and HORN, G. W. (1962), *Trans. AVS Vac. Symp.* **9**, 273.
FARNSWORTH, H. E. (1925), *Phys. Rev.* **25**, 41.
FARNSWORTH, H. E. (1963), *Appl. Phys. Letters* **2**, 199.
FARNSWORTH, H. E. and TUUL, J. (1959), *J. Phys. Chem. Solids* **9**, 48.
FARNSWORTH, H. E., SCHLIER, R. E., GEORGE, T. H. and BURGER, R. M. (1958), *J. Appl. Phys.* **29**, 1150.
FARRAR, E., MCINTYRE, R. A. and YORK, D. (1964), *Proc. ASTM Conf. on Mass Spectrometry* **12**, 194.

FEAKES, F. and TORNEY, F. L. (1963), *Trans. AVS Vac. Symp.* **10**, 257.
FEAKES, F., TORNEY, F. L. and BROCK, F. J. (1965), *Gauge Calibration Study in Extreme High Vacuum*, NASA Report CR 167, STAR Index N65-17126.
FEDORENKO, N. V. (1954), *Zh. Tekh. Fiz.* **24**, 784.
FEDORENKO, N. V. (1959), *Sov. Phys. Uspekhi* **2**, 526.
FEDORENKO, N. V., FILIPENKO, L. G. and FLAKS, I. P. (1960), *Sov. Phys. Tech. Phys.* **5**, 45.
FEUERSTEIN, S. and RICE, L. (1966), *Trans. Metallurgical Soc. AIME* **236**, 1674.
FEUERSTEIN, S., RICE, L. and CONRAD, H. (1964), *Appl. Phys. Lett.* **4**, 154.
FIELD, F. H. and FRANKLIN, J. L. (1957), *Electron Impact Phenomena and the Properties of Gaseous Ions*, Academic Press, New York.
FIRSOV, O. B. (1958a), *Sov. Phys. JETP* **6**, 534.
FIRSOV, O. B. (1958b), *Soviet Physics JETP* **7**, 308.
FISCHBECK, H. J. (1966), *Phys. Status Solidi* **15**, 387.
FISCHER, E. (1965), *J. Vac. Sci. and Technol.* **2**, 142.
FISCHER, G. E. and MACK, R. A. (1965), *J. Vac. Sci. and Technol.* **2**, 123.
FITE, W. L. (1962), in *Atomic and Molecular Processes* (ed. D. R. Bates; Academic Press, New York), p. 421.
FITE, W. L. and BRACKMANN, R. T. (1958a), *Phys. Rev.* **112**, 1141.
FITE, W. L. and BRACKMANN, R. T. (1958b), *Phys. Rev.* **112**, 1151.
FITE, W. L., SMITH, A. C. H. and STEBBINGS, R. F. (1962), *Proc. Roy Soc. (London)* **A268**, 527.
FITE, W. L., STEBBINGS, R. F. and BRACKMANN, R. T. (1959), *Phys. Rev.* **116**, 356.
FITE, W. L., STEBBINGS, R. F., HUMMER, D. G. and BRACKMANN, R. T. (1960), *Phys. Rev.* **119**, 663.
FLECKEN, F. A. and NÖLLER, H. G. (1961), *Trans. AVS Vac. Symp.* **8**, 58.
FLUIT, J. M., ROL, P. K. and KISTEMAKER, J. (1963), *J. Appl. Phys.* **34**, 690.
FOMIN, O. K., TIKHOMIROV, M. V. and TUNITSKII, N. N. (1965), *Sov. Phys.Tech. Phys.* **9**, 1114.
FONER, S. N., MAUER, F. A. and BOLZ, L. H. (1959), *J. Chem. Phys.* **31**, 546.
FORTH, H. J. (1965), *Le Vide* **20**, 343.
FOSTER, P. K. (1960), *Nature*, **188**, 399.
FOWLER, H. A. and FARNSWORTH, H. E. (1958), *Phys. Rev.* **111**, 103.
FOX, R. E. (1959), *Proc. Joint Conf. on Advances in Mass Spectrometry*, London, p. 397.
FOX, R. E. (1960), *J. Chem. Phys.* **33**, 200.
FOX, R. E. and KNOLL, J. S. (1960), *Trans. AVS Vac. Symp.* **7**, 364.
FRANK, R. C. and THOMAS, J. E. (1960), *J. Phys. Chem. Solids* **16**, 144.
FRANK, R. C., SWETS, E. D. and LEE, R. W. (1961), *J. Chem. Phys.* **35**, 1451.
FRASER, R. G. J. (1931), *Molecular Rays* (Cambridge University Press).
FREDERICKS, W. J. and COOK, C. J. (1961), *Phys. Rev.* **121**, 1693.
FREDERICKS, W. J. and COOK, C. J. (1963), *J. Phys. Soc. Japan* **18**, Suppl. 2, 281.
FRENKEL, J. (1924), *Zeit. fur. Phys.* **26**, 117.
FREYTAG, J. P. and SCHRAM, A. (1963), *Suppl. Nuovo Cimento* **1**, 405.
FRISCH, K. and STERN, O. (1933), *Zeit. fur Phys.* **84**, 430.
FUJIWARA, K., HAYAKAWA, K. and MIYAKE, S. (1966), *Jap. J. Appl. Phys.* **5**, 295.
FULS, E. N., JONES, P. R., ZIEMBA, F. P. and EVERHART, E. (1957), *Phys. Rev.* **107**, 704.
FUMI, F. G. and TOSI, M. P. (1964), *J. Phys. Chem. Solids* **25**, 31.

G

GABOR, D. (1962), *Brit. Patent* 887251, Jan. 17, 1962.
GADZUK, J. W. (1967), *Phys. Rev.* **153**, 759.
GAFNER, G. (1964), *Surface Science* **2**, 534.
GARBE, S. (1963), *Vakuum Technik* **12**, 201.
GARBE, S. and CHRISTIANS, K. (1962), *Vakuum Technik* **11**, 9.
GARBE, S., KLOPFER, A. and SCHMIDT, W. (1960), *Vacuum* **10**, 81.
GARWIN, E. L. (1963), in *Cryogenic Technology*, (ed. R. W. Vance), p. 332. (Wiley, New York).
GAVRILYUK, V. M. (1961), *Proc. Acad. Sci. USSR: Phys. Chem. Section* **141**, 938.
GAVRILYUK, V. M. and MEDVEDEV, V. K. (1963), *Sov. Phys. Solid State* **4**, 1737.
GAY, W. L. and HARRISON, D. E. (1964), *Phys. Rev.* **135**, A1780.
GERMER, L. H. (1965), *Scientific American* **212**, No. 3, 32.
GERMER, L. H. (1966), *Surface Science* **5**, 147.
GERMER, L. H. and MACRAE, A. U. (1962), *J. Appl. Phys.* **33**, 2923.
GERMER, L. H., GOLDSZTAUB, S., ESCARD, J., DAVID, G. and DEVILLE, J. P. (1966), *Compt. Rend.* **B262**, 1059.
GEYER, K. H. (1942), *Ann. der Physik* **42**, 241.
GIBBONS, M. D. and STOUT, V. L. (1959), *Proc. 19th Ann. M.I.T. Conf. on Phys. Electronics*, p. 117 (Cambridge, Mass.).
GIBSON, J. B., GOLAND, A. N., MILGRAM, M. and VINEYARD, G. H. (1960), *Phys. Rev.* **120**, 1229.
GILBEY, D. M. (1962), *J. Phys. Chem. Solids* **23**, 1453.
GILBREATH, W. P. and WILLIAMS, D. P. (1966), *J. Vac. Sci. and Technol.* **3**, 316.
GIMPEL, I. and RICHARDSON, O. (1943), *Proc. Roy. Soc. (London)* **A182**, 17.
GIORDMAINE, J. A. and WANG, T. C. (1960), *J. Appl. Phys.* **31**, 463.
GLASSTONE, S., LAIDLER, K. J. and EYRING, H. (1941), *The Theory of Rate Processes* (McGraw Hill, New York).
GOERZ, D. J. (1960), *Trans. AVS Vac. Symp.* **7**, 65.
GOETHERT, B. H. (1964), *Le Vide* **19**, 373.
GOLDMAN, D. T. and SIMON, A. (1958), *Phys. Rev.* **111**, 383.
GOMBAS, P. (1956), *Handbuch der Phys.* **36**, 109.
GOMER, R. (1953), *Rev. Sci. Instr.* **24**, 993.
GOMER, R. (1958), *J. Chem. Phys.* **29**, 441.
GOMER, R. (1959), *Disc. Far. Soc.* **28**, 23.
GOMER, R. (1961), *Field Emission and Field Ionization* (Harvard University Press).
GOMER, R. (1966), *Disc. Far. Soc.* **41**, 14.
GOMOYUNOVA, M. V. and LETUNOV, N. A. (1965), *Sov. Phys. Solid State* **7**, 316.
GOMOYUNOVA, M. V. and LETUNOV, N. A. (1966), *Sov. Phys. Solid State* **7**, 2427.
GOOD, R. H. and MÜLLER, E. W. (1956), *Handbuch der Physik*, **21**, 176.
GOODMAN, F. O. (1964), *Proc. 4th Int. Symp. Rarefied Gas Dynamics*, Vol. 2, p. 366.
GOODMAN, F. O. (1965), *J. Phys. Chem. Solids* **26**, 85.
GOODMAN, F. O. (1967), *Phys. Rev.* **164**, 1113
GOODMAN, F. O. and WACHMAN, H. Y. (1966), *M.I.T. Fluid Dynamics Res. Lab. Report*. No. 66-1 (Cambridge, Mass.). See also *J. Chem. Phys.* **46**, 2376, (197).
GOODRICH, G. W. and WILEY, W. C. (1961), *Rev. Sci. Instr.* **32**, 846.
GORMAN, J. K. and NARDELLA, W. R. (1962), *Vacuum* **12**, 19.

GOSSELIN, C. M. and BRYANT, P. J. (1965), *J. Vac. Sci. and Technol.* **2**, 293.
GOULD, C. L. and DRYDEN, R. A. (1961), *Trans. AVS Vac. Symp.* **8**, 369.
GRANDE, P. E., WATTERS, R. L. and HUDSON, J. B. (1966), *J. Vac. Sci. and Technol.* **3**, 329.
GRANT, W. A. and CARTER, G. (1965), *Vacuum* **15**, 477.
GRANT, W. A. and CARTER, G. (1966), *Vacuum* **16**, 485.
GRANT, W. A. and CARTER, G. (1967), *Brit. J. Appl. Phys.* **18**, 527.
GRANT, H. L. and DAVEY, J. E. (1963), *Rev. Sci. Instr.* **34**, 587.
GREEN, E. F., MOURSUND, A. L. and ROSS, J. (1966), *Advances in Chemical Physics*, **10**, 135.
GREENWOOD, G. W. and SPEIGHT, M. V. (1963), *J. Nuclear Materials* **10**, 140.
GRENIER, G. E. and STERN, S. A. (1966), *J. Vac. Sci. and Technol.* **3**, 334.
GRIGOR'EV, A. M. (1959), *Instr. and Exp. Tech.*, p. 870.
GRIGSON, C. W. B., NIXON, W. C. and TOTHILL, F. (1966), *Proc. 6th Int. Cong. for Electron Microscopy*, Vol. 1, p. 157 (Kyoto).
GROSZKOWSKI, J. (1965a), *Bull. Acad. Polon. Sci. Ser. Sc. Tech.* **13**, 57 (397).
GROSZKOWSKI, J. (1965b), *Bull. Acad. Polon. Sci. Ser. Sc. Tech.* **13**, 43 (261).
GROSZKOWSKI, J. (1965c), *Bull. Acad. Polon, Sci. Ser. Sc. Tech.* **13**, 15 (177).
GROSZKOWSKI, J. (1966a), *Bull. Acad. Polon. Sci., Ser. Sc. Tech.* **14**, 75 (551).
GROSZKOWSKI, J. (1966b), *Bull. Acad. Polon. Sci., Ser. Sc. Tech.* **14**, 87 (563).
GROSZKOWSKI, J. (1966c), *Bull. Acad. Polon. Sci., Ser. Sc. Tech.* **14**, 169 (1023).
GRYZINSKI, M. (1963), *Proc. 3rd Int. Conf. on the Physics of Electronic and Atomic Collisions* (London), p. 226.
GRYZINSKI, M. (1965), *Phys. Rev.* **138**, A336.

H

HAAS, C. (1960), *J. Phys. Chem. Solids.* **15**, 108.
HAAS, T. W. and JACKSON, A. G. (1966), *J. Vac. Sci. and Technol.* **3**, 168
HABER, J. and STONE, F. S. (1963), *Trans. Far. Soc.* **59**, 192.
HACHENBERG, O. and BRAUER, W. (1959), *Advances in Electronics and Electron Physics* **11**, p. 413.
HAEFER, R. (1953a), *Acta Physica Austriaca* **7**, 251.
HAEFER, R. (1953b), *Acta Physica Austriaca* **7**, 52.
HAEFER, R. (1954), *Acta Physica Austriaca* **9**, 200.
HAEFER, R. A. and HENGEVOSS, J. (1960), *Trans. AVS Vac. Symp.* **7**, 67.
HAGSTRUM, H. D. (1951), *Rev. Mod. Phys.* **23**, 185.
HAGSTRUM, H. D. (1953), *Rev. Sci. Instr.* **24**, 1122.
HAGSTRUM, H. D. (1954a), *Phys. Rev.* **96**, 325.
HAGSTRUM, H. D. (1954b), *Phys. Rev.* **96**, 336.
HAGSTRUM, H. D. (1956a), *Phys. Rev.* **104**, 317.
HAGSTRUM, H. D. (1956b), *Phys. Rev.* **104**, 672.
HAGSTRUM, H. D. (1956c), *Phys. Rev.* **104**, 1516.
HAGSTRUM, H. D. (1961), *Phys. Rev.* **123**, 758.
HAGSTRUM, H. D. (1966), *Phys. Rev.* **150**, 495.
HAGSTRUM, H. D. and D'AMICO, C. (1960), *J. Appl. Phys.* **31**, 715.
HAGSTRUM, H. D. and BECKER, G. E. (1966), *Phys. Rev. Lett.* **16**, 230.
HAGSTRUM, H. D., TAKEISHI, Y. and PRETZER, D. D. (1965), *Phys. Rev.* **139**, A526.

HALLER, F. B. (1964), *Rev. Sci. Instr.* **35**, 1356.
HALSEY, G. D., (1961), *Trans. AVS Vac. Symp.* **8**, 119.
HANCOX, R. R. (1932), *Phys. Rev.* **42**, 864.
HANLE, W. and RIEDE, D. (1952), *Zeit. fur Phys.* **133**, 537.
HANLEY, T. E. (1948), *J. Appl. Phys.* **19**, 583.
HANSEN, N. (1962), *Vakuum Technik* **11**, 70.
HANSEN, N. and LITTMAN, W. (1965), *Trans. 3rd Int. Vac. Cong.*, Vol. 2, p. 465 (Stuttgart).
HARRA, D. J. and HAYWARD, W. H. (1967), *Suppl. Nuovo Cimento* **5**, No. 1, 56.
HARRIS, L. A. (1968), *J. Appl. Phys.* **39**, 1419.
HARRISON, D. E. and MAGNUSON, D. G. (1961), *Phys. Rev.* **122**, 1421.
HARRISON, D.E., CARLSTON, C.E. and MAGNUSON, G. D. (1965) *Phys. Rev.* **139**, A737.
HARRISON, D. E., JOHNSON J. P. and LEVY, N. S. (1966), *Appl. Phys. Lett.* **8**, 33.
HARROWER, G. A. (1956a), *Phys. Rev.* **102**, 340.
HARROWER, G. A. (1956b), *Phys. Rev.* **104**, 52.
HARTLEY, B. M. and SWAN, J. B. (1966), *Phys. Rev.* **144**, 295.
HARTMAN, T. E. (1963), *Rev. Sci. Instr.* **34**, 1190.
HASTED, J. B. (1962), in *Atomic and Molecular Processes* (Academic Press; Ed. D. R. Bates) p. 696.
HASTED, J. B. (1964), *Physics of Atomic Collisions* (Butterworths, London).
HAYAKAWA, T. (1957), *Bull. Chem. Soc. Japan* **30**, 124, 236, 243, 332, 337.
HAYASHI, C. (1966), *J. Vac. Sci. and Technol.* **3**, 286.
HAYGOOD, J. D. (1963), *J. Phys. Chem.* **67**, 2061.
HAYWARD, D. O. and TRAPNELL, B. M. W. (1964), *Chemisorption* (Butterworths, London).
HAYWARD, D. O., TAYLOR, N. and TOMPKINS, F. C. (1966), *Disc. Far. Soc.* **41**, 75.
HAYWARD, W. H. and JEPSEN, R. L. (1962), *Trans. AVS Vac. Symp.* **9**, 459.
HAYWARD, W. H., JEPSEN, R. L. and REDHEAD, P. A. (1963), *Trans. AVS Vac. Symp.* **10**, 228.
HEDDLE, D. W. O. (1962), *J. Quant. Spect. Radiation Transfer* **2**, 349.
HELMER, J. C. and JEPSEN, R. L. (1961), *Proc. I.R.E.* **49**, 1920.
HENDERSON, W. G., MARK, J. T. and GEIGER, C. S. (1959), *Trans. AVS Vac. Symp.* **6**, 170.
HENGEVOSS, J. (1965) *Trans. 3rd Int. Vac. Cong.* Vol. 1, p. 51.
HENGEVOSS, J. and HUBER, W. K. (1961), *Trans. AVS Vac. Symp.* **8**, 1332.
HENGEVOSS, J. and TRENDELENBURG, E.A. (1963), *Z. Naturforsch.* **18a**, 558.
HERB, R. G. (1960), *Adv. in Vac. Sci. and Technol.* **1**, 45.
HERB, R. G., PAULY, T. and FISCHER, K. J. (1963), *Bull. Am. Phys. Soc.* **8**, 336.
HEROLD, D. (1965), *Zeit. fur angew Physik* **20**, 113.
HEROLD, D. and SCHÖNHEIT, E. (1965), *Zeit. fur angew Physik* **20**, 102.
HERRING, C. and NICHOLS, M. H. (1949), *Rev. Mod. Phys.* **21**, 185.
HERRON, J. T., ROSENSTOCK, H. M. and SHIELDS, W. R. (1965), *Nature* **206**, 611.
HERSH, C. K. (1961), *Molecular Sieves* (Reinhold, New York).
HERZBERG, G. (1944), *Atomic Spectra and Atomic Structure* (Dover, New York).
HERZBERG, G. (1950), *Molecular Spectra and Molecular Structure, I. Spectra of Diatomic Molecules* (Van Nostrand, New York).
HERZOG, L. F., ESKEW, T. J. and ERWIN, R. L. (1961), *Trans. AVS Vac. Symp.* **8**, 581.

HICKMAN, K. C. D. (1961), *Trans. AVS Vac. Symp.* **8**, 307.
HICKMOTT, T. W. (1960a), *J. Chem. Phys.* **32**, 810.
HICKMOTT, T. W. (1960b), *J. Appl. Phys.* **31**, 128.
HICKMOTT, T. W. (1965), *J. Vac. Sci. and Tech.* **2**, 257.
HICKMOTT, T. W. and EHRLICH, G. (1958), *J. Phys. Chem. Solids* **5**, 47.
HIGATSBERGER, M. J., DEMOREST, H. L. and NIER, A. O. (1954), *J. Appl. Phys.* **25**, 883.
HINRICHS, C. H. and DONALDSON, E. E. (1968), *J. Chem. Phys.* (in press).
HIRSCHFELDER, J. O., CURTISS, C. F. and BIRD, R. B. (1964), *Molecular Theory of Gases and Liquids* (Wiley, New York).
HIRTH, J. P. and POUND, G. M. (1963), *Condensation and Evaporation; Nucleation and Growth Kinetics* (MacMillan, New York).
HOBSON, J. P. (1956), *Can. J. Phys.* **34**, 1089.
HOBSON, J. P. (1959), *Can. J. Phys.* **37**, 300.
HOBSON, J. P. (1961a), *Trans. AVS Vac. Symp.* **8**, 26.
HOBSON, J. P. (1961b), *Vacuum* **11**, 16.
HOBSON, J. P. (1961c), *J. Chem. Phys.* **34**, 1850.
HOBSON, J. P. (1963), *Brit. J. Appl. Phys.* **14**, 544.
HOBSON, J. P. (1964), *J. Vac. Sci. and Technol.* **1**, 1.
HOBSON, J. P. (1965a), *Can. J. Phys.* **43**, 1934.
HOBSON, J. P. (1965b), *Can. J. Phys.* **43**, 1941.
HOBSON, J. P. (1966), *J. Vac. Sci. and Technol.* **3**, 281.
HOBSON, J. P. (1967), in *The Solid Gas Interface*, (Ed. E. A. Flood; Marcel Dekker, New York), Vol. 1, p. 447.
HOBSON, J. P. and ARMSTRONG, R. A. (1963), *J. Phys. Chem.* **67**, 2000
HOBSON, J. P. and EARNSHAW, J. (1967), *J. Vac. Sci. and Technol.* **4**, 257.
HOBSON, J. P. and EDMONDS, T. (1963), *Can. J. Phys.* **41**, 827.
HOBSON, J. P. and REDHEAD, P. A. (1958), *Can. J. Phys.* **36**, 271.
HOBSON, J. P. and REDHEAD, P. A. (1968), *Trans. 4th Int. Vac. Congr.* (*Manchester*) (to be published).
HOCH, H. (1963), *Proc. 2nd Eur. Vac. Symp.* p. 305.
HOCH, H. (1966), *Le Vide* **21**, 221.
HOENIG, M. O. (1964), *Adv. in Cryogenic Eng.* **9**, 482.
HOLKEBOER, D. H. (1963), *Trans. AVS Vac. Symp.* **10**, 292.
HOLLAND, L. (1960), *Trans. AVS Vac. Symp.* **7**, 168.
HOLLAND, L. (Ed.) (1965a), *Thin Film Microelectronics; The Preparation and Properties of Components and Circuit Arrays* (Chapman & Hall, London).
HOLLAND, L. (1965b), *Brit. J. Appl. Phys.* **16**, 1053.
HOLLAND, L. and LAURENSON, L. (1963a), *Nature* **199**, 274.
HOLLAND, L. and LAURENSON, L. (1963b), *Suppl. Nuovo Cimento No.* 1, 470.
HOLLAND, L. and LAURENSON, L. (1964), *Vacuum* **14**, 325.
HOLLAND, L., LAURENSON, L. and ALLEN, P. G. W. (1961), *Trans. AVS Vac. Symp.* **8**, 208.
HOLLAND, L., LAURENSON, L. and FULKER, M. J. (1966), *Vacuum* **16**, 663.
HOLLAND, L. and PRIESTLAND, C. (1966), *Nature* **209**, 274.
HOLLIDAY, J. E. and STERNGLASS, E. J. (1957), *J. Appl. Phys.* **28**, 1189.
HOLLIDAY, J. E. and STERNGLASS, E. J. (1959), *J. Appl. Phys.* **30**, 1428.
HOLSCHER, A. A. (1964), *J. Chem. Phys.* **41**, 579.

HOLZL, J. (1965), *Zeit. fur Phys.* **184**, 50.
HONIG, J. M. (1954), *Ann. N.Y. Acad. Sci.* **58**, 741.
HONIG, R. E. and HOOK, H. O. (1960), *R.C.A. Review* **21**, 360.
HONIG, W. and PARZEN, P. (1955), *I.R.E. Conv. Rec.* **3** (Pt. 3), 3.
HOOVERMAN, R. H. (1963), *J. Appl. Phys.* **34**, 3505.
HOPKINS, B. J. and PENDER, K. R. (1966), *Surface Science* **5**, 155.
HOUSTON, J. M. (1956), *Bull. Am. Phys. Soc.* **1**, 301.
HUANG, A. B. (1965), *J. Vac. Sci. and Technol.* **2**, 6.
HUBER, W. K. (1963a), *Vacuum* **13**, 399.
HUBER, W. K. (1963b), *Vacuum* **13**, 469.
HUBER, W. K. (1966), *Le Vide* **21**, 202.
HUDSON, J. B. and WATTERS, R. L. (1966), *IEEE Trans. on Instr. and Measurement* **IM-15**, 94.
HUGHES, A. L. and du BRIDGE, L. A. (1932), *Photoelectric Phenomena* (McGraw Hill, New York).
HUNT, A. L., DAMM, C. C. and POPP, E. C. (1961), *J. Appl. Phys.* **32**, 1937.
HUNT, A. L., TAYLOR, C. E. and OMOHUNDRO, J. E. (1963a), *Adv. in Cryogenic Eng.* **8**, 100.
HUNT, A. L., DAMM, C. C. and POPP, E. C. (1963b), *Adv. in Cryogenic Eng.* **8**, 110.
HURLBERT, R. C. and KONECNY, J. O. (1961), *J. Chem. Phys.* **34**, 655.
HURLBUT, F. C. (1957), *J. Appl. Phys.* **28**, 844.
HURLBUT, F. C. and BECK, D. E. (1959), Univ. of California, Eng. Proj. Rep. HE-150-166.
HURLBUT, F. C. and MANSFIELD, R. J. (1963), *Adv. in Cryogenic Eng.* **8**, 46.
HUXLEY, L. G. H. and CROMPTON, R. W. (1962), in *Atomic and Molecular Processes* (Academic Press; Ed. D. R. Bates), p. 335.

I

INKLEY, F. A. (1965), *Vacuum* **15**, 401.
IRELAND, J. V. and GILBODY, H. B. (1963), *3rd Int. Conf. on Physics of Electronic and Atomic Collisions* (London), p. 666.
ISHII, H. and NAKAYAMA, K. (1961), *Trans. AVS Vac. Symp.* **8**, 519.
ISHIKAWA, K. (1965), *Jap. J. Appl. Phys.* **4**, 461.

J

JACKSON, A. G. and HAAS, T. W. (1967), *J. Vac. Sci. and Technol.* **4**, 42.
JAECKEL, R. and TELOY, E. (1961), *Trans. AVS Vac. Symp.* **8**, 406.
JAMES, L. H. and CARTER, G. (1962), *Trans. AVS Vac. Symp.* **9**, 502.
JAMES, L. H., LECK, J. H. and CARTER, G. (1964), *Brit. J. Appl. Phys.* **15**, 681.
JANSEN, C. G. J. and VENEMA, A. (1959), *Vacuum* **9**, 219.
JEPSEN, R. L. (1961), *J. Appl. Phys.* **32**, 2619.
JEPSEN, R. L., FRANCIS, A. B., RUTHERFORD, S. L. and KIETZMANN, B. E. (1960), *Trans. AVS Vac. Symp.* **7**, 45.
JEPSEN, R. L., MERCER, S. L. and CALLAGHAN, M. J. (1959), *Rev. Sci. Instr.* **30**, 377.
JOHNSON, J. B. and MCKAY, K. G. (1953), *Phys. Rev.* **91**, 582.
JOHNSON, K. I. and KELLER, D. V. (1967a), *J. Vac. Sci. and Technol.* **4**, 115.
JOHNSON, K. I. and KELLER, D. V. (1967b) *J. Appl. Phys.* **38**, 1896.

JOHNSON, R. A. (1966), *Phys. Rev.* **145**, 423.
JOHNSON, V. R. and VAUGHN, G. W. (1956), *J. Appl. Phys.* **27**, 1173.
JONES, H. A. and LANGMUIR, I. (1927), *Gen. Elec. Rev.* **30**, 310.
JONKER, J. L. H. (1951), *Philips Res. Report* **6**, 372.
JONKER, J. L. H. (1957), *Philips Res. Report* **12**, 249.
JOST, W. (1952), *Diffusion in Solids, Liquids, Gases* (Academic, New York).

K

KAGANER, M. G. (1957), *Proc. Acad. Sci. USSR. Phys. Chem. Sect.* **116**, 603.
KAGANSKII, M. G., KAMINSKII, D. L. and KLYUCHAREV, A. N. (1964), *Sov. Phys. Tech. Phys.* **9**, 815.
KAMINKER, D. M. and FEDORENKO, N. V. (1955), *Zh. Tekh. Fiz.* **25**, 2239.
KAMINSKY, M. (1965), *Atomic and Ionic Impact on Metal Surfaces* (Academic, New York).
KANTER, H. (1957), *Ann. der Physik* **20**, 144.
KANTER, H. (1961), *Phys. Rev.* **121**, 461.
KANTER, H. (1964), *Brit. J. Appl. Phys.* **15**, 555.
KANTER, H. (1967), *Appl. Phys. Lett.* **10**, 73.
KARPUZOV, D. S., EL'TEKOV, V. A. and YURASOVA, V. E. (1967), *Sov. Phys. Solid State* **8**, 1726.
KEESOM, W. H. and SCHWEERS, J. (1941), *Physica* **8**, 1020.
KELLY, B. T. (1966), *Irradiation Damage to Solids* (Pergamon, Oxford).
KENDALL, B. R. F. (1962), *J. Sci. Instr.* **39**, 267.
KENDALL, B. R. F. (1965), *J. Vac. Sci. and Technol.* **2**, 1.
KENNARD, E. H. (1938), *Kinetic Theory of Gases* (McGraw Hill, New York).
KENTY, C. (1953), *Proc. 13th Ann. M.I.T. Conf. on Phys. Elect.* p. 138 (Cambridge, Mass.).
KERVALISHVILI, N. A. and ZHARINOV, A. V. (1966), *Sov. Phys. Tech. Phys.* **10**, 1682.
KESSEL, Q. C., RUSSEK, A. and EVERHART, E. (1965) *Phys. Rev. Lett.* **14**, 484.
KEYWELL, F. (1955), *Phys. Rev.* **97**, 1611.
KHAN, I. H., HOBSON, J. P. and ARMSTRONG, R. A. (1963), *Phys. Rev.* **129**, 1513.
KHASHABA, S. and MASSEY, H. S. W. (1958), *Proc. Phys. Soc. (London)* **71**, 574.
KIEFFER, L. J. and DUNN G. H. (1966), *Rev. Mod. Phys.* **38**, 1.
KIENEL, G. and WUTZ, M. (1966), *Vakuum Technik* **15**, 40.
KINCHIN, G. H. and PEASE, R. S. (1955), *Reports on Prog. in Phys.* **18**, 1.
KINDALL, S. M. and WANG, E. S. J. (1962), *Trans. AVS Vac. Symp.* **9**, 243.
KINDL, B. (1963), *Suppl. Nuovo Cimento* **1**, 646.
KING, D. A. (1966), *Disc. Far. Soc.* **41**, 63.
KINGDON, K. H. (1923), *Phys. Rev.* **21**, 408.
KIRCHNER, F. and KIRCHNER, H. (1956), *Zeit angew Phys.* **8**, 478.
KISELEV, A. V. and POSHKUS, D. P. (1963), *Trans. Far. Soc.* **59**, 176.
KISLIUK, P. (1957), *J. Phys. Chem. Solids* **3**, 95.
KISLIUK, P. (1958), *J. Phys. Chem. Solids* **5**, 78.
KISLIUK, P. J. (1959a), *J. Chem. Phys.* **30**, 174.
KISLIUK, P. J. (1959b), *J. Chem. Phys.* **31**, 1605.
KISLIUK, P. (1961), *Phys. Rev.* **122**, 405.

KISTEMAKER, J. (1945), *Physica* **11**, 277.
KLEIN, H. J. (1965), *Zeit. fur Phys.* **188**, 78.
KLEINT, CH. (1960), *Exp. Tech. der Phys.* **8**, 193.
KLEINT, CH. (1962), *Proc. Symp. on Electron and Vac. Phys.* Hungarian Acad. of Sciences, Budapest, p. 365.
KLEINT, CH. (1963), *Ann. Phys.* **10**, 309.
KLEINT, CH. and MOLDENHAUER, W. (1965), *Trans. 3rd Int. Vac. Cong.* Vol. 2, p. 475 (Stuttgart).
KLEMPERER, O. and SHEPHERD, J. P. G. (1963), *Advances in Physics* **12**, 355.
KLEMPERER, O. and THIRLWELL, J. (1966), *Solid State Communications* **4**, 15.
KLOPFER, A. (1961a), *Vakuum Technik* **10**, 113.
KLOPFER, A. (1961b), *Trans. AVS Vac. Symp.* **8**, 439.
KLOPFER, A. (1963), *Proc. 2nd European Vac. Symp.*, p. 271.
KLOPFER, A. and ERMRICH, W. (1959a), *Vakuum Technik* **8**, 162.
KLOPFER, A. and ERMRICH, W. (1959b), *Trans. AVS Vac. Symp.* **6**, 297.
KLOPFER, A. and ERMRICH, W. (1960), *Adv. in Vac. Sci. and Technol.* **1**, 427.
KLOPFER, A. and SCHMIDT, W. (1960), *Vacuum* **10**, 363.
KNAUER, W. (1962), *J. Appl. Phys.* **33**, 2093.
KNAUER, W. and LUTZ, M. A. (1963), *Appl. Phys. Lett.* **2**, 109.
KNAUER, F. and STERN, O. (1929), *Zeit. fur Phys.* **53**, 766.
KNOR, Z. (1963), *Czech. Jour. of Phys.* **13**, 302.
KNUDSEN, M. (1915), *Ann. der Physik* **48**, 1113.
KNUDSEN, M. (1934), *Kinetic Theory of Gases* (Wiley, New York).
KNUDSEN, M. (1950), *The Kinetic Theory of Gases* (Wiley, New York).
KODERA, K. and ONISHI, Y. (1959), *Bull. Chem. Soc. Japan* **32**, 356.
KOEDAM, M. (1959), *Physica* **25**, 742.
KOEDAM, M. and HOOGENDOORN, A. (1960), *Physica* **26**, 351.
KOENIG, H. (1953), *Vacuum* **3**, 3.
KOLLATH, R. (1956), *Handbuch der Physik* **21**, 232.
KOMIYA, S., SATO, H. and HAYASHI, C. (1966), *J. Vac. Sci. and Technol.* **3**, 300.
KOPITZKI, K. and STIER, H. E. (1962), *Z. fur Naturforsch.* **17a**, 346.
KOPITZKI, K. and STIER, H. E. (1963), *Phys. Lett.* **4**, 232.
KORNELSEN, E. V. (1959), *Proc. 19th Phys. El. Conf. M.I.T.*, p. 156 (Cambridge, Mass.).
KORNELSEN, E. V. (1960), *Trans. AVS Vac. Symp.* **7**, 29.
KORNELSEN, E. V. (1961), *Trans. AVS Vac. Symp.* **8**, 281.
KORNELSEN, E. V. (1964), *Can. J. Phys.* **42**, 364.
KORNELSEN, E. V. (1965), *Trans. of the 3rd Int. Vacuum Congress*, Vol. I, p. 65 (Stuttgart).
KORNELSEN, E. V. (Unpublished).
KORNELSEN, E. V. and DOMEIJ, B. (1966), *J. Vac. Sci. and Technol.* **3**, 20.
KORNELSEN, E. V. and SINHA, M. K. (1966), *Appl. Phys. Lett.* **9**, 112.
KORNELSEN, E. V. and SINHA, M. K. (1968), *J. Appl. Phys.* (in press).
KORNELSEN, E. V., BROWN, F., DAVIES, J. A., DOMEIJ, B. and PIERCY, G. R. (1964), *Phys. Rev.* **136**, A849.
KOTOWSKI, J. W. and TISCHER, K. M. (1964), *Vakuum Technik* **13**, 214.
KRAMER, I. R. and PODLASEK, S. (1963), *Acta Metallurgica* **11**, 70.
KRAUS, T. (1963), *Trans. AVS Vac. Symp.* **10**, 77.

KRUYER, S. (1955), *Proc. K. Ned. Akad. Wetensch.* B **58**, 73.
KUCHAI, S. A. and RODIN, A. M. (1958), *Soviet J. Atomic Energy* **4**, 277.
KUYATT, C.E., SIMPSON, J. A. and MIELCZAREK, S. R. (1965), *Phys. Rev.* **138**, A385.
KUZ'MIN, A. A. (1963), *Inst and Exp. Tech.*, No. 3, p. 497.

L

LAEGREID, N. and WEHNER, G. K. (1961), *J. Appl. Phys.* **32**, 365.
LAFFERTY, J. M. (1951), *J. Appl. Phys.* **22**, 299.
LAFFERTY, J. M. (1960), *Trans. AVS Vac. Symp.* **7**, 97.
LAFFERTY, J. M. (1961a), *J. Appl. Phys.* **32**, 424.
LAFFERTY, J. M. (1961b), *Trans. AVS Vac. Symp.* **8**, 460.
LAFFERTY, J. M. (1962), *Trans. AVS Vac. Symp.* **9**, 438.
LAFFERTY, J. M. (1963), *Rev. Sci. Instr.* **34**, 467.
LAFFERTY, J. M., DAVIS, W. D. and FAVREAU, L. J. (1967), *Study of Hot-Cathode Magnetron Ion Gauge*, NASA Report CR-701, STAR Index No. N67-15670.
LANDER, J. J. (1950), *Rev. Sci. Instr.* **21**, 672.
LANDER, J. J. (1965), in *Progress in Solid State Chemistry*, Vol. II, p. 26 (Ed. H. Reiss; Pergamon Press, Oxford).
LANDER, J. J. and MORRISON, J. (1962), *J. Chem. Phys.* **37**, 729.
LANDER, J. J. and MORRISON, J. (1963a), *J. Appl. Phys.* **34**, 1403.
LANDER, J. J. and MORRISON, J. (1963b), *J. Appl. Phys.* **34**, 3517.
LANDER, J. J. and MORRISON, J. (1964), *J. Appl. Phys.* **35**, 3593.
LANDER, J. J. and MORRISON, J. (1967), *Surface Science* **6**, 1.
LANE, G. H. and EVERHART, E. (1960), *Phys. Rev.* **120**, 2064.
LANGE, W. J. (1960), *Proc. 20th Ann. Conf. on Phys. Electronics. M.I.T.*, p. 226 (Cambridge, Mass.).
LANGE, W. J. (1965), *J. Vac. Sci. and Techhol.* **2**, 74.
LANGE, W. J. and ERIKSEN, D. P. (1966), *J. Vac. Sci. and Technol.* **3**, 303.
LANGE, W. J. and RIEMERSMA, H. (1961), *Trans. AVS Vac. Symp.* **8**, 167.
LANGE, W. J. and SINGLETON, J. H. (1966), *J. Vac. Sci. and Techhol.* **3**, 319.
LANGE, W. J., SINGLETON, J. H. and ERIKSEN, D. P. (1966), *J. Vac. Sci. and Technol.* **3**, 338.
LANGMUIR, I. (1915), *J. Am. Chem. Soc.* **37**, 417
LANGMUIR, I. (1918), *J. Am. Chem. Soc.* **40**, 1361.
LAPONSKY, A. B. and WHETTEN, N. R. (1960), *Phys. Rev.* **120**, 801.
LARGE, L. N. (1963), *Proc. Phys. Soc. (London)* **81**, 1101.
LASSETTRE, E. A. and FRANCIS, S. A. (1964), *J. Chem. Phys.* **40**, 1208.
LAW, J. T. (1962), *Trans. AVS Vac. Symp.* **8**, 104.
LAW, R. T. (1958), *J. Chem. Phys.* **28**, 511.
LAWSON, R. W. (1962), *J. Sci. Instr.* **39**, 281.
LAWSON, R. W. and WOODWARD, J. W. (1967). *Vacuum* **17**, 205.
LAZAREV, B. G., BOROVIK, E. S., FEDOROVA, M. F. and TSIN, N. M. (1957), *Ukrain Fiz. Zhur.* **2**, 175.
LeBLANC, L. J., FARRELL, J. S. and JUENKER, D. W. (1964), *J. Opt. Soc. Am.* **54**, 956.
LECK, J. H. (1964), *Pressure Measurement in Vacuum Systems* (Chapman and Hall, London).
LEHMANN, C. and SIGMUND, P. (1966), *Phys. Stat. Sol.* **16**, 507.

LEIBY, C. C. and CHEN, C. L. (1960), *J. Appl. Phys.* **31,** 268.
LENNARD-JONES, J. E. and DEVONSHIRE, A. F. (1936a), *Proc. Roy. Soc. (London)* **A156,** 6.
LENNARD-JONES, J. E. and DEVONSHIRE, A. F. (1936b), *Proc. Roy. Soc. (London)* **A156,** 29.
LEONAS, V. B. (1963), *Zh. Prikladnoi Mekhaniki i Tekhnicheskoi Fiziki No* **6,** 124. NASA Technical Translation TTF-265 (1965).
LEVENSON, L. L., MILLERON, N. and DAVIS, D. H. (1963), *Le Vide* **18,** 42.
LEWIN, G. (1965), *Fundamentals of Vacuum Science and Technology*, (McGraw-Hill, New York).
LICHTMAN, D. (1963), Private Communication.
LICHTMAN, D. (1964), *J. Vac. Sci. and Technol.* **1,** 23.
LICHTMAN, D. (1965a), *J. Vac. Sci. and Technol.* **2,** 70.
LICHTMAN, D. (1965b), *J. Vac. Sci. and Technol.* **2,** 91.
LICHTMAN, D. and KIRST, T. R. (1966), *Phys. Lett.* **20,** 7.
LICHTMAN, D. and MCQUISTAN, R. B. (1965), *Progress in Nuclear Energy, Series IX* **4**(2), 95 (Pergamon Press).
LICHTMAN, D., MCQUISTAN, R. B. and KIRST, T. R. (1966), *Surface Science*, **5,** 120.
LIFSHITZ, E. M. (1956), *Soviet Phys. JETP* **2,** 73.
LINDHARD, J. (1964), *Phys. Lett.* **12,** 126.
LINDHARD, J. (1965), *Mat. Fys. Medd. Dan. Vid. Selsk.* **34,** No. 14.
LINDHARD, J. and SCHARFF, M. (1961), *Phys. Rev.* **124,** 128.
LINDHARD, J., NIELSEN, V., SCHARFF, M. and THOMSEN, P. V. (1963a), *Mat. Fys. Medd. Dan. Vid. Selsk.* **33,** No. 10.
LINDHARD, J., SCHARFF, M. and SCHIØTT, H. E. (1963b), *Mat. Fys. Medd. Dan. Vid. Selsk.* **33,** No. 14.
LITTLE, J. G., QUINN, C. M. and ROBERTS, M. W. (1964), *J. of Catalysis* **3,** 57.
LITTLE, R. P. and WHITNEY, W. T. (1962), *Trans. AVS Vac. Symp.* **9,** 456.
LLOYD, O. (1966), *Brit. J. Appl. Phys.* **17,** 357.
LOCKWOOD, G. J., HELBIG, H. F. and EVERHART, E. (1963), *Phys. Rev.* **132,** 2078.
LOGAN, R. M. and STICKNEY, R. E. (1966), *J. Chem. Phys.* **44,** 195.
LOVE, H. M. and WILSON, J. R. (1967), *Can. J. Phys.* **45,** 225.
LUTZ, H. and SIZMANN, R. (1963), *Phys. Lett.* **5,** 113.
LYE, R. G. (1955), *Phys. Rev.* **99,** 1647.
LYE, R. G. and DEKKER, A. J. (1957), *Phys. Rev.* **107,** 977.

M

MACLENNAN, D. A. (1966), *Phys. Rev.* **148,** 218.
MACRAE, A. U. (1963), *Science* **139,** 379.
MACRAE, A. U. (1964a), *Surface Science* **1,** 319.
MACRAE, A. U. (1964b), *Surface Science* **2,** 522.
MACRAE, A. U. and GERMER, L. H. (1962), *Phys. Rev. Lett.* **8,** 489.
MACRAE, A. U. and GERMER, L. H. (1963), *Ann. New York Academy of Sciences* **101,** 627.
MADEY, T. E. and YATES, J. T. (1967), *Suppl. Nuovo Cimento* **5,** No. 2, 483.
MADEY, T. E., YATES, J. T. and STERN, R. C. (1965), *J. Chem. Phys.* **42,** 1372.
MAGNUSON, G. D. and CARLSTON, C. E. (1963a), *J. Appl. Phys.* **34,** 3267.

MAGNUSON, G. D. and CARLSTON, C. E. (1963b), *Phys. Rev..* **129**, 2403.
MAGNUSON, G. D. and CARLSTON, C. E. (1963c), *Phys. Rev.* **129**, 2409.
MALIAKAL, J. C., LIMON, P. J., ARDEN, E. E. and HERB, R. G. (1964), *J. Vac. Sci. and Technol.* **1**, 54.
MAMYRIN, B. A. (1967), *Instr. and Exp. Tech.*, No. 4, p. 900.
MANES, M. and GRANT, R. J. (1963), *Trans. AVS Vac. Symp.* **10**, 122.
MARGENAU, H. (1939), *Rev. Mod. Phys.* **11**, 1.
MARK, J. T. and DREYER, K. (1959), *Trans. AVS Vac. Symp.* **6**, 176.
MARK, H. and SOMMERS, R. D. (1962), *Adv. in Cryogenic Engineering* **8**, 93
MARMET, P. and KERWIN, L. (1960), *Can. J. Phys.* **38**, 787.
MARMET, P. and MORRISON, J. D. (1961), *J. Chem. Phys.* **35**, 746.
MARMET, P. and MORRISON, J. D. (1962), *J. Chem. Phys.* **36**, 1238.
MARTIN, D. W., LANGLEY, R. A., HOOPER, J. W., HARMER, D. S. and MCDANIEL, E. W. (1963), *Proc. 3rd Int. Conf. on Electronic and Atomic Collisions* (London), p. 679.
MARTIN, G. and LEWIN, G. (1966), *J. Vac. Sci. and Technol.* **3**, 6.
MARTON, L., LEDER, L. B. and MENDLOWITZ, H. (1955), *Advances in Electronics and Electron Physics* **7**, 183.
MARTYNENKO, YU. V. (1965a), *Sov. Phys. Solid State* **6**, 1581.
MARTYNENKO, YU. V. (1965b), *Sov. Phys. Solid State* **6**, 2827.
MARTYNENKO, YU. V. (1966), *Phys. Stat. Sol.* **15**, 767.
MARYOTT, A. A. and BUCKLEY, F. (1953), NBS Circular 537. (U.S. Dept of Commerce, National Bureau of Standards).
MASHKOVA, E. S. and MOLCHANOV, V. A. (1965a), *Sov. Phys. Solid State* **6**, 2792.
MASHKOVA, E. S. and MOLCHANOV, V. A. (1965b), *Sov. Phys. Tech. Phys.* **10**, 449.
MASHKOVA, E. S., MOLCHANOV, V. A., PARILIS, E. S. and TURAEV, N. YU. (1965), *Phys. Lett.* **18**, 7.
MASHKOVA, E. S., MOLCHANOV, V. A., SOSHKA, V. and FARUK, M. A. (1966), *Sov. Phys. Solid State* **7**, 2371.
MASON, E. A. and VANDERSLICE, J. T. (1958), *J. Chem. Phys.* **29**, 361.
MASON, E. A. and VANDERSLICE, J. T. (1962), in *Atomic and Molecular Processes* (ed. D. R. Bates; Academic Press, New York), p. 663.
MASSEY, H. S. W. (1956), *Handbuch der Physik* **36**, 307.
MASSEY, H. S. W. and BURHOP, E. H. S. (1952), *Electronic and Ionic Impact Phenomena* (Clarendon Press, Oxford).
MASSEY, H. S. W. and MOHR, C. B. O. (1934), *Proc. Roy. Soc.* **A144**, 188.
MAURICE, L. and SAGOT, S. (1964), *Le Vide* **19**, 109.
MEDVED, D. B. (1958), *J. Chem. Phys.* **28**, 870.
MEDVED, D. B. and STRAUSSER, Y. E. (1965), *Adv. in Electronics and Electron Physics* **21**, 101.
MEINKE, C. and REICH, G. (1963), *Vakuum Technik* **12**, 79.
MENZEL, D. (1965), *Surface Science* **3**, 424.
MENZEL, D. and GOMER, R. (1964a), *J. Chem. Phys.* **41**, 3311.
MENZEL, D. and GOMER, R. (1964b), *J. Chem. Phys.* **41**, 3329.
METCALFE, R. A. and TRABERT, F. W. (1961), *Trans. AVS Vac. Symp.* **8**, 1211.
METSON, G. H. (1951), *Brit. J. Appl. Phys.* **2**, 46.
MEYER, L. (1956), *Phys. Rev.* **103**, 1593.

MEYER, L. and GOMER, R. (1958), *J. Chem. Phys.* **28**, 617.
MEYER, E. A. and HERB, R. G. (1967), *J. Vac. Sci. and Technol.* **4**, 63.
MEYER, V. D., SKERBELE, A. and LASSETTRE, E. A. (1965), *J. Chem. Phys.* **43**, 805
MIKHAILOV, G. V. and KULAKOV, YU. A. (1963), *Inst. and Exp. Tech.*, p. 1136.
MILLER, C. F. and SHEPARD, R. W. (1961), *Vacuum* **11**, 58.
MILLER, T. M. and GEIGER, K. A. (1962), *Trans. AVS Vac. Symp.* **9**, 270.
MILLERON, N. (1965), *Trans. 3rd Int. Vac. Cong.*, Vol. 2, p. 189 (Stuttgart).
MILLERON, N. (1967) *IEEE Trans. on Nucl. Sciences*, NS-14, 794.
MIMEAULT, V. J. and HANSEN, R. S. (1963), *Vacuum* **13**, 229.
MINTURN, R. E., DATZ, S. and TAYLOR, E. H. (1960), *J. Appl. Phys.* **31**, 876.
MIZUSHIMA, Y. and ODA, Z. (1959) *Rev. Sci. Instr.* **30**, 1037.
MOE, D. E. and PETSCH, O. H. (1958) *Phys. Rev.* **110**, 1358.
MOESTA, H. and RENN, R. (1957), *Vakuum Technik* **6**, 35.
MOISEIWITSCH, B. L. (1956), *Proc. Phys. Soc. (London)* **A69**, 653.
MOLCHANOV, V. A. and SOSHKA, V. (1965), *Sov. Phys. Tech. Phys.* **10**, 741.
MOLCHANOV, V. A. and TEL'KOVSKII, V. O. (1962), *Izvestia A.N.* **26**, 1359.
MOORE, B. C. (1962), *Trans. AVS Vac. Symp.* **9**, 212.
MOORE, B. C. (1964), *J. Vac. Sci. and Technol.* **1**, 10.
MOORE, B. C. (1965), *J. Vac. Sci. and Technol.* **2**, 211.
MOORE, G. E. (1959), *J. Appl. Phys.* **30**, 1086.
MOORE, G. E. (1961), *J. Appl. Phys.* **32**, 1241.
MOORE, G. E. and UNTERWALD, F. C. (1964a), *J. Chem. Phys.* **40**, 2626.
MOORE, G. E. and UNTERWALD, F. C. (1964b), *J. Chem. Phys.* **40**, 2639.
MOORE, R. W. (1961), *Trans. AVS Vac. Symp.* **8**, 426.
MORGAN, G. H. and EVERHART, E. (1962), *Phys. Rev.* **128**, 667.
MORGULIS, N. D. and GORODETSKII, D. A. (1956), *Soviet Physics JETP* **3**, 535.
MORGULIS, N. D. and MARCHENKO, R. I. (1960), *Instr. and Exp. Tech.*, p. 795.
MORRISON, J. A. and TUZI, Y. (1965), *J. Vac. Sci. and Technol.* **2**, 109.
MORRISON, J. L. and ROBERTS, J. K. (1939), *Proc. Roy. Soc. (London)* **A173**, 1
MOSER, H. and POLTZ, H. (1957), *Zeit. fur Instrumentkunde* **65**, 43.
MOTT, N. F. and MASSEY, H. S. W. (1965), *The Theory of Atomic Collisions* (Clarendon Press, Oxford).
MOTT-SMITH, H. M. and LANGMUIR, I. (1926), *Phys. Rev.* **28**, 727.
MOURAD, W. G., PAULY, T. and HERB, R. G. (1964), *Rev. Sci. Instr.* **35**, 661.
MUEHLHAUSE, C. O., GANOCZY, M. and KUPIEC, C. *IEEE Trans. on Nuc. Sc.* **12**, 478.
MULLEN, L. O. and JACOBS, R. B. (1962), *Trans. AVS Vac. Symp.* **9**, 220.
MÜLLER, D. (1965), *Zeit. fur Phys.* **188**, 326.
MÜLLER, K. G. (1960), *Vakuum Technik* **9**, 13.
MURAKAMI, Y. and OKAMOTO, H. (1963), *Trans. AVS Vac. Symp.* **10**, 93.
MYERS, H. P. (1952), *Proc. Roy. Soc.* (London) **A215**, 329.

Mc

MCCARGO, M., DAVIES, J. A. and BROWN, F. (1963), *Can. J. Phys.* **41**, 1231.
MCCARROLL, B. and EHRLICH, G. (1963), *J. Chem. Phys.* **38**, 523.
MCCRACKEN, G. M. (1965), *Vacuum* **15**, 433.
MCCRACKEN, G. M. and PASHLEY, N. A. (1966), *J. Vac. Sci. and Technol.* **3**, 96.

MCDANIEL, E. W. (1964), *Collision Phenomena in Ionized Gases* (Wiley, New York).
MCDOWELL, C. A. (Ed.) (1963a), *Mass Spectrometry* (McGraw Hill, New York).
MCDOWELL, M. R. C. (Ed.) (1963b), *Proc. 3rd Int. Conf. on the Physics of Electronic and Atomic Collisions* (London).
MCGOWAN, W. and KERWIN, L. (1960), *Can. J. Phys.* **38**, 567.
MCKAY, K. G. (1948), *Advances in Electronics* **1**, 65.
MCRAE, E. G. (1966), *J. Chem. Phys.* **45**, 3258.
MCRAE, E. G. and CALDWELL, C. W. (1964), *Surface Science* **2**, 509.

N

NAKHODKIN, N. G. and MEL'NIK, P. V. (1964), *Sov. Phys. Solid State* **5**, 1779.
NARCISI, R. S., BRUBAKER, W. M., POEHLMANN, H. C., FEDCHENKO, R. P. and WIENS, F. B. (1962), *Trans. AVS Vac. Symp.* **9**, 232.
NAZAROV, A. S., IVANOVSKII, G. F. and KOUZNETSOV, M. V. (1965), *Trans. 3rd Int. Vac. Cong.*, Vol. 2, p. 663 (Stuttgart).
NEAL, R. B. (1965), *J. Vac. Sci. and Technol.* **2**, 149.
NECHAI, E. P., POPOV, K. V. and PANENKOVA, L. S. (1960), *Physics of Metals and Metallography* **10**, No. 6, 45.
NELSON, R. S. (1965), *Phil. Mag.* **11**, 291.
NELSON, R. S. (1966), *Phil. Mag.* **14**, 637.
NELSON, R. S. (1967), *Phil. Mag.* **15**, 845.
NELSON, R. S., THOMPSON, M. W. and MONTGOMERY, H. (1962), *Phil. Mag.* **7**, 1385.
NESMEYANOV, A. N. (1963), *Vapour Pressure of the Chemical Elements* (Elsevier, Amsterdam).
NICHOLS, D. K. and VAN LINT, V. A. J. (1966), *Adv. in Solid State Physics* **18**, 1.
NOBLE, R. and JACOB, L. (1956), *Nature* **178**, 814.
NÖLLER, H. G. (1966), *Handbook of Vacuum Physics*, Vol. 1, p. 321 (ed. A. H. W. Beck).
NORTON, F. J. (1957), *J. Appl. Phys.* **28**, 34.
NORTON, F. J. (1961), *Trans. AVS Vac. Symp.* **8**, 8.
NOTTINGHAM, W. B. (1947), 7th Ann. Conf. on Physical Electronics, M.I.T. (Cambridge, Mass.).
NOTTINGHAM, W. B. (1954), *Trans. AVS Vac. Symp.* **1**, 76.
NOTTINGHAM, W. B. (1956), *Handbuch der Physik* **21**, 1.
NOTTINGHAM, W. B. (1961), *Trans. AVS Vac. Symp.* **8**, 494.

O

OEN, O. S. (1965), *Phys. Lett.* **19**, 358.
OEN, O.S. and ROBINSON, M. T. (1963), *Appl. Phys. Lett.* **2**, 83.
OEN, O. S., HOLMES, D. K. and ROBINSON, M. T. (1963), *J. Appl. Phys.* **34**, 302.
OGURI, T. (1963), *J. Phys. Soc. Japan* **18**, 1280.
OKAMOTO, H. and MURAKAMI, Y. (1967), *Vacuum* **17**, 79.
OLDAL, E. and TAHY, P. (1965), *Trans 3rd Int. Vac. Cong.*, Vol. 2, p. 617 (Stuttgart).
OPPENHEIMER, J. R. (1928), *Phys. Rev.* **32**, 361.
OSHCHEPKOV, P. K., SKVORTSOV, B. N., OSANOV, B. A. and SIPRIKOV, I. V. (1960), *Instr. and Exp. Tech.*, p. 611.
OUTLAW, R. A. (1966), *J. Vac. Sci. and Technol.* **3**, 352.

OWENS, C. L. (1965), *J. Vac. Sci. and Technol.* **2**, 104.

P

PACK, J. L., VOSHALL, R. E. and PHELPS, A. V. (1962), *Phys. Rev.* **127**, 2084.
PALLUEL, P. (1947), *Compt. Rend.* **224**, 1492.
PARCEL, R. W., CLAUSS, F. J., O'HARA, C. F. and YOUNG, W. C. (1963), *Trans. AVS Vac. Symp.* **10**, 3.
PARILIS, E. S. and KISHINEVSKII, L. M. (1961), *Sov. Phys. Solid State* **3**, 885.
PARILIS, E. S. and TURAEV, N. YU. (1965), *Sov. Phys. Doklady* **10**, 212.
PARK, R. L. (1966), *J. Appl. Phys.* **37**, 295.
PARK, R. L. and FARNSWORTH, H. E. (1964a), *Surface Science* **2**, 527.
PARK, R. L. and FARNSWORTH, H. E. (1964b), *J. Appl. Phys.* **35**, 2220.
PATY, L. (1959), *Instr. and Exp. Tech.*, p. 863.
PEACH, G. (1966), *Proc. Phys. Soc. (London)* **87**, 381.
PENNING, F. M. and NIENHUIS, K. (1949), *Philips Tech. Rev.* **11**, 116.
PEROVIC, B. (Ed.) and TOSIC, D. (Ed.) (1965), *Proc. 7th Int. Conf. on Phenomena in Ionized Gases* (Belgrade).
PÉTERMANN, L. A. (1963a), *Proc. 2nd European Vac. Symp.*, p. 157.
PÉTERMANN, L. A. (1963b), *Suppl. Nuovo Cimento* **1**, 601.
PÉTERMANN, L. (1965), *Brevet Francais* 1,405,264.
PÉTERMANN, L. A. and BAKER, F. A. (1965), *Brit. J. Appl. Phys.* **16**, 487.
PHILIPP, H. R. and EHRENREICH, H. (1964), *J. Appl. Phys.* **35**, 1416.
PIERCY, G. R., BROWN, F., DAVIES, J. A. and MCCARGO, M. (1963), *Phys. Rev. Lett.* **10**, 399.
PIEROTTI, R. A. and HALSEY, G. D. (1959), *J. Phys. Chem.* **63**, 680.
PIERRE, J. (1961), *Le Vide* **16**, 18.
PINSKER, Z. G. (1953), *Electron Diffraction* (Butterworths, London).
PINSON, J. D. and PECK, A. W. (1962), *Trans. AVS Vac. Symp.* **9**, 406
PISANI, C. and DELLA PORTA, P. (1967), *Suppl. Nuovo Cimento* **5**, No. 1, 261.
PLUMLEE, R. H. and SMITH, L. P. (1950), *J. Appl. Phys.* **21**, 811.
PODGURSKI, H. H. and DAVIS, F. N. (1960), *Vacuum* **10**, 377.
POLLEY, M. H., SCHAEFFER, W. D. and SMITH, W. R. (1953), *J. Phys. Chem.* **57**, 469.
POWELL, C. J. (1965), *Phys. Rev. Lett.* **15**, 852.
POWELL, C. J. and SWAN, J. B. (1959), *Phys. Rev.* **115**, 869.
POWELL, C. J. and SWAN, J. B. (1960), *Phys. Rev.* **118**, 640.
POWER, B. D. (1966), *High Vacuum Pumping Equipment* (Reinhold, New York).
POWERS, D. and WHALING, W. (1962), *Phys. Rev.* **126**, 61.
PRESNYAKOV, L., SOBELMAN, I. and VAINSHTEIN, L. (1966), *Proc. Phys. Soc. (London)* **89**, 511.
PROPST, F. M. and PIPER, T. C. (1967), *J. Vac. Sci. and Technol.* **4**, 53.
PTUSHINKSII, YU. G. (1958), *Sov. Phys. Tech. Phys.* **3**, 1302.
PTUSHINKSII, YU. G. and CHUIKOV, B. A. (1967), *Surface Science* **6**, 42.

R

RABINOWICZ, E. (1965), *Friction and Wear of Materials* (Wiley, New York).
RAETHER, H. (1965), *Ergebn. exact. Naturw.* **38**, 84.
RAMSAUER, P. (1921a), *Ann. Phys.* **64**, 513.
RAMSAUER, P. (1921b), *Ann. Phys.* **66**, 546.

RAMSAUER, P. (1923), *Ann. Phys.* **72,** 345.
RAMSEY, N. F. (1956), *Molecular Beams* (Clarendon Press, Oxford).
RAPP, D. (1963), *Proc. 11th ASTM Mass Spectrometry Conf.* p. 2.
RAPP, D. and BRIGLIA, D. D. (1965), *J. Chem. Phys.* **43,** 1480.
RAPP, D. and ENGLANDER-GOLDEN, P. (1965), *J. Chem. Phys.* **43,** 1464.
RAPP, R. A., HIRTH, J. P. and POUND, G. M. (1960), *Can. J. Phys.* **38,** 709.
RAPP, R. A., HIRTH, J. P. and POUND, G. M. (1961), *J. Chem. Phys.* **34,** 184.
READ, P. L. (1963), *Vacuum* **13,** 271.
REDHEAD, P. A. (1958), *Can. J. Phys.* **36,** 255.
REDHEAD, P. A. (1959a), *Can. J. Phys.* **37,** 1260.
REDHEAD, P. A. (1959b), *Trans. AVS Vac. Symp.* **6,** 12.
REDHEAD, P. A. (1960a), *Trans. AVS Vac. Symp.* **7,** 108.
REDHEAD, P. A. (1960b), *Rev. Sci. Instr.* **31,** 343.
REDHEAD, P. A. (1961), *Trans. Far. Soc.* **57,** 641.
REDHEAD, P. A. (1962a), *Vacuum* **12,** 267.
REDHEAD, P. A. (1962b), *Vacuum* **12,** 203.
REDHEAD, P. A. (1962c), *Proc. Symp. on Electron and Vacuum Physics* (Hungary), p. 89.
REDHEAD, P. A. (1963), *Vacuum* **13,** 253.
REDHEAD, P. A. (1964a), *Can. J. Phys.* **42,** 886.
REDHEAD, P. A. (1964b), *Appl. Phys. Lett.* **4,** 166.
REDHEAD, P. A. (1965a), Private Communication.
REDHEAD, P. A. (1965b), *Can. J. Phys.* **43,** 1001.
REDHEAD, P. A. (1966), *J. Vac. Sci. and Technol.* **3,** 173.
REDHEAD, P. A. (1967a), *J. Vac. Sci. and Technol.* **4,** 57.
REDHEAD, P. A. (1967b), *Suppl. Nuovo Cimento* **5,** No. 2, 586.
REDHEAD, P. A. (1967c), *Can. J. Phys.* **45,** 1791.
REDHEAD, P. A. and HOBSON, J. P. (1965), *Brit. J. Appl. Phys.* **16,** 1555.
REDHEAD, P. A., KORNELSEN, E. V. and HOBSON, J. P. (1962a), *Can. J. Phys.* **40,** 1814.
REDHEAD, P. A., HOBSON, J. P. and KORNELSEN, E. V. (1962b), *Adv. in Elect. and Electron Physics* **17,** 323.
REED, R. I. (1962), *Ion Production by Electron Impact* (Academic Press, New York).
REICH, G. (1960), *Trans. AVS Vac. Symp.* **7,** 112.
REICH, G. (1961), *Vakuum Technik* **10,** 109.
REICH, G. (1966), *Zeit. fur. Instrumentkunde* **74,** 254.
REICH, G. and MEINKE, C. (1966), *J. Vac. Sci. and Technol.* **3,** 302.
REIKHRUDEL, E. M., CHERNETSKII, A. V., MIKHNEVICH, V. V. and VASIL'EVA, I. A. (1952), *Zh. Tekh Fiz.* **22,** 1945. Available in English as R.A.E. Library Translation No. 490.
REIKHRUDEL, E. M. and SHERETOV, E. P. (1966a), *Sov. Phys. Tech. Phys.* **10,** 972.
REIKHRUDEL, E. M. and SHERETOV, E. P. (1966b), *Rad. Eng. and Electronic Phys.* No. 3, p. 449.
REIKHRUDEL, E. M. and SMIRNITSKAYA, G. M. (1958), *Radiofizika* **1**(2), 36.
REIKHRUDEL, E. M., SMIRNITSKAYA, G. V. and SHERETOV, E. P. (1962), *Rad. Eng. and Electronic Physics,* **7,** 1672.
REYNOLDS, J. H. (1956), *Rev. Sci. Instr.* **27,** 928.
RHODIN, T. N. (1950), *J. Amer. Chem. Soc.* **72,** 5691.

RHODIN, T. N. and ROVNER, L. H. (1960), *Trans. AVS Vac. Symp.* **7**, 228.
RICCA, F. (1967), *Suppl. Nuovo Cim.* **5**, No. 2, 339.
RICCA, F. and BELLARDO, A. (1967), *Zeit fur Phys. Chem.* **52**, 318.
RICCA, F. and MEDANA, R. (1964), *La Ricerca Scientifica* **34**(II-A), 4, 617.
RICCA, F., BELLARDO, A. and MEDANA, R. (1966), *La Ricerca Scientifica* **36**, 460.
RICCA, F., MEDANA, R. and SAINI, G. (1965), *Trans. Far. Soc.* **61**, 1492.
RICCA, F., MEDANA, R. and BELLARDO, A. (1967), *Z. fur Phys. Chem.* **52**, 276.
RICCA, F., NASINI, A. G. and SAINI, G. (1962), *J. of Catalysis* **1**, 458.
RICHARDS, P. I. and HAYS, E. E. (1950), *Rev. Sci. Instr.* **21**, 99.
RICHARDSON, D. M. and STREHLOW, R. A. (1963), *Trans. AVS Vac. Symp.* **10**, 97.
RICHMAN, J. and HOOD, C. B. (1962), *Trans. AVS Vac. Symp.* **9**, 282.
RIDDOCH, A. and LECK, J. H. (1958), *Proc. Phys. Soc. (London)* **72**, 467.
RIGBY, L. J. (1964), *Can. J. Phys.* **42**, 1256.
RIGBY, L. J. (1965), *Can. J. Phys.* **43**, 1020.
RITCHIE, R. H. (1957), *Phys. Rev.* **106**, 874.
RIVIÈRE, J. C. and ALLINSON, J. D. (1964), *Vacuum* **14**, 97.
RIVIÈRE, J. C., THOMPSON, J. B., READ, J. E. and WILSON, I. (1965), *Vacuum* **15**, 353.
ROBERTS, J. K. (1935), *Proc. Roy. Soc. (London)* **A152**, 477.
ROBERTS, R. W. (1963), *Brit. J. Appl. Phys.* **14**, 537.
ROBERTS, R. W. and VANDERSLICE, T. A. (1963), *Ultrahigh Vacuum and its Applications* (Prentice-Hall, New Jersey).
ROBINS, J. L. (1962), *Trans. AVS Vac. Symp.* **9**, 510.
ROBINS, J. L. (1963), *Can. J. Phys.* **41**, 1385.
ROBINS, J. L. and SWAN, J. B. (1960), *Proc. Phys. Soc.* (London) **76**, 857.
ROBINSON, C. F. and HALL, L. G. (1956), *Rev. Sci. Instr.* **27**, 504.
ROBINSON, C. F. and SHARKEY, A. G. (1958), *Rev. Sci. Instr.* **29**, 250.
ROBINSON, M. T. (1962), *Appl. Phys. Lett.* **1**, 49.
ROBINSON, N. W. (1960), *Adv. in Vac. Sci. and Tech.* **2**, 616.
ROBINSON, N. W. and BERZ, F. (1959), *Vacuum* **9**, 48.
ROBINSON, M. T. and OEN, O. S. (1963), *Phys. Rev.* **132**, 2385.
ROEHRIG, J. R. and SIMONS, J. C. (1961), *Trans. AVS Vac. Symp.* **8**, 511.
ROGERS, K. W. (1963), *Trans. AVS Vac. Symp.* **10**, 84.
ROGERS, W. A., BURITZ, R. S. and ALPERT, D. (1954), *J. Appl. Phys.* **25**, 868.
ROL, P. K., FLUIT, J. M. and KISTEMAKER, J. (1960a), *Physica* **26**, 1000.
ROL, P. K., FLUIT, J. M. and KISTEMAKER, J. (1960b), *Physica* **26**, 1009.
ROOTSAERT, W. J. M., VAN REIJEN, L. L. and SACHTLER, W. M. H. (1962), *J. of Catalysis* **1**, 416.
ROSENBERG, D. and WEHNER, G. K. (1962), *J. Appl. Phys.* **33**, 1842.
ROSENBERG, P. (1939), *Rev. Sci. Instr.* **10**, 131.
ROSS, S. and OLIVIER, J. P. (1964), *On Physical Adsorption* (Interscience, New York).
ROSS, J. R. H. and ROBERTS, M. W. (1965), *J. Catalysis* **4**, 620.
ROTHE, E. W. (1964), *J. Vac. Sci. and Technol.* **1**, 66.
ROUSSEL, J., THIBAULT, J. J. and NANOBOFF, A. (1965), *Le Vide* **20**, 249.
ROZGONYI, G. A. (1966), *J. Vac. Sci. and Technol.* **3**, 187.
ROZGONYI, G. A., POLITO, W. J. and SCHWARTZ, B. (1966), *Vacuum* **16**, 121.
RUBET, L. (1966), *Le Vide* **21**, 227.
RUDBERG, E. (1936), *Phys. Rev.* **50**, 138.
RUEDL, E. and KELLY, R. (1965), *J. Nuclear Materials* **16**, 89.

RUSSEK, A. (1963), *Phys. Rev.* **132**, 246.
RUSSEK, A. and THOMAS, M. T. (1958), *Phys. Rev.* **109**, 2015.
RUTHERFORD, S. L. (1963), *Trans. AVS Vac. Symp.* **10**, 185.

S

SALISBURY, J. W., GLASER, P. E., STEIN, B.A. and VONNEGUT, B. (1964), *J. Geophys. Res.* **69**, 235.
SAMS, J. R., CONSTABARIS, G. and HALSEY, G. D. (1960), *J. Phys. Chem.* **64**, 1689.
SANDS, A. and DICK, S. M. (1966), *Vacuum* **16**, 691.
SANTELER, D. J. (1959), *Trans. AVS Vac. Symp.* **6**, 129.
SANTELER, D. J., HOLKEBOER, D. H., JONES, D. W. and PAGANO, F. (1966), *Vacuum Technology and Space Simulation*, NASA SP-105. STAR Index No. N66-36129.
SCHÄFER, K. and GERSTACKER, H. (1956), *Z. Elektrochem* **60**, 874.
SCHÄFER, K. and TEGGERS, H. (1953), *Z. Elektrochem* **57**, 747.
SCHAGEN, P. (1965), *Brit. J. Appl. Phys.* **16**, 293.
SCHIFF, L. I. (1949), *Quantum Mechanics* (McGraw-Hill, New York).
SCHISSEL, P. O. (1962), *J. Appl. Phys.* **33**, 2659.
SCHISSEL, P. O. and TRULSON, O. C. (1965), *J. Chem. Phys.* **43**, 737.
SCHITTKO, F. J. (1963), *Vacuum* **13**, 525.
SCHLIER, R. E. (1958), *J. Appl. Phys.* **29**, 1162.
SCHMIDLIN, F. W., HEFLINGER, L. O. and GARWIN, F. L. (1962), *Trans. AVS Vac. Symp.* **9**, 197.
SCHRAM, A. (1967), *Compt. Rend.* **264**, 1248.
SCHRAM, B. L. (1966), *Physica* **32**, 197.
SCHRAM, B. L., BOERBOOM, A. J. H. and KISTEMAKER, J. (1966), *Physica* **32**, 185.
SCHRAM, B. L., BOERBOOM, A. J. H., KLEINE, W. and KISTEMAKER, J. (1965a), *Proc. 7th Int. Conf. on Ionization Phenomena in Gases* (Belgrade) Vol. I, p. 170.
SCHRAM, B. L., DE HEER, F. J., VAN DER WIEL, M. J. and KISTEMAKER, J. (1965b), *Physica* **31**, 94.
SCHUEMANN, W. C. (1963), *Rev. Sci. Instr.* **34**, 700.
SCHUEMANN, W. C., DE SEGOVIA, J. L. and ALPERT, D. (1963), *Trans. AVS Vac. Symp.* **10**, 223.
SCHUHMANN, S. (1962), *Trans. AVS Vac Symp.* **9**, 463.
SCHULTZ, A. A. and POMERANTZ, M. A. (1963), *Phys. Rev.* **130**, 2135.
SCHULZ, G. J. (1957), *J. Appl. Phys.* **28**, 1149.
SCHULZ, G. J. (1959), *Phys. Rev.* **116**, 1141.
SCHULZ, G. J. (1962), *Phys. Rev.* **125**, 229.
SCHULZ, G. J. (1963), *Phys. Rev. Lett.* **10**, 104.
SCHÜTZE, H. J. and EHLBECK, H. W. (1961), *Trans. AVS Vac. Symp.* **8**, 451.
SCHÜTZE, H. J. and STORK, F. (1962a), *Vakuum Technik* **11**, 133.
SCHÜTZE, H. J. and STORK, F. (1962b), *Trans. AVS Vac. Symp.* **9**, 431.
SCHÜTZE, W. and BERNHARD, F. (1956), *Zeit. fur Physik* **145**, 44.
SCHWARZ, H. (1951), *Arch. Tech. Mess.* Sept. V 1341-2.
SCHWARZ, H. (1952), *Arch. Tech. Mess.* Jan V 1341-3; March V 1341-4; May V 1341-5.
SCHWARZ, H. (1960), *Arch. Tech. Mess.* Dec. V 1341-6.

SCHWARZ, H. (1961), *Arch. Tech. Mess.* Feb. V 1341-7; March V 1341-8.
SEARS, G. W. and HUDSON, J. B. (1963), *J. Chem. Phys.* **38**, 2025.
SEATON, M. J. (1962), *Atomic and Molecular Processes* (Academic Press, New York; Ed. D. R. Bates), p. 374.
SEATON, M. J. (1966), *Proc. Phys. Soc.* (London) **88**, 801.
SEITZ, F. and KOEHLER, J. S. (1956), *Adv. in Sol. State Phys.* **2**, 305.
SEWELL, P. B. and COHEN, M. (1965), *Appl. Phys. Lett.* **7**, 32.
SHAPIRO, E. (1952), *J. Am. Chem. Soc.* **74**, 5233.
SHAW, M. L. (1966), *Rev. Sci. Instr.* **37**, 113.
SHELTON, H. (1957), *Phys. Rev.* **107**, 1553.
SHEN, H., PODLASEK, S. E. and KRAMER, I. R. (1966), *Acta Metallurgica* **14**, 341.
SHERIDAN, W. F., OLDENBERG, D. and CARLETON, N. P. (1961), *2nd Int. Conf. on Physics of Electronic & Ionic Collisions* (University of Colorado), p. 440.
SHUL'MAN, A. R. and BAZHANOVA, N. P. (1967), *Sov. Phys. Solid State* **8**, 2235.
SIDDIQI, M. M. and TOMPKINS, F. C. (1962), *Proc. Roy. Soc.* (London) **A268**, 452.
SILSBEE, R. H. (1957), *J. Appl. Phys.* **28**, 1246.
SILVERNAIL, W. L. (1955), Ph.D. Dissertation, University of Missouri, Columbia, Mo.
SIMONOV, V. A. and MILESHKIN, A. G. (1962), *Cesk. Cas. Fys.* **12(A)**, 653.
SINGLETON, J. H. (1963), *Trans. AVS Vac. Symp.* **10**, 267.
SINGLETON, J. H. (1966a), *J. Chem. Phys.* **45**, 2819.
SINGLETON, J. H. (1966b), *J. Vac. Sci. and Technol.* **3**, 354.
SINGLETON, J. H. (1967a), *J. Vac. Sci. and Technol.* **4**, 103.
SINGLETON, J. H. (1967b), *J. Chem. Phys.* **47**, 73.
SINGLETON, J. and LANGE, W. J. (1965), *J. Vac. Sci. and Technol*, **2**, 93.
SKERBELE, A. and LASSETTRE, E. A. (1966), *J. Chem. Phys.* **44**, 4066.
SLOANE, R. H. and WATT, C. S. (1948), *Proc. Phys. Soc.* (London) **61**, 217.
SLUYTERS, T. J. M. and KISTEMAKER, J. (1959), *Physica* **25**, 1389.
SMEATON, G. P. and CARTER, G. (1966), *J. Vac. Sci. and Technol.* '**3**, 208.
SMEATON, G. P., CARTER, G. and LECK, J. H. (1962), *Trans. AVS Vac Symp.* **9**, 491.
SMETANA, F. O. and CARLEY, C. T. (1966), *J. Vac. Sci. and Technol.* **3**, 47.
SMIRNITSKAYA, G. V. and REIKHRUDEL, E. M. (1957), *Rad. Eng. and Electronic Phys.* **2(10)**, 123.
SMITH, D. G. (1966), *J. Sci. Instr.* **43**, 270.
SMITH, H. R. (1959), *Trans. AVS Vac. Symp.* **6**, 140.
SMITH, J. N. (1964), *J. Chem. Phys.* **40**, 2520.
SMITH, J. N. and FITE, W. L. (1962), *Proc. 3rd Symp. Rarefied Gas Dynamics*, Vol. 1, p. 430.
SMITH, H. I. and GUSSENHOVEN, M. S. (1965), *J. Appl. Phys.* **36**, 2326.
SMITH, J. N. and SALTSBURG, H. (1964), *J. Phys. Chem.* **40**, 3585.
SMITH, P. T. (1930), *Phys. Rev.* **36**, 1293.
SMITH, R. V., EDMONDS, D. K., BRENTARI, E. G. F. and RICHARDS, R. J. (1964), *Adv. in Cryogenic Eng.* **9**, 88.
SMITHELLS, C. J. (1962), *Metals—Reference Handbook*. Vol. II, p. 580 (Butterworths, London).
SNOEK, C., GEBALLE, R., VAN DER WEG, W. F., ROL, P. K. and BIERMAN, D. J. (1965), *Physica* **31**, 1553.

SNOEK, C. and KISTEMAKER, J. (1965), *Adv. in Electronics and Electron Physics* **21**, 67.
SNOUSE, T. W. and HAUGHNEY, L. C. (1966), *J. Appl. Phys.* **37**, 700.
SOO, S. L. and HUANG, A. B. (1962), *Trans. AVS Vac. Symp.* **9**, 443.
SOUTHERN, A. L., WILLIS, W. R. and ROBINSON, M. T. (1963), *J. Appl. Phys.* **34**, 153.
SPALVINS, T. and BUCKLEY, D. H. (1966), *J. Vac. Sci. and Technol.* **3**, 107.
SPANGENBERG, K. R. (1948), *Vacuum Tubes* (McGraw-Hill, New York).
SPARNAAY, M. J. (1958), *Physica* **24**, 751.
STECKELMACHER, W. (1965), *J. Sci. Instr.* **42**, 63.
STEDEFORD, J. B. H. and HASTED, J. B. (1955), *Proc. Roy. Soc.* **A227**, 466.
STEELE, W. A. and HALSEY, G. D. (1954), *J. Chem. Phys.* **22**, 979.
STEELE, W. A. and HALSEY, G. D. (1955), *J. Phys. Chem.* **59**, 57.
STEHBERGER, K. H. (1928), *Ann Physik* **86**, 825.
STEINRISSER, F. (1967), *J. Vac. Sci. and Technol.* **4**, 44.
STEPHENS, J. B. (1966), NASA Report No. CR-74654. STAR Index N66-24511.
STERN, O. (1929), *Naturwiss.* **17**, 391.
STERN, R. M. (1964), *Appl. Phys. Letters* **5**, 218.
STERN, R. M. (1965), 150th Meeting, Colloid & Surface Chemistry, Am. Chem. Soc., (Atlantic City, N.J.).
STERN, S. A., MULLHAUPT, J. T., HEMSTREET, R. A. and DI PAOLO, F. S. (1965), *J. Vac. Sci. and Technol.* **2**, 165.
STERN, S. A., HEMSTREET, R. A. and RUTTENBUR, D. M. (1966), *J. Vac. Sci. and Technol.* **3**, 99.
STERNGLASS, E. J. (1954), *Phys. Rev.* **95**, 345.
STERNGLASS, E. J. (1957), *Phys. Rev.* **108**, 1.
STEVENSON, D. L. (1959), *Trans .AVS Vac. Symp.* **6**, 134.
STEVENSON, D. P. (1955), *J. Chem. Phys.* **23**, 203.
STEVENSON, D. P. and HIPPLE, J. A. (1942), *Phys. Rev.* **62**, 237.
STEWART, D. T. (1956), *Proc. Phys. Soc.* (London) **A69**, 437.
STICKNEY, W. W. and DAYTON, B. B. (1963), *Trans. AVS Vac. Symp.* **10**, 105.
STONE, F. S. (1964), *Proc. Conf. on Physical Chemistry of Solid Surfaces* (Madrid), p. 109.
STOUT, V. L. and GIBBONS, M. D. (1955), *J. Appl. Phys.* **26**, 1488.
STUART, R. V. and WEHNER, G. K. (1962), *J. Appl. Phys.* **33**, 2345.
STUART, R. V. and WEHNER, G. K. (1964), *J. Appl. Phys.* **35**, 1819.
STUBER, F. A. (1965), *J. Chem. Phys.* **42**, 2639.
STUDIER, M. H. (1963), *Proc. 11th ASTM Mass Spectrometry Conf.* p. 143.
SUGATA, E., KIM, H. and ISHII, S. *Proc. 6th Int. Cong. for Electron Microscopy*, Vol. 1, p. 245 (Kyoto).
SWAN, J. B. (1964), *Phys. Rev.* **135**, A1467.
SWANSON, N. and POWELL, C. J. (1966), *Phys. Rev.* **145**, 195
SWETS, D. E., LEE, R. W. and FRANK, R. C. (1961) *J. Chem. Phys.* **34**, 17.

T

TAKEISHI, Y. and HAGSTRUM, H. D. (1965), *Phys. Rev.* **137**, A641.
TATE, J. T. and SMITH, P. T. (1932), *Phys. Rev.* **39**, 270.
TATE, J. T. and SMITH, P. T. (1934), *Phys. Rev.* **46**, 773.

TAYLOR, L. H., LONGLEY, W. W. and BRYANT, P. J. (1965), *J. Chem. Phys.* **43**, 1184.
TELKOVSKY, W. G. (1957), *Proc. 3rd Int. Conf. on Phenomena in Ionized Gases* (Venice), p. 1079.
TERENIN, A. and SOLONITZIN, YU. (1959), *Disc. Farad. Soc.* **28**, 28.
THEODOROU, D. G., MCINTOSH, R. O., CONKLIN, T. H. and EARL, K. C. (1966), *Vacuum* **16**, 237.
THIEME, O. (1932), *Zeit. fur Phys.* **78**, 412.
THOMAS, C. H. (1925), *Phys. Rev.* **26**, 739.
THOMAS, L. B. (1961), Private Communication to Goodman (1964).
THOMAS, L. B. (1965), Private Communication to Wachman (1966).
THOMAS, L. B. and BROWN, R. E. (1950), *J. Chem. Phys.* **18**, 1367.
THOMAS, L. B. and OLMER, F. G. (1943), *J. Amer. Chem. Soc.* **65**, 1036.
THOMAS, L. B. and SCHOFIELD, E. B. (1955), *J. Chem. Phys.* **23**, 861.
THOMAS, A. M., JOHNSON, D. P. and LITTLE, J. W. (1962), *Trans. AVS Vac. Symp.* **9**, 468.
THOMPSON, M. W. (1961), *Proc. 5th Int. Conf. on Ionization Phenomena in Gases* (*Munich*), p. 83.
THOMPSON, M. W. (1963), *Phys. Lett.* **6**, 24.
THOMPSON, M. W. and NELSON, R. S. (1962), *Phil. Mag.* **7**, 2015.
TIEDEMA, T. J., DE JONG, B. C. and BURGERS, W. G. (1960), *Proc. K. Ned. Akad. Wetensch.* **B63**, 422.
TITRAN, R. H. and HALL, R. W. (1966), NASA-TN D-3222, STAR Index N66-15485.
TODD, B. J. (1955), *J. Appl. Phys.* **26**, 1238.
TODD, B. J. (1956), *J. Appl. Phys.* **27**, 1209.
TODD, B. J., LINEWEAVER, J. L. and KERR, J. T. (1960), *J. Appl. Phys.* **31**, 51.
TOM, T. and JAMES, B. D. (1966), *J. Vac. Sci. and Technol.* **3**, 300.
TOMINAGA, G. (1965), *Jap. J. Appl. Phys.* **4**, 129.
TOMPKINS, F. C. (1967), in *The Solid-Gas Interface*, Vol. II, p. 765 (Marcel Dekker, New York).
TORNEY, F. L. and FEAKES, F. (1963), *Rev. Sci. Instr.* **34**, 1041.
TOWNES, C. H. and SCHAWLOW, A. L. (1955), *Microwave Spectroscopy* (McGraw Hill, New York).
TOWNSEND, J. S. and BAILEY, V. A. (1922), *Phil. Mag.* **43**, 593.
TRENDELENBURG, E. A. (1963), *Ultrahochvakuum* (Braun, Karlsruhe).
TSAREV, B. M. (1965), *Rad. Eng. and Electron. Phys.* **9**, 1341.
TSEITLIN, A. B. (1965), *Inst. & Exp. Tech.*, No. 4, p. 1204.
TUCKER, C. W. (1964), *J. Appl. Phys.* **35**, 1897.
TUCKER, C. W. (1965), *Surface Science* **3**, 427.
TUCKER, C. W. and NORTON, F. J. (1960), *J. Nucl. Materials* **2**, 329.
TULINOV, A. F., KULIKAUSKAS, V. S. and MALOV, M. M. (1965), *Phys. Lett.* **18**, 304.
TURNER, F. T. and HOGAN, W. H. (1966), *J. Vac. Sci. and Technol.* **3**, 252.
TUZI, Y. (1962), *J. Phys. Soc.* (*Japan*) **17**, 218.
TYUTIKOV, A. M. and SHUBA, YU. A. (1960), *Optics and Spectroscopy* **9**, 332.

U

USTINOV, YU. K. and IONOV, N. I. (1966), *Sov. Phys. Tech. Phys.* **10**, 1607.
UTTERBACK, N. G. and GRIFFITH, T. (1966), *Rev. Sci. Instr.* **37**, 866.

V

VALDRÈ, U., PASHLEY, D. W., ROBINSON, E. A., STOWELL, M. J., ROUTLEDGE, K. J. and VINCENT, R. (1966), 6th Int. Cong. for Electron Microscopy, Vol. I, p. 155 (Kyoto).
VAN DINGENEN, W. and VAN ITTERBEEK, A. (1939), Physica 6, 49.
VAN ITTERBEEK, A., HELLEMANS, R. and VAN DAEL, W. (1964), Physica 30, 324.
VAN OOSTROM, A. G. J. (1961), Trans. AVS Vac. Symp. 8, 443.
VAN OOSTROM, A. G. J. (1966), Phil. Res. Rep. Suppl. No. 1.
VARADI, P. F. (1960), Trans. AVS Vac. Symp. 7, 149
VARADI, P. F. (1961), Trans. AVS Vac. Symp. 8, 73.
VARNERIN, L. J. and CARMICHAEL, J. H. (1955), J. Appl. Phys. 26, 782.
VARNERIN, L. J. and CARMICHAEL, J. H. (1957), J. Appl. Phys. 28, 913.
VASIL'EVA, M. N. and REIKHRUDEL, E. M. (1962), Sov. Phys. Tech. Phys. 7, 528.
VEKSLER, V. I. (1965), Sov. Phys. Solid State 7, 500.
VEKSLER, V. I. (1966), Sov. Phys. JETP 22, 65.
VENEMA, A. (1959), Vacuum 9, 54.
VENEMA, A. and BANDRINGA, M. (1958), Phil. Tech. Rev. 20, 145.
VINOGRADOV, M. I. and RUDNITSKII, E. M. (1966), Inst. and Exp. Tech., p. 377.
VOLOSOK, V. I. and CHIRIKOV, B. V. (1957), Sov. Phys. Tech. Phys. 2, 2437.
VON ARDENNE, M. (1956), Tabellen der Elektronenphysik, Ionenphysik und Ubermikroskopie (Veb Deutscher Verlag der Wissenschaften, Berlin).
VON KOCH, H. and LINDHOLM, E. (1961), Arkiv. för Fysik 19, 123.
VON ROOS, O. (1957), Zeit. fur Phys. 147, 210.
VON ZAHN, U. (1963), Rev. Sci. Instr. 34, 1.

W

WACHMAN, H. Y. (1962), Am. Rocket Soc. Journal 32, 2.
WACHMAN, H. Y. (1966), J. Chem. Phys. 45, 1532.
WAGENER, J. S. (1958), Proc. 4th National Conf. on Tube Techniques, p. 1 (New York).
WAGENER, S. (1951), The Oxide-Coated Cathode (Chapman & Hall, London).
WAHBA, M. and KEMBALL, C. (1953), Trans. Far. Soc. 49, 1351.
WALDSCHMIDT, E. (1954), Metall. 8, 749.
WARGO, P., HAXBY, B. V. and SHEPHERD, W. G. (1956), J. Appl. Phys. 27, 1311.
WASA, K. and HAYAKAWA, S. (1965), J. Phys. Soc. Japan 20, 1219.
WATERS, P. M. (1958), Phys. Rev. 109, 1466.
WEAST, R. C. (1966), Handbook of Chemistry and Physics, Chemical Rubber Co., Cleveland.
WEBER, R. E. and CORDES, L. F. (1966), Rev. Sci. Instr. 37, 112.
WEBER, R. E. and PERIA, W. T. (1967), J. Appl. Phys. 38, 4355.
WEHNER, G. K. (1955a), Adv. in Electronics and Electron Physics 7, 239.
WEHNER, G. K. (1955b), J. Appl. Phys. 26, 1056.
WEHNER, G. K. (1957), Phys. Rev. 108, 35.
WEHNER, G. K. (1959), J. Appl. Phys. 30, 1762.
WEHNER, G. K. and ROSENBERG, D. (1960), J. Appl. Phys. 31, 177.
WEINBAUM, S. (1935), J. Chem. Phys. 3, 547.
WEINREICH, O. A. (1951), Phys. Rev. 82, 573.

WEISSLER, G. L. (1956), *Handbuch. der Phys.* **21**, 304.
WEISSLER, G. L., SAMSON, J. A. R., OGAWA, M. and COOK, G. R. (1959), *J. Opt. Soc. Amer.* **49**, 338.
WELLS, E. R. and POSTMA, H. (1962), *Trans. AVS Vac. Symp.* **9**, 266.
WEXLER, S. (1958), *Rev. Mod. Phys.* **30**, 402.
WHETTEN, N. R. (1964), *J. Appl. Phys.* **35**, 3279.
WHETTEN, N. R. (1965), *J. Vac. Sci. and Technol.* **2**, 84.
WHETTEN, N. R. and LAPONSKY, A. B. (1957a), *Phys. Rev.* **107**, 1521.
WHETTEN, N. R. and LAPONSKY, A. B. (1957b), *J. Appl. Phys.* **28**, 515.
WHETTEN, N. R. and LAPONSKY, A. B. (1959), *J. Appl. Phys.* **30**, 432.
WHETTEN, N. R. and YOUNG, J. R. (1959), *Rev. Sci. Instr.* **30**, 472.
WHITE, F. A., SHEFFIELD, J. C. and DAVIS, W. D. (1961), *Nucleonics* **19**(8), 58.
WILEY, W. C. and HENDEE, C. F. (1962), *IRE Trans. Nucl. Sci.* **9**(3), 103.
WILLIAMS, C. E. and BEAMS, J. W. (1961), *Trans. AVS Vac. Symp.* **8**, 295.
WILSON, K. R., KWEI, G. H., NORRIS, J. A., HERM, R. R., BIRELY, J. H. and HERSCHBACH, D. R. (1964), *J. Chem. Phys.* **41**, 1154.
WINDSOR, E. E. (1963), *Trans. 2nd Eur. Vac. Symp.*, p. 278.
WINSLOW, P. M. and MCINTYRE, D. V. (1966), *J. Vac. Sci. and Technol.* **3**, 54.
WINTERS, H. F., DENISON, D. R., BILLS, D. G. and DONALDSON, E. E. (1963), *J. Appl. Phys.* **34**, 1810.
WINTERS, H. F., HORNE, D. E. and DONALDSON, E. E. (1964), *J. Chem. Phys.* **41**, 2766.
WOLSKY, S. P. and ZDANUK, E. J. (1960), *Trans. AVS Vac. Symp.* **7**, 282.
WOOD, E. A. (1964), *J. Appl. Phys.* **35**, 1306.
WOOD, R. W. (1915), *Phil. Mag.* **30**, 300.
WRIGHT, R. W. (1941), *Phys. Rev.* **60**, 465.

Y

YATES, J. T., MADEY, T. E. and PAYN, J. K. (1967), *Suppl. Nuovo Cimento* **5**, No. 2, 558.
YEUNG, T. H. Y. *Proc. Phys. Soc.* (London) **71**, 341.
YOUNG, D. M. and CROWELL, A. D. (1962), *Physical Adsorption of Gases* (Butterworths, London).
YOUNG, J. R. (1956a), *J. Appl. Phys.* **27**, 926.
YOUNG, J. R. (1956b), *J. Appl. Phys.* **27**, 1.
YOUNG, J. R. (1959), *J. Appl. Phys.* **30**, 1671.
YOUNG, J. R. (1960), *J. Appl. Phys.* **31**, 921.
YOUNG, J. R. (1963a), *Rev. Sci. Instr.* **34**, 374.
YOUNG, J. R. (1963b), *Rev. Sci. Instr.* **34**, 891.
YOUNG, J. R. (1966a), *J. Vac. Sci. and Technol.* **3**, 345.
YOUNG, J. R. (1966b), *Rev. Sci. Instr.* **37**, 1414.
YOUNG, J. R. and HESSION, F. P. (1963), *Trans. AVS Vac. Symp.* **10**, 234.
YOUNG, J. R. and HESSION, F. P. (1964), *J. Vac. Sci. and Technol.* **1**, 65.
YOUNG, J. R. and WETTEN, N. R. (1961), *Rev. Sci. Instr.* **32**, 453.
YURASOVA, V. E., PLESHIVTSEV, N. V. and ORFANOV, I. V. (1960), *Sov. Phys. JETP* **10**, 689.

Z

ZAHL, H. and ELLETT, A. (1931), *Phys. Rev.* **38**, 977.

ZANDBERG, E. Ya and IONOV, N. I. (1959), *Sov. Phys. Uspekhi* **2**, 255.
ZAPHIROPOULOS, R. and DE TADDEO, D. (1965), *Trans. 3rd Int. Vac. Cong.*, Vol 2, p. 373 (Stuttgart).
ZINGERMAN, Ya. P., ISCHUK, V. A. and KRUTILINA, T. A. (1966), *Soviet Physics Solid State* **7**, 2078.
ZOLLWEG, R. J. (1964), *Surface Science* **2**, 409.
ZSCHEILE, H. (1966), *Phys. Stat. Sol.* **14**, K15.
ZWANZIG, R. W. (1960), *J. Chem. Phys.* **32**, 1173.

APPENDIX A
RECENT REFERENCES

Listed below are some recent papers which are relevant to the discussion in this book, but appeared too late to be included in the text.

CHAPTER 2

ARMAND, G. and LAPUJOULADE, J. (1967), 'Le facteur pre-exponentiel en cinetique de desorption gaz-solide', *Surface Science* **6**, 345.

ARMAND, G. (1966), 'Molecular flow in the transition region. Validity of the diffusion equation', *Le Vide* **21**, 408.

BACHMAN, L. and SHIN, J. J. (1966), 'Measurement of the sticking coefficient of Ag and Au in UHV', *J. Appl. Phys.* **37**, 242.

BARRER, R. M. (1966), 'Specificity in physical adsorption', *J. Colloid & Interface Science* **21**, 415.

BELL, A. A. and GOMER, R. (1966), 'Adsorption of carbon monoxide on tungsten. Abundances, dipole moments, and sticking coefficients', *J. Chem. Phys.* **44**, 1065.

BERGSNOV-HANSEN, B. and PASTERNAK, R. A. (1966), 'Low pressure study of heterogeneous hydrogen-deuterium exchange at steady state', *J. Chem. Phys.* **45**, 1199.

CHAMBERS, C. M. and KINZER, E. T. (1966), 'Higher dimensional crystal models. A theory of thermal accommodation coefficients', *Surface Science* **4**, 33.

COCHRANE, H., WALKER, P. L., DIETHORN, W. S. and FRIEDMAN, H. C. (1967), 'Xenon adsorption on graphitized carbon blacks over a wide coverage range', *J. Colloid and Interface Science* **24**, 405.

DALGARNO, A. and DAVISON, W. D. (1966), 'The calculations of Van der Waals interactions', *Adv. in Atomic and Molecular Physics* **2**, 1.

DEGRAS, D. A. (1967), 'Capture of carbon monoxide on nickel (110) and polycrystalline tungsten', *Suppl. Nuovo Cimento* **5**, No. 2, 408.

EBISUZAKI, Y., KASS, W. J. and O'KEEFFE, M. (1967), 'Diffusion and solubility of hydrogen in single crystals of nickel and nickel vanadium alloy', *J. Chem. Phys.* **46**, 1378.

FARRELL, G. and CARTER, G. (1967), 'Diffusive properties in a solid during tempering', *Vacuum* **17**, 15.

GADZUK, J. W. (1967), 'Theory of atom-metal interactions. I. Alkali atom adsorption', *Surface Science* **6**, 133.

GADZUK, J. W. (1967), 'Theory of atom-metal interactions. II. One electron transition matrix elements', *Surface Science* **6**, 159.

GAVRILYUK, V. M., MEDVEDEV, V. K. and SMEREKA, T. P. (1967), 'Adsorption of nitrogen on the (113) plane of tungsten single crystals', *Sov. Phys.—Solid State* **8**, 1983.

GOTTWALD, B. A. and HAUL, R. (1968), 'Adsorptionskinetik. III. Ermittlung von adsorptions—Isothermen aus molekularstromungs-versuchen', *Surface Science* **10**, 76.

GRIFFITH, R. and PRYDE, J. A. (1968), 'Degassing rate constant for dilute solution of nitrogen in tantalum', *Trans. Farad. Soc.*, **64**, 507.

HAGENA, O. F. (1967), 'Velocity distribution measurements of molecular beams scattered from solid surfaces', *Appl. Phys. Lett.* **9**, 385.

HANSEN, N. (1967), 'Evaluation of the sticking probability of gases chemisorbed on metal films without measurement of gas pressure', *Suppl. Nuovo Cim.* **5**, No. 2, 389.

HARRIS, L. B., HUDSON, J. B. and ROSS, S. (1967), 'Low-pressure physical adsorption and electron microscope study of the surface of annealed pyrolite graphite', *J. Phys. Chem.* **71**, 377.

HAUL, R. and GOTTWALD, B. A. (1966), 'Adsorptionskinetik. I. Messung der verweilzeit adsorbierter teilschen durch molekular-stromungs-versuche', *Surface Science* **4**, 321.

HAUL, R. and GOTTWALD, B. A. (1966), 'Adsorptionskinetik. II. Verweilzeiten adsorbierter edelgasatome an Pyrexglas-oberflachen', *Surface Science* **4**, 334.

HAYEK, K., FARNSWORTH, H. E. and PARK, R. L. (1968), 'Interaction of oxygen, carbon monoxide and nitrogen with (001) and (110) faces of molybdenum', *Surface Science* **10**, 429.

HAYWARD, D. O., KING, D. A. and TOMPKINS, F. C. (1967), 'Sticking probabilities, heats of adsorption, and redistribution processes of nitrogen on tungsten films at 195 and 290°K', *Proc. Roy. Soc.* **A297**, 305.

HENGEVOSS, J. and TRENDELENBURG, E. A. (1967), 'Continuous cryotrapping of hydrogen by argon in temperature range between 4.2 and 15°K', *Vacuum* **17**, 495.

HORGAN, A. M. and KING, D. A. (1968), 'Adsorption efficiency in collision of molecular oxygen with clean tungsten surfaces', *Nature* **217**, 60.

KIRKENDALL, T. D., VARADI, P. F. and DOOLITTLE, H. D. (1966), 'Some effects of surface layers on the degassing properties of copper', *J. Vac. Sci. and Technol.* **3**, 214.

KNOR, Z. and MÜLLER, E. W. (1968), 'A refined model of the metal surface and its interaction with gases in the field ion microscope', *Surface Science* **10**, 21.

KOUPTSIDIS, J. and MENZEL, D. (1967), 'Akkommodation der edelgase an reinen wolfram-und molybdanoberflachen bei zimmertemperatur', *Ber. Bun. Ges.* **71**, 720.

LEE, R. W. and FRY, D. L. (1966), 'A comparative study of the diffusion of hydrogen in glass', *Phys. Chem. Glass* **7**, 19.

LEVENSON, L. L. (1967), 'Condensation coefficients of argon, krypton and xenon measured with a quartz microbalance at 4.2°K', *Suppl. Nuovo Cim.* **5**, No. 2, 321.

MAY, J. W., GERMER, L. H. and CHANG, C. C. (1966), 'Coadsorption of oxygen and carbon monoxide upon a (110) tungsten surface', *J. Chem. Phys.* **45**, 2383.

MAZHUGA, V. V. (1966), 'Collisions of an atom with a solid surface', *Doklady Akad. Nauk SSSR* **167**, 1012.

MCCAROLL, B. (1967), 'Chemisorption and oxidation: Oxygen on tungsten', *J. Chem. Phys.* **46**, 863.

MCKEE, C. S. and ROBERTS, M. W. (1967), 'Probability of gas adsorption on metal films,' *Trans. Farad. Soc.* **63**, 1418.

MESSER, G. and MOLIÈRE, K. (1966), 'Determination of the thermal accommodation coefficients of inert gases on tungsten', *Vakkum-Technik* **15**, 224.

MIMEAULT, V. J. and HANSEN, R. S. (1966), 'Flash desorption and isotopic mixing of hydrogen and deuterium adsorbed on tungsten, iridium, rhodium', *J. Chem. Phys.* **45**, 2240.

PÉTERMANN, L. A. (1967), 'Bond strength and mean time of adsorption for active gases on metals', *Suppl. Nuovo Cim.* **5**, No. 2, 364.

ROBERTSON, A. J. B. (1966), 'Catalytic reactions on metal surfaces at very low gas pressures', *Vacuum* **16**, 289.

SALTSBURG, H. and SMITH, J. N. (1966), 'Molecular beam scattering from the (111) plane of silver', *J. Chem. Phys.* **45**, 2175.

SCHRAM, A. (1967), 'Physical adsorption of argon on nickel between 78°K and 120°K', *Suppl. Nuovo Cim.* **5**, No. 2, 291.

SMITH, C. G. and LEWIN, G. (1966), 'Free molecular conductance of a cylindrical tube with wall sorption', *J. Vac. Sci. and Technol.* **3**, 92.

STERN, S. A. and DIPAOLO, F. S. (1967), 'The adsorption of atmospheric gases on molecular sieves at low pressures and temperatures. The effect of pre-adsorbed water', *J. Vac. Sci. and Technol.* **4**, 347.

TEUBNER, W. (1967), 'Molecular flow of gases through tubes taking account of particle-wall interactions', *Vakuum-Technik* **16**, 69.

TEUBNER, W. (1967), Experimental evaluation of deviations from Knudsen's Law for stationary flow', *Vakuum-Technik* **16**, 95.

TOMINAGA, G. (1967), 'Mean adsorption time of oil molecules used in vacuum techniques', *Suppl Nuovo Cim.* **5**, No. 2, 274.

VOSTROV, G. A. and BOL'SHAKOV, O. I. (1966), 'Helium passage through glasses', *Inst. and Exp. Tech.* No. 2, p. 382.

YU, J. S. and SOO, S. L. (1966), 'Interaction of gases with a condensed phase', *J. Vac. Sci. and Technol.* **3**, 11.

ZINGERMAN, Ya. P. and ISCHUK, V. A. (1967), 'Investigation of the process of sorption of O_2 on 100 and 110 faces of W single crystal by the electron-stimulated desorption effect', *Sov. Phys. —Solid State* **9**, 623.

ZINGERMAN, Ya. P. and ISCHUK, V. A. (1967), 'Mechanism of the sorption of oxygen on the (100) face of a tungsten single crystal', *Sov. Phys.—Solid State* **8**, 2394.

CHAPTER 3

DE HEER, F. J., SCHUTTEN, J. and MOUSTAFA, H. (1966), 'Ionization and electron capture cross-sections for helium ions incident on noble and diatomic gases between 10–150 keV', *Physica* **32**, 1793.

DE HEER, F. J., SCHUTTEN, J. and MOUSTAFA, H. (1966), 'Ionization and electron capture cross-sections for protons incident on noble and diatomic gases between 10–140 keV', *Physica* **32**, 1766.

LOTZ, W. (1967), 'Electron impact ionization cross-sections and ionizing rate coefficients for atoms and ions', *Astrophys. Jour.*, Suppl. No. 128, **14**, 207.

LOTZ, W. (1967), 'An empirical formula for the electron-impact ionization cross-section', *Zeit. für Phys.* **206**, 205.

MOISEIWITSCH, B. L. and SMITH, S. J. (1968), 'Electron impact excitation', *Rev. Mod. Phys.* **40**, 238.

SUNSHINE, G., AUBREY, B. B. and BEDERSON, B. (1967), 'Absolute measurements of total cross-sections for the scattering of low-energy electrons by atomic and molecular oxygen', *Phys. Rev.* **154**, 1.

CHAPTER 4

ABROYAN, I. A., EREMEEV, M. A. and PETROV, N. N. (1967), 'Excitation of electrons in solids by relatively slow atomic particles', *Sov. Phys.—Uspekhi* **10**, 332.

AKHMETOVA, B. G., PLETS, YU. M. and TULINOV, A. F. (1967), 'Scattering of 5—40 keV protons by molybdenum single crystals', *Sov. Phys.—JEPT* **24**, 1108.

ANDERSON, G. S. (1967), 'Etching rate of an ion-bombarded tungsten (110) surface', *J. Appl. Phys.* **38**, 1989.

ANDERSON, J. and ESTRUP, P. J. (1968), 'Effect of electrons on NH_3 adsorbed on a W(100) surface', *Surface Science* **9**, 463.

AUSLENDER, V. I. and MICHENKOV, G. B. (1967), 'Outgassing of metal surfaces by electron bombardment', *Sov. Phys.—Tech Phys.* **11**, 1535.

BAUER, E. (1967), 'Multiple scattering versus superstructures in low energy electron diffraction', *Surface Science* **7**, 351.

CHUTJIAN, A. (1967), 'Coherence area in low energy electron diffraction', *Phys. Letters* **24A**, 615.

DATSIEV, M. I. and IONOV, N. I. (1967), 'Desorption from adsorbed layers on copper by slow electrons', *Sov. Phys.—Tech. Phys.* **12**, 821.

DAVIES, J. A., DENHARTOG, J. and WHITTON, J. L. (1968), 'Channeling of MeV projectiles in tungsten and silicon', *Phys. Rev.* **165**, 345.

DEGRAS, D. A. and LECANTE, J. (1967), 'Interaction of low energy electrons and adsorbed CO', *Suppl. Nuovo. Cim.* **5**, No. 2, 598.

ERENTS, K. and CARTER, G. (1967), 'Investigation into the mechanism of trapping of inert gas ions in polycrystalline tungsten', *Vacuum* **17**, 215.

ERENTS, K., LAWSON, R. P. W. and CARTER, G. (1967), 'Thermal re-emission of inert gas atoms in tungsten and gold', *J. Vac. Sci. and Technol.* **4**, 252.

ERENTS, K., NAVINSEK, B. and CARTER, G. (1967), 'Sputtering of trapped gases from tungsten', *Brit. J. Appl. Phys.* **18**, 587.

ESTRUP, P. J. and ANDERSON, J. (1967), 'Characterization of chemisorption by LEED', *Surface Science* **8**, 101.

EVDOKIMOV, I. N., MASHKOVA, E. S. and MOLCHANOV, V. A. (1967), 'On a new method of observing defects annealing in crystals', *Phys. Letters* **25A**, 619.

FARNSWORTH, H. E. and HAYEK, K. (1967), 'Investigation of surface bombardment damage by LEED', *Surface Science* **8**, 35.

GERSTNER, J. and CUTLER, P. H. (1968), 'Slow electron scattering from one-dimensional crystal models', *Surface Science* **9**, 198.

GOTO, K. and ISHIKAWA, K. (1968), 'Secondary electron emission from diffusion pump oils. II δ-η analysis for DC-705', *Jap. J. Appl. Phys.* **7**, 227.

GRANT, W. A. and CARTER, G. (1967), 'Thermal desorption of inert gases ionically pumped into glass', *Phys. Chem. Glass* **8**, 35.

GRANT, W. A. and CARTER, G. (1967), 'Ion bombardment induced emission of gas from glass', *Brit. J. Appl. Phys.* **18**, 527.

IONA, F., LEVER, R. F. and GUNN, J. B. (1968), 'Comments on "Multiple scattering versus superstructures in low energy electron diffraction" by Bauer, E.', *Surface Science* **9**, 468.

ISHIKAWA, K. and GOTO, K. (1967), 'Secondary electron emission from diffusion pump oils I', *Japan J. Appl. Phys.* **6**, 1329.

IVANOVSKII, G. F., RADZHABOV, T. D. and ZAGORSKAYA, T. N. (1967), 'Mechanism of sorption of inert gas ions by titanium', *Sov. Phys.—Tech. Phys.* **11**, 1096.

JACKSON, A. G. and HOOKER, M. P. (1967), 'A LEED study of CO and CO_2 adsorption on Mo(110)', *Surface Science* **6**, 297.

KAMBE, K. (1967), 'Theory of low energy electron diffraction. I. Application of cellular method to monatomic layers', *Z. Naturf.* **22a**, 322.

KHAN, J. M., POTTER, D. L., WORLEY, R. D. and SMITH, H. P. (1967), 'Characteristic x-ray production in single crystals (Al, Cu, W) by proton bombardment. II. Protons of 250–1560 KeV', *Phys. Rev.* **163**, 81.

KLOPFER, A. (1967), 'Interactions of electrons with adsorbed water', *Suppl. Nuovo Cim.* **5**, No. 2, 606.

KONJEVIC, R., GRANT, W. A. and CARTER, G. (1967), 'Thermal desorption of argon ionically pumped into glass', *Vacuum* **17**, 501.

KUNZE, D., PETERS, O. and SAUERBREY, G. (1967), 'Polymerization of adsorbed hydrocarbons by electron bombardment', *Z. angew Phys.* **22**, 69.

LAWSON, R. P. W. and CARTER, G. (1968), 'Inert gas ion bombardment induced work function changes in polycrystalline tungsten and gold ribbon', *Vacuum* **18**, 205.

LYON, H. B. and SOMORJAI, G. A. (1967), 'Low-energy electron diffraction study of clean (100), (111), and (110) faces of Pt.', *J. Chem. Phys.* **46**, 2539.

MCCRACKEN, G. M., BARTON, R. S. and DILLON, W. (1967), 'Electron induced desorption of gas from stainless steel', *Suppl. Nuovo Cim.* **5**, No. 1, 146.

MCRAE, E. G. (1967), 'Self-consistent multiple scattering approach to interpretation of low energy electron diffraction', *Surface Science* **8**, 14.

MENZEL, D. (1968), 'Mass spectrometric study of inelastic interaction of slow electrons with adsorbed particles', *Z. Naturf.* **23a**, 330.

POLLARD, J. H. and DANFORTH, W. E. (1967), 'Desorption of surface atoms by emerging high energy electrons', *Appl. Phys. Letters* **11**, 9.

SANDSTROM, D. R., LECK, J. H. and DONALDSON, E. E. (1967), 'Electron bombardment desorption: CO adsorbed on W', *J. Appl. Phys.* **38**, 2851.

SCHIØTT, H. E. (1966), 'Range-energy relations for low energy ions', *Mat. Fys. Meddr.* **35**, No. 9.

SHINODA, G., KATO, S., FUJIMOTO, S. and KOBAYASHI, H. (1966), 'Apparatus for alternative observation of low energy electron diffraction and photo-electric emission', *Rev. Sci. Instr.* **37**, 533.

SMITH, D. P. (1967), 'Scattering of low energy noble gas ions from metal surfaces', *J. Appl. Phys.* **38**, 340.

TAYLOR, N. J. (1966), 'A LEED study of the epitaxial growth of copper on the (110) surface of tungsten', *Surface Science* **4**, 161.

THARP, L. N. and SCHEIBNER, E. J. (1967), 'Energy spectra of inelastically scattered electrons and LEED studies of tungsten', *J. Appl. Phys.* **38**, 3320.

VAN DER WEG, W. F. and KISTEMAKER, J. (1968), 'Characteristic x-ray production from metals by 90 keV argon ion bombardment,' *Phys. Letters* **26A**, 592.

WEIJSENFELD, C. H. (1967), 'Yield, energy and angular distribution of sputtered atoms', *Philips Res. Report* Supp. 2.
WHITTON, J. L. (1967), 'Channelling in gold', *Can. J. Phys.* **45**, 1947.
WILSON, R. G. (1966), 'Vacuum thermionic work functions of polycrystalline Be, Ti, Cr, Fe, Ni, Cu, Pt and 304 S.S.', *J. Appl. Phys.* **37**, 2261.
WINTERS, H. F. and KAY, E. (1967), 'Gas incorporation into sputtered films', *J. Appl. Phys.* **38**, 3928.
ZINGERMAN, YA. P. and ISCHUK, V. A. (1968), 'Mechanism of electron-stimulated desorption of oxygen from the surface of tungsten', *Sov. Phys.—Solid State* **9**, 2638.
Proceedings of the International Conference on Atomic Collision and Penetration Studies with Energetic (keV) Ion Beams (Chalk River, 1967), *Can. J. Phys.* **46**, 449–782 (1968)

CHAPTER 5

WACLAWSKI, B. J., HUGHEY, L. R. and MADDEN, R. P. (1967), 'Effect of oxygen adsorption on the photo-electron yield from tungsten in the vacuum ultraviolet', *Appl Phys. Lett.* **10**, 305.

CHAPTER 6

JOHNSON, K. I. and KELLER, D. V. (1967), 'Factors affecting the adhesion of titanium and molybdenum couples', *J. Vac. Sci. and Technol.* **4**, 115.

CHAPTER 7

BURROUS, C. N., LIEBER, A. J. and ZAVIANTSEFF, V. I. (1967),' Detection efficiency of a continuous channel electron multiplier for positive ions', *Rev. Sci. Instr.* **38**, 1477.
CHUBB, J. N. and GOWLAND, L. (1967), 'Experience with a pulsed gas flow technique for calibrating vacuum gauges', *Vacuum* **17**, 449.
DAVIS, W. D. (1968), 'Ultrahigh vacuum gauge calibration', *J. Vac. Sci. and Technol.* **5**, 23.
ELLIOT, K. W. T., WOODMAN, D. M. and DADSON, R. S. (1967), 'A study of the Knudsen and McLeod gauge methods for the calibration of high vacuum gauges', *Vacuum* **17**, 439.
FLETCHER, B. and WATTS, J. F. (1967), 'Design and construction of an absolute precision gauge calibration system', *Vacuum* **17**, 445.
GIORGI, T. A. and PISANI, C. (1966), 'On the concept of "pumping speed" in gettered and cryopumped systems', *Vacuum* **16**, 669.
HOLLAND, L. and PRIESTLAND, C. (1966), 'Experiments on the molecular flow of gas through tubes and vacuum system components using a B-A ionization gauge', *Vacuum* **16**, 601.
HOOD, C. B. and BARNES, C. B. (1966), 'Relationship between pumping speed, capture probability and chamber geometry in a spherical cryopumped vacuum chamber', *J. Vac. Sci. and Technol.* **2**, 302.

KOZLOV, V. F., KOLOT, V. Ya. and DOVBNYA, A. N. (1965), 'Low energy ion counter', *Inst. & Exp. Tech.* No. **6**, 1385.

MEINKE, C. and REICH, G. (1967), 'Comparison of static and dynamic calibration methods for ionization gauges', *J. Vac. Sci. and Technol.* **4**, 356.

MILLER, J. R. (1967), 'Extension of the Moser and Plotz McLeod gauge from 10^{-10} to 10^{-2} torr', *Vacuum* **17**, 387.

MOORE, B. C., BERGQUIST, L. E., CAMARILLO, R. G. and LARSON, R. (1965), 'The effect of localized gas densities on vacuum ionization gauges', *Trans. 3rd. Int. Vac. Cong.*, Vol. 2, p. 251 (Stuttgart).

MORRISON, C. F. (1967), 'Reference-transfer method of vacuum gauge calibration', *J. Vac. Sci. and Technol.* **4**, 246.

NAKAYAMA, K., AKIYAMA, Y. and HASHIMOTO, H. (1968), 'Capillary depression in McLeod gauges', *Vacuum* **18**, 65.

PERDIJK, H. J. R. (1967), 'A compilation of gas reactions as observed in electron tubes', *Suppl. Nuovo Cim.* **5**, No. 1, 73.

PTUSHINSKII, Y. G. and CHUIKOV, B.A. (1967), 'State of oxygen desorbed from the tungsten surface', *Surface Science* **7**, 90.

STECKELMACHER, W. (1966), 'A review of the molecular flow conductance for systems of tubes and components and the measurement of pumping speed', *Vacuum* **16**, 561.

THOMAS, A. M. and CROSS, J. L. (1967), 'Micrometer U-tube manometers for medium-vacuum measurements', *J. Vac. Sci. and Technol.* **4**, 1.

VÖLTER, J. and BERNDT, H. (1967), 'Reduced sorption and reaction of molecular gases in a modified Bayard-Alpert gauge', *Suppl. Nuovo Cim.* **5**, No. 1, 234.

CHAPTER 8

ALBERT, M. J., ATTA, M. A. and GABOR, D. (1967), 'A new type of composite all-metal electron emitter for valves and electron-optical devices', *Brit. J. Appl. Phys.* **18**, 627.

BUCKINGHAM, J. D. and THORNE, P. N. (1965), 'A comparative study of the thermionic emission properties of rhenium when coated with either lanthanum hexaboride or thoria, *Trans. 3rd Czech. Conf. on Elect. and Vac. Phys.*, p. 375 (Prague 1965).

CESPIRO, Z. (1965), 'On the precision of measurement of the compression manometer', *Trans. 3rd Czech. Conf. on Elect. and Vac. Phys.*, p. 471 (Prague 1965).

CLEAVER, J. S. and ZAKRZEWSKI, W. H. (1968), 'A comparative study of modulated Bayard-Alpert gauges and an extractor gauge', *Vacuum* **18**, 73.

COMSA, G. (1965), 'On ion collection in a Bayard-Alpert gauge', *Trans. 3rd Czech. Conf. on Elect. and Vac. Phys.*, p. 489 (Prague 1965).

COMSA, G., 'Initial ion-energy distribution and the measurement of very low pressure', *Rev. Roum. Phys.* **11**, 901 (1966); *Rev. Roum. Phys.* **12**, 63 (1967).

COMSA, G. (1967), 'Collection of energetic ions in a Bayard-Alpert gauge', *Vacuum* **17**, 373.

GROSZKOWSKI, J. (1967), 'Ionization gauges with reduced electronic desorption effects', *Bull. Acad. Polon. Sci., Ser. Sc. Tech.* **15**, 347.

HELMER, J. C. and HAYWARD, W. H. (1966), 'Ion gauge for vacuum pressure measurements below 1×10^{-10} torr', *Rev. Sci. Instr.* **37**, 1652.

LAWSON, R. W. (1967), 'Anomalous ion currents in Bayard-Alpert ionization gauges', *Brit. J. Appl. Phys.* **18**, 1763.

NAKEO, I. F. (1966), 'On the characteristics of the hot cathode magnetron gauge for measurement of UHV', *J. Jap. Vac. Soc.* **9**, 49.

NICHIPOROVICH, G. A. (1966), 'Inverted magnetron gauge for measuring pressures of 10^{-12} to 10^{-4} torr', *Inst. and Exp. Tech.*, No. 5, p. 1215.

NICHIPOROVICH, G. A. (1966), 'Comparative calibration of a hot-cathode magnetron gauge and a cold-cathode inverted magnetron gauge in UHV', *Inst. and Exp. Tech.*, No. 6, p. 1440.

PELZ, D. and NEWTON, G. (1967), 'Pressure conversion constants for magnetron ionization gauges', *J. Vac. Sci. and Technol.* **4**, 239.

POPOV, YU. S., (1967), 'Low pressure cold-cathode Penning discharge', *Sov. Phys. —Tech. Phys.* **12**, 81.

REIKHRUDEL, E. M. and ISAKAEV, E. KH. (1966), 'Ignition of the discharge in a high vacuum Penning gauge', *Sov. Phys.—Tech. Phys.* **11**, 486.

REIKHRUDEL, E. M. and SHERETOV, E. P. (1966), 'Current of self-maintained discharge in high vacuum in crossed electric and magnetic fields', *Rad. Eng. and Electron. Phys.*, No. 3, p. 449.

SCHUURMAN, W. and BRANDT, B. (1965), 'Theory of current-voltage characteristics of a low pressure discharge in a planar magnetron', *Proc. 7th Int. Conf. on Phen. in Ionized Gases, Belgrade*, Vol. I, p. 506.

SCHUURMAN, W. (1967), 'Investigation of a low pressure Penning discharge', *Physica* **36**, 136.

VAN OOSTROM, A. (1967), 'Total pressure measurements with ionization gauges and their limiting factors', *Vakuum-Technik* **16**, 159.

VAN OOSTROM, A. (1967), 'Modulation of Bayard-Alpert ionization gauges with grid end-caps', *J. Sci. Instr.* **44**, 927.

VISSER, J. (1967), 'A hot-cathode magnetron ionization gauge with photo-current suppressor', *Vacuum* **17**, 73.

CHAPTER 9

BLAUTH, E. W. (1968), 'Trends in partial pressure measurement', *Vakuum-Technik* **17**, 90.

BOERS, A. L. (1968), 'Some remarks on gas analysis with a molecular sieve', *Vacuum* **18**, 81.

DAWSON, P. H. and WHETTEN, N. R. (1968), 'Ion storage in three-dimensional, rotationally symmetric, quadrupole fields. I. Theoretical treatment', *J. Vac. Sci. and Technol.* **5**, 1.

HOCH, H. (1967), 'Total and partial pressure measurements in the range 2×10^{-10} to 10^{-12} torr', *Vakuum-Technik* **16**, 8.

PATY, L. (1965), 'Processes on the condensation surface at very low temperature', *Trans. 3rd Czech. Conf. on Elect. and Vac. Phys.*, p. 525 (Prague 1965).

RETTINGHAUS, G. (1967), 'Nachweis niedriger partialdrucke mit dem ionenkafig', *Z. angew Phys.* **22**, 321.

TORNEY, F. L. (1967), 'Cold cathode quadrupole mass spectrometer', *Rev. Sci. Instr.* **38**, 1404.

Appendix A

CHAPTER 10

HOLLAND, L., FULKER, M. J. and LAURENSON, L. (1967), 'The influence of the pre-evacuation cycle on the gas atmosphere in a Penning pump system, *Suppl. Nuovo Cim.* **5**, No. 1, 242.

HOPKINS, B. J. (1968), 'Ultrahigh vacua', *Contemporary Physics* **9**, 115.

CHAPTER 11

BAEVA, N. N., DANILOVA, N. P. and SHELNIKOV, A. I. (1966), 'Cryopump for obtaining ultrahigh vacuum', *Inst. & Exp. Tech.*, p. 204, Nov.-Dec. 1966.

BELL, P. R., MOORE, E. C. and WYRICK, J. (1967), 'Starting sputter-ion pumps and the outgassing of metal surfaces', *Vacuum* **17**, 87.

BIGUENET, C., SHROFF, A. M., MATHIEZ, P. and BOROSSAY, J. (1968), 'Clean titanium wire for UHV sublimation pumps', *Le Vide* **23**, 15.

BIRYUKOVA, N. E., VINOGRADOV, M. I., DANILOV, K. D. and SHISHLOVSKII, S. S. (1968), 'Cooled titanium sorption pump', *Inst. & Exp. Tech.* No. 2, p. 610.

ESPE, W., HYBL, C. and LHOTSKY, O. (1965), 'Adsorption of gases and vapours on molecular sieves', *Proc. 3rd Czech. Conf. on Elect. and Vac. Phys.*, p. 551 (Prague 1965).

GREGORY, G. L., BRADFORD, J. M. and MUGLER, J. P. (1966), Vacuum capabilities of the 150 cu ft space facility at the Langley Research Centre', *Am. Inst. Aeronautics J.*, p. 154.

HOPKINS, B. J. and PENDER, R. R. (1967), 'A liquid helium cooled trap for UHV systems', *J. Sci. Instr.* **44**, 73.

RADZHABOV, T. D. and IVANOVSKII, G. F. (1967), 'Ion pumping with continuous renewal of the sorbent surface', *Sov. Phys.—Tech. Phys.* **11**, 1539.

SCHULZE, H. (1967), 'Investigation of the back streaming from oil diffusion pumps', *Vakuum-Technik* **16**, 102.

CHAPTER 12

BLAHNIK, C. E. (1967), 'Differentially pumped ultrahigh vacuum system bakeable to 900°C', *J. Vac. Sci. and Technol.* **4**, 378.

ROZGONYI, G. A. and SOSNIAK, J. (1968), 'Influence of gauge operation on background gases during high vacuum mass spectrometry', *Vacuum* **18**, 1.

SUBJECT INDEX

Accelerators, 433
Accommodation coefficient, thermal, 59–63, 69–71
Adhesion of materials at very low pressures, 243–245
Adsorption,
   area of adsorbents (Table), 51
   chemical adsorption spectrometers, 340, 355–361
   chemisorption properties of metals and semiconductors, 29–37
   electron impact on adsorbed gas, 167–181
   isotherms, 39–54
   of mixtures, 83–88, 355–365
   on bonded-adsorbent cryopumps, 387–389
   on cryopumps with flat surfaces, 389–391
   on loose adsorbents, 385
   physical adsorption spectrometers, 340, 355, 361–365
Area of adsorbents (Table), 51
Atoms,
   see Molecules

Background,
   see Residual
Backscattering,
   see Collisions
Baking,
   see also Outgassing,
   of glass systems, 375–376
   of metal system, 377, 424
   outgassing procedures after baking, 378
   procedures, 369–370, 373–378
Barkhausen-Kurtz oscillations, 308–309
Binding states,
   of ionically trapped inert gases, 211–213
   multiple, in chemisorption, 36, 37, 355–361
Blears effect, 280–281
Bonded-adsorbent pumps, 387–389
Born-Mayer constants, for rare gases, 182

Calibration,
   of desorption spectrometers, 355

   of gauges, extensions to lower pressure, 257–262
   of uhv gauges, 253–263
Cathodes,
   see also Filaments,
   gas interactions at hot cathodes, 275–280, 336, 425–427
   other than thermionic, 301
   thermionic, 299–303
   thermionic, rates of evaporation, 301–303
Channelling, ions in solids, 218–221
Characteristic energy loss, of electrons, 152–159
Chemical adsorption,
   see Chemisorption and Adsorption
Chemical reactions,
   at filaments, 275–280, 336, 425–427
   in gas phase, 21
Chemisorption,
   activation energies of desorption, 31, 80–81
   adsorption properties of metals and semi-metals, 29–37
   energies, 29–37
   heats, 27–38
   in getter pumps, 399–404
   in ion pumps, 404–405, 411–416
   isotherms, 50–54
   replacement, 86–88, 360
   spectrometer, 355–361
   sticking coefficient, 66–68
Cleaning, of single crystals, 146–147
Coefficient,
   accommodation, 59–63, 69–71
   backscattering of electrons, total, 144, 159–167
   condensation, 63–66, 383, 393, 396
   friction, 246–248
   recombination of positive ions, 131–133
   recombination, dissociative (Table), 132
   rediffusion, of electrons, 157–159
   reflection, of ions at surfaces, 189–191
   second virial (Table), 19
   sticking, see Sticking probability,
   surface diffusion, 82

p*

Subject Index

Cold-cathode ionization gauges,
  see Gauges
Cold-cathode ion pumps,
  see Pumps
Collisions,
  charged particles with surfaces, 134–233
  electron-adsorbed molecule, 167–181
  electron-molecule, elastic, 111–114
  electron-molecule, exciting, 117–120
  electron-molecule, ionizing, 114–117
  electron-surface, 134–181
  ion-molecule, charge transfer, 123–125
  ion-molecule, elastic, 120–123, 353, 354
  ion-molecule, exciting, 123, 125–128
  ion-molecule, ionizing, 125–128
  ion-surface, 181–229
  molecule-molecule, energetic, 120–123
  molecule-molecule, reactive, 21
  molecule-molecule, thermal, 18–22
  molecule-surface, angular distribution, 55–59, 72
  molecule-surface, theory, 68–72
  molecules, energetic, with surfaces, 181–229
  molecules, thermal, with surfaces, 54–73
  phenomena at various projectile energies (Table), 187
Condensation,
  coefficient, 63–66, 383, 393, 396
  in cryopumps, 383–399
  in space simulation chambers, 396–399
Creep, of materials at very low pressures, 248–250
Cross-section, collision,
  molecule-molecule, 20
  collision processes in gases (Table), 112
  elastic, atom-atom, 120–123
  electron-impact desorption, 168–172
  excitation, electron-atom, 117–120
  ionization, electron-atom, 114–117
  ionization, ion-atom, 125–128
  reaction, molecule-molecule, 21
  recombination, 131–133
Current measurements,
  general problem of, 294–295
  with conversion-scintillation detectors, 298–299
  with dc amplifiers, 295
  with electron multipliers, 295–298, 320, 342, 344, 354
Cryogenic pumps,
  see Pumps
Cryopumps,
  see Pumps
Cryosorption pumps,
  see Pumps
Cryotrapping of mixtures of gases, 83–86, 384–385

Degassing,
  See Outgassing
Desorption,
  activation energies of, 31, 80–81
  by electron impact, 167–181
  by photons, 239–242
  from adsorbed layers in uhv systems, 369–370, 420, 422
  in cryopumps, 383–399
  mean time of sojourn, 73–77, 386
  of molecules from cathodes, 301–303
  of molecules from surfaces, 73–83, 239–242, 355–361, 369–370, 420, 422
  spectra, of chemisorbed species, 36, 355–361
  spectra, of rare gases in metals, 211–212
  thermal, first order, 77–80, 360–361, 362, 369–370
  thermal, second order, 80–81, 361
Desorption spectrometers,
  chemical, 340, 355–361
  general, 355
  physical, 340, 355, 361–365
  sensitivity, 361
Diffraction, electron,
  general considerations, 135–139
  LEED, 135–139, 145–151
  RHEED or HEED, 139
Diffusion,
  constants, gases in metals, 91
  constants, gases in non-metals, 90
  during outgassing, 370, 371–378
  examples of, in uhv practice, 105–109
  gases in solids, 88–91, 370, 371–378
  in a cylinder, 95
  in a semi-infinite solid, 92–93
  in a slab, 93–94
  in a sphere, 95
  in vacuum chamber wall, 99, 371–378
  modifications to idealized solutions, 100–105
  surface, 28, 81–83, 386
Dispersion energies between molecules, 12–16
Dissociation energies of molecules (Table), 16

Electronic desorption,
  see Electron-impact desorption
Electron emission,
  Auger, electron induced, 159
  Auger, ion induced, 223–226
  photoelectric, 234–239, 304, 306, 315, 319, 321, 322, 327
  secondary, 159–167
  thermionic, 299–303
Electron-impact desorption,
  during outgassing, 172–176, 307–308, 378
  from adsorbed layers, 168–172

Subject Index

Electon-impact desorption—*cont.*
  from surfaces of unknown state, 174–176
  from thick layers, 172–174
  in accelerators, 433
  in gauges, 290–294, 315, 316, 323, 326, 327, 329, 337, 420–422
  in partial pressure gauges, 355, 420–422
  ion current from, 315–316, 324–325, 329, 337
  theory of, 176–181
Electron multipliers,
  see Multipliers, electron
Electrons, collisions in gases,
  elastic collisions with atoms, 111–114
  exciting collisions with atoms, 117–120
  ionizing collisions with atoms, 114–117
  recombination with positive ions, 131–133
Electrons, collisions at surfaces,
  Auger, 159
  bombardment during outgassing, 172–176, 307–308, 378
  characteristic energy losses, 152–159
  desorption of gas by electrons, 167–181, 307–308, 378
  diffraction, 135–139, 145–151
  ejection by ions at surfaces, 222–229
  elastic reflection, 135–151
  energies of photoeclectrons, 238–239
  energies of scattered, 134–167
  LEED, 135–139, 145–151
  rediffused, 157–159
  scattering from surfaces, 134–167
  secondary, emission of, 159–167
  secondary, in ionization gauges, 308, 331
  RHEED or HEED, 139
  theory of electron impact desorption, 176–181
Electron space charge,
  see Space charge
Electrostatic energies between molecules, 12, 16–18
Emission,
  electrons, by ion impact, 222–229
  ions, from hot surfaces, 229–233
  photoelectric, 234–239
  photoelectric, energy distribution of electrons, 238–239
  photoelectric, yield from x-rays, 289
  secondary electron, 159–167
Emissivity in cryogenic pumps, 394
Energies,
  backscattered electrons, 134–167
  characteristic losses, of electrons, 152–159
  chemisorption, 29–37, 355, 360, 361
  desorption, in uhv systems, 369–370
  dispersion, 12–16
  dissociation (Table), 16
  electrostatic, between molecules, 12, 16–18

  induction, 12, 16–18, 27
  interaction between molecules, > 1 eV, 181–185
  molecule-molecule, 11–18
  molecule-surface, 22–29
  photoelectrons, 238–239
  physisorption, 22–29, 355
  physisorption, on heterogeneous surfaces, 26–27, 362
  physisorption, variation with location on single crystal, 25
  surface diffusion, in physisorption, 28
  valence, 12
  Van der Waals, 12–16
  vaporization and sublimation (Table), 25
Entrapment of positive ions in solids, 209–216
Errors in pressure measurement,
  in Bayard-Alpert gauges, 304
  in ionization gauges, 323–329
  in partial pressure gauges, 354
Evaporation from cathodes, 301–303
Fatigue of materials at very low pressures, 248–250
Field emission, use of in gauges, 340, 365–366
Filaments,
  see also Cathodes,
  chemical reactions at, 275–280, 336, 425–427
  in ionization gauges, 309, 317
Flash filament,
  see Desorption spectrometers
Fracture of materials at very low pressures, 248–250
Friction of materials at very low pressures, 246–248
Gas admission to uhv systems, 427–430
Gaskets, 430, 431
Gauges, partial pressure,
  see also Desorption spectrometers; Mass spectrometers,
  desorption spectrometers, chemical, 340, 355–361
  desorption spectrometers, physical, 340, 355, 361–365
  electron-impact desorption in, 293–294
  errors, 354
  mass spectrometers, examples of uhv, 345–352, 419
  mass spectrometers, general characteristics, 340, 342, 346
  using field emission, 340, 365–366
  using work function changes, 340, 365–366
Gauges, total pressures,
  Bayard-Alpert, 305–313
  calibration, uhv gauges, 253–263
  cold-cathode, current-pressure relation, 332, 334–335

Gauges—*cont.*
  cold-cathode, magnetic, 329–336
  cold-cathode, oscillatory behaviour, 335
  comparison of types of gauge, 336–339
  conventional ionization, 304–306
  crossed-field, cold-cathode, 329–336
  electron-impact desorption in, 290–294, 315, 316, 323–329, 337, 420–422
  envelope potential, 308–309
  errors, 304, 323–329
  extractor, 315–317
  hot-cathode, 304–329
  hot-cathode with magnetic field, 319–322
  inverted-magnetron, 333–334
  magnetron, 331–333
  McLeod, 254–257
  modulation method, 312, 314, 325–327, 337
  non-linear sensitivity, 312, 319, 329–336, 338
  orbitron, 317–319
  oscillatory behaviour, 308, 329, 334–336
  outgassing of, 307–308, 378
  Penning, 329–331
  pumping in, 263–280, 308, 309, 325, 329, 336
  re-emission in, 263–280, 336, 420–424, 428
  residual currents, measurement of, 323–329
  sensitivity as a function of geometry, 309–311
  sensitivity as a function of operating conditions, 309, 318–319, 322, 331–338
  sensitivity, for different gases, 262–264
  space charge gauges, 322–323
  suppressor, 313–315
  suspended rotor, 304
  theory of cold-cathode gauges, 334–336
  trigger discharge, 330
Getters,
  getter-ion pumps, 411–416
  getter materials, 399–402
  initial sticking probability, 399–403; (Tables), 401, 402
  maximum capacity, 399–403; (Tables), 401
  types of pumps, 403–404
  use of, during system processing, 376, 378
Grids, non-sag, 307

Heats,
  see also Energies,
  chemisorption, 27–38
  chemisorption, correlation with metallic radius, 35
  chemisorption, hydrogen on evaporated films, 34
  chemisorption, oxygen on evaporated films, 32

  differential, 32
  isosteric, 31, 362
Hot-cathode ionization gauge,
  See Gauges

Impact,
  see Collisions
Induction energies between molecules, 12, 16–18, 27
Instability,
  in Bayard-Alpert gauges, 308
  in cold-cathode gauges, 329, 334–336
  in ion pumps, 407, 410, 411, 413
Internuclear distances (Tables), 16, 185
Ion collector,
  current, 304–339
  in Bayard-Alpert gauges, 305–313
  in ionization gauges, 304–339
  negative collector currents, 315
Ion pumps, 404–416
Ionic space charge,
  see Space charge
Ions, collisions at surfaces,
  channelling in solids, 218–221
  electron ejection by, 222–229
  energies of interaction, 181–185
  entrapment in solids, 209–216
  penetration in solids, 216–218
  reflection at surfaces, 188–192
  radiation damage by, 204–208
  re-emission, 215–216, 404–406, 424, 428
  sputtering, 192–204
  surface ionization, 230–233, 420–424
Ions, collisions in gases,
  collisions with molecules, charge transfer, 123–125
  collisions with molecules, elastic, 120–123, 353, 354
  collisions with molecules, exciting, 123, 125–128
  collisions with molecules, ionizing, 125–128
  created in electron-molecule collisions, 114–117, 168–172
  energies of interaction, 181–185
  fragments, in mass spectrometers, 341 (Table), 352
  photo-ionization, 128–131
  recombination of positive, 131–133
  scattering in mass spectrometers, 354
  trapping in mass spectrometers, 348, 352
Ion current,
  from electron-impact desorption, 315, 316, 324, 325, 329, 337
  in ionization gauges, 304–339
  in ion pumps, 404–416
  in mass spectrometers, 342, 344

Ion current—*cont.*
  residual current in gauges, 312, 313, 315, 323–329, 330, 332, 336
  true ion current in ionization gauges, 312, 315, 323, 324
Ionization,
  by electrons, 114–117, 168–172
  in partial pressure gauges, 340–355; (Table), 341
  in total pressure gauges, 304–339
  multiple, in mass spectrometers, 348, 352
  surface, 230–233, 420–424
Isotherms,
  adsorption, 39–54
  Dubinin–Radushkevich, 46–50
  Freundlich, 33, 50–54
  Langmuir, 50–54
  role in cryosorption pumps, 384–391
  Temkin, 33, 50–54

Leak,
  detectors, 429–430
  rates of residual gases, 423–427
  terms in pressure equations, 263–275, 423–427
Low energy electron diffraction (LEED), 135–139, 145–151
  effect of target temperature, 150–151
  surface structure, O_2 on Ni, 148–150
Low pressure limit,
  of ionization gauges, 304–339
  of mass spectrometers, 342, 351
  of uhv pumps, 379–416
Lubrication of materials at very low pressures, 247–248

Magnetic fields,
  in cold-cathode gauges, 329–336
  in hot-cathode gauges, 319, 322–323
  in ion pumps, 406–412
  requirements for mass spectrometers, 344, 345–349, 351
Mass spectrometers,
  see also Gauges, partial pressure,
  accessibility of ionization volume, 344
  background currents, 344, 354
  construction and circuitry, 344
  examples of, 345–352, 419
  fragment ions in, 341 (Table), 352
  general characteristics, 342, 346
  interference from ion scattering, 354
  magnet requirements, 344, 345–349, 351
  magnetic deflection, 180°, 346, 348
  magnetic sector, 60°, 346, 347
  magnetic sector, 90°, 345, 346, 419
  mass discrimination, 354
  mass numbers typically observed, 341 (Table), 354–355, 419
  measurement problems, 352–354
  monopole, 346, 349
  multiple ionization in, 348, 352
  non-linearity at high pressures, 353
  omegatron, 351–352, 419
  operation of, 340–345, 346
  outgassing of, 343, 378, 421–423
  peak shape, 343, 350
  properties of (Table), 346
  quadrupole, 346, 349, 419
  resolving power, 343, 346
  scanning speed, 345
  sensitivity, 342, 346, 354
  time-of-flight, 344, 345, 346, 350–351
  trochoidal or cycloidal, 346, 348–349, 419
Materials,
  adhesion at low pressures, 243–245
  choice of, for uhv, 369
  creep at low pressures, 248–250
  fatigue at very low pressures, 248–250
  fracture at very low pressures, 248–250
  getter, 399–402
  lubrication at very low pressures, 247–248
  outgassing of glass, 107, 373–377
  outgassing of stainless steel, 108–109, 371–372
  pre-treatment of, 371–373
Mean free path,
  ion-atom, 353
  molecular collisions, 20
Measurement of pressure,
  see Pressure measurement;
  Gauges; Partial pressure measurement
Mechanical properties of materials at very low pressures, 243–250
Modulation method in ionization gauges, 312, 314, 325–327, 337
Molecular sieves,
  bonding to surfaces, 387–389
  in cryosorption pumps, 385–387
Molecules, collisions in gases,
  elastic collisions with molecules, 18–22, 120–123
  charge transfer collisions with ions, 123–125
  elastic collisions with ions, 120–123, 353, 354
  excitation by ions, 123, 125–128
  ionization by electrons, 114–117
  ionization by ions, 125–128
Molecules, collisions at surfaces,
  energetic molecules, 181–229
  sputtering by ions, 192–204
  theory, 68–73
  thermal energies, 54–68
  thermodynamics, 37–54
Molecules, energies of interaction,
  molecule-molecule energies, 11–16
  molecule-surface energies, 22–29

Subject Index

Moments,
  electric dipole (Table), 17
  electric quadrupole (Table), 17
Multipliers, electron,
  current measurement with, 295–298, 320, 342, 344, 345
  noise in, 296–297, 342
Noise,
  in electron multipliers, 296–297, 342
  in field emission gauges, 365

Orbits,
  electrons in ionization gauges, 310, 317–319
  ions in ionization gauges, 310–311
Oscillations,
  see also Instabilities,
  in ion pumps, 407, 410–413
  in total pressure gauges, 308, 329, 334–336
Outgassing,
  by baking, 369–370, 373–378, 424
  by electron bombardment, 172–176, 307–308, 378
  by ion bombardment, 378
  examples of, in uhv, 105–109, 373–378
  of glass, 107, 373–377
  of grids in ionization gauges, 172–176, 307–308, 378
  of mass spectrometers, 343, 378, 421–423
  of materials for uhv, 369, 371–373
  of stainless steel, 108–109, 371–372, 377
  of water during baking, 373
  procedures after baking, 378

Partial pressure measurement,
  see also Gauges, partial pressure; Mass spectrometers; Desorption spectrometers
  fragmentation, 341 (Table), 352
  gauges using field emission, 340, 365–366
  gauges using work function changes, 340, 365–366
  general problems, 340, 352
  interference from ion scattering, 354
  ion trapping, 348, 352
  mass discrimination, 354
  mass numbers typically observed, 341 (Table), 354, 417–420, 420–434
  multiple ionization, 348, 352
  non-linearity at high pressures, 353
  variations of transmission probability, 353
Penetration of ions in solids, 216–218
Permeation,
  constants, gases in metals, 102
  constants, gases in non-metals, 101
  of gases through chamber walls, 99, 100, 107–109, 424, 425
Photo-absorption, 128–131
Photodesorption, 239–242
Photo-electric emission, 234–239
Photo-ionization, 128–131
Photons,
  interaction with adsorbed gases, 239–242
  photo-adsorption, 128–131
  photo-electric emission, 234–239, 289, 304, 306, 315, 319, 321, 322, 327, 433
  photo-ionization, 128–131
Physical adsorption,
  see also Physisorption
Physisorption,
  desorption spectrometers, 361–365
  energies of, 22–29, 355, 362
  in cryosorption and cryogenic pumps, 383–399
  isotherms, 39–50
  on heterogeneous surfaces, 26–27
Polarizability (Table), 17
Potential,
  see Energies
Pressure,
  equations, 263–275, 423–427
  gauges, see Gauges, total pressure; Gauges, partial pressure
  local, in uhv systems, 424
  vapour, 37–39
Pressure measurement,
  absolute, 253–257
  current-pressure relation in cold-cathode gauges, 332, 334–335
  errors in gauges, 304, 323–329, 354
  in non-uniform environments, 281–287
  partial pressure, 340–366
  total pressure, 304–339
Probability, sticking,
  see Sticking probability
Processing techniques,
  baking procedures, 369–370, 373–378, 424
  degassing procedures, 307–308, 378
  for uhv systems, 369–378
  material pre-treatment, 371–373
Pumping,
  at hot surfaces, 275–280
  during system processing, 376
  in desorption spectrometers, 357, 363
  in gauges, 263–280, 308, 309, 325, 329, 336
  in tubing, 280–281
  of typical uhv systems, 417–434
  speed, of pumps, 379–416
  speed, of uhv gauges (Table), 268
Pumps,
  comparison of properties, 414–415
  cryogenic, 384, 391–399, 414–415, 430, 432
  cryopumps, general, 383–385, 432

Subject Index

Pumps—cont.
  cryosorption, 384, 385–391, 414–415, 432
  diffusion, 380–383, 414–415
  for uhv, 379–416, 419, 430
  getter, 399–404, 414–415
  getter-ion, 411–416
  in space simulation chambers, 396–399
  ion, 404–416
  molecular drag, 379–380
  orbitron, 412–416
  speed of, 379–416
  sputter-ion, 406–411, 414–415
  turbo-molecular, 379–380, 414–415

Radiation,
  see also Photons,
  damage, 204–208
  in space simulation chambers, 396–399
Recombination, positive ions, 131–133
Rediffused electrons, 157–159
Re-emission,
  in gauges, 263–280, 336, 420–423, 424, 428
  in ion pumps, 404–411, 424, 428
  in leak detectors, 429–430
  of ionically trapped gas, 215–216
Reflection,
  electron, elastic at surfaces, 135–151
  ions, at surfaces, 188–192
  molecules, at surfaces, 55–59
  x-ray, 287, 314
Replacement,
  in chemisorption, 86–88, 360
  in physisorption, 86
  of ionically trapped gas, 203–204
Residual,
  conditions, achieved with diffusion pumps, 382
  conditions, for thin film deposition, 429,
  conditions, in small glass and metal systems, 417–430; (Table), 419
  gas, desorption spectra, 355–361
  gases, fragment ions of (Table), 341
  gases, in uhv systems, 341, 402, 403, 417–427; (Table), 419
  processes, in small glass and metal systems, 420–427
  gas leak rates, 343, 373–378, 387, 420–427
Residual currents,
  electron-impact desorption, 315–316, 324–325, 329, 337
  in total pressure gauges, 323–329
  soft x-ray photo-emission, 287–290
Residual gas analysers,
  see Gauges, partial pressure; Mass spectrometers.
Resolving power of mass spectrometers, 343, 346

Scattering,
  see Collisions
Scintillation detectors, 298–299
Secondary emission of electrons, 159–167
Sensitivity,
  non-linear, in gauges, 312, 319, 329–336, 338
  of gauges, for different gases, 262–264
  of gauges, for different geometries, 309–311
  of gauges, for different operating conditions, 309, 318–319, 322, 331–338
  of mass spectrometers, 342, 346, 354
Sinks, of gas,
  in gauges, 263–280, 308, 309, 325, 329, 336
  in tubing, 280–281
Sojourn time, of a molecule on a surface, 73–77, 386
Solubilities,
  of gases in getters, 399–403
  of gases in metals, 97, 98
  of gases in non-metals, 96
Sorption pumps,
  see Pumps
Sources of gas,
  during outgassing, 369–378
  in gauges, 263–280, 424–427
  in pumps, 420–427
  in small glass and metal systems, 417–430
  in tubing, 280–281
Space charge,
  electron, in mass spectrometers, 348, 352
  gauges depending on, 322–323
  in ionization gauges, 319, 322–323, 329, 330, 334–336, 354
  ion, in mass spectrometers, 352, 353
Space simulation chambers, 396–399, 434
Spectra,
  in mass spectrometers, 340–355, 417–430
  thermal desorption, for chemisorbed gases, 36, 355–361
  thermal desorption, for ionically trapped gases, 211–212
  thermal desorption, for physisorbed gases, 363–364
Spectrometer,
  see also Mass Spectrometer: Gauges, partial pressure; Desorption spectrometers
Speed, pumping,
  of uhv gauges, 263–280, 308, 309, 325, 329, 336; (Table), 268
  of uhv pumps, 379–416; (Table), 410
Sputtering,
  by energetic ions, 192–204
  sputter-ion pumps, 406–411
  re-emission of previously pumped gas, 203
  theory, 200–203

Sputtering—*cont.*
  yields, positive ions on metals, 193–198, 200, 202–203
Sticking probability,
  adsorption, 66–68, 72–73
  of ions in ion pumps, 404–416
  on getters, 399–403
  positive ions on surfaces, 210, 213
Storage rings, 433
Sublimation energies, 25
Surface,
  area, of absorbents, 51
  collisions of electrons with, 134–181
  collisions of ions with, 181–229
  collisions and interactions of molecules with, 37–88, 181–229
  collisions of photons with, 234–242
  desorption of gas from, 73–83, 167–181, 211-212, 239–242, 355–361, 369–370, 383–399, 420, 422; see also Desorption
  diffusion, 81–83, 381, 386
  ionization, 229-233, 420–424
  molecule-surface energies, 22–29
  phase in uhv systems, 423–424
  structure observed in LEED, 148–150
Symbols, index of, 4–7
Systems, examples of uhv,
  conventional systems, moderate size, 431–432
  cryogenic, 432–433
  double-walled, 431–432
  gauge calibration, characteristics of, 256
  glass systems, small, 417–430
  large systems, 430–434
  metal systems, small, 417–430
  plasma physics research, 433–434
  space simulation, 396–399, 434
  storage rings, 433
Sytems, outgassing of,
  degassing procedures after baking, 378
  double-walled, 378
  glass, 375–376
  metal, 108–109, 371–372, 377–378

Thermal description,
  see also Desorption,
  first order, 77–80, 210–214, 360–361, 362, 369–370
  second order, 80–81, 361
Thermal transpiration, 284–287
Thermionic emission,
  of electrons, 299–303
  of ions, 229–233
Thin film deposition, 429
Total pressure gauges,
  see Pressure measurement;
  Gauges, total pressure
Traps,
  for diffusion pumps, 380–383, 386
  required for various pumps, 314–315
  surface diffusion in, 381
  using molecular sieves, 385

Vacuum envelopes, 419
Vacuum firing of materials, 371–373
Valence energies, between molecules, 12
Van der Waals energies, 12–16
Vaporization energies, 25
Vapour pressure, role in cryogenic pumps, 384, 391–392, (Tables) 37–39
Vapour stream effect, 255
Virial coefficient (Table), 19
Wear, at low pressures, 246–248
Welding, cold, at low pressures, 243–245
Work function,
  changes, due to adsorbed gas, 143
  gauges using work function changes, 340, 365–366
  ionic, 229–233
X-rays,
  limit in gauges, 304–315, 319–329, 337–338, 344, 349, šee also Residual currents
  photo-electric yield, 234–239, 289, 304, 306, 315, 319, 321, 322, 327, 433
  reflection of, 287, 314
  reverse x-ray effect, 309, 315, 328–329

AUTHOR INDEX

A

Abrahamson, 14, 15, 121, 182, 185, 186
Abrikosova, 244
Abroyan, 227, 470
Ackley, 292
Adams, 241, 242
Adamson, 48
Afrosimov, 127, 128, 191
Ageev, 68, 86, 88
Agishev, 68, 86, 87, 88, 102, 231
Akhmetova, 470
Akiyama, 473
Albert, 473
Alexeff, 322
Allen, F. G., 232
Allen, J. S., 295
Allen, P.G.W., 403, 426
Allinson, 424
Almén, 194, 195, 196, 197, 200, 201, 202, 214, 215
Alpert, 1, 90, 96, 101, 107, 209, 210, 261, 263, 268, 273, 275, 278, 279, 292, 302, 304, 305, 306, 324, 386, 424, 426
Altemose, 90, 96, 101
Amdur, 60, 121
Ames, 391
Anderson, G. S., 198, 199, 470
Anderson, H. U., 265
Anderson, J., 143, 470
Anderson, P. A., 1
Andreen, 220, 221
Apgar, 281
Apker, 355
Appelt, 325, 326
Arden, 412, 413
Arifov, 226
Armand, 467
Armstrong, 43, 45, 47, 49, 140, 141, 142, 143, 146, 294, 362, 366
Ash, 75, 76
Asundi, 115
Atkinson, 323
Atta, 473
Aubry, 254, 257, 470
Auslender, 470

B

Bächler, 65, 382, 395, 405, 408
Bachman, 467
Baeva, 475
Bailey, J. R., 341, 352
Bailey, V. A., 111
Baker, 78, 171, 215, 268, 272, 278, 353
Ball, 217
Ballance, 397
Balwanz, 384, 397
Bance, 406, 410
Bandringa, 288, 309, 417
Banks, 323
Bannock, 385
Barnes, C. B., 472
Barnes, G., 265, 268
Baronetzky, 280
Barrer, 28, 75, 76, 82, 89, 90, 91, 96, 101, 467
Barrington, 410, 429
Barton, 369, 471
Bas, 364
Basalaeva, 369
Bashanova, 139
Batanov, 227
Bates, 123, 125, 127
Batzer, 244
Bauer, E., 148, 149, 150, 470
Bauer, S. H., 384, 387, 388
Bayard, 1, 305, 306, 324
Beams, 304, 379
Beattie, 12, 17
Beck, A. H., 333
Beck, D. E., 58
Becker, E. J., 81, 278, 279
Becker, J. A., 66, 67, 68, 72, 81, 223, 278, 279
Becker, W., 379, 380
Bederson, 470
Beeck, 34
Beeler, 217
Behrisch, 182, 193
Bell, A. A., 467
Bell, P. R., 475
Bellardo, 44, 45, 46, 47, 51
Belyakov, 102
Bennette, 169

Bennewitz, 265
Bergquist, 473
Bergsnov-Hansen, 467
Bernardini, 175, 242, 433
Berndt, 473
Bernhard, 296, 298
Bernstein, 19, 21
Besco, 217
Bierman, 128
Biguenet, 475
Billington, 204
Bills, 210, 230, 232, 267, 268, 270, 273, 302, 413
Biondi, 131, 132, 385, 386
Bird, 12, 14, 15, 17, 18, 19
Birely, 21
Biryukova, 475
Björkqvist, 221
Blahnik, 378, 432
Blankenfeld, 166
Blauth, 280, 342, 474
Bleakney, 114, 115
Blears, 280, 372
Blodgett, 238
Boerboom, 115, 298
Boers, 474
Boersch, 155, 156, 157
Bøgh, 192, 220
Bohm, 153
Bohr, 122, 123, 183, 184
Bol'shakov, 469
Bolz, 66
Bond, 29, 30, 31, 33, 34, 35
Borossay, 475
Borovik, 39, 384, 395, 396
Bowden, 246, 247
Brackmann, 84, 115, 116, 118, 119, 123, 124, 127
Bradford, 475
Brandes, 81, 278, 279
Brandt, B., 474
Brandt, W., 202
Brauer, 134, 159, 161, 163, 167
Brennan, 32, 33, 35, 40, 51
Brentari, 393
Briglia, 117
Brinkman, 207
Brisbane, 333
Brock, 332, 333, 334, 338, 432
Brombacher, 304
Bronshtein, 160, 163, 164, 165, 166
Brown, E., 271
Brown, F., 191, 217, 218, 219, 220
Brown, R. D., 246
Brown, R. E., 60
Brubaker, 388
Bruce, 194, 195, 196, 197, 200, 201, 202, 214, 215
Bruining, 157, 166

Brunauer, 45, 46
Brunnée, 188, 189, 226, 227, 228, 294
Bryant, 28, 48, 50, 244, 330, 331, 332, 333, 419, 425, 426, 427
Buck, 232
Buckingham, 299, 300, 473
Buckley, D. H., 248
Buckley, F., 17
Buckman, 250
Buffham, 393
Bultemann, 346, 349, 419, 421
Bunshah, 244
Bureau, 260, 262
Burger, 147
Burgers, 91
Burhop, 61, 110, 117, 119, 138
Buritz, 90, 96, 101, 107, 209, 256, 261, 302
Burke, 118
Burns, 166, 167
Burrous, 472
Burton, 246
Burtt, 203, 210
Buser, 425
Busol, 391, 395, 396

C.

Cairns, 236, 237
Calder, 109, 371
Caldwell, 140
Callaghan, 385
Camarillo, 473
Campbell, 87, 88
Carbone, 121, 127
Caren, 394
Carleton, 119, 125, 210, 267, 268
Carley, 259
Carlston, 196, 197, 199, 226, 228, 229
Carmichael, 203, 210, 215, 216, 269, 271, 386, 405, 406
Carr, 253
Carslaw, 99
Carter, 79, 182, 203, 204, 208, 209, 210, 211, 212, 215, 216, 265, 266, 267, 268, 270, 271, 272, 308, 467, 470, 471
Casimir, 244
Caswell, 390, 391, 428, 429
Cavaleru, 215
Cespiro, 473
Chambers, 467
Chang, 468
Chen, 90, 96, 101
Chernetskii, 335
Chirikov, 323, 336
Choumoff, P. S., 257, 396
Choumoff, S., 257
Christensen, 391
Christian, 260
Christians, 374

Chubb, 66, 395, 432, 472
Chuikov, 81, 473
Chutjian, 470
Clarke, 130
Clausing, P., 55, 75, 76
Clausing, R. E., 400, 401, 402
Clauss, 247, 248
Clay, 317, 338
Cleaver, 473
Close, 211
Cobic, 210, 265, 266, 268
Cochrane, 467
Cohen, 135, 139, 150
Coleman, 386, 387
Colgate, 121
Colligon, 203, 210, 213, 214
Comeaux, 197, 199
Comes, 236
Compton, 114, 229
Comsa, 215, 273, 310, 473
Conklin, 419
Conn, 319
Conrad, 250
Constabaris, 23, 40, 43, 51, 74, 75
Cook, C. J., 140, 153
Cook, G. R., 130
Cordes, 230
Coy, 392, 393
Craggs, 115
Craig, 348, 406, 410
Cramer, 121, 122
Crank, 92, 94
Crawford, 204
Crawley, 376, 377
Crompton, 114
Cross, 473
Crowell A. D., 24, 25, 40, 45
Crowell, C. R., 366
Cunningham, 394
Curtiss, 12, 14, 15, 17, 18, 19
Cutler, 142, 470
Czavinsky, 122

D.

Dacey, 28
Dadson, 472
Daglish, 319
Dahl, 191
Dalgarno, 124, 467
Dallos, 408
D'Amico, 139
Damm, 385, 391, 431
Danforth, 169, 471
Danilov, 475
Danilova, 475
Das, 109
Datsiev, 470
Datz, 21, 57, 58, 191, 230, 231, 232

Davey, 387
David, 138, 140
Davies, 191, 217, 218, 219, 220, 470
Davis, D. H., 55
Davis, F. N., 254, 255
Davis, J. C.. 142
Davis, W. D., 174, 256, 261, 265, 270, 271, 293, 295, 297, 342, 343, 345, 346, 347, 354, 355, 377, 419, 420, 424, 425, 426, 472
Davison, 467
Dawson, J. P., 65, 384, 393
Dawson, P. H., 474
Dayton, 65, 100, 103, 104, 105, 282, 284, 285
de Boer, 75, 76, 78, 166
Degras, 78, 84, 170, 467, 470
de Heer, 114, 469
de Jong, 91
Dekker, 159, 160, 162, 163
Delbart, 254
Delchar, 29, 143
de Leeuw, 110
Delgmann, 346, 349, 419, 421
della Porta, 67, 400, 402
Demorest, 298
Denhartog, 470
Denison, 230, 270, 273, 293, 302, 413
Denisov, 160, 163
de Poorter, 60
Deryagin, 244
de Segovia, 278, 279, 292, 426
de Taddeo, 431
Deville, 138, 140
Devonshire, 69
de Vries, 255
Dick, 388, 389
Diels, 369
Dienes, 204
Diethorn, 467
Dilke, 34
Dillon, J. A., 147
Dillon, W., 471
di Paolo, 44, 51, 384, 385, 469
Ditchburn, 128
Dobretsov, 163, 230, 231
Dohmann, 265
Domeij, 55, 191, 217, 218, 219, 220, 221, 284, 356, 426
Donaldson, 170, 230, 232, 241, 242, 270, 273, 293, 302, 471
Doolittle, 468
Dore, 372
Dorman, 115
Douglas, 412, 413
Dovbnya, 473
Dowden, 34
Drain, 43, 49
Drawin, 120
Dreyer, 434
Dryden, 404

Dubinin, 46
DuBridge, 234
Duckworth, 342
Duffy, 298
Dumas, 335
Dunkelman, 236
Dunn, 115, 116, 117, 310
Dunning, 26
Dupp, 195, 196, 202
Dushman, 20, 91, 97, 98, 102, 106, 112, 305, 319, 379, 381, 383

E.

Earl, 419
Earnshaw, 50, 273, 372, 376, 423, 424, 426
Eberhagen, 27, 366
Ebisuzaki, 467
Edmonds, D. K., 393
Edmonds, T., 58, 216, 255, 256, 258, 268, 284, 285
Eggleton, 60
Ehlbeck, 288, 289, 290
Ehrenreich, 289
Ehrlich, 25, 28, 29, 33, 34, 37, 65, 68, 72, 79, 80, 143, 265, 356, 357, 360, 362
Eisinger, 68, 274
Eley, 34
Ellett, 57
Elliot, 472
Elsworth, 391, 401
El'tekov, 191
Emmett, 46
Endow, 44, 46, 47
Englander-Golden, 114, 263, 265
Englert, 183, 206
Eremeev, 470
Erents, 79, 203, 204, 211, 212, 470
Erginsoy, 183, 206
Eriksen, 256, 261, 330, 418, 419, 424
Eriksson, 220
Ermrich, 169, 401, 403, 411
Erwin, 347
Escard, 138, 140
Eschbach, 75, 90, 96
Eskew, 347
Espe, 98, 101, 369, 384, 475
Estermann, 55
Estrup, 143, 470
Evans, 297
Evdokimov, 228, 470
Everhart, 121, 122, 123, 125, 127, 128, 129, 191
Evett, 232
Eyring, 77

F.

Fano, 128
Farkass, 431

Farnsworth, 144, 145, 146, 147, 148, 149, 468, 470
Farrar, 346, 419
Farrell, G., 467
Farrell, J. S., 289
Faruk, 192
Favreau, 265
Feakes, 331, 332, 333, 334, 338, 390, 432
Fedchenko, 388
Fedorenko, 121, 125, 127, 128, 191
Fedorova, 396
Feuerstein, 250
Field, 117
Filipenko, 121
Firsov, 121, 122, 123, 183, 191, 192
Fischbeck, 167
Fischer, E., 433
Fischer, G. E., 171, 176, 242, 433
Fischer, K. J., 317
Fite, 57, 84, 115, 116, 117, 118, 119, 123, 124, 126, 127
Flaks, 121
Flecken, 376
Fletcher, 472
Flinn, 393
Fluit, 195, 196, 200, 201, 202
Fomin, 232
Foner, 66
Forth, 395
Foster, 91
Fowler, H. A., 144
Fowler, W. E., 236
Fox, 115, 116, 215, 216, 267, 269
Fraiman, 165
Francis, A. B., 410
Francis, S. A., 117
Frandsen, 384
Frank, 90, 91
Franklin, 117
Fraser, 55
Fredericks, 140, 153
Frenkel, 78
Freytag, 309
Friedman, 467
Frisch, 69
Fry, 468
Fujimoto, 471
Fujiwara, 151
Fulker, 413, 475
Fuls, 127, 128, 191
Fumi, 183

G.

Gabor, 317, 473
Gadzuk, 69, 70, 467
Gafner, 142
Gaines, 265, 268
Ganoczy, 242

Garbe, 170, 173, 174, 175, 279, 374
Garwin, E. L., 396
Garwin, F. L., 83, 84, 85
Gavrilyuk, 72, 73, 468
Gay, 190
Geballe, 128
Geiger, C. S., 434
Geiger, J., 155, 156, 157
Geiger, K. A., 431
George, 147
Germer, 138, 140, 145, 146, 147, 148, 150, 151, 468
Gerstacker, 61
Gerstner, 470
Geyer, 166
Gibbons, 400
Gibson, 183, 191, 201, 206, 207
Gilbey, 69, 70
Gilbody, 127
Gilbreath, 245
Gilcrest, 394
Gimpel, 157
Giordmaine, 284
Giorgi, 67, 215, 402, 472
Glaser, 245
Glasstone, 77
Goerz, 386
Goethert, 396, 397
Goland, 183, 191, 201, 206, 207
Goldman, 195, 202
Goldsztaub, 138, 140
Gombas, 183
Gomer, 28, 33, 37, 62, 63, 67, 68, 83, 169, 170, 176, 179, 180, 181, 309, 467
Gomoyunova, 157, 158, 163
Good, 365
Goodman, 25, 28, 62, 69, 70, 71
Goodrich, 297, 298
Gordeev, 128
Gorman, 102
Gorodetskii, 144
Gosselin, 330, 331, 332, 333, 419, 425, 426, 427
Goto, 470, 471
Gottwald, 468
Gould, C. L., 404
Gould, P. R., 431
Govier, 369
Gowland, 395, 432, 472
Graham, 40, 51
Grande, 350
Grant, H. L., 387
Grant, R. J., 385
Grant, W. A., 79, 182, 203, 204, 208, 209, 270, 470, 471
Greene, 19, 21
Greenwood, 208
Greer, 372
Gregory, 475

Grenier, 64, 388
Griffing, 127
Griffith, R., 468
Griffith, T., 254
Grigor'ev, 304, 334
Grigson, 135
Grishin, 39
Grishina, 39
Groszkowski, 290, 309, 317, 326, 473
Grove, 76
Gryzinski, 120
Guildner, 60
Gunn, 471
Gussenhoven, 244
Gutshall, 244

H.

Haas, C., 91
Haas, T. W., 381, 402
Haber, 239
Hachenberg, 134, 159, 161, 163, 167
Haefer, 280, 333, 334
Hagena, 467
Hagstrum, 117, 139, 181, 188, 189, 190, 222, 223, 225, 226, 294, 355
Hall, L. G., 348
Hall, R. W., 250
Haller, 386
Halsey, 23, 24, 26, 39, 40, 42, 43, 49, 51, 74, 75
Hancox, 57
Hanle, 111
Hanley, 299, 300
Hansen, N., 43, 45, 47, 49, 67, 468
Hansen, R. S., 275, 277, 469
Hanwell, 246, 247
Harden, 348
Harmer, 126
Harra, 401, 402
Harris, L. A., 159
Harris, L. B., 468
Harrison, 190, 193, 201, 228, 229
Harrower, 140, 159
Hartley, 155
Hartman, 292
Hashimoto, 473
Hasted, 110, 111, 113, 114, 115, 117, 118, 119, 121, 123, 124, 126, 129, 130, 131, 353
Haughney, 199
Haul, 468
Haxby, 166
Hayakawa, K., 151
Hayakawa, S., 335
Hayakawa, T., 25
Hayashi, 331, 409
Hayek, 468, 470
Hayes, 32, 33, 35
Haygood, 65, 83, 84, 85, 384

Hays, 298
Hayward, D. O., 32, 33, 34, 35, 52, 53, 54, 83, 468
Hayward, W. H., 261, 327, 328, 401, 402, 473
Heddle, 131
Heflinger, 83, 84, 85
Helbig, 125
Hellemans, 47
Hellwig, 155, 156, 157
Helmer, 334, 335, 473
Hemstreet, 44, 51, 384, 385, 387, 388, 389
Henault, 393
Hendee, 297
Henderson, 434
Hengevoss, 39, 83, 84, 280, 363, 364, 365, 468
Hennes, 236
Herb, 317, 318, 319, 338, 404, 417, 413
Herm, 21
Herold, 298
Herring, 142, 143
Herron, 232, 233
Herschbach, 21
Hersh, 384
Herzberg, 12, 16, 129
Herzog, L. F., 347
Herzog, R. F. K., 429
Hession, 330, 374, 419, 421
Hetherington, 250
Hickman, 382
Hickmott, 52, 81, 273, 274, 275, 276, 277, 278, 374, 375, 420
Higatsberger, 298
Hines, 220, 221
Hinrichs, 170
Hipple, 115
Hirschfelder, 12, 14, 15, 17, 18, 19
Hirth, 66
Hobson, 2, 26, 27, 43, 44, 45, 47, 48, 49, 50, 58, 66, 68, 75, 78, 84, 140, 142, 144, 146, 210, 216, 255, 256, 258, 265, 267, 268, 270, 273, 274, 284, 285, 294, 304, 313, 314, 315, 325, 326, 331, 332, 333, 334, 335, 338, 355, 359, 362, 369, 370, 372, 376, 385, 386, 387, 389, 405, 417, 423, 424, 426
Hoch, 293, 359, 419, 474
Hoenig, 396
Hogan, 396
Holkeboer, 282, 416
Holland, 164, 173, 285, 383, 391, 401, 403, 413, 426, 431, 472, 475
Holliday, 158, 159
Holmes, 218
Holscher, 143
Holzl, 144
Honig, J. M., 48
Honig, R. E., 38, 39
Honig, W., 335
Hood, 397, 398, 472
Hoogendoorn, 199
Hook, 38, 39
Hooker, 471
Hooper, 126
Hooverman, 317
Hopkins, 143, 475
Horgan, 468
Horn, 431
Houston, 319
Huang, 48, 49, 423
Huber, 340, 363, 364, 365, 419, 425
Hudda, 25, 28
Hudson, 66, 346, 349, 350, 468
Hughes, 234
Hughey, 472
Hummer, 123, 124, 127
Hunt, 385, 390, 391, 431
Hurlbert, 91
Hurlbut, 55, 56, 58, 397
Huxley, 114
Hybl, 384, 475

I.

Imbusch, 157
Inkley, 386, 387
Iona, 471
Ionov, 68, 86, 87, 88, 102, 231, 470
Iosifescu, 215, 273
Ireland, 127
Isakaev, 474
Ischuk, 143, 469, 472
Ishii, 169, 170, 181, 255
Ishikawa, 266, 267, 268, 269, 271, 309, 470, 471
Ivanovskii, 403, 471, 475

J.

Jackson, 381, 402, 471
Jacob, 288
Jacobs, 395
Jaeckel, 75, 274, 369
Jaeger, 99
James, B. D., 410
James, L. H., 203, 271, 272
Jansen, 254
Jeffers, 384, 387, 388
Jepsen, 261, 327, 328, 334, 335, 385, 410
Jespersgaard, 218, 220
Johnson, D. P., 254
Johnson, J. B., 166
Johnson, J. P., 201
Johnson, K. I., 243, 472
Johnson, R. A., 183
Johnson, V. R., 247
Jones, D. W., 416
Jones, H. A., 300

Author Index

Jones, P. R., 127, 128, 191
Jonker, 164
Jordan, 121
Jost, 89
Juenker, 289

K.

Kaganer, 46
Kaganskii, 336
Kambe, 471
Kaminker, 127
Kaminskii, 336
Kaminksy, 59, 60, 61, 63, 181, 188, 193, 222, 226, 231, 232
Kanter, 139, 157, 159
Karpuzov, 191
Kass, 467
Kato, 471
Kay, 472
Kees, 265, 268
Keesom, 49
Keller, D. V., 243, 472
Keller, J. M., 260, 262
Kelly, B. T., 204, 205
Kelly, R., 208
Kemball, 34
Kendall, 175, 271, 345, 346, 351, 419
Kennard, 283
Kenty, 242
Kerr, 173
Kervalishvili, 335
Kerwin, 116, 265
Kessel, 128
Keywell, 194, 195
Khan, I. H., 140, 142, 146, 294
Khan, J. M., 471
Khashaba, 118
Khozinksii, 226
Kieffer, 115, 116, 117, 310
Kienel, 396
Kietzmann, 410
Kim, 169, 170, 181
Kinchin, 202
Kindall, 384
Kindl, 400
King, 468
Kingdon, 310, 317
Kinzer, 467
Kirchner, F., 365
Kirchner, H., 365
Kirkendall, 468
Kirst, 169, 170
Kiselev, 42, 78
Kishinevskii, 228, 229
Kisliuk, 53, 68, 72, 141, 143
Kistemaker, 114, 115, 125, 188, 195, 196, 200, 201, 202, 253, 255, 298, 471
Klein, 228
Kleine, 298

Kleint, 361, 365
Klemperer, 134, 152, 153, 154
Klipping, 65, 395
Klopfer, 274, 279, 280, 321, 322, 338, 341, 346, 351, 374, 401, 403, 411, 419, 422, 423, 426, 471
Klyucharev, 336
Knauer, F., 55
Knauer, W., 334
Knoll, 215, 216, 267, 269, 406
Knor, 385, 468
Knudsen, 55, 283, 284
Kobayashi, 471
Kodera, 40
Koedam, 199
Koehler, 203, 204, 207
Koenig, 365
Kollath, 134, 159, 161
Kolot, 473
Komiya, 409
Konecny, 91
Konjevic, 471
Kopotzki, 198
Kornelsen, 2, 55, 191, 201, 203, 204, 209, 210, 211, 212, 213, 217, 218, 219, 220, 267, 268, 270, 271, 284, 304, 355, 356, 359, 364, 376, 400, 401, 409, 411, 419, 426
Kotowski, 299
Kouptsidis, 468
Kouznetsov, 403
Kovalenko, 395, 396
Kozlov, 473
Kramer, 250
Kraus, 105
Krebs, 296
Krutilina, 143
Kruyer, 78
Kuchai, 215
Kulakov, 403
Kulikauskas, 192, 221
Kunze, 471
Kupiec, 242
Kurepa, 115
Kuyatt, 119
Kuz'min, 403
Kwei, 21

L.

Laegreid, 193, 194, 195, 196
Lafferty, 20, 91, 97, 98, 102, 106, 112, 265, 299, 300, 305, 319, 320, 321, 338, 379, 381
Laidler, 77
Lander, 1, 136, 139, 140, 142, 150, 151, 304
Lane, 121, 122, 123
Lange, 240, 261, 288, 289, 327, 330, 338, 346, 348, 375, 386, 418, 419, 424, 427
Langley, 126
Langmuir, 42, 229, 275, 300, 310

Laponsky, 160, 166, 167
Lapujoulade, 467
Large, 226, 227
Larson, 473
Laslett, 260, 262
Lassettre, 117
Laubert, 202
Laurenson, 164, 173, 391, 401, 403, 413, 426, 475
Lavrov, 227
Law, 67, 232, 278
Lawrence, 125
Lawson, 351, 403, 470, 471, 474
Lazarev, 384, 396
LeBlanc, 289
Lecante, 170, 470
Leck, 22, 54, 203, 210, 213, 214, 215, 254, 260, 265, 266, 268, 270, 271, 272, 302, 304, 305, 308, 471
Leder, 134, 152
Lee, 90, 468
Lehmann, 201
Leiby, 90, 96, 101
Lennard-Jones, 69
Leonas, 71
Letunov, 157, 158, 163
Levenson, 55, 468
Lever, 471
Levy, 201
Lewin, 109, 369, 371, 434, 469
Lhotsky, 475
Lichten, 128
Lichtman, 22, 168, 169, 170, 171, 175, 176, 232, 347, 355, 419, 422, 426
Lieber, 472
Lifshitz, 244
Limon, 412, 413
Lindhard, 192, 202, 217, 218, 219, 220
Lindholm, 117
Lineweaver, 173
Ling, 48
Little, J. G. 87, 88
Little, J. W., 254
Little, R. P., 365
Littman, 67
Lloyd, 323
Lockwood, 125
Logan, 72
Longley, 28, 48, 50, 331, 332, 333
Lothrop, 292
Lotz, 469
Love, 143
Lutz, H., 217
Lutz, M.A., 334
Lye, 160. 162, 166
Lyon, 471

M.

Mack, 171, 176, 242, 433
MacLennan, 223
MacRae, 137, 145, 147, 148, 149, 150, 151
Madey, 37, 141, 143, 169
Madden, 472
Magnuson, 193, 196, 197, 199, 226, 228, 229
Magyar, 191
Mahadevan, 197, 199
Maliakal, 412, 413
Malter, 242, 433
Malov, 192, 221
Mamyrin, 297
Manes, 385
Mansfield, 397
Marchenko, 304
Margenau, 22
Mark, H., 390, 432
Mark, J. T., 434
Marmet, 116, 117, 175
Martin, D. W., 126
Martin, G., 434
Marton, 134, 152
Martynenko, 195, 202
Maryott, 17
Mashcher, 65, 395
Mashkova, 192, 228, 470
Mason, 121, 122
Massey, 20, 61, 110, 114, 115, 117, 118, 119, 138
Mathiez, 475
Matskevich, 163
Mauer, 66
Maurice, 380
Maxted, 34
May, 468
Mazhuga, 468
Medana, 44, 45, 46, 47, 51, 68
Medved, 226, 239
Medvedev, 73, 468
Meinke, 255, 259, 473
Melfi, 317, 338
Mel'nik, 239
Mendlowitz, 134, 152
Menzel, 150, 169, 170, 176, 179, 180, 181, 468, 471
Mercer, 385
Messer, 469
Metcalfe, 431
Metson, 1, 304, 313
Meyer, E. A., 317, 318, 319, 338
Meyer, L., 49, 62, 63
Meyer, V. D., 117
Michel, 155, 156, 157
Michenkov, 470
Melczarek, 119
Mikhailov, G. V., 403
Mikhailov, I. F., 384, 396
Mikhnevich, 335
Mileshkin, 403
Milgram, 183, 191, 201, 206, 207

Miller, C. F., 101
Miller, J. R., 473
Miller, T. M., 431
Milleron, 55, 372, 386, 388, 431
Mimeault, 275, 277, 469
Minturn, 21, 230
Miyake, 151
Mizushima, 292
Moe, 127
Moesta, 265
Mohr, 20
Moiseiwitsch, 122, 470
Molchanov, 192, 200, 228, 470
Moldenhauer, 361
Molière, 469
Montgomery, 207
Moore, B. C., 55, 173, 284, 286, 393, 396, 473
Moore, E. C., 475
Moore, G. E., 57, 58, 81, 86, 88, 91, 97, 170, 172, 278, 282
Moore, R. W., 282
Morgan, 127, 129, 191
Morgulis, 144, 304
Morrison, C. F., 473
Morrison, J., 140, 150, 151
Morrison, J. A., 43, 45, 62, 283
Morrison, J. D., 115, 116, 117, 175
Morrison, J. L., 60
Moser, 255
Mott, 110, 114, 115
Mott-Smith, 310
Mourad, 318
Moursund, 19, 21
Moustafa, 469
Movnin, 227
Muehlause, 242
Mugler, 475
Mullen, 395
Müller, D., 75, 76
Müller, E. W., 365, 468
Müller, K. G., 309
Mullhaupt, 44, 51, 384, 385
Murakami, 364, 365, 382
Myers, 157

Mc.

McCargo, 191, 217, 219, 220
McCarrol, 125
McCarroll, 65, 468
McCracken, 391, 403, 404, 419, 471
McDaniel, 110, 123, 124, 126, 127, 129
McDowell, C. A., 342
McDowell, M. R. C., 110
McGowan, 265
McIntosh, 419
McIntyre, D. V., 244
McIntyre, R. A., 346, 419

McKay, 134, 166
McKee, 469
McQuestan, 22, 168, 170, 175, 176
McRae, 140, 142, 471

N.

Nakayama, 255, 473
Nakeo, 474
Nakhodkin, 239
Nanoboff, 384
Narcisi, 388
Nardella, 102
Nasini, 87, 88
Navinsek, 470
Nazarov, 403
Neal, 135
Nechai, 91
Nelson, 192, 197, 198, 204, 207, 221
Nesmeyanov, 39
Newton, 474
Nichiporovich, 474
Nichols, D. K., 182
Nichols, M. H., 142, 143
Nicholson, 75, 76
Niedrig, 157
Nielsen, K. O., 220
Nielsen, V., 202
Nienhuis, 329
Nier, 298
Nightingale, 372
Nikolaev, 395
Nixon, 135
Noble, 288
Nöller, 376, 381
Norris, 21
Norton, 91, 100, 101, 208
Nottingham, 1, 135, 254, 267, 304, 309

O.

Oda, 292
Odinstov, 228
Oen, 207, 208, 217, 218, 221
Ogawa, 130
Oguri, 68
O'Hara, 247, 248
Okamoto, 364, 365, 382
O'Keeffe, 467
Oldal, 361
Oldenberg, 119
Olivier, 48
Olmer, 61
Omohundro, 390, 391
Onishi, 40
Öpik, 128
Oppenheimer, 115
Orfanov, 191, 199
Osanov, 297

Oshchepkov, 297
Outlaw, 431
Owens, 260

P.

Pack, 111
Pagano, 416
Palluel, 157
Panenkova, 91
Panov, 128
Parcel, 247, 248
Parilis, 192, 228, 229
Park, 145, 146, 147, 148, 468
Parzen, 335
Pashley, D. W., 135
Pashley, N. A., 403
Pasternak, 44, 46, 47, 467
Paty, 304, 474
Pauley, 317, 318
Payn, 169
Peach, 120
Pease, 202
Peck, 396, 397, 398
Pelz, 474
Pender, 143, 475
Penning, 329
Perdijk 473
Peria, 159
Perovic, 110
Pétermann, 170, 171, 268, 278, 372, 469
Peters, 471
Petrov, 470
Petsch, 127
Phelps, 111
Philipp, 289
Piercy, 191, 217, 218, 219, 220
Pierotti, 24
Pierre, 309
Pines, 153
Pinsker, 136, 139
Pinson, 396, 397, 398
Piper, 152
Pisani, 400, 472
Pleshivtsev, 199
Plets, 470
Plumlee, 175
Podgurski, 254, 255
Podlasek, 250
Poehlmann, 388
Polder, 244
Polito, 418, 419
Pollard, I. E., 395, 432
Pollard, J. H., 471
Polley, 40
Poltz, 255
Pomerantz, 166
Popov, K. V., 91
Popov, Yu. S., 474

Popp, 385, 391, 431
Poshkus, 42, 78
Postma, 434
Potter, 471
Pound, 66
Powell, 154, 155, 156
Power, 379, 381, 382, 384, 406, 410, 413
Powers, 218
Presnyakov, 120
Pretzer, 223
Priestland, 285, 472
Propst, 152
Pryde, 468
Ptushinksii, 66, 81, 473

Q.

Quinn, 87, 88

R.

Rabinowicz, 246
Radushkevich, 46
Radzhabov, 471, 475
Raether, 134, 152
Rakhimov, 226
Ramsauer, 111, 113
Ramsey, 284
Rapp, D., 114, 117, 124, 263, 265
Rapp, R. A., 66
Read, J. E., 391, 404, 419
Read, P. L., 385, 424
Redhead, 2, 36, 66, 68, 78, 79, 80, 169, 170, 171, 174, 176, 177, 178, 179, 180, 181, 210, 265, 267, 275, 290, 291, 292, 293, 304, 309, 312, 313, 314, 315, 316, 325, 326, 327, 328, 331, 333, 334, 335, 338, 348, 353, 355, 357, 359, 360, 366, 376, 389
Reed, 117
Reich, 255, 259, 280, 304, 325, 359, 473
Reikhrudel, 293, 334, 335, 474
Renn, 265
Rettinghaus, 474
Reynolds, 173, 346, 347, 419, 421, 424, 427
Rhodin, 51, 268, 278
Ricca, 24, 28, 44, 45, 46, 47, 51, 68, 87, 88
Rice, 250
Richards, P. I., 298
Richards, R. J., 393
Richardson, D. M., 364
Richardson, O., 157
Richman, 397, 398
Ricketson, 392, 393
Riddoch, 22, 302
Riede, 111
Riemersma, 240
Rigby, 86, 87, 88, 358
Ritchie, 153
Rivière, 391, 404, 419, 424

Author Index

Roberts, J. K., 60, 87, 88
Roberts, M. W., 46, 87, 88, 469
Roberts, R. W., 147, 369, 379
Roberts, W. K., 397
Robertson, 469
Robins, 86, 88, 154, 155, 174, 293
Robinson, C. F., 299, 348
Robinson, E. A., 135
Robinson, M. T., 195, 199, 200, 207, 208, 215, 217, 218
Robinson, N. W., 50
Rodin, 215
Roehrig, 256, 262
Rogers, K. W., 103, 104
Rogers, W. A., 90, 96, 101, 107
Rol, 128, 195, 196, 200, 201, 202, 255
Rootsaert, 36
Rosenberg, D., 193, 194, 196, 197
Rosenberg, P., 255
Rosenstock, 232, 233
Ross, J., 19, 21
Ross, J. R. H., 46
Ross, S., 48, 467
Rothe, 255
Rotter, 296
Roussel, 384
Routledge, 135
Rovner, 268, 278
Rozgonyi, 418, 419, 428, 429, 475
Rubet, 380
Rudberg, 153
Rudnitskii, 409
Ruedl, 208
Ruiz, 298
Russek, 127, 128, 229
Rutherford, 407, 410
Ruttenbur, 51, 387, 388, 389

S.

Sachtler, 36
Sagot, 380
Saini, 68, 87, 88
Salisbury, 245
Saltsburg, 57, 58, 469
Sams, 23, 40, 74, 75
Samson, 130, 236, 237
Sands, 388, 389
Sandstrom, 471
Santeler, 286, 287, 384, 396, 397, 416, 434
Sato, 409
Sauerbrey, 471
Sauermann, 429
Schäfer, 61, 64
Schaeffer, 40
Schagen, 301
Scharff, 202, 218
Scharmann, 195, 196, 202
Schawlow, 17

Scheibner, 471
Schiff, 12
Schiøtt, 218, 471
Schissel, 279
Schittko, 369
Schlier, 68, 147, 278, 365
Schmidlin, 83, 84, 85
Schmidt, 279, 341, 351, 374
Schofield, 60, 71
Schönheit, 298
Schram, A., 309, 469
Schram, B. L., 44, 47, 48, 114, 115, 298
Schuchinsky, 166
Schuemann, 278, 279, 292, 313, 426
Schuhmann, 256, 258
Schultz, 166
Schulz, 117, 119, 120, 152, 265
Schulze, 475
Schutten, 469
Schütze, H. J., 288, 289, 290, 309
Schütze, W., 298
Schuurman, 474
Schwartz, 418, 419
Schwarz, 304
Schweers, 49
Searcy, 60
Sears, 66
Seaton, 116, 118, 120
Segal, 163, 164
Seitz, 203, 204, 207
Sewell, 135, 139, 150
Shapiro, 300
Sharevskii, 395
Sharkey, 299
Shaw, 265
Sheffield, 297
Shelton, 143
Shen, 250
Shepard, 101
Shepherd, J. P. G., 134, 152, 153, 154
Shepherd, W. G., 166
Shelnikov, 475
Sheretov, 293, 334, 335, 474
Sheridan, 119
Shields, 232, 233
Shin, 467
Shinoda, 471
Shishlovskii, 475
Shoulders, 378, 432
Shroff, 475
Shuba, 235
Shul'man, 139
Siddiqi, 86, 87
Sigmund, 201
Silsbee, 201, 207
Silvernail, 62
Simon, 195, 202
Simonov, 403
Simons, J. C., 256, 262

Simons, J. H., 121, 122
Simpson, 119
Singer, 384
Singleton, 40, 43, 68, 87, 88, 268, 275, 276, 277, 278, 279, 292, 327, 330, 338, 375, 382, 418, 419, 420, 424, 427
Sinha, 203, 204, 212, 213
Siprikov, 297
Sizmann, 217
Skerbele, 117
Skvortsov, 297
Sloane, 300
Sluyters, 125
Smeaton, 215, 216
Smereka, 468
Smetana, 259
Smirnitskaya, 335
Smith, A. C. H., 123, 124
Smith, C. G., 469
Smith, D. G., 298
Smith, D. P., 471
Smith, H. I., 244
Smith, H. P., 471
Smith, H. R., 382
Smith, J. N., 57, 58, 469
Smith, L. P., 175
Smith, P. T., 111, 114, 115
Smith, R. V., 393
Smith, S. J., 470
Smith, W. R., 40
Smithells, 91, 98
Snoek, 128, 188, 191
Snouse, 199
Sobelman, 120
Solonitzin, 239, 240
Somorjai, 471
Sommers, 390, 432
Soo, 423, 469
Sormani, 67
Soshka, 192
Sosniak, 475
Southern, 195, 199, 200
Spalvins, 248
Spangenberg, 353
Sparnaay, 244
Speight, 208
Spicer, 238
Spitzer, 304
Stebbings, 118, 119, 123, 124, 127
Steckelmacher, 304, 473
Stedeford, 124
Steele, 26, 39, 42, 43, 49, 51
Stehberger, 157
Stein, 245
Steinrisser, 375, 408
Stephens, 432
Stern, O., 55, 57, 69
Stern, R. C., 37
Stern, R. M., 143, 147
Stern, S. A., 44, 51, 64, 384, 385, 387, 388, 389, 469
Sternglass, 157, 158, 159, 228
Stevenson, D. L., 382
Stevenson, D. P., 34, 115
Stewart, 119
Stickney, R. E., 72
Stickney, W. W., 65, 282
Stier, 199
Stockmayer, 12, 17
Stone, F. S., 239, 240
Stone, J., 121
Stork, 290, 309
Stout, 400
Stowell, 135
Strayer, 169
Strausser, 226
Strehlow, 364
Stuart, 193, 196, 198, 199
Stuber, 115
Studier, 351
Sugata, 169, 170, 181
Sullivan, 425
Sunshine, 470
Swan, 154, 155
Swanson, L. W., 169
Swanson, N., 156
Swets, 90

T.

Tabor, 246
Tahy, 361
Takeishi, 223
Tarbell, 397
Tate, 114, 115
Taylor, C. E., 390, 391
Taylor, E. H., 57, 58, 230, 231, 232
Taylor, L. H., 28, 48, 50, 244
Taylor, N., 53, 54, 471
Teale, 391
Teggers, 64
Tel'kovskii, 200
Telkovsky, 226
Teller, 46
Teloy, 274
Terenin, 239, 240
Teubner, 469
Tharp, 471
Theodorou, 419
Thibault, 384
Thieme, 119
Thirlwell, 154
Thomas, A. M., 254, 473
Thomas, C. H., 288
Thomas, J. E., 91
Thomas, L. B., 60, 61, 62, 70, 71
Thomas, M. T., 127
Thompson, J. B., 391, 404, 419, 424

Author Index

Thompson, M. W., 195, 198, 199, 201, 204, 207
Thomsen, 202
Thomson, 87, 88
Thorne, 473
Tiedema, 91
Tikhomirov, 232
Tischer, 299
Titran, 250
Todd, 95, 96, 173, 373, 374
Tom, 410
Tominaga, 76, 469
Tompkins, 27, 29, 53, 54, 60, 86, 87, 468
Torney, 331, 332, 333, 334, 338, 390, 432, 474
Tosi, 183
Tosic, 110
Tothill, 135
Townes, 17
Townsend, 111
Trabert, 431
Trapnell, 32, 33, 34, 35, 52, 83
Trendelenburg, 84, 203, 271, 369, 379, 468
Trulson, 279
Tsarev, 301
Tseitlin, 419, 425
Tsin, 396
Tucker, 91, 138, 150, 208
Tulinov, 192, 221, 470
Tunitskii, 232
Turaev, 192
Turnover, 396
Tuul, 145
Tuzi, 62, 283, 420
Tyutikov, 235

U.
Uggerhøj, 192, 220
Unterwald, 81, 86, 88, 91, 97, 279
Ustinov, 68, 86, 87, 88
Utterback, 254

V.
Valdrè, 135
Van Dael, 47
Vanderslice, J. T., 121, 122
Vanderslice, T. A., 345, 369, 379
Van der Weg, 128, 471
Van der Wiel, 114
Van Dingenen, 44
Van Itterbeek, 44, 47
Van Lint, 182
Van OOstrom, 143, 309, 311, 359, 474
Van Reijen, 36
Van Voorhis, 114
Varadi, 372, 468
Varnerin, 210, 215, 216, 269, 405, 406

Vasil'eva, I. A., 335 335
Vasil'eva, M. N., 335
Vaughn, 247
Veinshtein, 120
Veksler, 188, 189, 191
Venema, 254, 288, 309, 381, 417
Venus, 280
Vincent, 135
Vineyard, 183, 191, 201, 204, 206, 207
Vinogradov, 409, 475
Visser, 474
Volosok, 323, 336
Volter, 473
von Ardenne, 101
von Koch, 117
Vonnegut, 245
von Roos, 189, 228
von Zahn, 349
Voshall, 111
Vostrov, 469

W.
Wachman, 59, 62, 69, 71
Waclawski, 472
Wade, 304
Wagener, J. S., 67
Wagener, S., 300
Wahba, 34
Waldschmidt, 97, 98, 102
Walker, 467
Wang, E. S. J., 384
Wang, T. C., 284
Wargo, 166
Wasa, 335
Waters, 228
Watt, 300
Watters, 346, 349, 350
Watts, 472
Weast, 107, 108
Weber, 159, 230
Wehner, 192, 193, 194, 195, 196, 197, 198, 199, 201
Weijsenfeld, 472
Weinbaum, 122
Weinreich, 299
Weissler, 128, 130, 234, 237
Wells, 434
Wenning, 236
Wexler, 66
Whaling, 218
Wheeler, 292
Whetten, 160, 165, 166, 167, 427, 428, 474
White, 297
Whitney, 365
Whitton, 470, 472
Wiens, 388
Wiley, 297, 298
Williams, C. E., 379
Williams, D. P., 245

Willis, 195, 199, 200
Wilson, I., 391, 404, 419, 424
Wilson, J. R., 143
Wilson, K. R., 21
Wilson, R. G., 472
Windsor, 385
Winslow, 244
Winters, 230, 270, 273, 293, 302, 472
Wolsky, 196
Wood, E. A., 136, 137, 139, 145
Wood, R. W., 55
Woodman, 472
Woodward, 403
Worley, 471
Wright, 230
Wutz, M., 396
Wyrick, 475

X, Y, Z.

Yadav, 124
Yarwood, 211
Yates, 37, 141, 143, 169
Yeung, 132
York, 346, 419
Young, D. M., 24, 25, 40, 45
Young, J. R., 159, 170, 172, 267, 268, 269, 278, 295, 330, 338, 374, 419, 421, 423, 425, 427, 428
Young, R. L., 394
Young, W. C., 247, 248
Yu, 469
Yuferov, 391
Yurasova, 191, 199
Zabritski, 412, 413
Zagorskaya, 471
Zahl, 57
Zakrzewski, 473
Zandberg, 231
Zaphiropoulos, 431
Zaviantseff, 472
Zdanuk, 196
Zharinov, 335
Ziemba, 127, 128, 191
Zierman, 394
Zingerman, 143, 469, 472
Zollweg, 141, 142
Zscheile, 228
Zwanzig, 69